Practical
Bioinformatics

Dedication

This book is dedicated to my mother, Ruth Agostino, who tolerated my smelly biology and chemistry experiments in the basement of our house, and the endless number of muddy clothes and shoes from my frequent explorations of the woods near my home. I owe my love of exploration and discovery to you.

Practical
Bioinformatics

Michael Agostino

Garland Science
Taylor & Francis Group

NEW YORK AND LONDON

Vice President: Denise Schanck
Senior Editor: Gina Almond
Assistant Editor: David Borrowdale
Development Editor: Mary Purton
Production Editor: Ioana Moldovan
Typesetter and
Senior Production Editor: Georgina Lucas
Copy Editor: Jo Clayton
Proofreader: Sally Huish
Illustrations: Oxford Designers & Illustrators
Cover Design: Andrew Magee
Indexer: Medical Indexing Ltd

ISBN 978-0-8153-4456-8

Front cover image:
Chapter 8 of this book focuses on protein analysis. One example in this chapter is the superimposition of the black swan and Atlantic cod fish lysozyme structures (see Section 8.6). This allows the viewer to see the impact, or lack thereof, of the amino acid differences on the structures of these distantly related proteins. The book cover shows the amino acid sequence of this swan lysozyme (UniProt accession number P00717), repeated many times to fill the page, and is combined with the structure of the lysozyme protein (PDB identifier 1gbs).

About the author:
Michael Agostino received his PhD in Molecular Biology from Roswell Park Memorial Institute, a division of SUNY at Buffalo, New York. His thesis characterized the unusual structure and evolution of sea urchin histone genes. Postdoctoral work included the development of a molecular assay for DNA strand scission agents used in chemotherapy. In 1984, he moved to the University of North Carolina at Chapel Hill where he co-developed a vector trap for gene enhancers. Other work included the creation of a synthetic gene and an *E. coli* blue-white reporter gene assay for HIV protease activity. In 1991 he formally switched careers to bioinformatics by joining GlaxoSmithKline. There, he provided sequence analysis, consulting, user-support, and training for the Glaxo scientists. In 1996 he moved to Genetics Institute, where he was appointed manager of a bioinformatics department. This group was responsible for the sequence analysis and database of a high-throughput effort to identify, express, and patent the human genes that encode secreted proteins. Presently, he provides bioinformatics analysis and end-user support for multiple sites of the Pfizer Research organization. He is also an adjunct professor in the Biology Department at Merrimack College, North Andover, Massachusetts (USA).

Library of Congress Cataloging-in-Publication Data
Agostino, Michael J.
  Practical bioinformatics / Michael Agostino.
    p. cm.
  ISBN 978-0-8153-4456-8 (alk. paper)
  1. Nucleotide sequence--Data processing. 2. Bioinformatics. I. Title.
  QP625.N89A39 2013
  572.8'6330285--dc23
                    2012017992

Published by Garland Science, Taylor & Francis Group, LLC, an informa business, 711 Third Avenue, 8th floor, New York, NY 10017, USA, and 3 Park Square, Milton Park, Abingdon, OX14 4RN, UK.

Printed in the United States of America

15 14 13 12 11 10 9 8 7 6 5 4 3 2 1

Garland Science
Taylor & Francis Group

Visit our Website at http://www.garlandscience.com

Preface

Although bioinformatics is a relatively new scientific discipline, it has become quite broad in definition. It is often described as including diverse topics such as the analysis of microarrays and the accompanying statistics, protein structure prediction, and pathway and protein interaction analysis. Of course, computer programming, database development, and even hardware design are included in the field. *Practical Bioinformatics* is focused on the fundamental skills of bioinformatics: the analysis of DNA, RNA, and protein sequences. The chapters take the reader through a commonly asked question, "What can I learn about this sequence?" The only requirement is access to the Internet and a web browser; no other software is required.

This book is designed as an introduction to bioinformatics sequence analysis for biology and biochemistry majors. There are many published books that teach about detailed algorithms, sophisticated programs, and advanced interpretation of data. Although these are excellent sources of information, many biologists and biochemists are not prepared for, nor do they need, the depth and detail of these texts. Instead, they need the practical knowledge and skills to analyze sequences. They are asking questions such as "Which tools do I use?" "What settings should I use?" "What database should I search?" "What do these results mean?" "How do I export this information?" *Practical Bioinformatics* addresses these questions, and many more, in 12 easily read chapters.

Concepts will be introduced within each chapter and then demonstrated through the analysis of problems using selected gene/protein examples. Adequate background, details, illustrations, and references will be provided to insure that readers understand the fundamentals and can do further reading if desired. Along the way, interesting genes, phenotypes, mutations, and biology will be introduced but not discussed extensively or analyzed. These topics are purposefully left open so they may easily be turned into literature searches, analysis problems, or senior projects for the ambitious student. Just thinking about these problems and how to analyze them will instill the habit of identifying topics needing exploration.

The best way to learn this material is by "doing." Readers of this book will learn the concepts by performing many analysis problems. To get the most out of this book, readers should perform most, if not all, of the analysis steps and recreate the figures for themselves. By the time readers finish the book, they will have significant experience in sequence analysis problems, approaches, and solutions. They should then be ready to perform many analysis steps on their own, and tackle more advanced books on the subject.

A common error when approaching a sequence analysis problem is to use powerful analysis software with little understanding of how it works or how to interpret the output. Web forms and software can completely hide the details. This text will emphasize the proper use of established analysis software and the need to evaluate new tools. There are literally hundreds of bioinformatics tools available and no book could possibly contain or instruct on all the tools that are available. However, the repeated experience of performing guided analysis problems will teach the reader to be critical of bioinformatics software and to use proper positive and negative controls when testing unfamiliar tools. When this book is finished, readers will have both the practical knowledge and experience to address their own problems, and take advantage of the mountains of genetic data being generated today.

I would like to thank the staff and associates of Garland Science for their tremendous support during the process of writing this book. Thanks to Gina Almond who believed in the project from the very beginning and was never short on enthusiasm, David Borrowdale for guiding the book through the many steps, and Mary Purton for her infinite patience during the editing process. My thanks go to Ioana Moldovan, Georgina Lucas, Jo Clayton, and Sally Huish for their tremendous attention to detail and style during the final editing. Special thanks go to Oxford Designers & Illustrators for numerous illustrations. Thanks to Josephine Modica-Napolitano who gave me my first job in teaching, and the students at Merrimack College; together they put me on the path of writing this book. My special thanks to Donald J. Mulcare, my undergraduate advisor, for advice, encouragement, and my first real taste of what it is like to be a scientist. My years in industry would not have been the same without knowing the members of the "Dream Team:" Yuchen Bai, Sreekumar Kodangattil, Ellen Murphy, Padma Reddy, and Wenyan Zhong. They are the best sequence analysts I know. Additional thanks go to Maryann Whitley and Steve Howes for providing a calm and steady leadership at Pfizer. Many thanks to my daughter Becky, who inspires me to be better every day. Finally, this book would not have been possible without the years of support, encouragement, and love from my wife, Nan. Dreams can come true.

Instructor Resources Website

Accessible from www.garlandscience.com, the Instructor Resource Site requires registration and access is available only to qualified instructors. To access the Instructor Resource Site, please contact your local sales representative or email science@garland.com.

The images in *Practical Bioinformatics* are available on the Instructor Resource Site in two convenient formats: PowerPoint® and JPEG, which have been optimized for display. The resources may be browsed by individual chapter or a search engine. Figures are searchable by figure number, figure name, or by keywords used in the figure legend from the book.

Answers to end of chapter questions/exercises are available on the Instructor Resource Site.

Resources available for other Garland Science titles can be accessed via the Garland Science Website.

PowerPoint is a registered trademark of Microsoft Corporation in the United States and/or other countries.

Acknowledgments

The author and publisher of *Practical Bioinformatics* gratefully acknowledge the contributions of the following reviewers in the development of this book:

Enrique Blanco	University of Barcelona, Spain
Ron Croy	Durham University, UK
John Ferguson	Bard College, USA
Laurie Heyer	Davidson College, USA
Torgeir Hvidsten	Umeå University, Sweden
Ian Kerr	University of Nottingham, UK
Daisuke Kihara	Purdue University, USA
Peter Kos	Biological Research Centre of the Hungarian Academy of Sciences, Hungary
Jean-Christophe Nebel	Kingston University London, UK
Samuel Rebelsky	Grinnell College, USA
Rebecca Roberts	Ursinus College, USA
Hugh Shanahan	Royal Holloway, University of London, UK
Shin-Han Shiu	Michigan State University, USA
Shaneen Singh	Brooklyn College, USA
Alan Ward	Newcastle University, UK

Contents

CHAPTER 1

Introduction to Bioinformatics and Sequence Analysis

Key concepts

- The scope of bioinformatics
- The origins and growth of DNA databasess
- Evidence of evolution from bioinformatics
- Example sequence analysis and displays using human Factor IX

1.1 INTRODUCTION

We are witnessing a revolution in biomedical research. Although it has been clear for decades that exploring the genetics of biological systems was crucial to understanding them, it was far too expensive and complex to consider obtaining genetic sequences for that exploration. But now, acquiring genetic sequences is affordable and simple, and data are being generated at unprecedented rates. The heart of understanding all this sequence lies in bioinformatics sequence analysis, and this book serves as an introduction to this powerful study of DNA, RNA, and protein sequence.

Bioinformatics concerns the generation, visualization, analysis, storage, and retrieval of large quantities of biological information. The generation of biomedical data, including DNA sequence, in its raw form does not involve bioinformatics skills. But in order for that sequence to be usable, it must be analyzed, annotated, and reformatted to be suitable for databases. These are all bioinformatics activities. Many of these activities can be automated, but their development and support come from someone with skills or experience in bioinformatics.

Once the data have been made available, how do you analyze the data? Is there text like DNA and protein sequence files? If yes, it should be presented in a way to allow interpretation or easy input into programs for analysis. Or is there so much information that data are represented graphically? This form of data reduction is quite powerful and without it we would be staring at pages and pages of sequence without, literally, seeing the big picture.

Some analysis is manual, ranging from looking at the individual nucleotides or amino acids, to submitting sequence to a program that transforms the sequence into another form. This could include the location of features such as functional domains, modification sites, and coding regions. Often, analysis includes the searching of databases for the purposes of comparison or discovery, and this will be the primary activity for a number of chapters. Much of the content of this book is concerned with analysis.

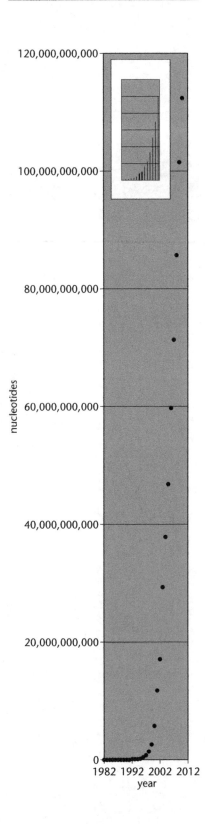

> **ⓘ Floppy disk databases**
> In the early years of GenBank, if you wanted access to the database you ordered a handful of floppy disks that were delivered in the mail.

Storage is usually not a responsibility of those who will analyze the sequences. However, the creation of properly structured databases or storage forms so data can be queried and retrieved is essential for the analysts to do their work. Sequence files and other forms of data can be decades old or just created yesterday. But unless you can retrieve them easily, the value decreases quickly. "Easily" is not just describing the speed of the computers and connections delivering the information to you, although this can be extremely important. It also includes the steps to access and query the stored data. The ideal approach is often a Web form with easily understood options, online help, and results pages rich with hypertext. Bioinformatics was one of the first areas of science to embrace the Web as a vehicle for disseminating information and we'll be using many Web pages in this book.

Finally, bioinformatics activities often involve large quantities of data. Even if you are focusing on a single gene, you still may have mountains of data that are connected to this single sequence. With a good database or software tool, you may only be aware of the quantities yet not overwhelmed with details that don't interest you. Still, it can't be emphasized enough that one of the biggest challenges facing the field of bioinformatics is the absolute deluge of information and how to generate, visualize, analyze, store, and retrieve these data.

1.2 THE GROWTH OF GenBank

How much data are we talking about? One way to answer this is to describe the amount of DNA sequence data in public databases. GenBank is a huge repository run by the US National Center for Biotechnology Information (**NCBI**). The inset in **Figure 1.1** shows the steady growth in the early years of GenBank but the rate of growth has been rapid since then. As of early 2011, there are over 126 billion nucleotides in this standard division of GenBank from over 380,000 organisms. If this were not impressive enough, there are an additional 91 billion nucleotides in the whole genome shotgun (a type of sequencing) division, the section of GenBank dedicated to unfinished large sequencing efforts. If a DNA sequence is considered "public information," it is deposited in GenBank, the DNA Data Bank of Japan (DDBJ), or the database of the European Bioinformatics Institute (EBI). The contents of these three databases are synchronized. In terms of disk space, the database is over 500 Gigabytes in size.

1.3 DATA, DATA, EVERYWHERE

Where are all the data coming from? The quick answer is everywhere! In recent years there has been a dramatic drop in prices and rapid advances in both sequencing technology and computing power. What was once too time-consuming and expensive is now very possible and affordable; biological sequence generation is now commonplace. A major driver for the advances being realized today is the **Human Genome Project**. Even though the completion of the sequencing of the human genome was announced in 2001, the analysis of the data is ongoing and will take many years. These advances had to be coupled with dramatic improvements in computers and the drop in cost for processing power, memory, and storage. Of course, the Human Genome Project and all the spin-offs are only possible because of simultaneous advances in bioinformatics.

This intersection of sequencing technologies, computational power, and advances in bioinformatics has made DNA sequencing quite routine and paved the way for many bold and ambitious projects. Projects now come from scientists

Figure 1.1 GenBank growth. Plotted is the size of GenBank in nucleotides versus the years from 1982 to the first three months of 2011. The inset shows data for years 1982–1994, not visible on the larger plot. From the GenBank Release Notes of Release 184, ncbi.nlm.nih.gov.

in numerous fields of biology, medicine, agriculture, ecology, history, energy, and forensics, just to name a few. Here are some prominent examples.

- The 1000 Genomes Project (www.1000genomes.org). An effort to sequence the genomes of 1000 people to identify genetic variants that affect 1% of the human population. In addition to providing insights to genetic disorders and health risks, the history of human migrations is being revealed. In recent years, people have proposed that the number of human genomes to sequence for this project grow to be 10,000 or higher.

- The 1001 *Arabidopsis thaliana* Genomes Project (www.1001genomes.org). *Arabidopsis* is a widely used plant model due to its habitat diversity, genetics, and ease of manipulation. This genome project aims to study the genomes of 1001 strains that differ in phenotype including adaptation to growth in a wide variety of conditions. Project scientists and those in the *Arabidopsis* community are able to grow huge numbers of genetically identical plants and can vary the environment at will to challenge and observe the underlying genetic elements which define these strains.

- The Genome 10K Project (genome10k.soe.ucsc.edu). An effort to sequence the genomes of 10,000 vertebrate species, one from every genus. Along with all the other genomes sequenced, this project will make a tremendous impact on understanding the relationship between organisms. We can only guess what will be discovered from these animals, having so much in common with us but with such diverse physiologies and phenotypes, and occupying such a wide range of habitats.

- The i5k Initiative (www.arthropodgenomes.org/wiki/i5K). An effort to sequence the genomes of 5000 insects and arthropods. Many insects are either pests, carriers of disease, or beneficial to agriculture and man. More knowledge of their biochemical pathways will surely result in new avenues of control, utilization, and fascination.

- **Metagenomics.** This is a broad term covering the sequencing of DNA samples from the environment as well as from biomedical sources. For example, sequencing has led to the identification of the hundreds of bacterial species inhabiting our skin, mouth, and digestive system. The populations that live on and within us vary with our health state and are clearly linked to our physiology (as we are to theirs). The NCBI lists almost 350 metagenomics projects (www.ncbi.nlm.nih.gov/genomes/lenvs.cgi) that are either at the beginning stages or completed. These projects each generate anywhere from thousands to millions of sequences.

- Cancer Genome Atlas. This is a massive project (cancergenome.nih.gov) where thousands of specimens from all the major cancer types and their matched normal controls will have their RNAs and many of their genes sequenced.

- EST generation. **ESTs** (expressed sequence tags) are small samples of transcribed genes and a quick avenue for discovering the genes expressed in tissues or organisms. Clones are generated and sequenced by the thousands. There are at least 72 million EST sequences in GenBank.

- The Barcode of Life (www.barcodeoflife.org). Distinguishing closely related species is often difficult, even for taxonomy experts. For example, there are approximately 11,000 species of ants. How can you easily tell them apart? The Barcode of Life project aims to identify a DNA "signature" for each species in the world using a 648 base pair sequence of the cytochrome c oxidase 1 gene. The five-year goal is to have sequences from 500,000 species. Nice examples of consumer use of this information include the identification of illegal fishing of endangered species and illegal logging activities.

- The NCBI lists over 1700 eukaryotic genome sequencing projects (www.ncbi. nlm.nih.gov/genomes/leuks.cgi), over 11,000 microbial genome projects (www.ncbi.nlm.nih.gov/genomes/lproks.cgi), and over 3100 viral genomes.

ⓘ Tumbling costs
According to Eric Lander, director of the Genome Biology Program at the Broad Institute, it now costs about $20 to sequence the *Escherichia coli* genome, sequencing each of the 4.7 million nucleotides twenty times to ensure accuracy.

ⓘ **Keep flossing**
Human microbiome studies have sampled bacteria from skin from all over your body, the gut flora (of course), even your navel (nicknamed "bellybutton biodiversity"). According to a *Nature Reviews Microbiology* Editorial, dental plaque is very dense with bacteria. The number of bacteria in a single gram is equivalent to the number of people who have ever lived.

There are also private sequencing efforts where the data are not always released to the public yet the parties acquiring the data still have to cope with the huge amount of sequence generated by these projects.

- Firms such as pharmaceutical and biotechnology companies are contracting other companies to generate sequence from patients, animals, important crops, plants, cell lines, tumors, and pathogens. They are also doing **deep sequencing** of complementary DNA (cDNA) libraries to identify rarely expressed genes. These efforts are being used to develop products such as new drugs, crops, and diagnostic kits.

- In response to an infectious disease, the genomes of suspected pathogens are being sequenced. For example, in 2011 there was a major pathogenic *Escherichia coli* outbreak in Europe that eventually killed several dozen people. In 2010 there was a cholera outbreak in Haiti following the devastating earthquake there. In both cases, the genomes of the causative bacteria were sequenced to better understand the pathogens and learn how to treat the diseases. Literally tens of thousands of human immune deficiency virus (HIV) genomes have been sequenced. As the price drops, medical sequencing will probably become more commonplace for diagnosis of individuals in the general population.

There are many "smaller" projects that are contributing to the public data growth. There is a division of the NCBI Website (www.ncbi.nlm.nih.gov/popset) that only contains population studies (PopSet): collections of sequences from many members of the same species. For example, there are PopSets for spiders (102), rabbits (179), squirrels (83), skunks (94), robins (114), and ants (94). Within these records you can find the sequence of a single gene from hundreds or thousands of individuals.

Smaller still in size, but not importance, are the efforts to understand a single gene or gene family. This analysis, often originating in an individual laboratory or academic department, is often very detailed and associated with publications. These analysis studies are at the heart of understanding how genes function. Many automated annotation efforts absolutely depend on these manual and long-term projects to serve as reference sequences.

Further examples of human genome sequencing

Personal genome sequencing

Families with a common last name have often cooperated to establish links by common ancestors. Now some are using sequencing from the Y chromosome (inherited from father to son), mitochondria (passed from mothers to their children), or both with the specific purpose of establishing or verifying these links. Some have already uncovered unknown connections between families that would not have been possible to identify without the DNA sequence. Companies have formed that specialize in these kinds of sequence analysis services. They can provide partial family histories for adoptees, provide information concerning paternity, and even identify the presence of the so-called "warrior gene" (*MAOA*), a gene variant associated with aggressive responses to threats.

There are companies that offer the sequencing of your entire genome and the accompanying analysis as a service. As the cost comes down (estimated to drop as low as $1000 per genome) and the predictive value of genes goes up, you can expect more people to have their genomes sequenced.

Paleogenetics

This is a relatively new field, made possible by vast improvements in the isolation and amplification of DNA from ancient biological specimens. Scientists are now able to ask genetic questions of ancient times in history. For example, Schuenemann and colleagues sequenced DNA from the remains of people who were fourteenth century victims of the Black Death that swept through Europe.

ⓘ **A long journey**
In 2011, there were surprising reports of a mountain lion being seen in Connecticut, not the current habitat of these large cats. Shortly after these reports, a 140-pound male mountain lion was struck and killed by a car on a Connecticut highway. For the preceding several years, scientists using the DNA found in scat and hair samples had been tracing its movement from South Dakota, Minnesota, and Wisconsin, making the journey to Connecticut of at least 1500 miles.

Their work shows that *Yersinia pestis*, the agent most probably causing the Black Death, was present but is a different strain to the one found today.

A spectacular display of **paleogenetics** is the sequencing of the Neanderthal genome. The DNA was obtained from bones thousands of years old and carefully sequenced by Richard E. Green and colleagues in the laboratory of Svante Pääbo. The analysis of these data has just begun but has already yielded interesting findings about our ancient relatives. Early work examined the language gene, *FOXP2*, investigating their ability to speak. It was also discovered that Neanderthals had a *MCR1* gene variant that leads to red hair. Comparisons between our genome and that of the Neanderthal reveal that approximately 2.5% of Neanderthal DNA sequence is in our genome, indicating that our ancestors interbred. Very recently, a 41,000-year-old bone from a new human ancestor (**hominin**) was discovered in Siberia and their genome indicates that they contributed a small amount of sequence to present-day Melanesians (people of the islands northeast of Australia).

Focused medical genomic studies

Genetic testing is a well-established hospital procedure for carrier or prenatal testing, diagnostics, and newborn screening for common genetic disorders. The formation of companies that specialize in gene sequencing for establishing genetic risks has some parties concerned that testing without reason or access to qualified counseling can lead to fear or poorly informed life decisions. Testing positive does not mean that you definitely have or will develop a disorder and a negative test does not guarantee that you will not develop the disorder. Others are concerned that disclosure of a positive test to an employer or insurance company may lead to negative consequences.

There are a number of studies in which patients had their DNA sequenced and analyzed for the purpose of identifying the molecular basis of genetic disorders. In the laboratory of David Galas, genome sequences were obtained from the parents and their two children who inherited separate and different recessive genetic disorders. One child was born with Miller syndrome, which causes facial and limb development abnormalities, and the other child was born with primary ciliary dyskinesia. The latter is characterized by the malfunction of microscopic cilia in the respiratory tract. Through careful analysis of the sequences, the disorders were narrowed down to four possible genes. Another finding from this study was an accurate measurement of the mutation rate per generation. Each child in this family was born with 70 mutations (sequences different from either parent), which was lower than that estimated for human generations using other methods.

In another study, a scientist, James Lupski, used his own DNA to identify the molecular basis of the disease Charcot-Marie-Tooth neuropathy that affected him and other members of his family. By sequencing his own genomic DNA, candidates for the cause of his disease were identified. More directed sequencing of the DNA of family members confirmed the mutations responsible for his family's disorder.

In a final example, newspaper reporters received a Pulitzer Prize for an article describing a team effort at The Medical College and Children's Hospital of Wisconsin where the genome sequence of a sick child was determined to assist in the diagnosis of his unusual disease. The Medical College has started a program where physicians can nominate medical cases where knowing the patient's genomic sequence may help, and at least six patients are in the queue to have their DNA sequenced.

1.4 THE SIZE OF A GENOME

How much data is generated when a genome is sequenced? The genome size and gene number generally increase with the complexity of the organism, but there are some surprises. *E. coli,* the object of research for decades and resident in our

Table 1.1 The size of genomes

Species	Genome size (10^6 nucleotides)	Number of genes
Escherichia coli	4.7	4300
Saccharomyces cerevisiae	12	6700
Drosophila melanogaster	169	13,900
Danio rerio	1500	26,000
Homo sapiens	3200	21,000
Zea mays	3200	63,000
Oryza sativa	488	57,000

Source: The Ensembl Genome Browser (www.ensembl.org) April 2012.

digestive system, has 4300 genes in 4.7 million nucleotide pairs. *Saccharomyces cerevisiae*, a single-celled yeast used in cooking and fermentation, has an incrementally larger genome and number of genes (**Table 1.1**).

Multicellular organisms show an increase in these numbers. The common fruit fly, *Drosophila melanogaster*, has almost 14,000 genes in 169 million base pairs. The genome of the vertebrate zebrafish, *Danio rerio*, is almost tenfold larger, yet only contains 26,000 protein-coding genes. The human genome is approximately 3.2 billion nucleotides long and contains approximately 21,000 protein-coding genes and at least 12,000 noncoding genes. Each mammal genome sequenced in the projects listed above will generate approximately the same amount of partially processed data as seen in human genome analysis.

Plants have complex genomes, reflecting a history of genetic duplications that far surpass the number seen in vertebrates. As a result they often have large genomes and gene numbers; maize (*Zea mays*) has 63,000 genes in a genome of size comparable to that of mammals while rice (*Oryza sativa*) has over 57,000 genes in a much smaller genome.

1.5 ANNOTATION

Of course, if all we had were file after file of just DNA sequence, we would learn little about the object of our sequencing efforts. The true value is realized when the DNA or protein sequence is described to tell us about genetic or protein elements, structures, similarities, functions, and predictions associated with these sequences. Collectively, these details are referred to as **gene annotation**. Like bioinformatics, annotation is a broad term and has different meanings to different people. Here, it is used to describe details such as where a gene starts and ends; similarities to other genes and proteins based on database searches; places that are known to vary; translation start and stop sites; places where the protein is predicted to be or is modified; association with a phenotype or disorder; and ties to other analysis or publications.

Annotation efforts are a big part of any genome or gene project and, depending on the size of the project, can be either manual or automated. When a genome sequence is finished, the annotation of the hundreds or thousands of genes has to be automated. Bioinformatics experts join together a **"pipeline"** of software tools that systematically analyzes each region of the genome, identifies genes, and then determines the details of those genes. The fields of bioinformatics and gene analysis would be at a near standstill without these pipelines, and this form of analysis is both powerful and very accurate.

An automated process cannot perform every conceivable analysis, however. The developers of the pipeline choose the questions to be asked and this analysis

 Economic impact of the Human Genome Project

The Human Genome Project cost the US government $3.8 billion yet the return on that investment has been incredible. According to a report by Battelle Technology Partnership Practice, the breakthroughs in technology and information spawned the birth and growth of both companies and academic laboratories, followed by the creation of products and services. In 2010 alone, this generated $67 billion in US output, supporting 310,000 jobs and $20 billion in personal income. Since the Human Genome Project started, over $49 billion in taxes have been paid to the US government from these genomic-related activities.

provides valuable, but basic, information about these newly discovered genetic elements. Automated efforts will miss details that the software is not trained to recognize. Importantly, automated annotation is not always updated. Annotations entered in a database when a gene is newly discovered may never be updated. If other members of this gene family later appear in the database, there may be no link between the older sequence and these new, more fully described sequences. The description on the older file may be frozen in time.

1.6 WITNESSING EVOLUTION THROUGH BIOINFORMATICS

In the history of life, there have been countless times when a gene's sequence has randomly mutated with a concomitant change in the encoded protein structure and function. Some of these new functions imparted advantages to the organism and were retained for future generations. Deleterious mutations were quickly eliminated from the population. Other changes were neutral and, because they caused no harm, may or may not have been retained. Genes have been duplicated again and again, with each copy continuing to evolve, leading to large gene families and new functions. The path from unicellular to multicellular organisms, and the development of tissue, organs, and limbs, also increased genetic complexity, visible today in higher organisms. Throughout this book there will be numerous demonstrations where the fields of genomics and bioinformatics will show these steps in evolution.

Recent evolutionary changes to plants and animals

About 10,000 years ago, humans began to change from a hunter-gatherer lifestyle to practicing agriculture. Seeds were collected and kept from consumption for planting in the ground in the vicinity of their dwellings. By selecting seeds of plants with superior characteristics, ancient varieties of plants grew taller, produced bigger seeds, produced more nuts or fruit, and resisted inclement weather or disease. We can barely recognize the ancestral plants because this selection process has transformed their appearance so dramatically. However, their DNA sequence reveals the evolution.

The same applies to domesticated animals. Recent sequencing of the dog genome reveals their origins from wolves and places the time of domestication much earlier than plant domestication. Over time, we have transformed dogs into breeds with strikingly different phenotypes. They are all clearly dogs, but the size range, accomplished by careful breeding by humans, is astonishing. An adult Chihuahua weighs no more than 6 pounds and can be as small as six inches high, while a Saint Bernard can reach 180 pounds. Just based on weight, this variation is equivalent to a small human newborn and an adult man.

Other animals have also been bred for specific traits: cows (increased milk production), horses (speed or strength), sheep (wool quantity and quality), poultry (more breast meat), and fish (speed of maturation). Through the sequencing and study of their genes, genetic screening and manipulation may prove to be a more direct route to desired phenotypes. These studies are taking place now.

1.7 LARGE SOURCES OF HUMAN SEQUENCE VARIATION

One of the contributing reasons for the sharp decline in the cost of sequencing the human genome is that the first sequence to be obtained stands as a template to guide the assembly and analysis of subsequent genomes. These newer, so-called **re-sequencing** efforts do not require the many weeks of assembly and problem solving seen during the first genome sequencing. However, there are still considerable differences seen between individual people.

 The Japanese Warrior Crab

The Japanese Warrior Crab (*Heikea japonica*) has on its back an uncanny resemblance to an artistic portrait of a Samurai warrior. Over hundreds of years, any captured crabs not looking like a warrior were kept for the market (and no longer reproduced) while those resembling the warriors were returned to the sea.

First, there are **single nucleotide polymorphisms (SNPs)**. The entire human genome is approximately 3.2 billion nucleotide pairs long, and there are approximately 3 million nucleotides that differ when you compare the genomes of two people. These common differences are found in about 1% of the population. Many of these differences have no apparent impact on the function of the genome, while others disrupt gene regulation, or change coding regions resulting in altered amino acid sequences. People studying the genomes of tumors find many SNPs arise within the tumor. Some of these may be responsible for the cancer state, while others accumulate independently of the biochemical changes necessary to become a cancer cell.

There are also tremendous differences between genomes due to **copy number variations (CNVs)**. Comparing your DNA sequence to that of the human "standard" genome, there are thousands of DNA segments which range from 1000 to several million nucleotides in length, and they are either present, present in multiple copies, or absent from your genome.

Kimberly Pelak and colleagues did a fascinating study published in 2010 where they sequenced the genomes of 20 people, 10 of which had hemophilia A. Although they were faced with the many differences between individuals, they were able to identify the mutations causing hemophilia in 6 of the 10 patients. Surprisingly, they found that "on average, each genome carries 165 homozygous protein-truncating or stop loss variants in genes representing a diverse set of pathways." Of the 21,000 protein-coding genes, almost 0.8% of our genes are unable to be translated to full-length proteins, essentially "knocking out" many of these genes.

1.8 RECENT EVOLUTIONARY CHANGES TO HUMAN POPULATIONS

Since the emergence from Africa, humans have migrated to all continents except Antarctica (**Box 1.1**).

Box 1.1 The author's DNA

My mother's ancestors walked out of Africa perhaps 50,000 years ago. I don't know their exact path, or how they got across rivers or mountains, or survived winters. For countless generations they fanned out from the Middle East across Europe, displacing, and eventually driving to extinction, the Neanderthals who had inhabited these lands for several hundred thousand years. But not before the Neanderthals contributed a small amount to their gene pool, shown recently by careful analysis of both modern human and Neanderthal genomic DNA sequences. My ancestors eventually crossed Eastern Europe and settled in what is now called Lithuania. The evidence for this narrative is contained in my DNA.

A partnership between National Geographic and IBM (The Genographic Project) aims to establish the migratory paths of modern humans through the collection and sequencing of DNA from many tens of thousands of volunteers. I am one of those volunteers and had a small section of my mitochondrial DNA sequenced. This snippet of DNA tracks with my mother's side of the family as the mitochondrial DNA is only contributed through females. My DNA sequence is seen in **Figure 1**. The handful of nucleotides that vary from a reference sequence, shown in a lighter shade, indicate that I am in the "T haplogroup," which places my ancestors on the migratory path described above. With time and more analysis, perhaps more details will be filled in.

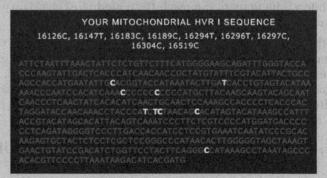

Figure 1 The author's DNA. The sequence shown is from the author's mitochondrial DNA. The sequence was provided by The Genographic Project, www.nationalgeographic.com/genographic.

Along the way, in response to the environment, they have changed their diet and lifestyle. Here are a few examples along with the genetic changes associated with these adaptations. Many may have occurred during the last 40,000 years, or since the more recent start of agriculture. Included are the official gene names or symbols when known.

- Skin color. Humans near the equator have retained a darker skin color to block damaging ultraviolet light. However, people closer to the Earth's poles need and have paler skin allowing them to make enough light-induced vitamin D. Sequence variation in a number of genes, such as *SLC24A5*, appears to be responsible for this skin pigmentation.

- Lactose tolerance. It is estimated that as recently as 8000 years ago, goats and cattle were domesticated and their milk was consumed by humans, especially at times of poor crop yields. This practice was probably a contributing factor to preventing starvation in the historically frequent famines. Normally, the ability to digest lactose rapidly decreases after early childhood, resulting in considerable intestinal discomfort after consuming milk. But a mutation arose which resulted in the persistence of lactase expression into adulthood, allowing milk consumption without side effects. Lactose tolerance quickly spread through the European population and a sequence variation near the promoter of the lactase gene, *LCT*, appears to be responsible. Interestingly, a different set of promoter sequence variations arose independently in pastoral African populations, reaching 90% of the Tutsi population.

- Digestion of starch. Like the digestion of lactose, there have been selective pressures for the increased ability to survive on high-starch diets. Amylase is an enzyme found in saliva and provides the first steps in the digestion of starch. It has been found that populations that consume a lot of starch have high copy numbers of the amylase (*AMY1A*) gene while populations with low-starch diets have fewer amylase genes. More amylase gene copies results in higher amylase expression, especially in saliva, conferring an advantage for digesting food.

- Malaria resistance and sickle cell anemia. Malaria is a tropical disease caused by a *Plasmodium* blood infection. It attacks hundreds of millions of people each year and is fatal to tens of thousands. Sickle cell anemia is a disease where mutations in the hemoglobin B gene (*HBB*) lead to misshapen red blood cells. In addition to carrying less oxygen, the crescent-shaped cells cause problems of poor circulation such as pain and organ damage. However, you are more likely to survive malaria if you carry one copy of the sickle cell disease gene and, over multiple generations, mutations of the sickle cell trait have spread rapidly through the populations most at risk for malaria.

- Life at high altitude. There are several human populations that live and thrive at extremely high altitudes and have high red blood cell counts in response to the low oxygen levels. Yi and colleagues identified a transcription factor gene, *EPAS1*, as having a SNP present in most high-altitude Tibetans but mostly absent from Han Chinese living at low elevation. The Tibetan population split from the Han population less than 3000 years ago. Interestingly, *EPAS1* expression rises in response to low oxygen levels.

1.9 DNA SEQUENCE IN DATABASES

In earlier sections, the major drivers of database growth were described. With this growth and wealth of data comes the ability to address long-term questions such as finding molecular evidence of evolution, and examples of this were also described. Genomic and cDNA sequences are chiefly responsible for the flood of information into GenBank and the basics of DNA sequence assembly and cDNA

The black blood of Uro Indians

There is a legend that the Uro Indians of Peru had "black blood" which helped them survive at the cold and high altitudes. Although the legend may not be true it is interesting that the story comes from a time of limited biochemistry knowledge yet blood color, and therefore oxygen-binding hemoglobin, is connected to this legend.

synthesis will be described here and explained in more detail in Chapter 5. The same principles of **genome sequence assembly** apply to both established and next-generation sequencing methods. Importantly, bioinformatics plays a key role in assembling the millions of sequences into contiguous pieces of genome. As we search databases and come across pieces of DNA sequence, it is important to appreciate the origins of those fragments, both from a scientific point of view and source of pride in human ingenuity.

Genomic DNA assembly

Most genomes are millions of nucleotides long, far surpassing the length of sequence generated by current sequencing technologies. So genomic sequencing efforts have all involved breaking chromosomal DNA into pieces and then working with the smaller fragments. Once the fragments are in a suitable format and size, the DNA sequence is determined and bioinformatics software assembles the fragments into long contiguous stretches with the goal of assembling the genome sequence from end to end.

Now, it is important to remember that when chromosomal DNA is isolated, you are not working with just one copy. DNA is obtained from something abundant, for example cells grown in culture, a whole organ, or even a whole organism or flask of organisms. Since you are not working with a single cell, you are isolating the DNA from many millions of cells and therefore have millions of copies of each gene.

It is also important to realize that you are sequencing random pieces of DNA. The approach you took to randomly fragment the chromosomal DNA generated many different beginnings and endings. That is, you are running thousands or millions of sequencing reactions at once and they correspond to regions all over the genome. Furthermore, your sequencing reactions are not generating the entire sequence of a gene; sequencing only generates short stretches. Finally, you are starting at random places in gene A, gene B, gene C, and so on. After you are done with the sequencing reactions, you have DNA sequence from the beginning and end of the genes, and everywhere in between. Because you had multiple copies of each gene in the original sequencing reactions, you have overlapping copies of sequence. But all of this is required to get any level of accuracy of sequence.

To understand the strength in the randomness described above, let's start with an analogy. Imagine taking a piece of paper on which two sentences are written and with scissors, cutting the page into those individual sentences. Consider these two adjacent sentences,

```
Here comes a fox. The fox jumps over the lazy dog.
```

and the pieces that you generated with scissors, deliberately put out of order:

```
The fox jumps over the lazy dog.
Here comes a fox.
```

If you had no knowledge of the original order of sentences and were asked to assemble these as they were before, you would be at a loss since there is no overlap between sentences. There is some hint of sentence order if you consider them individually because you understand English, but you can't be absolutely sure about the order when you try to reassemble the sentences. But if you had multiple copies of the sentences, and many pieces of sentences cut at random places, and pieces that spanned the two sentences, you could assemble, with confidence, the order of the sentences and place a consensus assembly (built from the agreement between words) underneath the fragments:

```
Here comes
    comes a fox. The f
          fox. The fox jumps over the
                          over the lazy dog.
                              the lazy dog.
Here comes a fox. The fox jumps over the lazy dog.
```

This analogy is close to the approach and solution to sequencing and assembling genomic DNA. The overlapping nature allows us to confidently determine the relationship among fragments. The unique words (here, comes, jumps, over, lazy, dog) are analogous to genes, scattered about. Genomic DNA has repetitive elements, much like the repeating words (fox, the), also distributed unevenly. These can be a problem unless you have adjacent unique sequences (words).

Now look at a small stretch of DNA sequence taken from a rat gene:

```
ACAAGGGACTAGAGAAACCAAAACGAAAGGTGCAGAAGGGGAAACAGATGCAGAAAGCATCTGGAGACAA
```

Let's "sequence" multiple copies, starting from random locations, in overlapping pieces, and build a consensus, shown below it:

```
ACAAGGGACTAGAGAAACCAAAA
        AGAAACCAAAACGAAAGGTGCAGAA
                AACGAAAGGTGCAGAAGGGGAAACAGATGCAGA
                        GAAGGGGAAACAGATGCAGAAAGCATCT
                                        AGAAAGCATCTGGAGACAA
ACAAGGGACTAGAGAAACCAAAACGAAAGGTGCAGAAGGGGAAACAGATGCAGAAAGCATCTGGAGACAA
```

Like the two-sentence analogy, the overlapping fragments allowed you to order the pieces. Note that there are multiple places of "AAA" which could confuse the assembly process but the adjacent sequence allowed the correct assembly.

Many early genome assemblies aimed for approximately sixfold coverage (overlapping regions six sequences deep) but with the next generation of sequencing machines, 20–100-fold coverage is now commonplace. Even so, errors in assembly can and do occur, often because of scattered repetitive regions ranging from hundreds to thousands of nucleotides long. These repeats are nearly identical in sequence and can be indistinguishable from each other.

For reasons that are not always clear, some regions of DNA cannot be isolated very easily, do not clone at high efficiency, and/or cannot be sequenced very accurately. This results in regions that are underrepresented in the multifold coverage or not represented at all. These "holes" can be mapped and addressed through alternative means, but nevertheless represent a hurdle in generating genomic sequence. You will often see these holes as a long string of Ns (NNNNNNNNNNNNNNNNNNNNNNNNNN) as placeholders where the scientists know the length of a fragment, based on the distance between flanking markers, but not its content.

Remember, DNA is double-stranded. When you sequence DNA randomly, you could be sequencing the complementary strand, which gives you twice the sequence to consider when trying to pull all your fragments together to build the consensus. In practice, sometimes the sequence of a complementary strand is more easily obtained, so having the other strand's sequence is a benefit and simply contributes to the redundancy you need for accuracy.

Finally, you cannot forget the huge bioinformatics contribution for the assembly of genomic DNA sequence. That is, if you have to assemble literally millions of randomly generated DNA sequences, each fragment ranging from 25 to 900 nucleotides long, you must use a computer program to accomplish this difficult goal. Going back to the two-sentence analogy above, imagine trying to assemble the sentences of a 23-volume encyclopedia, with millions of words in each volume, and 20–100 copies of those 23 volumes. Assembling the original version of the human genome sequence required the assembly of over 27 million fragments (approximately fivefold coverage) and literally weeks of computing time on one of the largest known computers at that time. Complete understanding of the complex computer programs that accomplished this feat is beyond the scope of this book, but it is valuable to appreciate that bioinformatics was key to this historic event. With the above description as a background, be prepared that the genomic DNA sequence in our databases may:

- be very long pieces (contiguous stretches or contigs) but often, many small fragments;
- contain regions of unknown sequence;
- contain mistakes in sequence;
- be assembled incorrectly;
- contain either strand of the double helix in the database unless it is described in more detail (like a gene);
- be represented multiple times in the database.

cDNA in databases—where does it come from?

A huge contribution to the sequence in DNA databases is cDNA. As more thoroughly explained in Chapter 5, messenger RNA (mRNA) is quite unstable and using enzymes to convert this polymer into stable DNA is the preferred approach for cloning and sequencing of these transcripts. cDNA analysis is critical to the understanding of gene expression and function so, as a result, this form of DNA is very prominent in the analysis in this book. The basic steps in cDNA synthesis are described below.

There are approximately 21,000 human protein-encoding genes. Around 8000 are ubiquitously expressed (that is, transcribed and translated) in all tissues and have functions in common with all cells: DNA replication, energy metabolism, regulation of transcription, translation, and so on. The expression of the other 13,000 genes is thought to be somewhat specific between the cell types, tissues, developmental states, or any circumstances that make a cell type or condition unique. For example, genes important for the function of blood should only be expressed in blood cells, and genes important for liver function should only be found in liver. A direct approach to identifying liver-specific genes would be to isolate all the proteins in a sample of liver and identify them by sequencing. This is technically challenging and is beyond the expertise of many laboratories.

Another approach is to isolate and study the mRNAs in liver. These encode the proteins found in liver and you could generate a list of genes that appear to be liver-specific. However, studying mRNA is technically challenging, as mRNA is very labile so only the most meticulous handling will prevent it from breaking down quickly. A technically easier approach to studying mRNA is to make a complementary DNA copy of the mRNA and then sequence this copy. Complementary DNA, or cDNA, is very stable, easily handled, and sequenced with little difficulty. You still have the challenge of isolating and properly handling mRNA from cells, but once you have cDNA, your success is almost guaranteed. A brief explanation of mRNA and cDNA synthesis will help you understand what you are looking at in DNA databases. This will also be covered in different detail in Chapter 5.

Can measuring mRNA allow good estimates of protein levels? Be aware that not all transcripts are translated. We'll learn in Chapter 10 that there are possibly thousands of genes that have RNA as their end product and are never translated. Furthermore, translation rates of transcripts vary considerably so there may not be a direct correlation between protein levels and the abundance of an mRNA.

mRNA is a long polymer with a "cap" at the beginning (5P end, pronounced "five prime") and tail of AAAs (**poly(A) tail**) at the 3P end (**Figure 1.2A**). cDNA

(i) "Tissue-specific" expression

The transcripts of most genes can be found in more than one tissue. Rather than think of genes as tissue-specific, it may be more accurate to think that gene expression is more "selective" for one tissue or cell type. Nevertheless, it is very common to refer to genes as being tissue-specific.

synthesis is begun by mixing the mRNA with the components necessary for the synthesis of cDNA. These components include the individual nucleotides (represented here as A, T, G, and C) and an enzyme called reverse transcriptase. As the name of this enzyme might suggest, it makes cDNA out of RNA. The enzyme requires a starting point: it needs a short stretch of DNA, called a primer, already sitting on the mRNA as a place to begin cDNA synthesis. Reverse transcriptase, like other DNA-synthesizing enzymes, can only begin at the 3P end of the primer. If you add a primer of poly(T) DNA to the reaction, it will base-pair with the poly(A) tail in the opposite orientation (Figure 1.2B).

Reverse transcription will then begin at the 3P end of the poly(T) and extend the DNA synthesis toward the 5P end of the mRNA (Figure 1.2C). The mRNA is then removed from the reaction, leaving only single-stranded cDNA behind (Figure 1.2D). The second strand of DNA now needs to be synthesized, but what will act as the 3P primer for this reaction? Multiple solutions have been devised but an early way to prime the reaction was to allow the cDNA to fold back on itself and self-prime (Figure 1.2E). The second strand synthesis was then completed (Figure 1.2F). Subsequent steps then clone the cDNA into vectors suitable for cloning and sequencing.

cDNA synthesis does have limitations. Synthesis of the first strand is not always efficient and the reverse transcriptase may fall off the mRNA before reaching the 5P end of the message. Since much of the early cDNA synthesis started with poly(T) priming, cDNAs in databases are often biased toward the 3P end. Reverse transcriptase is also error-prone so there may be mistakes in the sequence.

One type of cDNA mentioned above in Section 1.3 is called an Expressed Sequence Tag or EST. ESTs are often less than 500 nucleotides long even though they were derived from mRNAs thousands of nucleotides in length. What they lack in length is balanced out by quantity: ESTs are synthesized and sequenced in very high numbers (thousands). The result is often a very thorough sampling of the mRNAs expressed in that cell line, tissue, or organ. If nothing is done to normalize the mRNA population, the cDNA synthesized will be proportional to the abundance of the various mRNAs found in those cells. That is, abundant mRNAs will give rise to most of the ESTs, and rare mRNAs will give rise to rare ESTs or none at all. When you randomly pick ESTs to sequence, by chance you will sequence the

Figure 1.2 Synthesis of a cDNA from an mRNA. (A) mRNA. (B) A DNA primer is attached to the poly(A) tail. (C) Reverse transcriptase extends the cDNA to the 5P end of the mRNA. (D) cDNA after removal of the mRNA. (E) Formation of a "hairpin" at the 3P end of the cDNA acts as a primer for synthesis of the complementary strand (F).

(A)

    mRNA 5P-GAAUUCACGUGGGAAUUCGCAGCAAAAUGAUGCAUAGCUCGCUGAUAGCUUUGAUAAAAAAAAAAAAAAA-3P

(B)

    mRNA 5P-GAAUUCACGUGGGAAUUCGCAGCAAAAUGAUGCAUAGCUCGCUGAUAGCUUUGAUAAAAAAAAAAAAAAA-3P
                                                                  3P-**TTTTTTTTTTT**-5P

(C)

    mRNA 5P-GAAUUCACGUGGGAAUUCGCAGCAAAAUGAUGCAUAGCUCGCUGAUAGCUUUGAUAAAAAAAAAAAAAAA-3P
    cDNA 3P-**CTTAAGTGCACCCTTAAGCGTCGTTTTACTACGTATCGAGCGACTATCGAAACTATTTTTTTTTTTTTTT**-5P

(D)

    cDNA 3P-**CTTAAGTGCACCCTTAAGCGTCGTTTTACTACGTATCGAGCGACTATCGAAACTATTTTTTTTTTTTTTT**-5P

(E)

                      GT
                    C   GAATTC-3P
                    A   CTTAAGCGTCGTTTTACTACGTATCGAGCGACTATCGAAACTATTTTTTTTTTTTTTT-5P
                      CC

(F)

                      GT
                    C   GAATTCGCAGCAAAATGAUGCATAGCTCGCTGATAGCTTTGATAAAAAAAAAAAAAAA-3P
                    A   CTTAAGCGTCGTTTTACTACGTATCGAGCGACTATCGAAACTATTTTTTTTTTTTTTT-5P
                      CC

abundant cDNAs repeatedly. So scientists will sequence ESTs by the thousands to find those rare EST sequences. ESTs will be discussed further in Chapter 5.

1.10 SEQUENCE ANALYSIS AND DATA DISPLAY

The following example illustrates a very simple sequence analysis problem. As the analysis progresses, the display of data changes, demonstrating some of the variety of styles that you will see in this book.

Figure 1.3 shows the mRNA transcript from a human gene called Factor IX (pronounced "factor nine"). The Factor IX gene encodes a protein critical to the cascade of proteins that respond and work together to properly clot blood. Transcript sequences are conventionally shown in databases and in many publications as cDNA sequences, using "T" instead of "U." In this form, the sequence is mostly uninformative, not providing any details except for general impressions about the nucleotide content and length (the sequence in Figure 1.3 is 2802 nucleotides long).

But what if you knew two simple rules: protein-coding regions begin with "ATG" and end with "TAA," "TGA," or "TAG." **Figure 1.4** shows bold and underlined the

Figure 1.3 The sequence of the mRNA for human Factor IX. In GenBank and many other databases, sequence files are given unique identification numbers called "accession numbers." This sequence is from GenBank accession number NM_000133.

```
ACCACTTTCACAATCTGCTAGCAAAGGTTATGCAGCGCGTGAACATGATCATGGCAGAATCACCAGGCCT
CATCACCATCTGCCTTTTAGGATATCTACTCAGTGCTGAATGTACAGTTTTTCTTGATCATGAAAACGCC
AACAAAATTCTGAATCGGCCAAAGAGGTATAATTCAGGTAAATTGGAAGAGTTTGTTCAAGGGAACCTTG
AGAGAGAATGTATGGAAGAAAGTGTAGTTTTGAAGAAGCACGAGAAGTTTTTGAAAACACTGAAAGAAC
AACTGAATTTTGGAAGCAGTATGTTGATGGAGATCAGTGTGAGTCCAATCCATGTTTAAATGGCGGCAGT
TGCAAGGATGACATTAATTCCTATGAATGTTGGTGTCCCTTTGGATTTGAAGGAAAGAACTGTGAATTAG
ATGTAACATGTAACATTAAGAATGGCAGATGCGAGCAGTTTTGTAAAAATAGTGCTGATAACAAGGTGGT
TTGCTCCTGTACTGAGGGATATCGACTTGCAGAAAACCAGAAGTCCTGTGAACCAGCAGTGCCATTTCCA
TGTGGAAGAGTTTCTGTTTCACAAACTTCTAAGCTCACCCGTGCTGAGACTGTTTTTCCTGATGTGGACT
ATGTAAATTCTACTGAAGCTGAAACCATTTTGGATAACATCACTCAAAGCACCCAATCATTTAATGACTT
CACTCGGGTTGTTGGTGGAGAAGATGCCAAACCAGGTCAATTCCCTTGGCAGGTTGTTTTGAATGGTAAA
GTTGATGCATTCTGTGGAGGCTCTATCGTTAATGAAAAATGGATTGTAACTGCTGCCCACTGTGTTGAAA
CTGGTGTTAAAATTACAGTTGTCGCAGGTGAACATAATATTGAGGAGACAGAACATACAGAGCAAAAGCG
AAATGTGATTCGAATTATTCCTCACCACAACTACAATGCAGCTATTAATAAGTACAACCATGACATTGCC
CTTCTGGAACTGGACGAACCCTTAGTGCTAAACAGCTACGTTACACCTATTTGCATTGCTGACAAGGAAT
ACACGAACATCTTCCTCAAATTTGGATCTGGCTATGTAAGTGGCTGGGGAAGAGTCTTCCACAAAGGGAG
ATCAGCTTTAGTTCTTCAGTACCTTAGAGTTCCACTTGTTGACCGAGCCACATGTCTTCGATCTACAAAG
TTCACCATCTATAACAACATGTTCTGTGCTGGCTTCCATGAAGGAGGTAGAGATTCATGTCAAGGAGATA
GTGGGGGACCCCATGTTACTGAAGTGGAAGGGACCAGTTTCTTAACTGGAATTATTAGCTGGGGTGAAGA
GTGTGCAATGAAAGGCAAATATGGAATATATACCAAGGTATCCCGGTATGTCAACTGGATTAAGGAAAAA
ACAAAGCTCACTTAATGAAAGATGGATTTCCAAGGTTAATTCATTGGAATTGAAAATTAACAGGGCCTCT
CACTAACTAATCACTTTCCCATCTTTTGTTAGATTTGAATATATACATTCTATGATCATTGCTTTTTTCTC
TTTACAGGGGAGAATTTCATATTTTACCTGAGCAAATTGATTAGAAAATGGAACCACTAGAGGAATATAA
TGTGTTAGGAAATTACAGTCATTTCTAAGGGCCCAGCCCTTGACAAAATTGTGAAGTTAAATTCTCCACT
CTGTCCATCAGATACTATGGTTCTCCACTATGGCAACTAACTCACTCAATTTTCCCTCCTTAGCAGCATT
CCATCTTCCCGATCTTCTTTGCTTCTCCAACCAAAACATCAATGTTTATTAGTTCTGTATACAGTACAGG
ATCTTTGGTCTACTCTATCACAAGGCCAGTACCACACTCATGAAGAAGAACACAGGAGTAGCTGAGAGG
CTAAAACTCATCAAAAACACTACTCCTTTTCCTCTACCCTATTCCTCAATCTTTTACCTTTTCCAAATCC
CAATCCCAAATCAGTTTTTCTCTTTCTTACTCCCTCTCTCCCTTTTACCCTCCATGGTCGTTAAAGGAG
AGATGGGGAGCATCATTCTGTTATACTTCTGTACACAGTTATACATGTCTATCAAACCCAGACTTGCTTC
CGTAGTGGAGACTTGCTTTTCAGAACATAGGGATGAAGTAAGGTGCCTGAAAAGTTTGGGGGAAAAGTTT
CTTTCAGAGAGTTAAGTTATTTTATATATATAATATATATATAAAATATATAATATACAATATAAATATA
TAGTGTGTGTGTATGCGTGTGTGTAGACACACACGCATACACACATATAATGGAAGCAATAAGCCATTCT
AAGAGCTTGTATGGTTATGGAGGTCTGACTAGGCATGATTTCACGAAGGCAAGATTGGCATATCATTGTA
ACTAAAAAAGCTGACATTGACCCAGACATATTGTACTCTTTCTAAAAATAATAATAATAATGCTAACAGA
AAGAAGAGAACCGTTCGTTTGCAATCTACAGCTAGTAGAGACTTTGAGGAAGAATTCAACAGTGTGTCTT
CAGCAGTGTTCAGAGCCAAGCAAGAAGTTGAAGTTGCCTAGACCAGAGGACATAAGTATCATGTCTCCTT
TAACTAGCATACCCCGAAGTGGAGAAGGGTGCAGCAGGCTCAAAGGCATAAGTCATTCCAATCAGCCAAC
TAAGTTGTCCTTTTCTGGTTTCGTGTTCACCATGGAACATTTTGATTATAGTTAATCCTTCTATCTTGAA
TCTTCTAGAGAGTTGCTGACCAACTGACGTATGTTTCCCTTTGTGAATTAATAAACTGGTGTTCTGGTTC
AT
```

"ATG" and "TAA" triplets that bound the protein-coding region. You can now see that there are regions upstream and downstream of the coding region that do not code for protein. These are the 5P and 3P **untranslated regions (UTRs)**, respectively.

But these aren't perfect rules. If you closely look at the sequence, there are many other instances of "ATG" and "TAA." However, there are some additional constraints to consider. ATG can appear multiple times in a gene sequence, but often (but not always—this is biology!) the first ATG is used to start the coding region, as in this sequence. There are no other ATGs upstream of the one indicated in Figure 1.4.

The protein-coding region is read three nucleotides at a time, starting at the ATG. These are called codons. The coding sequence can now be formatted to show all the codons (**Figure 1.5**). Although straightforward, this grouping step is completely dependent on the sequence being accurate. If the sequence was incorrect and a single nucleotide was missing or inserted, the grouping of three would be completely wrong from that point onward. A mistake involving two nucleotides would also be incorrect. However, if the insertion or deletion were a multiple of three, an event such as this would only have obvious consequences at the point of change, as all the other codons would be correct.

```
ACCACTTTCACAATCTGCTAGCAAAGGTT**ATG**CAGCGCGTGAACATGATCATGGCAGAATCACCAGGCCT
CATCACCATCTGCCTTTTAGGATATCTACTCAGTGCTGAATGTACAGTTTTTCTTGATCATGAAAACGCC
AACAAAATTCTGAATCGGCCAAAGAGGTATAATTCAGGTAAATTGGAAGAGTTTGTTCAAGGGAACCTTG
AGAGAGAATGTATGGAAGAAAGTGTAGTTTTGAAGAAGCACGAGAAGTTTTTGAAAACACTGAAAGAAC
AACTGAATTTTGGAAGCAGTATGTTGATGGAGATCAGTGTGAGTCCAATCCATGTTTAAATGGCGGCAGT
TGCAAGGATGACATTAATTCCTATGAATGTTGGTGTCCCTTTGGATTTGAAGGAAAGAACTGTGAATTAG
ATGTAACATGTAACATTAAGAATGGCAGATGCGAGCAGTTTTGTAAAAATAGTGCTGATAACAAGGTGGT
TTGCTCCTGTACTGAGGGATATCGACTTGCAGAAAACCAGAAGTCCTGTGAACCAGCAGTGCCATTTCCA
TGTGGAAGAGTTTCTGTTTCACAAACTTCTAAGCTCACCCGTGCTGAGACTGTTTTTCCTGATGTGGACT
ATGTAAATTCTACTGAAGCTGAAACCATTTTGGATAACATCACTCAAAGCACCCAATCATTTAATGACTT
CACTCGGGTTGTTGGTGGAGAAGATGCCAAACCAGGTCAATTCCCTTGGCAGGTTGTTTTGAATGGTAAA
GTTGATGCATTCTGTGGAGGCTCTATCGTTAATGAAAAATGGATTGTAACTGCTGCCCACTGTGTTGAAA
CTGGTGTTAAAATTACAGTTGTCGCAGGTGAACATAATATTGAGGAGACAGAACATACAGAGCAAAAGCG
AAATGTGATTCGAATTATTCCTCACCACAACTACAATGCAGCTATTAATAAGTACAACCATGACATTGCC
CTTCTGGAACTGGACGAACCCCTTAGTGCTAAACAGCTACGTTACACCTATTTGCATTGCTGACAAGGAAT
ACACGAACATCTTCCTCAAATTTGGATCTGGCTATGTAAGTGGCTGGGGAAGAGTCTTCCACAAAGGGAG
ATCAGCTTTAGTTCTTCAGTACCTTAGAGTTCCACTTGTTGACCGAGCCACATGTCTTCGATCTACAAAG
TTCACCATCTATAACAACATGTTCTGTGCTGGCTTCCATGAAGGAGGTAGAGATTCATGTCAAGGAGATA
GTGGGGGACCCCATGTTACTGAAGTGGAAGGGACCAGTTTCTTAACTGGAATTATTAGCTGGGGTGAAGA
GTGTGCAATGAAAGGCAAATATGGAATATATACCAAGGTATCCCGGTATGTCAACTGGATTAAGGAAAAA
ACAAAGCTCACT**TAA**TGAAAGATGGATTTCCAAGGTTAATTCATTGGAATTGAAAATTAACAGGGCCTCT
CACTAACTAATCACTTTCCCATCTTTTGTTAGATTTGAATATATACATTCTATGATCATTGCTTTTTCTC
TTTACAGGGGAGAATTTCATATTTTACCTGAGCAAATTGATTAGAAAATGGAACCACTAGAGGAATATAA
TGTGTTAGGAAATTACAGTCATTTCTAAGGGCCCAGCCCTTGACAAAATTGTGAAGTTAAATTCTCCACT
CTGTCCATCAGATACTATGGTTCTCCACTATGGCAACTAACTCACTCAATTTTCCCTCCTTAGCAGCATT
CCATCTTCCCGATCTTCTTTGCTTCTCCAACCAAAACATCAATGTTATTAGTTCTGTATACAGTACAGG
ATCTTTGGTCTACTCTATCACAAGGCCAGTACCACACTCATGAAGAAAGAACACAGGAGTAGCTGAGAGG
CTAAAACTCATCAAAAACACTACTCCTTTTCCTCTACCCTATTCCTCAATCTTTTACCTTTTCCAAATCC
CAATCCCCAAATCAGTTTTTCTCTTTCTTACTCCCTCTCTCCCTTTTACCCTCCATGGTCGTTAAAGGAG
AGATGGGGAGCATCATTCTGTTATACTTCTGTACACAGTTATACATGTCTATCAAACCCAGACTTGCTTC
CGTAGTGGAGACTTGCTTTTCAGAACATAGGGATGAAGTAAGGTGCCTGAAAAGTTTGGGGGAAAAGTTT
CTTTCAGAGAGTTAAGTTATTTTATATATATAATATATATATAAAATATATAATATACAATATAAATATA
TAGTGTGTGTGTATGCGTGTGTGTGTAGACACACACGCATACACACATATAATGGAAGCAATAAGCCATTCT
AAGAGCTTGTATGGTTATGGAGGTCTGACTAGGCATGATTTCACGAAGGCAAGATTGGCATATCATTGTA
ACTAAAAAAGCTGACATTGACCCAGACATATTGTACTCTTTCTAAAAATAATAATAATAATGCTAACAGA
AAGAAGAGAACCGTTCGTTTGCAATCTACAGCTAGTAGAGACTTTGAGGAAGAATTCAACAGTGTGTCTT
CAGCAGTGTTCAGAGCCAAGCAAGAAGTTGAAGTTGCCTAGACCAGAGGACATAAGTATCATGTCTCCTT
TAACTAGCATACCCCGAAGTGGAGAAGGGGTGCAGCAGGCTCAAAGGCATAAGTCATTCCAATCAGCCAAC
TAAGTTGTCCTTTTCTGGTTTCGTGTTCACCATGGAACATTTTGATTATAGTTAATCCTTCTATCTTGAA
TCTTCTAGAGAGTTGCTGACCAACTGACGTATGTTTCCCTTTGTGAATTAATAAACTGGTGTTCTGGTTC
AT
```

Figure 1.4 Applying two rules for describing the human Factor IX mRNA sequence. Those two rules are (a) coding regions begin with "ATG" and (b) coding regions end with one of three terminator sequences, "TAA," "TGA," or "TAG." Two of the many possible matches to these rules are in bold and underlined.

```
ACCACTTTCACAATCTGCTAGCAAAGGTT
ATG CAG CGC GTG AAC ATG ATC ATG GCA GAA TCA CCA GGC CTC ATC ACC ATC TGC CTT TTA GGA TAT CTA CTC AGT
GCT GAA TGT ACA GTT TTT CTT GAT CAT GAA AAC GCC AAC AAA ATT CTG AAT CGG CCA AAG AGG TAT AAT TCA GGT
AAA TTG GAA GAG TTT GTT CAA GGG AAC CTT GAG AGA GAA TGT ATG GAA GAA AAG TGT AGT TTT GAA GAA GCA CGA
GAA GTT TTT GAA AAC ACT GAA AGA ACA ACT GAA TTT TGG AAG CAG TAT GTT GAT GGA GAT CAG TGT GAG TCC AAT
CCA TGT TTA AAT GGC GGC AGT TGC AAG GAT GAC ATT AAT TCC TAT GAA TGT TGG TGT CCC TTT GGA TTT GAA GGA
AAG AAC TGT GAA TTA GAT GTA ACA TGT AAC ATT AAG AAT GGC AGA TGC GAG CAG TTT TGT AAA AAT AGT GCT GAT
AAC AAG GTG GTT TGC TCC TGT ACT GAG GGA TAT CGA CTT GCA GAA AAC CAG AAG TCC TGT GAA CCA GCA GTG CCA
TTT CCA TGT GGA AGA GTT TCT GTT TCA CAA ACT TCT AAG CTC ACC CGT GCT GAG ACT GTT TTT CCT GAT GTG GAC
TAT GTA AAT TCT ACT GAA GCT GAA ACC ATT TTG GAT AAC ATC ACT CAA AGC ACC CAA TCA TTT AAT GAC TTC ACT
CGG GTT GTT GGT GGA GAA GAT GCC AAA CCA GGT CAA TTC CCT TGG CAG GTT GTT TTG AAT GGT AAA GTT GAT GCA
TTC TGT GGA GGC TCT ATC GTT AAT GAA AAA TGG ATT GTA ACT GCT GCC CAC TGT GTT GAA ACT GGT GTT AAA ATT
ACA GTT GTC GCA GGT GAA CAT AAT ATT GAG GAG ACA GAA CAT ACA GAG CAA AAG CGA AAT GTG ATT CGA ATT ATT
CCT CAC CAC AAC TAC AAT GCA GCT ATT AAT AAG TAC AAC CAT GAC ATT GCC CTT CTG GAA CTG GAC GAA CCC TTA
GTG CTA AAC AGC TAC GTT ACA CCT ATT TGC ATT GCT GAC AAG GAA TAC ACG AAC ATC TTC CTC AAA TTT GGA TCT
GGC TAT GTA AGT GGC TGG GGA AGA GTC TTC CAC AAA GGG AGA TCA GCT TTA GTT CTT CAG TAC CTT AGA GTT CCA
CTT GTT GAC CGA GCC ACA TGT CTT CGA TCT ACA AAG TTC ACC ATC TAT AAC AAC ATG TTC TGT GCT GGC TTC CAT
GAA GGA GGT AGA GAT TCA TGT CAA GGA GAT AGT GGG GGA CCC CAT GTT ACT GAA GTG GAA GGG ACC AGT TTC TTA
ACT GGA ATT ATT AGC TGG GGT GAA GAG TGT GCA ATG AAA GGC AAA TAT GGA ATA TAT ACC AAG GTA TCC CGG TAT
GTC AAC TGG ATT AAG GAA AAA ACA AAG CTC ACT TAA
TGAAAGATGGATTTCCAAGGTTAATTCATTGGAATTGAAAATTAACAGGGCCTCTCACTAACTAATCACTTTCCCATCTTTTGTTAGATTTGAATATATACA
TTCTATGATCATTGCTTTTTCTCTTTACAGGGGAGAATTTCATATTTTACCTGAGCAAATTGATTAGAAAATGGAACCACTAGAGGAATATAATGTGTTAGG
AAATTACAGTCATTTCTAAGGGCCCAGCCCTTGACAAAATTGTGAAGTTAAATTCTCCACTCTGTCCATCAGATACTATGGTTCTCCACTATGGCAACTAAC
TCACTCAATTTTCCCTCCTTAGCAGCATTCCATCTTCCCGATCTTCTTTGCTTCTCCAACCAAAACATCAATGTTTATTAGTTCTGTATACAGTACAGGATC
TTTGGTCTACTCTATCACAAGGCCAGTACCACACTCATGAAGAAACAACACAGGAGTAGCTGAGAGGCTAAAACTCATCAAAAACACTACTCCTTTTCCTCT
ACCCTATTCCTCAATCTTTTACCTTTTCCAAATCCCAATCCCCAAATCAGTTTTTCTCTTTCTTACTCCCTCTCTCCCTTTTACCCTCCATGGTCGTTAAAG
GAGAGATGGGGAGCATCATTCTGTTATACTTCTGTACACAGTTATACATGTCTATCAAACCCAGACTTGCTTCCGTAGTGGAGACTTGCTTTTCAGAACATA
GGGATGAAGTAAGGTGCCTGAAAAGTTTGGGGGAAAAGTTTCTTTCAGAGAGTTAAGTTATTTTATATATATAATATATATATAAAATATATAATATACAAT
ATAAATATATAGTGTGTGTGTATGCGTGTGTGTAGACACACACGCATACACACATATAATGGAAGCAATAAGCCATTCTAAGAGCTTGTATGGTTATGGAGG
TCTGACTAGGCATGATTTCACGAAGGCAAGATTGGCATATCATTGTAACTAAAAAAGCTGACATTGACCCAGACATATTGTACTCTTTCTAAAAATAATAAT
AATAATGCTAACAGAAAGAAGAGAACCGTTCGTTTGCAATCTACAGCTAGTAGAGACTTTGAGGAAGAATTCAACAGTGTGTCTTCAGCAGTGTTCAGAGCC
AAGCAAGAAGTTGAAGTTGCCTAGACCAGAGGACATAAGTATCATGTCTCCTTTAACTAGCATACCCCGAAGTGGAGAAGGGTGCAGCAGGCTCAAAGGCAT
AAGTCATTCCAATCAGCCAACTAAGTTGTCCTTTTCTGGTTTCGTGTTCACCATGGAACATTTTGATTATAGTTAATCCTTCTATCTTGAATCTTCTAGAGA
GTTGCTGACCAACTGACGTATGTTTCCCTTTGTGAATTAATAAACTGGTGTTCTGGTTCAT
```

Figure 1.5 Coding regions are read as triplets. The human Factor IX mRNA with the start and termination codons bold and underlined. The coding region has been divided into the three-base codons. The 5P and 3P untranslated regions (5P and 3P UTRs, respectively) appear before and after the coding region.

Scanning by eye, you can see that there are no other terminator codons—TAA, TAG, or TGA—within this coding region. But there are other ATG triplets, including two just downstream of the first one. All of these codons are translated into a polypeptide chain according to the genetic code.

If you wanted to, you could count the 462 codons to deduce the length of the protein as 461 amino acids (the terminator codon does not encode an amino acid). There are many software programs to do this for you, but you should also get into the habit of examining a sequence by eye as well. After all, software only finds things it is designed to look for but you may notice something that is not yet described.

There are programs that take DNA sequence as input and translate this into an amino acid sequence using the genetic code (to be covered later, in Chapter 4). **Figure 1.6** shows this translation below each codon using the one-letter code for the amino acids. This figure is a little complex since it includes both the nucleotides, some grouped as three-letter codons, and one-letter amino acids. It has its value, though; for example, you can see that there can be multiple codons for each amino acid. Methionine (M) is always ATG, and tryptophan (W) is always TGG, but valine (V) can be GTG, GTT, GTA, or GTC.

Figure 1.7, which shows just the amino acid sequence, is a much simpler figure to study. If you knew the biochemical properties of the amino acids, you might recognize regions that are hydrophilic or hydrophobic. Based on the sequence you see, regions of amino acids that tend to fold into helical structures or sheets might be noticed. Or you might recognize certain groups of amino acids that often have attached sugar groups. These structural features may tell a story about

ACCACTTTCACAATCTGCTAGCAAAGGTT

```
ATG CAG CGC GTG AAC ATG ATC ATG GCA GAA TCA CCA GGC CTC ATC ACC ATC TGC CTT TTA GGA TAT CTA CTC AGT
 M   Q   R   V   N   M   I   M   A   E   S   P   G   L   I   T   I   C   L   L   G   Y   L   L   S

GCT GAA TGT ACA GTT TTT CTT GAT CAT GAA AAC GCC AAC AAA ATT CTG AAT CGG CCA AAG AGG TAT AAT TCA GGT
 A   E   C   T   V   F   L   D   H   E   N   A   N   K   I   L   N   R   P   K   R   Y   N   S   G

AAA TTG GAA GAG TTT GTT CAA GGG AAC CTT GAG AGA GAA TGT ATG GAA GAA AAG TGT AGT TTT GAA GCA CGA
 K   L   E   E   F   V   Q   G   N   L   E   R   E   C   M   E   E   K   C   S   F   E   E   A   R

GAA GTT TTT GAA AAC ACT GAA AGA ACA ACT GAA TTT TGG AAG CAG TAT GTT GAT GGA GAT CAG TGT GAG TCC AAT
 E   V   F   E   N   T   E   R   T   T   E   F   W   K   Q   Y   V   D   G   D   Q   C   E   S   N

CCA TGT TTA AAT GGC GGC AGT TGC AAG GAT GAC ATT AAT TCC TAT GAA TGT TGG TGT CCC TTT GGA TTT GAA GGA
 P   C   L   N   G   G   S   C   K   D   D   I   N   S   Y   E   C   W   C   P   F   G   F   E   G

AAG AAC TGT GAA TTA GAT GTA ACA TGT AAC ATT AAG AAT GGC AGA TGC GAG CAG TTT TGT AAA AAT AGT GCT GAT
 K   N   C   E   L   D   V   T   C   N   I   K   N   G   R   C   E   Q   F   C   K   N   S   A   D

AAC AAG GTG GTT TGC TCC TGT ACT GAG GGA TAT CGA CTT GCA GAA AAC CAG AAG TCC TGT GAA CCA GCA GTG CCA
 N   K   V   V   C   S   C   T   E   G   Y   R   L   A   E   N   Q   K   S   C   E   P   A   V   P

TTT CCA TGT GGA AGA GTT TCT GTT TCA CAA ACT TCT AAG CTC ACC CGT GCT GAG ACT GTT TTT CCT GAT GTG GAC
 F   P   C   G   R   V   S   V   S   Q   T   S   K   L   T   R   A   E   T   V   F   P   D   V   D

TAT GTA AAT TCT ACT GAA GCT GAA ACC ATT TTG GAT AAC ATC ACT CAA AGC ACC CAA TCA TTT AAT GAC TTC ACT
 Y   V   N   S   T   E   A   E   T   I   L   D   N   I   T   Q   S   T   Q   S   F   N   D   F   T

CGG GTT GTT GGT GGA GAA GAT GCC AAA CCA GGT CAA TTC CCT TGG CAG GTT GTT TTG AAT GGT AAA GTT GAT GCA
 R   V   V   G   G   E   D   A   K   P   G   Q   F   P   W   Q   V   V   L   N   G   K   V   D   A

TTC TGT GGA GGC TCT ATC GTT AAT GAA AAA TGG ATT GTA ACT GCT GCC CAC TGT GTT GAA ACT GGT GTT AAA ATT
 F   C   G   G   S   I   V   N   E   K   W   I   V   T   A   A   H   C   V   E   T   G   V   K   I

ACA GTT GTC GCA GGT GAA CAT AAT ATT GAG GAG ACA GAG CAT ACA GAG CAA AAG CGA AAT GTG ATT CGA ATT ATT
 T   V   V   A   G   E   H   N   I   E   E   T   E   H   T   E   Q   K   R   N   V   I   R   I   I

CCT CAC CAC AAC TAC AAT GCA GCT ATT AAT AAG TAC AAC CAT GAC ATT GCC CTT CTG GAA CTG GAC GAA CCC TTA
 P   H   H   N   Y   N   A   A   I   N   K   Y   N   H   D   I   A   L   L   E   L   D   E   P   L

GTG CTA AAC AGC TAC GTT ACA CCT ATT TGC ATT GCT GAC AAG GAA TAC ACG AAC ATC TTC CTC AAA TTT GGA TCT
 V   L   N   S   Y   V   T   P   I   C   I   A   D   K   E   Y   T   N   I   F   L   K   F   G   S

GGC TAT GTA AGT GGC TGG GGA AGA GTC TTC CAC AAA GGG AGA TCA GCT TTA GTT CTT CAG TAC CTT AGA GTT CCA
 G   Y   V   S   G   W   G   R   V   F   H   K   G   R   S   A   L   V   L   Q   Y   L   R   V   P

CTT GTT GAC CGA GCC ACA TGT CTT CGA TCT ACA AAG TTC ACC ATC TAT AAC AAC ATG TTC TGT GCT GGC TTC CAT
 L   V   D   R   A   T   C   L   R   S   T   K   F   T   I   Y   N   N   M   F   C   A   G   F   H

GAA GGA GGT AGA GAT TCA TGT CAA GGG GAT AGT GGG GGA CCC CAT GTT ACT GAA GTG GAA GGG ACC AGT TTC TTA
 E   G   G   R   D   S   C   Q   G   D   S   G   G   P   H   V   T   E   V   E   G   T   S   F   L

ACT GGA ATT ATT AGC TGG GGT GAA GAG TGT GCA ATG AAA GGC AAA TAT GGA ATA TAT ACC AAG GTA TCC CGG TAT
 T   G   I   I   S   W   G   E   E   C   A   M   K   G   K   Y   G   I   Y   T   K   V   S   R   Y

GTC AAC TGG ATT AAG GAA AAA ACA AAG CTC ACT TAA
 V   N   W   I   K   E   K   T   K   L   T   Stop
```

TGAAAGATGGATTTCCAAGGTTAATTCATTGGAATTGAAAATTAACAGGGCCTCTCACTAACTAATCACTTTCCCATCTTTTGTTAGATTTGAATATATACA
TTCTATGATCATTGCTTTTTCTCTTTACAGGGGAGAATTTCATATTTTACCTGAGCAAATTGATTAGAAAATGGAACCACTAGAGGAATATAATGTGTTAGG
AAATTACAGTCATTTCTAAGGGCCCAGCCCTTGACAAAATTGTGAAGTTAAATTCTCCACTCTGTCCATCAGATACTATGGTTCTCCACTATGGCAACTAAC
TCACTCAATTTTCCCTCCTTAGCAGCATTCCATCTTCCCGATCTTCTTTGCTTCTCCAACAAAACATCAATGTTTATTAGTTCTGTATACAGTACAGGATC
TTTGGTCTACTCTATCACAAGGCCAGTACCACACTCATGAAGAAAGAACACAGGAGTAGCTGAGAGGCTAAAACTCATCAAAAACACTACTCCTTTTCCTCT
ACCCTATTCCTCAATCTTTTACCTTTTCCAAATCCCAATCCCCAAATCAGTTTTTCTCTTTCTTACTCCCTCTCTCCCTTTTACCCTCCATGGTCGTTAAAG
GAGAGATGGGGAGCATCATTCTGTTATACTTCTGTACACAGTTATACATGTCTATCAAACCCAGACTTGCTTCCGTAGTGGAGACTTGCTTTTCAGAACATA
GGGATGAAGTAAGGTGCCTGAAAAGTTTGGGGGAAAAGTTTCTTTCAGAGAGTTAAGTTATTTTATATATATAATATATATATAAAATATATAATATACAAT
ATAAATATATAGTGTGTGTGTATGCGTGTGTGTAGACACACACGCATACACACATATAATGGAAGCAATAAGCCATTCTAAGAGCTTGTATGGTTATGGAGG
TCTGACTAGGCATGATTTCACGAAGGCAAGATTGGCATATCATTGTAACTAAAAAAGCTGACATTGACCCAGACATATTGTACTCTTTCTAAAAATAATAAT
AATAATGCTAACAGAAAGAAGAGAACCGTTCGTTTGCAATCTACAGCTAGTAGAGACTTTGAGGAAGAATTCAACAGTGTGTCTTCAGCAGTGTTCAGAGCC
AAGCAAGAAGTTGAAGTTGCCTAGACCAGAGGACAGCATAAGTATCATGTCTCCTTTAACTAGCATACCCCGAAGTGGAGAAGGGTGCAGCAGGCTCAAAGGCAT
AAGTCATTCCAATCAGCCACCTAAGTTGTCCTTTTCTGGTTTCGTGTTCACCATGGAACATTTTGATTATAGTTAATCCTTCTATCTTGAATCTTCTAGAG
AGTTGCTGACCAACTGACGTATGTTTCCCTTTGTGAATTAATAAACTGGTGTTCTGGTTCAT

Figure 1.6 Coding-region triplets are translated into amino acids. Each three-base codon can be translated into an amino acid using the genetic code. The one-letter representations for the amino acids appear in this figure (for example, "M" stands for methionine, "Q" stands for glutamine, and so on).

Figure 1.7 The protein sequence of human Factor IX. Here is the translation of the coding region appearing in Figure 1.6. This protein is 461 amino acids long.

```
MQRVNMIMAESPGLITICLLGYLLSAECTVFLDHENANKILNRPKRYNSGKLEEFVQGNL
ERECMEEKCSFEEAREVFENTERTTEFWKQYVDGDQCESNPCLNGGSCKDDINSYECWCP
FGFEGKNCELDVTCNIKNGRCEQFCKNSADNKVVCSCTEGYRLAENQKSCEPAVPFPCGR
VSVSQTSKLTRAETVFPDVDYVNSTEAETILDNITQSTQSFNDFTRVVGGEDAKPGQFPW
QVVLNGKVDAFCGGSIVNEKWIVTAAHCVETGVKITVVAGEHNIEETEHTEQKRNVIRII
PHHNYNAAINKYNHDIALLELDEPLVLNSYVTPICIADKEYTNIFLKFGSGYVSGWGRVF
HKGRSALVLQYLRVPLVDRATCLRSTKFTIYNNMFCAGFHEGGRDSCQGDSGGPHVTEVE
GTSFLTGIISWGEECAMKGKYGIYTKVSRYVNWIKEKTKLT
```

the function of this protein. Luckily, we have bioinformatics programs that will tell us about these biochemical properties, structures, and modifications. Protein analysis, along with structures and their visualization, will appear in Chapter 8.

This simple example illustrates how analysis of a raw sequence (Figure 1.3) can be broken down into steps and additional information can be extracted through the application of distinct rules. Knowing the intermediate steps makes you more aware of the dependencies. The next two examples show how sequences can be compared.

The Factor IX transcript, analyzed above, is a product of the Factor IX gene that is over 38,000 nucleotides long. A single nucleotide mutation, changing a G to a T at coordinate 25,531, results in hemophilia B, a severe bleeding disorder (**Figure 1.8**).

Single nucleotide changes elsewhere in the gene are quite common but tolerated, either because they do not change the protein sequence or do not disrupt splicing or any form of regulation, or do not make changes to the protein that biochemically compromise the function of Factor IX. A comparison between the human and chimpanzee Factor IX proteins (**Figure 1.9**) illustrates that conservative variation in at least one amino acid position can be tolerated in primates.

The Factor IX protein sequence can be divided into domains with different biochemical properties and functions. These domains interact with each other and with other proteins to function properly. They attain a three-dimensional structure via folding although they are depicted in a linear form in **Figure 1.10**. In this figure, the domains of interest are gray boxes while other amino acids, not described here, are in white boxes. Domain one, at the N-terminus of the protein, functions to direct the protein to the endoplasmic reticulum of liver cells, from where it is secreted into the blood. When the protein is secreted, this first domain is cleaved off by a protein called signal peptidase. Twelve glutamic acid residues in domain two (also called the "Gla" domain) are modified by the enzyme gamma-carboxylase to become gamma-carboxyglutamic acid residues. Domain three is an "epidermal growth factor (EGF)-like" domain that binds calcium. Skipping ahead, domain five is a peptidase domain that cleaves another protein in the clotting cascade, the X protein. This protease function only becomes active once the Factor IX protein is cleaved into two peptides, and this cut occurs in domain four. That is why domain four is called the "activation peptide." This cleavage results in two polypeptides, the Factor IX light chain consisting of domains two and three, and the Factor IX heavy chain consisting of domain five. The heavy and light chains remain covalently linked to each other by a disulfide bond between two cysteines. To transform from precursor to functional protein, Factor IX interacts with at least four other molecules, as described above.

Finally, here are two more views of the human Factor IX gene showing the context within genomic DNA. In **Figure 1.11A**, the entire 38,000-nucleotide gene is

Figure 1.8 Hemophilia B mutation. A simple pairwise alignment between a small portion of the normal sequence and a sequence from a hemophilia B patient is shown. A vertical bar links positions where the two sequences are identical. A single nucleotide change in the 38,059 base pair human Factor IX gene can cause hemophilia B. The sequence shown in this figure spans the location of the G-to-T mutation found at gene coordinate 25,531 in GenBank record K02402.

```
Normal     GATGCCAAACCAGGTCAATTCCCTTGGCAGGTACTTTATACTGATGGTGTGTCAAAACTG
           |||||||||||||||||||||||||||||||| |||||||||||||||||||||||||||
Mutation   GATGCCAAACCAGGTCAATTCCCTTGGCAGTTACTTTATACTGATGGTGTGTCAAAACTG
```

```
Query     1    MQRVNMIMAESPGLITICLLGYLLSAECTVFLDHENANKILNRPKRYNSGKLEEFVQGNL    60
               MQRVNMIMAESPGLITICLLGYLLSAECTVFLDHENANKILNRPKRYNSGKLEEFVQGNL
Sbjct     1    MQRVNMIMAESPGLITICLLGYLLSAECTVFLDHENANKILNRPKRYNSGKLEEFVQGNL    60

Query    61    ERECMEEKCSFEEAREVFENTERTTEFWKQYVDGDQCESNPCLNGGSCKDDINSYECWCP   120
               ERECMEEKCSFEEAREVFENTERTTEFWKQYVDGDQCESNPCLNGGSCKDDINSYECWCP
Sbjct    61    ERECMEEKCSFEEAREVFENTERTTEFWKQYVDGDQCESNPCLNGGSCKDDINSYECWCP   120

Query   121    FGFEGKNCELDVTCNIKNGRCEQFCKNSADNKVVCSCTEGYRLAENQKSCEPAVPFPCGR   180
               FGFEGKNCELDVTCNIKNGRCEQFCKNSADNKVVCSCTEGYRLAENQKSCEPAVPFPCGR
Sbjct   121    FGFEGKNCELDVTCNIKNGRCEQFCKNSADNKVVCSCTEGYRLAENQKSCEPAVPFPCGR   180

Query   181    VSVSQTSKLTRAETVFPDVDYVNSTEAETILDNITQSTQSFNDFTRVVGGEDAKPGQFPW   240
               VSVSQTSKLTRAETVFPDVDYVNSTEAETILDNITQSTQSFNDFTRVVGGEDAKPGQFPW
Sbjct   181    VSVSQTSKLTRAETVFPDVDYVNSTEAETILDNITQSTQSFNDFTRVVGGEDAKPGQFPW   240

Query   241    QVVLNGKVDAFCGGSIVNEKWIVTAAHCVETGVKITVVAGEHNIEETEHTEQKRNVIRII   300
               QVVLNGKVDAFCGGSIVNEKWIVTAAHCV+TGVKITVVAGEHNIEETEHTEQKRNVIRII
Sbjct   241    QVVLNGKVDAFCGGSIVNEKWIVTAAHCVDTGVKITVVAGEHNIEETEHTEQKRNVIRII   300

Query   301    PHHNYNAAINKYNHDIALLELDEPLVLNSYVTPICIADKEYTNIFLKFGSGYVSGWGRVF   360
               PHHNYNAAINKYNHDIALLELDEPLVLNSYVTPICIADKEYTNIFLKFGSGYVSGWGRVF
Sbjct   301    PHHNYNAAINKYNHDIALLELDEPLVLNSYVTPICIADKEYTNIFLKFGSGYVSGWGRVF   360

Query   361    HKGRSALVLQYLRVPLVDRATCLRSTKFTIYNNMFCAGFHEGGRDSCQGDSGGPHVTEVE   420
               HKGRSALVLQYLRVPLVDRATCLRSTKFTIYNNMFCAGFHEGGRDSCQGDSGGPHVTEVE
Sbjct   361    HKGRSALVLQYLRVPLVDRATCLRSTKFTIYNNMFCAGFHEGGRDSCQGDSGGPHVTEVE   420

Query   421    GTSFLTGIISWGEECAMKGKYGIYTKVSRYVNWIKEKTKLT    461
               GTSFLTGIISWGEECAMKGKYGIYTKVSRYVNWIKEKTKLT
Sbjct   421    GTSFLTGIISWGEECAMKGKYGIYTKVSRYVNWIKEKTKLT    461
```

Figure 1.9 Alignment of human (Query) and chimpanzee (Subject (Sbjct)) Factor IX proteins. Instead of a vertical bar to signify identity, as seen in Figure 1.8, this alignment places identical amino acids between the two sequences. There is only one amino acid difference over the 461 amino acid length (see if you can spot it). This change is biochemically conservative and so the space between the E (glutamic acid) and the D (aspartic acid) is indicated by "+" rather than a blank space indicating "no identity." The chimpanzee protein sequence is from the NCBI RefSeq sequence file NP_001129063.1 and the human sequence is from file NP_000124.1.

shown as the black arrow (labeled *F9*) near the center of the figure. Factor IX has several genetic neighbors including steroid-5-alpha-reductase, alpha polypeptide 1 pseudogene 1 (*SRD5A1P1*). A **pseudogene** is a sequence derived from a known functional gene but which is somehow defective structurally. There are over 15,000 pseudogenes in the human genome. These may represent opportunities for new functional genes to arise, or be defective genes that are destined to accumulate mutations until they are no longer recognizable as once being similar to any known gene. Another genomic neighbor is *MCF2* which encodes a protein capable of transforming normal tissue culture cells into a cancerous state. Note that the *MCF2* gene is over 126,000 nucleotides long and has the opposite orientation of the Factor IX gene, as indicated by the direction of the arrows. Genes can be found on both DNA strands and their sizes vary tremendously.

Figure 1.11B shows the location of the Factor IX gene on the X chromosome. Chromosomes are one long piece of genomic DNA with features that allow their identification and orientation. The X chromosome is 155 million nucleotides long, with a constriction near the center called the centromere. Throughout the chromosome are regions that, upon staining, show dark and light bands. Factor IX is one of 1500 genes and pseudogenes on this chromosome.

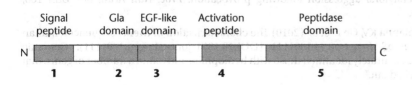

Figure 1.10 Factor IX protein domains. The Factor IX protein contains five major domains (gray boxes), each with specific functions in the molecule. They are depicted as boxes on a line representing the length of the protein, from N-terminus (N) to the C-terminus (C).

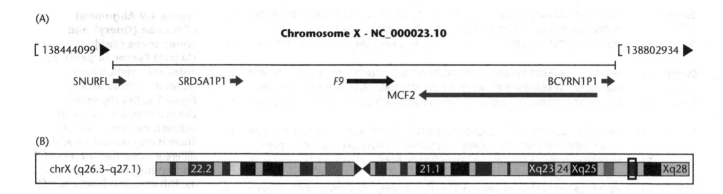

Figure 1.11 The position of the Factor IX gene on chromosome X. (A) Detail of the part of chromosome X that contains the Factor IX gene (the locus). The Factor IX gene is shown as a black arrow pointing to the right, 5P to 3P. It is labeled using the official gene symbol, "F9." Note that the genes shown differ in size and orientation. This screenshot is taken from the NCBI Gene database. (B) The human X chromosome with the banding pattern similar to what is seen with a light microscope. On the far right, the boxed area indicates the approximate region shown in (A). This screenshot is taken from the University of California at Santa Cruz Genome Browser.

1.11 SUMMARY

This chapter serves as an introduction to sequence analysis, which is, perhaps, the first specialty within bioinformatics and has become the cornerstone of interpreting the deluge of data that we are experiencing today. As this chapter described, sequence data come from animals, plants, and microbes and from research and medicine. In the coming chapters you will acquire the practical skills of using sequence analysis tools, giving you access to this wealth of information.

FURTHER READING

Anonymous (2011) Microbiology by numbers. *Nat. Rev. Microbiol.* 9, 628. A collection of interesting statistics about microbes.

Benson DA, Karsch-Mizrachi I, Lipman DJ et al. (2011) GenBank. *Nucleic Acids Res.* 39 (Database issue), D32–37.

Cordain L, Eaton SB, Sebastian A et al. (2005) Origins and evolution of the Western diet: health implications for the 21st century. *Am. J. Clin. Nutr.* 81, 341–354.

Feuk L, Carson AR & Scherer SW (2006) Structural variation in the human genome. *Nat. Rev. Genet.* 7, 85–97.

Gan X, Stegle O, Behr J et al. (2011) Multiple reference genomes and transcriptomes for *Arabidopsis thaliana. Nature* 477, 419–423.

Green RE, Krause J, Briggs AW et al. (2010) A draft sequence of the Neandertal genome. *Science* 328, 710–722.

Itan Y, Powell A, Beaumont MA et al. (2009) The origins of lactase persistence in Europe. *PLoS Comput. Biol.* 5, e1000491.

Johnson M, Gallagher K, Porter G et al. A baffling illness. Milwaukee Journal Sentinel, December 19, 2010. Sifting through the DNA haystack. Milwaukee Journal Sentinel, December 22, 2010. Embracing a risk. Milwaukee Journal Sentinel, December 26, 2010. An award-winning series of newspaper articles about the use of genomic DNA sequencing to help diagnose a patient.

Li J, Yang T, Wang L et al. (2009) Whole genome distribution and ethnic differentiation of copy number variation in Caucasian and Asian populations. *PLoS One* 4, e7958 (DOI: 10.1371/journal.pone.0007958).

Lupski JR, Reid JG, Gonzaga-Jauregui C et al. (2010) Whole-genome sequencing in a patient with Charcot-Marie-Tooth neuropathy. *New Engl. J. Med.* 362, 1181–1191.

McDermott R, Tingley D, Cowden J et al. (2009) Monoamine oxidase A gene (MAOA) predicts behavioral aggression following provocation. *Proc. Natl Acad. Sci. USA* 106, 2118–2123.

Pelak K, Shianna KV, Ge D et al. (2010) The characterization of twenty sequenced human genomes. *PLoS Genet.* 6, e1001111 (DOI:10.1371/journal.pgen.1001111). Genomic sequencing accurately identified those with hemophilia and found a number of genes that were "knocked out."

Perry GH, Dominy NJ, Claw KG et al. (2007) Diet and the evolution of human amylase gene copy number variation. *Nat. Genet.* 39, 1256–1260.

Pontius JU, Wagner L & Schuler GD (2003) UniGene: a unified view of the transcriptome. In The NCBI Handbook. Bethesda, MD: National Center for Biotechnology Information.

Reich D, Green RE, Kircher M et al. (2010) Genetic history of an archaic hominin group from Denisova Cave in Siberia. *Nature* 468, 1053–1060.

Roach JC, Glusman G, Smit AF et al. (2010) Analysis of genetic inheritance in a family quartet by whole-genome sequencing. *Science* 328, 636–639. The genomic sequencing of two healthy parents and their two children affected by genetic disorders.

Schuenemann VJ, Bos K, DeWitte S et al. (2011) Targeted enrichment of ancient pathogens yielding the pPCP1 plasmid of *Yersinia pestis* from victims of the Black Death. *Proc. Natl Acad. Sci. USA* 108, E746–752.

Sturm RA (2009) Molecular genetics of human pigmentation diversity. *Hum. Mol. Genet.* 18, R9–17. Includes a beautiful picture of 19 forearms showing the range in skin pigmentation.

Tishkoff SA, Reed FA, Ranciaro A et al. (2007) Convergent adaptation of human lactase persistence in Africa and Europe. *Nat. Genet.* 39, 31–40.

Tripp S & Grueber M (2011). Economic impact of the human genome project. Battelle Memorial Institute.

Yi X, Liang Y, Huerta-Sanchez E et al. (2010) Sequencing of 50 human exomes reveals adaptation to high altitude. *Science* 329, 75–78.

Zhu J, He F, Hu S & Yu J (2008) On the nature of human housekeeping genes. *Trends Genet.* 24, 481–484. Housekeeping genes are expressed in all cell types, taking care of basic functions such as transcription, translation, and cell division.

Internet resources

The NCBI has a collection of electronic textbooks available through their Website which is a good place to search for explanations of the biology and technology mentioned in this chapter: www.ncbi.nlm.nih.gov/books.

GenBank release notes: ftp://ftp.ncbi.nih.gov/genbank/gbrel.txt

ACAAGGGACTAGAGAAACCAAAA
AGAAACCAAAACGAAAGGTGCAGAA
AACGAAAGGTGCAGAAGGGGAAACAGATGCAGA
GAAGGGGAAACAGATGCAGAAAGCATC
AGAAAGCATC
ACAAGGGACTAGAGAAACCAAAACGAAAGGTGCAGAAGGGGAAACAGATGCAGAAAGCATC
CTAGAGAAACCAAAA
AGAAACCAAAACGAAAGGTGCAGAA
AACGAAAGGTGCAGAAGGGG
GAAGGGG

CHAPTER 2

Introduction to Internet Resources

Key concepts

- Searching the medical and scientific literature: PubMed, iHOP, OMIM
- Searching the patent literature
- Gene classifications: Ontology
- Sequence collections such as Gene and UniGene

2.1 INTRODUCTION

How do you first learn about an interesting topic in biology? Is it in a lecture? A book? From a newspaper article or a conversation? Or do you observe it yourself in the laboratory or during a walk in the woods? However you first learn about a topic, you may wish to learn more and the early steps in this journey should include a search of the medical and scientific literature. Here you find a virtual mountain of past observations, experiments, discussions, comparisons, speculations, and explanations about our world. The benefits of consulting past observations are too numerous to list but include saving you time by directing your next steps to understanding. Do you make more observations, or have they already taken place? Do you care to repeat them, or do you build on past accomplishments? You have entered the field of science because you are naturally curious, and want explanations, so why not learn from thousands who have walked before you?

This chapter introduces the tools and resources for obtaining information from published works and Internet databases. Here, and throughout the book, we will be exploring some of the best Websites in the world for genetic information. Our launch point is the Website for the National Center for Biotechnology Information (NCBI). This Website will be used extensively throughout this book and this introduction should give you the tools to work beyond the topics covered. The foundations of many observations are the genes that act alone or in concert to give rise to the phenotype, disease, behavior, or organism. In this chapter you will not be analyzing any sequences, but you will learn ways to find and better understand them. You will see a mix of locations and approaches with the ultimate goal of showing you a path to your objective.

2.2 THE NCBI WEBSITE AND ENTREZ

The NCBI Website is one of the major hubs of bioinformatics resources and innovation in the world, and provides a wealth of information. The NCBI home page

(**Figure 2.1**) welcomes you with quick access to commonly used tools or databases ("Popular Resources"), and links to collections of resources (left sidebar). The "Site Map" and "All Resources," also on the left sidebar, give you access to everything in one place and it is interesting to scroll through all that is offered here.

It is important to remember that this Website is not just a collection of hyperlinks. The scientific community is trying to cope with truly massive amounts of biological information and the scientists and computer experts at the NCBI have created and maintain many powerful solutions to the problems of data storage, retrieval, and display. Many other wonderful Websites will be introduced in this book, but the NCBI is certainly a major player in the world of bioinformatics.

A primary interface between you and this wealth of information is the Entrez (pronounced "on-tray"; it is French for "enter") retrieval system, the NCBI's own search engine. On most pages on the Website, there is a very simple text field along with a drop-down menu which launches your searches of any or all of the databases available through Entrez. This text field appears at the top of Figure 2.1. In this example, kibra, the name of a gene, has been entered. Hitting the return key or clicking the "Search" button will launch a very fast search of 38 NCBI databases.

When the page refreshes, the majority of the page is a large section listing numerous databases, each with a brief description. The number of hits to each database is displayed to the left (**Figure 2.2**). Some databases have zero hits but this is understandable. Unless there was an organism named "kibra" you would not expect to find any hits in the Taxonomy database, for example. But on display are a number of choices for information, sequences, and other forms of data.

Clicking on the question mark next to each brief database description provides a longer explanation. To explore the hits in an individual database, just click on the number. Note that on the Entrez results page seen in Figure 2.2, the drop-down menu of individual databases is no longer available. But if you hit the back button or navigate to any of the individual results (for example, Nucleotide) you will find the familiar drop-down menu. Many of the individual databases will be described in detail later in this book. In this section of the chapter, we will focus on "PubMed: biomedical literature citations and abstracts."

Figure 2.1 Home page for the NCBI (www. ncbi.nlm.nih.gov). At the top of the page is a drop-down menu where you can choose a specific database to search. In this example, all the NCBI databases will be searched with the term "kibra."

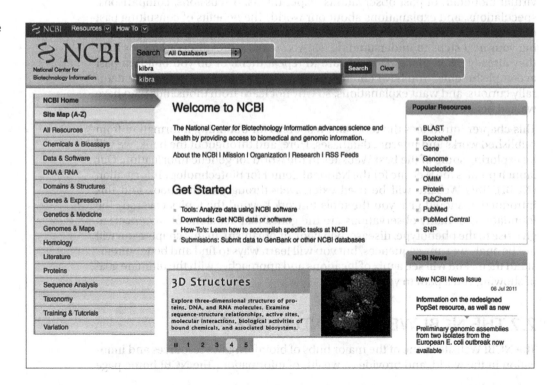

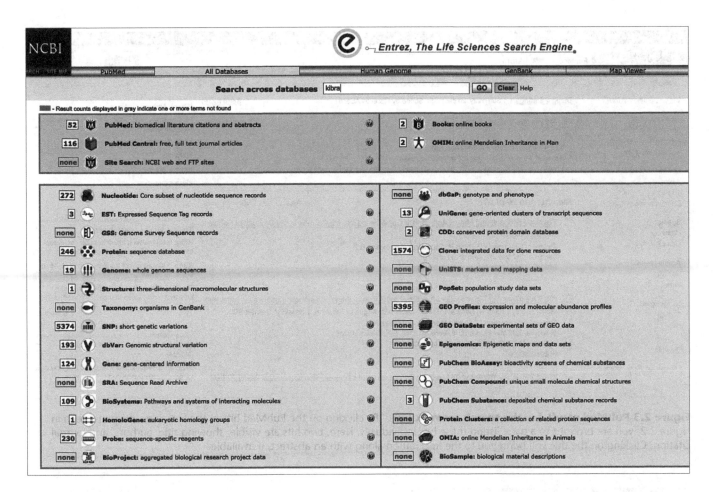

Entrez, The Life Sciences Search Engine.

2.3 PubMed

PubMed is a very large literature database, covering all major journals of biology and medicine. When you use Entrez to search PubMed, a number of searches are performed. First, your queries are used against an index of Medical Subject Headings (MeSH), which is a controlled vocabulary that describes the contents of a published paper. Queries are also used against indices from several fields including author. For example, searching with "white" will find papers on white blood cells or the *Drosophila* mutation called white, but also authors named White, too. You can specify which specific fields are searched, such as author and title/abstract, to eliminate many unwanted hits.

Clicking on the PubMed hits in Entrez brings you to a page providing more details: title, authors, and journal (**Figure 2.3**). The search term "kibra" found 52 hits in this example, sorted by date. In the upper-right corner of this figure, notice that related searches are suggested, including "kibra memory," "kibra lats," "hippo kibra," and "kibra expression." Clicking on these links will automatically return hits found with these terms, subsets of the original 52 since these hits must have both search terms present. Filters for the listed results, such as date, are found on the left side of the page.

The default display is 20 references to a page, which can save page-loading time by not providing long lists of hits should you have them. This display is also customizable. Clicking on the "Display Settings" hyperlink (upper left) opens a box (**Figure 2.4**). You can vary the information displayed, the citations per page, and the sorting order. Similar display settings are available on all of the Entrez results pages (for example, Nucleotide searches).

Back in the PubMed results (Figure 2.3), clicking on the hypertext title of the citation takes you to a page displaying the title, authors, and full abstract, if available (**Figure 2.5**). The information on this page can be exported using the "Send to:"

Figure 2.2 NCBI Entrez results page. The search for the term "kibra" performed in Figure 2.1 returned hits in many databases, including PubMed, Gene, OMIM, and UniGene. These specific databases will be explored in this chapter while others such as the Nucleotide, EST, Protein, and Structure databases will be covered in later chapters.

Search all of them or only what you need?

There are benefits for searching all the databases even if you are just looking for a nucleotide sequence. Scanning the results page, you may find it informative to see that there are only a few PubMed references, or that there are many. The same applies to all the other databases. So glance at the other hit numbers and let those bits of information provide you a little more background on your topic of interest.

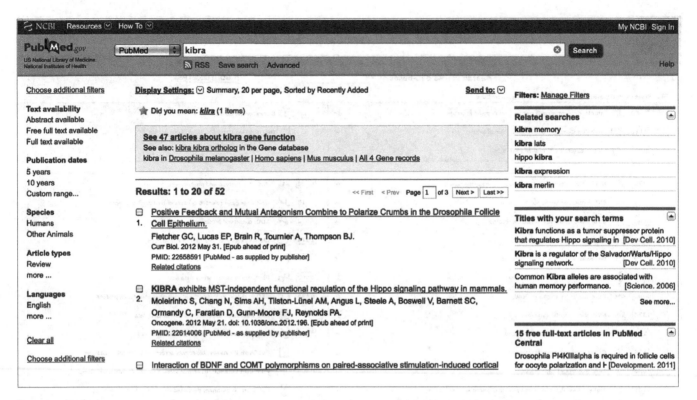

Figure 2.3 PubMed result for the search term "kibra." By clicking on the PubMed hits in the Entrez results page, seen in Figure 2.2, you are brought to a page listing those hits individually. Here, two hits are visible, showing title, authors, and journal citation. Clicking on the title will bring you to the full citation along with an abstract, if available.

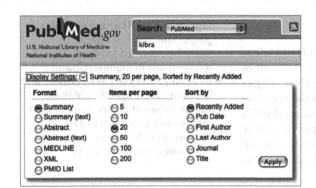

Figure 2.4 Display settings for PubMed. Each Entrez results subsection has a "Display Settings" menu, tailored for the data to be displayed. Shown here are the PubMed Display Settings where you can view the list of hits in multiple ways (Format), vary the number of citations listed per page (Items per page), and sort the results.

tool on the right, which opens a dialog box (Choose Destination). In the figure, this box partially obscures the link to the full article on the far right. This link is often the logo of a journal, in this case *Science* magazine.

The authors' names are hypertext (Figure 2.5) so if you want to search for all articles published by that author, just click on their name and the page refreshes, listing their articles. In addition, the author's name appears in the search window (**Figure 2.6A**). Along with the author's name (Papassotiropoulos A), PubMed inserts a field label ([Author]) to direct the search to author names, ignoring other fields. You may do this manually, but PubMed also provides a form to construct these and other types of searches, so there is no need to memorize the field labels. This form is available through the "Advanced search" link visible in this and earlier figures near the top of the page.

In Advanced Search (Figure 2.6B) a drop-down menu (under "Builder") offers many choices, only some of which are visible in this screenshot. You pick the field from the menu (Title/Abstract is this example), enter text in the window next to

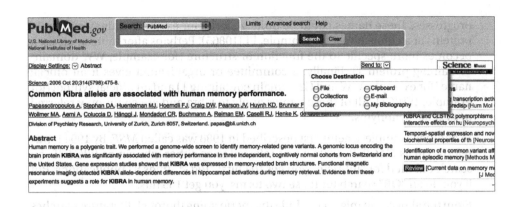

(A)

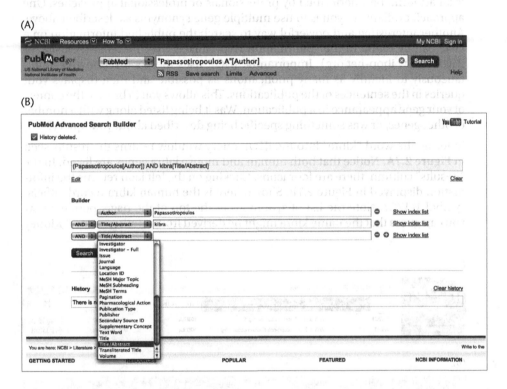

(B)

Figure 2.5 A single PubMed result. Clicking on any of the titles of the PubMed hits will take you to the full citation and an abstract, if available. The author names are hypertext; clicking on them will find other articles by that author. A "Send to:" menu, shown already open in this figure, allows you to export the citation in a variety of ways. There will also be a link to the full journal article (in this case, a link to *Science* magazine) if it is available online. Not all articles are free to view.

Figure 2.6 More specific searches in PubMed. (A) Clicking on the hypertext author name of a citation will create a more specific PubMed search. There is now a field description called "Author" in square brackets and this will direct PubMed to only display articles by this author. (B) Clicking on the "Advanced search" hypertext in the PubMed title bar takes you to a query builder where many specific fields of the database can be selected from a drop-down menu, and the Boolean terms "AND," "OR," and "NOT" can be selected from a separate menu (right). The "Limits" hyperlink, shown in this and earlier figures on PubMed, allows the placement of other constraints upon the search such as dates or language.

your choice, and the form automatically constructs your **query** in the text box at the top of the page. This form also offers a choice of "AND," "OR," and "NOT" to construct the logic for your approach (the default is "AND"). These choices are available next to the Builder drop-down menus. For example, let's say that there were two authors with the same name, "Papassotiropoulos A," and one has published articles about genes while the other has published papers about heart surgery. You may wish to construct "(Papassotiropoulos A[Author]) NOT surgery[Title/Abstract]" to generate a list of articles that are more specific.

2.4 GENE NAME EVOLUTION

Gene names can often be considered a moving target. If you search PubMed for "kibra" you find 41 hits. But using the official gene symbol, *WWC1*, given to this gene some time after it was first described, you only get 28 hits. If you do an "OR" search, kibra OR *WWC1*, you get 41 hits, suggesting that kibra is the term to use.

There are thousands of laboratories around the world and many laboratories make the same discoveries and publish their names for the same gene. Even the same laboratory could have multiple names for the same gene as their research progresses. These names could be quite generic (for example, kidney and brain protein) or a clone name (for example, FLJ10865). Perhaps after more analysis, it becomes more specific to the function or structure (for example, WW domain-containing protein 1). Finally, a committee or organization gives it an official name (for example, WW and C2 domain containing 1) and an official symbol (for example, *WWC1*). All of these names apply to kibra and are commonly referred to as gene "synonyms."

In another example, a gene first described in 1988 was called *MSF*. By 1994 it was referred to as *CACP*. By 2000 it was referred to as either lubricin or *PRG4*. If you search PubMed for "lubricin" you get 104 hits; "*PRG4*" finds 83 publications. But if you do an "OR" search for these two terms, you get 165 hits.

From the above, you might conclude that performing thorough literature searches is an art form, best performed by professionals or professional approaches. One approach available to you is to use multiple gene synonyms as described above. Another interesting and powerful way to search the published information on a gene can be found in the "information Hyperlinked Over Proteins" Website, or iHOP (www.ihop-net.org). Importantly, iHOP uses the gene synonyms simultaneously to identify as many publications as possible and then displays your queries in the sentences of the publications. This allows you to browse the context of your gene appearance in a publication. Was it being listed along with a number of other genes, or was something specific being described about your query?

Entering the word "kibra" into the iHOP query window returns the results seen in **Figure 2.7A**. Notice that both human and mouse kibra genes are listed. In the "Results" column, there are four icons. Clicking on the left icon returns the information displayed in Figure 2.7B. Shown here is the human kibra record, official symbol *WWC1*. Note the list of synonyms at the top of the page; perhaps now you can guess that the name kibra might be derived from another name, "Kidney

Figure 2.7 iHOP (information Hyperlinked Over Proteins) www.ihop-net.org. (A) Search results obtained by entering "kibra." (B) Clicking on the first rectangular icon in the "Results" column displays the details. At the top is a list of protein synonyms and below that are links to various databases. Below these hyperlinks are the individual sentences taken from PubMed citations where the search term (kibra) is bold and uppercase. Hypertext links to medical subject headings (MeSH terms, for example, transactivation) and other genes (for example, *HAX-1*) decorate these sentences.

(A)

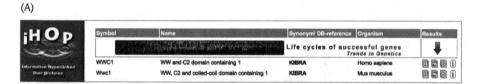

(B)

and <u>brain protein.</u>" There are links to common databases (for example, UniProt), some of which will be covered later in this book. Below these links are actual sentences taken from publications. MeSH terms and other gene names in these sentences are hyperlinked, allowing rapid navigation to topics and genes related to your query. To the right of the sentences are square icons for links to the entire abstracts of the publications, again marked with hypertext terms and names.

2.5 OMIM

Another database that can be searched by the NCBI Entrez query interface is OMIM (Online Mendelian Inheritance in Man). OMIM contains records on human genes and the many human disorders that are thought to be genetic in nature. It is manually authored and updated by staff at the Johns Hopkins University School of Medicine. From the NCBI home page, select OMIM from the Entrez drop-down menu and enter the query term "memory." Some of the results are displayed in **Figure 2.8**. Note that the "Limits" box (upper left) is checked and only the Title and Clinical Synopsis fields in the database are searched, two of the choices from this filter, which narrows the search from over 400 hits to 30. Limits do continue with subsequent searches until this box is unchecked, so be careful not to apply Limits by mistake. Of the hits shown in Figure 2.8, one gene is listed (kibra, the first hit) but the other hits are disorders. The symbols next to the OMIM record number (for example, 610533) designate a gene record with an associated phenotype (+), diseases where the gene defect is known (#), and mapped disorders where the gene is still unknown (%). Not shown is the fourth symbol (*) for genes not associated with a phenotype.

The thousands of records in OMIM contain basic information such as the chromosomal location of the genes and disorders (if known) and how the gene was isolated and identified (**Figure 2.9**). In disease records, information about clinical descriptions and findings, animal models, cellular pathways, and many other topics may be available, along with references on these topics. Records on common and major disorders such as diabetes can be dozens of pages long and easily have over 100 references. OMIM records are remarkable collections of biomedical data and should be explored whenever genetic information is needed.

> **OMIM record 100820**
> There is a disorder called "Achoo syndrome," characterized by sneezing in response to bright light, and it appears in the OMIM database. Here is a passage directly from the OMIM record:
> "Duncan (1995) pointed out public awareness of the ACHOO syndrome is much more widespread than one might guess, to the point that it has entered into the popular wisdom conveyed to preschoolers. In a best-selling children's book by Berenstain and Berenstain (1981), Papa and Mama Bear are taking Sister Bear and Brother Bear to their pediatrician, Dr. Grizzly, for a check-up. The cubs are expressing their apprehension about the possibility of injections when Papa Bear suddenly cuts loose with an explosive sneeze. 'Bless you!' said Mama. 'It's just this bright sunlight,' sniffed Papa. 'I never get sick.'"

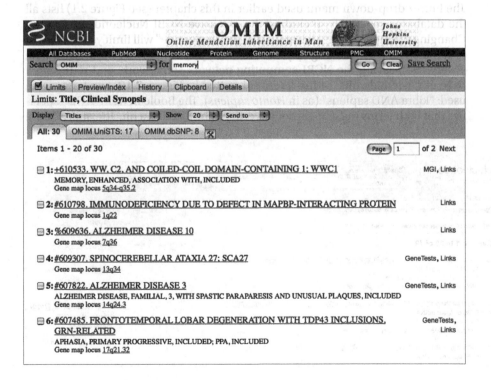

Figure 2.8 OMIM results using the search term "memory." By clicking on the "Limits" tab (visible near the top of the figure, on the left) the search can be restricted to specific fields in the records. In this case, the search was limited to "Title" and "Clinical Synopsis." This search resulted in 30 hits, some of which are shown in this figure.

Figure 2.9 OMIM detail of the first search result from Figure 2.8. Although the search was limited to "Title" and "Clinical Synopsis," the term "memory" can find this record because the "Other entities represented in this entry" field is also searched, and this subtitle appears in Figure 2.8. This is an OMIM record for a gene (kibra). For disease records, there are often many fields with extensive description of clinical findings and associated genetics.

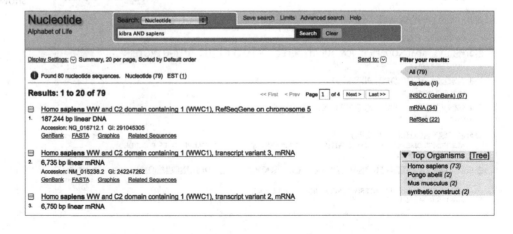

+610533

WW, C2, AND COILED-COIL DOMAIN-CONTAINING 1; WWC1

Alternative titles; symbols

KIDNEY AND BRAIN EXPRESSED PROTEIN; KIBRA

KIAA0869

Other entities represented in this entry:

MEMORY, ENHANCED, ASSOCIATION WITH, INCLUDED

HGNC Approved Gene Symbol: WWC1

Cytogenetic location: 5q34   *Genomic coordinates (GRCh37):* 5:167,719,064 - 167,899,307 (from NCBI)

Gene Phenotype Relationships

Location	Phenotype	Phenotype MIM number
5q34	[Memory, enhanced, association with]	

TEXT

Cloning

By sequencing clones obtained from a size-fractionated brain cDNA library, Nagase et al. (1998) cloned a partial WWC1 sequence, which they designated KIAA0869. RT-PCR ELISA detected highest expression in liver and kidney, intermediate expression in brain, lung, pancreas, and ovary, weak expression in heart and testis, and no expression in skeletal muscle and spleen.

Using dendrin (DDN; 610588) as bait in a yeast 2-hybrid screen of a brain cDNA library, followed by

2.6 RETRIEVING NUCLEOTIDE SEQUENCES

The Entrez drop-down menu used earlier in this chapter (see Figure 2.1) lists all the databases separately, including the extensive NCBI Nucleotide collection. Changing the menu from "All Databases" to "Nucleotide" will limit your search to only that database, speeding up your search (by just a bit, the NCBI search engine is very fast), saving the NCBI computing network some work, and returning a page of nucleotide hits. In the query submitted in **Figure 2.10**, two terms were used: "kibra AND sapiens" (as in *Homo sapiens*). The Boolean term "AND" specifies that both terms must be present in order to be listed as a hit. The query "kibra

Figure 2.10 Searching for nucleotide sequences at the NCBI Website. The Entrez drop-down menu has both nucleotide and protein sequences as choices. You can make the search quite specific using the "Advanced search" form or "Limits." You can also quickly construct queries that may work well enough for you to find the sequences you want. In this example, "kibra AND sapiens" demonstrates Boolean logic combining two terms to search all the fields. Using "sapiens" alone can lead to many thousands of hits that are from bacterial files because the search goes beyond the organism field.

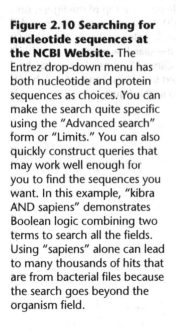

sapiens" generates the same results because listed terms are "AND'ed" together (the default). This approach should return fewer and more specific hits, but not always (as explained below). If you just use "kibra" as a query, 234 sequences are returned, while the "kibra AND sapiens" query returns 79 hits. In the first list, many sequences from other organisms are returned. In the more specific search, only a few hits are from nonhuman species, and they are listed on the right side of Figure 2.10 ("Top Organisms"). The annotation of many records often contains references to other organisms and this usually explains why hits from other organisms appear in search results.

Clicking on the hypertext line in the hit list (see Figure 2.10) takes you to that record. For example, **Figure 2.11** is a record that was listed as "WO 2005116204-A/350160: Double strand polynucleotides generating RNA interference" and you can see that this text was taken from the "DEFINITION" line. The default condition of the Entrez search uses your query against all fields in the nucleotide records. "Kibra" is found under the "COMMENT" field and "sapiens" is found in the "SOURCE" and "ORGANISM" fields. By using the "Advanced search" as we did before, you can limit your searches to specific fields.

This record was found using the terms "kibra patent." Under "COMMENT," "PN" stands for the patent publication number. This number (WO 2005116204-A/350160) can be used to find the patent application associated with this sequence.

Nucleotide searching will be extensively covered in chapters 3, 5, and 6.

2.7 SEARCHING PATENTS

Because they are legal documents, patent applications can contain very thorough descriptions of specific subjects. In the case of genes and proteins, patent applications can often act as excellent review articles containing many references

```
LOCUS       FW943634                19 bp    DNA     linear   PAT 19-APR-2011
DEFINITION  WO 2005116204-A/350160: Double strand polynucleotides generating
            RNA interference.
ACCESSION   FW943634
VERSION     FW943634.1  GI:329972699
KEYWORDS    WO 2005116204-A/350160.
SOURCE      Homo sapiens (human)
  ORGANISM  Homo sapiens
            Eukaryota; Metazoa; Chordata; Craniata; Vertebrata; Euteleostomi;
            Mammalia; Eutheria; Euarchontoglires; Primates; Haplorrhini;
            Catarrhini; Hominidae; Homo.
REFERENCE   1  (bases 1 to 19)
  AUTHORS   Naito,Y., Fujino,M., Oguchi,S. and Natori,Y.
  TITLE     Double strand polynucleotides generating RNA interference
  JOURNAL   Patent: WO 2005116204-A 350160 08-DEC-2005;
            RNAi Co Ltd
COMMENT     OS   Homo sapiens
            PN   WO 2005116204-A/350160
            PD   08-DEC-2005
            PF   11-MAY-2005 WO 2005IB001647
            PR   11-MAY-2004 JP 04P   232811
            PI   yuki naito,masato fujino,shinobu oguchi,yukikazu natori CC
            siRNA target sequence for KIBRA (NM_015238.1,3046-3064). FH   Key
            Location/Qualifiers.
FEATURES             Location/Qualifiers
     source          1..19
                     /organism="Homo sapiens"
                     /mol_type="unassigned DNA"
                     /db_xref="taxon:9606"
ORIGIN
        1 gacctgcagg cgacaagaa
```

Figure 2.11 A nucleotide record from NCBI's GenBank. The field names of a GenBank file appear on the far left of each line. Although there is a lot of information in this short record, it is easy to focus on the fields that interest you. For example, the first line says it is a 19 base pair (bp) linear DNA, and the PAT tag indicates that it is a sequence that appeared in a patent application. The DEFINITION lines are often descriptive and, in this case, indicate that this short sequence is used in RNA interference.

and interesting sequences. A drawback is that the legal language within a patent is not always an easy read. Nevertheless, you should not overlook patents as a source of information and sequences.

The World Intellectual Property Organization (WIPO) has a Website, www.wipo.int, which is easy to use and can be searched using words or patent numbers. The number associated with the kibra patent (see above) is the publication number. This number requires some consideration. "WO" ("World") indicates it is an international patent. Patents from countries begin with two-letter designations, for example Japan (JP) and the United States (US). The next four digits in the kibra patent represent the year of publication (2005). As you encounter patents on Websites and in databases, inconsistencies on how these patent numbers are displayed will be seen. You may see gaps between WO and the number, or slashes ("/"), so should a query not work, try again with a different format.

From the WIPO home page, click on the "Patents" link on the left sidebar of the Website (**Figure 2.12A**), then click on "PATENTSCOPE search" (www.wipo.int/patentscope) to construct and launch your query. The number found in the NCBI record (WO2005116204, no space) should be entered, but just the number 2005116204 would find the patent record also. The default section of the patent to be searched is the Front Page but other sections can be chosen by clicking on the tabs.

Figure 2.12B shows part of the results page; notice that multiple languages are available on the top of the Web page (visible in Figure 2.12B). Clicking on another language does an automated translation of the text, sometimes with imperfect results. Once in your language of choice, the tabs across the page provide access to the description and claims. Still, indexes are not always complete; using "kibra" as a WIPO Website query does not find this patent application. Persistence and multiple approaches are often needed to find what you seek. There are commercial databases of sequences from patents and these are more up to date than those available in the public domain.

When patent applications are filed, they can contain anywhere from a handful to thousands of sequences. Extraction of the sequences directly from the patent is labor-intensive and prone to error, and not something to be done routinely. But armed with a patent number, the nucleotide sequences at the NCBI Website could then be searched and those sequences associated with that patent number could be easily retrieved. However, there is a delay between the patent documents appearing in databases and the time when sequences appear in GenBank.

Figure 2.12 The World Intellectual Property Organization (WIPO) Website (www.wipo.int). You can search for patents using patent publication numbers like the one found in Figure 2.11. (A) The WIPO search engine (www.wipo.int/patentscope) can be found by clicking on the "Patents" link on the left sidebar. Click on the PATENTSCOPE link on that new page. (B) The results page with multiple languages available. The translation is automated and so it can be imperfect. For example, instead of saying "interference" (seen in the NCBI nucleotide record in Figure 2.11), the auto-translation says "interfere" in panel (B).

(A)

(B)

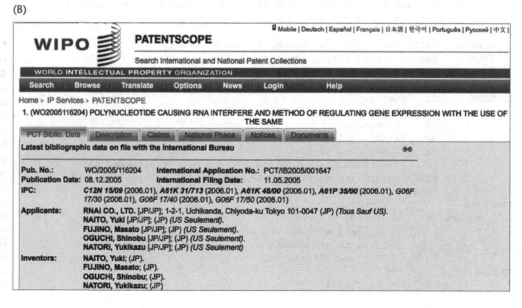

(A)

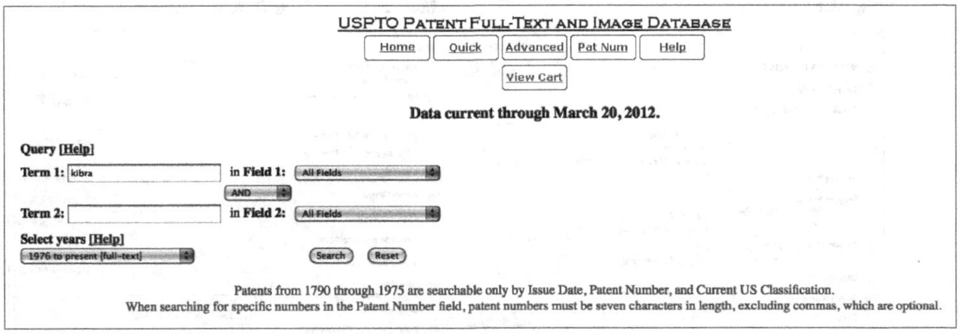

USPTO PATENT FULL-TEXT AND IMAGE DATABASE

Home   Quick   Advanced   Pat Num   Help

View Cart

Data current through March 20, 2012.

Query [Help]

Term 1: kibra   in Field 1: All Fields ▾

AND ▾

Term 2:   in Field 2: All Fields ▾

Select years [Help]

1976 to present [full-text] ▾   Search   Reset

Patents from 1790 through 1975 are searchable only by Issue Date, Patent Number, and Current US Classification.
When searching for specific numbers in the Patent Number field, patent numbers must be seven characters in length, excluding commas, which are optional.

(B)

PAT. NO.   Title
1 8,158,360 T Effects of inhibitors of FGFR3 on gene transcription
2 8,053,183 T Method of diagnosing esophageal cancer
3 7,993,836 T Genes affecting human memory performance
4 7,927,795 T Gene expression profiling in primary ovarian serous papillary tumors and normal ovarian epithelium
5 7,659,062 T Gene expression profiling of uterine serous papillary carcinomas and ovarian serous papillary tumors
6 7,067,633 T Targeting cellular entry, cell survival, and pathogenicity by dynein light chain 1/PIN in human cells

(C)

25 20090048266   Effects of Inhibitors of Fgfr3 on Gene Transcription
26 20090047216   Tumor vasculature markers and methods of use thereof
27 20080131887   Genetic Analysis Systems and Methods
28 20080108568   COMPOUNDS FOR IMPROVING LEARNING AND MEMORY
29 20080051360   PKC LIGANDS AND POLYNUCLEOTIDES ENCODING PKC LIGANDS
30 20070231822   Methods for the detection and treatment of cancer
31 20070105105   Surrogate cell gene expression signatures for evaluating the physical state of a subject
32 20070054268   Methods of diagnosis and prognosis of ovarian cancer

Figure 2.13 Searching for information at the US Patent and Trademark Office (USPTO), www.uspto.gov. Both issued patents and patent applications can be searched. (A) Using "kibra" as a search term in the patents database, only six issued patents are found (B). (C) Using the same term to search the patent applications database finds 48 patent applications, eight of which are shown here.

Another major Website to use for patents is from the US Patent and Trademark Office, www.uspto.gov. There is a lot of information on this Website and you have to find the "Search" hypertext on the home page. From there, at least two kinds of patent searches can be performed: searching patent applications and searching issued patents. There can be a considerable amount of time between application and issuance, and this will be demonstrated in **Figure 2.13**.

Figure 2.13A shows the query interface for issued patents. Note that you can search for patents that go back to the year 1790! As seen earlier at other Websites, drop-down menus allow you to search specific fields. In this example, the term "kibra" was used to search all fields of issued patents and the result is six patents shown in Figure 2.13B. These results are hypertext so you are one click away from viewing the patent. In Figure 2.13C, the search was repeated for patent applications (patents not issued yet) and 48 hits were returned, eight of which are shown in this figure. The difference in numbers makes sense in that kibra was only named in 2003 according to an iHOP search and not enough time has passed for these patent applications to be issued. Nevertheless, if you only searched issued patents, you would have missed 48 other documents. You may have to install a Web browser plugin to view patent figures or images. Trying a different browser will sometimes work too.

You can also search for patents at the European Patent Office (EPO), www.epo.org.

2.8 PUBLIC GRANTS DATABASE: NIH REPORTER

Searching patents gives you access to information about genes that can be so valuable that legal steps were taken to protect discoveries. Another source of gene information, some of which is not yet published, is public grant applications. The US National Institutes of Health (NIH) has made information on issued grants available through their NIH RePORTER Website, projectreporter.nih.gov.

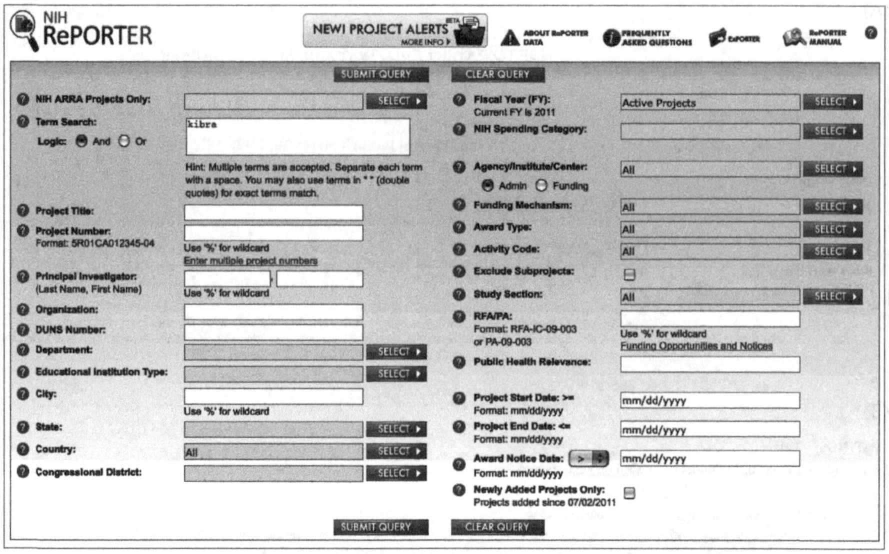

Figure 2.14 Query interface of the NIH RePORTER database (www.projectreporter.nih.gov).

You cannot see the entire grant but the abstract and associated publications will inform you about work that is underway which can potentially help you understand your gene of interest or identify potential collaborators. **Figure 2.14** shows the query interface, which provides many avenues for searching this database.

2.9 GENE ONTOLOGY

In addition to the multiple names possible for a single gene, there is still potential for more chaos and confusion. In the human genome alone, there are over 21,000 genes that encode proteins. What do they do? Where are they located? What chemicals or proteins do they associate with? The answers cannot be revealed in their names, although some names are more descriptive than others. The approach to making sense of these thousands of proteins is called **Gene Ontology**, and a major Website for this effort is found at www.geneontology.org. The introduction on their home page describes it well: "The Gene Ontology project is a major bioinformatics initiative with the aim of standardizing the representation of gene and gene product attributes across species and databases." As you can imagine, this is a massive project requiring tremendous organization, data analysis, and interpretation. We can only touch upon this effort and will do so with kibra to illustrate a simple query.

From the Gene Ontology home page, entering kibra into the simple text field and clicking on the "GO" button brings you to ten results, one of which is *Homo sapiens* (**Figure 2.15A**). This result shows that for human kibra, there are 14 known "associations," or ontologies, where there is some evidence linking kibra to a process, cellular component, or function. Clicking on the hypertext "14 associations" shows you this list (Figure 2.15B). For each ontology (for example, biological process) there can be multiple terms. For example, for this kibra result there are six different biological processes: cell migration, hippo signaling cascade, positive regulation of MAPK cascade, regulation of hippo signaling cascade, regulation of transcription, DNA-dependent, and transcription, DNA-dependent. Clicking on these terms takes you to a tree of larger categories that contain these terms. For example, cell migration is a form of cell motility, which is a form of locomotion, which is a form of localization, which is a biological process. Clicking on the "gene products" takes you to a listing of individual genes that have the same classification. For example, the proteins crumbs and merlin are also involved in the regulation of the hippo signaling cascade.

(A)

☐	WWC1	14 associations **protein** from *Homo sapiens*
	Protein **KIBRA**	BLAST

(B)

Accession, Term		Ontology	Qualifier	Evidence	Reference	Assigned by
☐ GO:0016477 : cell migration	6085 gene products / view in tree	biological process		IDA	PMID:18596123	UniProtKB
☐ GO:0035329 : hippo signaling cascade	181 gene products / view in tree	biological process		TAS	Reactome:REACT_118607	Reactome (via UniProtKB)
☐ GO:0043410 : positive regulation of MAPK cascade	1293 gene products / view in tree	biological process		IDA	PMID:18190796	UniProtKB
☐ GO:0035330 : regulation of hippo signaling cascade	46 gene products / view in tree	biological process		IMP	PMID:20159598	UniProtKB
☐ GO:0006355 : regulation of transcription, DNA-dependent	33779 gene products / view in tree	biological process		IEA With UniProtKB-KW:KW-0805	GO REF:0000037	UniProtKB
☐ GO:0006351 : transcription, DNA-dependent	38178 gene products / view in tree	biological process		IEA With UniProtKB-KW:KW-0804	GO REF:0000037	UniProtKB
☐ GO:0005737 : cytoplasm	166085 gene products / view in tree	cellular component		IDA	PMID:16684779	UniProtKB
				IDA	PMID:18190796	UniProtKB
☐ GO:0005829 : cytosol	22265 gene products / view in tree	cellular component		TAS	Reactome:REACT_118857	Reactome (via UniProtKB)
☐ GO:0005794 : Golgi apparatus	9127 gene products / view in tree	cellular component		IDA	PMID:18029348	HPA (via UniProtKB)
☐ GO:0005634 : nucleus	65458 gene products / view in tree	cellular component		IDA	PMID:16684779	UniProtKB
				IDA	PMID:18190796	UniProtKB
☐ GO:0048471 : perinuclear region of cytoplasm	2755 gene products / view in tree	cellular component		IDA	PMID:18596123	UniProtKB
☐ GO:0032587 : ruffle membrane	238 gene products / view in tree	cellular component		IDA	PMID:18190796	UniProtKB
☐ GO:0005515 : protein binding	54759 gene products / view in tree	molecular function		IPI With UniProtKB:P63167	PMID:16684779	UniProtKB
				IPI With UniProtKB:O95219	PMID:17994011	UniProtKB
				IPI With UniProtKB:Q05513	PMID:18190796	UniProtKB
				IPI With UniProtKB:Q8N3V7	PMID:18596123	UniProtKB
☐ GO:0003713 : transcription coactivator activity	1569 gene products / view in tree	molecular function		IDA	PMID:16684779	UniProtKB

So why is kibra classified in this manner? Notice that the Reference column in Figure 2.15B contains a PubMed Identifier (PMID) for a scientific reference, and there is the name of the principal source of this classification (for example, UniProtKB). UniProtKB is a fantastic database that will be covered in Chapter 4. In this example, the Gene Ontology Website drew upon the knowledge and analysis captured at this other Website to classify kibra.

The Evidence column in Figure 2.15B contains three-letter codes representing the type of evidence used to classify kibra. For example, the assignment of "cell migration" was Inferred from a Direct Assay (IDA), while the regulation of the hippo signaling cascade was Inferred from a Mutant Phenotype (IMP). These and the other codes for evidence are listed in **Figure 2.16**. Clicking on the evidence code on the Website will take you to their extensive help section with thorough explanations of each type of evidence along with examples. The tabular results as shown in Figure 2.15 can be filtered using these codes. For example, you can display only the results based on evidence from direct assays.

The evidence types listed in Figure 2.16 illustrate the varied forms of data that support or refute scientific conclusions. Throughout this book you will be gathering evidence through sequence analysis. You literally "play detective" and build a case classifying, naming, or otherwise describing sequences. This annotation process is sometimes straightforward but when it is not, you will need to pursue multiple avenues to gather evidence. This includes the use of literature and databases

Figure 2.15 The Gene Ontology project (www.geneontology.org). Gene Ontology uses a controlled list of terms to describe proteins, producing a very organized classification of proteins. (A) Entering "kibra" in the simple query form returns a list of proteins including the human form of kibra. (B) Clicking on the "14 associations" link shown in (A) brings up a list of terms associated with kibra function (for example, cell migration). Also displayed are Ontology (for example, cell migration is a "biological process" while protein binding is a "molecular function"), Evidence (how this was determined), and Reference (a publication that demonstrates this).

Experimental Evidence Codes

EXP	Inferred from Experiment
IDA	Inferred from Direct Assay
IEP	Inferred from Expression Pattern
IGI	Inferred from Genetic Interaction
IMP	Inferred from Mutant Phenotype
IPI	Inferred from Physical Interaction

Computational Analysis Evidence Codes

IBA	Inferred from Biological aspect of Ancestor
IBD	Inferred from Biological aspect of Descendant
IGC	Inferred from Genomic Context
IKR	Inferred from Key Residues
IRD	Inferred from Rapid Divergence
ISA	Inferred from Sequence
ISM	Inferred from Sequence Model
ISO	Inferred from Sequence Orthology
ISS	Inferred from Sequence or Structural Similarity
RCA	Inferred from Reviewed Computational Analysis

Other listings

TAS	Traceable Author Statement
NAS	Non-traceable Author Statement
IC	Inferred by Curator
ND	No biological Data available
IEA	Inferred from Electronic Annotation

where a lot of work has already been done for you. Internet Websites have many databases on specialized topics and gene families which are not covered by this book. For example, the ENZYME database (see **Box 2.1**) is an extensive resource on this class of proteins. Before struggling to gather the information you need for a project, search the Internet to determine if it has already been done for you.

2.10 THE GENE DATABASE

For the remainder of this chapter, we will return to the NCBI Website and visit two wonderful collections: the Gene and UniGene databases. Both collections are often used to retrieve DNA and protein sequences and we'll be using these databases later in the book for this purpose. Here, we will learn about their content.

From the Entrez interface (Figure 2.2), you can select a database called "Gene." Each record within Gene compiles information from numerous sources on a single gene. There are 105 separate Gene records with at least some mention of kibra

Box 2.1 Enzyme database

The Gene Ontology project provides a view of all gene products. An important subset, already possessing extensive descriptions and classification, are the enzymes. For a database that organizes the naming and classification of enzymes, go to enzyme.expasy.org. Here you can search the ENZYME database for enzymes based on names, Enzyme Commission (EC) numbers, and other terms. Enter "kinase" and you are shown many pages of different kinases. Clicking on their links shows the chemical reactions they catalyze, and links to their sequences and annotation.

and many of these genes are versions of kibra from various organisms. There are multiple human records found with the more specific query "kibra AND sapiens." But only one has the official gene symbol seen earlier (*WWC1*) and kibra listed as a synonym. The others have the word kibra somewhere in the annotation and are found by this search. Part of the human kibra Gene record is shown in **Figure 2.17**.

While this is not the only database where gene/protein-centric information has been compiled, it can serve as a launching point for many analysis projects. Most of the text in each record is hypertext linking you to other NCBI resources. Looking at Figure 2.17, at the top is a Summary of some basic facts about the gene,

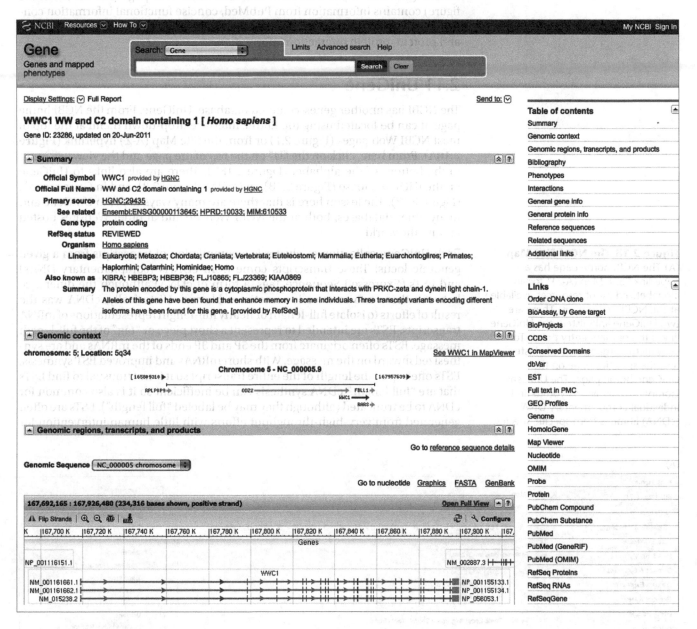

Figure 2.17 The NCBI Gene record for human kibra. Basic information for a gene is displayed along with links to related information from NCBI data sources on the right sidebar. This record was selected from the "Gene" hit list shown in Figure 2.2. The "Summary" section includes the official name for this gene, *WWC1* WW and C2 domain containing 1, along with a list of synonyms, including kibra. The "Genomic context" section shows kibra along with adjacent genes portrayed as arrows on the chromosome. The "Genomic regions, transcripts, and products" section is a map showing exons of kibra transcripts, with transcript and protein accession numbers on the left and right sides of the map, respectively. These accession numbers are unique identifiers for these molecules.

including synonyms that we saw earlier in the iHOP record. The official symbol is here (*WWC1*) along with the RefSeq (Reference Sequence) status; REVIEWED indicates that the sequence information in this record has been reviewed by a person, increasing the confidence about the gene structure and associated information. Many records in the Gene database were automatically generated and have not (yet) been reviewed. However, automated methods are quite powerful, accurate, and necessary when you are dealing with millions of genes.

The "Genomic context" section shows the position of the kibra gene (a dark arrow labeled WWC1) in relation to its neighbors on human chromosome 5. Below that are three different kibra gene transcripts represented as exons (boxes) and introns (thin lines). The bottom of the kibra Gene Web page (not shown in the figure) contains information from PubMed, concise functional information contributed by the scientific community (GeneRIF (Gene Reference into Function)), and protein–protein interaction information.

2.11 UniGene

The NCBI has another genes-centered database, UniGene. From the NCBI home page, it can be located using the Entrez interface drop-down menu at the top of most NCBI Web pages (Figure 2.1) or from the Site Map (A-Z) hyperlink (**Figure 2.18A**). From here, click on the "U" on the top of the page and the view will jump to the bottom of the alphabet (Figure 2.18B). There are also links to UniGene on the iHOP database (Figure 2.7B), in OMIM (Figure 2.8), and in Gene records (Figure 2.17). The lesson here is that there are many ways to get to UniGene and many other databases, both at the NCBI Website and at other Websites hosted around the world.

Each UniGene collection or cluster is a compilation of transcripts from a given genomic locus. These transcripts come from cDNAs (complementary DNAs) and ESTs (Expressed Sequence Tags, a form of cDNA) although the difference between these two sources is sometimes blurred. Historically, cDNA was the result of efforts to isolate full-length or nearly full-length representations of mRNA transcripts. ESTs are intended to represent a short piece or "tag" of the full-length message. ESTs often originate from the 5P and 3P ends of the mRNAs and are synthesized inward on the message. With short mRNAs and improved EST synthesis, ESTs often span the length of the entire transcript so it is not unusual to find ESTs that are "full length." cDNA synthesis can be inefficient so it is also common for cDNA to be truncated (although they may be labeled "full length"). ESTs are often generated from very-high-throughput efforts with little human intervention for

Figure 2.18 The NCBI Site Map.
(A) The NCBI home page has a hyperlink "Site Map (A-Z)" for an alphabetized list of resources available at the NCBI Website. (B) There are two UniGene links listed: the UniGene link is the primary entry point. To view all the organisms and clusters, find the statistics page on the UniGene home page. The UniGene Library Browser allows you to find individual complementary DNA (cDNA) libraries from specific species.

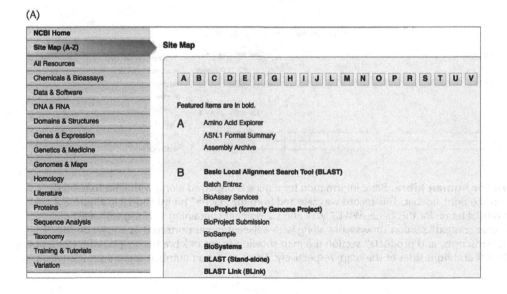

(A)

(B)

interpreting the sequence. At the other extreme, some cDNAs were very carefully synthesized, checked for accuracy, and fully annotated as part of a publication on that sequence. Regardless of these different possibilities, the transcripts of a chromosomal locus are in one place in the UniGene database.

From the home page of the UniGene database, click on the "UniGene Statistics" link and you navigate to a page (**Figure 2.19A**) displaying a long list of species, reflecting the widespread generation of ESTs. The first entry is from the cow, with over 45,000 loci. Some adjacent loci will eventually be joined because overlapping sequences will be found, linking distant exons to a single mRNA. Use of the latest technology to create and sequence cDNA is uncovering rarely transcribed loci, many of which will be shown not to encode proteins. The human entry shows over 129,000 loci, which far exceeds the 21,000 known protein-encoding genes. Guesses as to the identity of the other thousands of clusters range from regulatory RNAs to transcribed regions that have no function. The latter may represent nature's way of creating genes and function; first get transcribed and then find a role. This is an exciting and emerging area of molecular biology and bioinformatics so non-coding transcripts should not be overlooked.

(A)

UniGene Statistics			
Species	UniGene entries	Cutoff Date*	Release Date*
⊟ Chordata			
⊟ Mammalia			
Bos taurus (cow)	45178	Mar 18 2011	Nov 28 2011
Canis lupus familiaris (dog)	24459	May 3 2010	Jul 8 2010
Capra hircus (domestic goat)	38271	Nov 29 2011	Dec 20 2011
Equus caballus (horse)	6402	Mar 17 2011	Nov 4 2011
Homo sapiens (human)	129525	Mar 14 2012	Apr 11 2012
Macaca fascicularis (crab-eating macaque)	12546	Aug 28 2011	Nov 7 2011
Macaca mulatta (rhesus monkey)	7785	Mar 1 2011	Aug 8 2011
Monodelphis domestica (gray short-tailed opossum)	662	Oct 24 2010	Jan 31 2012
Mus musculus (mouse)	80414	Oct 20 2011	Mar 5 2012
Ornithorhynchus anatinus (platypus)	749	Nov 26 2009	Jul 27 2010
Oryctolagus cuniculus (rabbit)	7298	Mar 25 2012	Apr 3 2012
Ovis aries (sheep)	17318	Aug 27 2011	Nov 21 2011
Pan troglodytes (chimpanzee)	2412	Mar 15 2011	Mar 28 2011
Papio anubis (olive baboon)	11659	Mar 8 2011	Nov 7 2011
Peromyscus maniculatus (deer mouse)	11045	Mar 9 2010	Aug 3 2011
Pongo abelii (Sumatran orangutan)	7460	Nov 20 2010	Dec 27 2010
Rattus norvegicus (Norway rat)	66373	Jan 18 2012	Mar 2 2012
Sus scrofa (pig)	46613	Mar 1 2011	Apr 13 2011
Trichosurus vulpecula (silver-gray brushtail possum)	11405	Apr 3 2011	Aug 8 2011

(B)

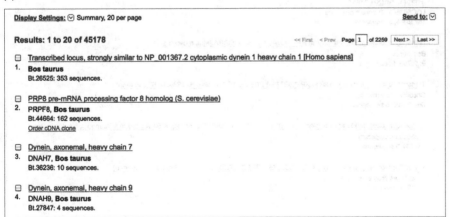

Figure 2.19 The UniGene database gathers all the transcription products of genomic loci into one place. (A) The statistics page lists the many species represented, each with thousands of UniGene "clusters." (B) A sampling of four cow transcribed loci. Notice that there are 45,178 UniGene clusters distributed over 2259 pages containing 20 records each.

Clicking on a species name brings you to the statistics of the collection. For example, the cow has approximately 1.4 million sequences assembled into almost 46,000 clusters. There are over 16,000 transcripts that are "singletons"; only a single cDNA sequence maps to each location. Clicking on the number of UniGene entries seen in Figure 2.19A takes you to a list of the individual clusters (Figure 2.19B) and you can see the number of sequences per cluster along with an automated identification. For example, the first cluster shows strong similarity to a known human gene and the second cluster is tentatively identified as the cow equivalent to a yeast (*Saccharomyces cerevisiae*) gene.

A search of the UniGene database using the term "kibra" returns results such as those seen in **Figure 2.20**. There are a total of 12 clusters found with kibra, from 12 different species, ranging from 228 sequences (human) to 1 (cow). If you wanted to clone the cow kibra gene, this single EST out of the 1.4 million cow sequences would be your ticket to success. These results also demonstrate the range of similarities that are calculated by the automated annotation of these sequences. The sequence identity is fairly certain if the term "similar" is used. The terms "strongly" and "moderately" reflect the degree of similarity to known genes.

The top of the UniGene cluster record for human kibra is shown in **Figure 2.21A**. Sequences from the cluster show high identity to known proteins, ranging from 100% (the human kibra) to 67% (*Danio rerio*). The next section in the UniGene record covers gene expression (Figure 2.21B); the origins of the cDNA or ESTs are used to determine where these sequences can be found. For example, if kibra ESTs were found in brain cDNA libraries, brain is listed as a possible location for the normal expression of kibra.

It is worth noting that this is not a perfect assumption. Certain tissues (for example, brain and testis) are known to express many genes in an apparent nonspecific

Figure 2.20 Searching UniGene.
In this example, "kibra" is used as a query and 12 clusters were found. Some are well characterized, for example the human, mouse, and *Drosophila* genes. The others are automatically annotated as loci that have similarity to kibra but more analysis is needed to verify if these are kibra genes.

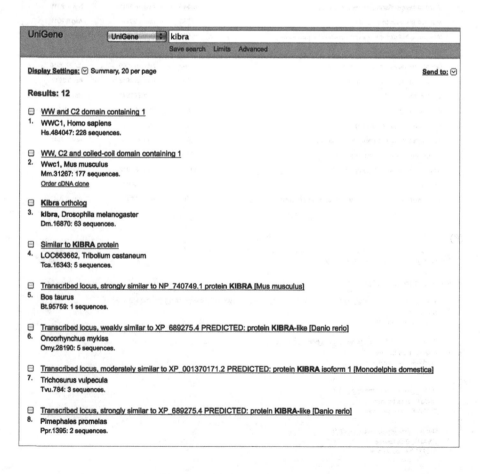

(A)

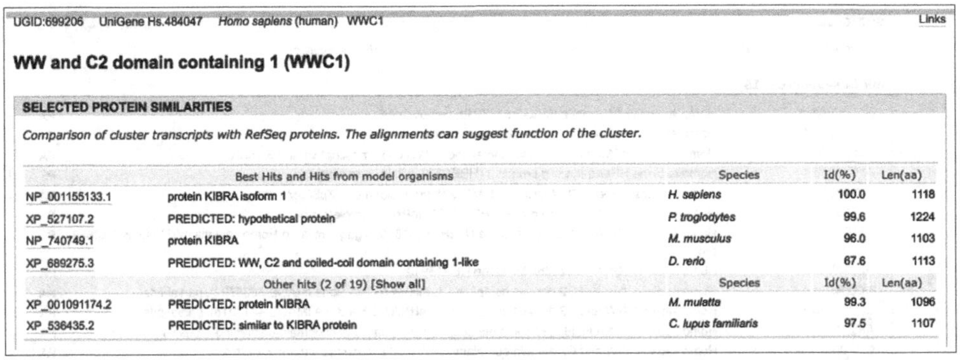

(B)

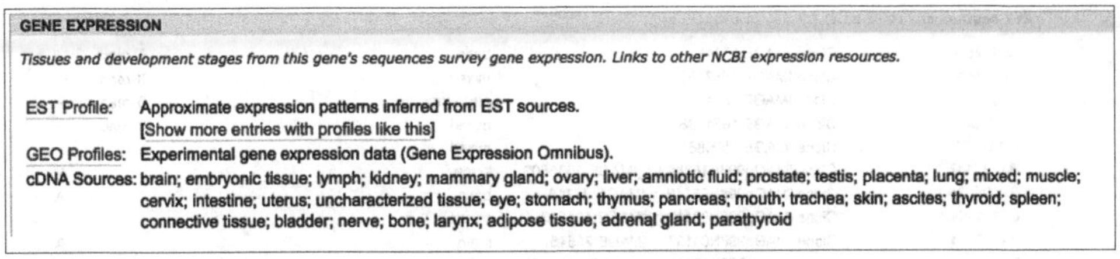

Figure 2.21 The UniGene record for human kibra. (A) A portion of the record, showing the best similarities to proteins from other organisms. This evidence supports the identity of this locus. (B) Another section of the record concerned with gene expression. The tissues of origin for the transcripts in this UniGene cluster are summarized as "cDNA Sources" and suggest widespread expression of the kibra transcripts. Clicking on "EST Profile" will give both a graphical and numerical presentation of these data.

manner. Nevertheless, tissues of expression are additional data points that can be gathered from these collections of cDNA in the UniGene database.

The next section of the UniGene database is the listing of sequences in the cluster (**Figure 2.22A**). The mRNAs are listed first. The left-hand column contains database **accession numbers**, unique identifiers in the database, which are hyperlinked to these sequences. The accession numbers that start with "NM_" are reference sequences, thought to be standards that are full length and without errors. Also included are brief descriptions and, to the right, a code indicating similarity to known proteins (P) and possession of a polyadenylation signal at the 3P end of the mRNA (A).

The list of ESTs includes hyperlinked accession numbers and brief descriptions. Those labeled "IMAGE" are from clones produced by the IMAGE consortium (Integrated Molecular Analysis of Genomes and their Expression). Members of this organization generated millions of clones, perhaps the world's largest collection, from many different tissues and made them available to everyone. A large number of clones were sequenced both at the 5P and 3P ends, and some were sequenced internally. As a result, their descriptions (second column) may indicate identical clone names. The annotation of the sequence file provides more information such as which end of the clone was sequenced (for example, 3P read). Another column lists the tissues that generated these clones. Figure 2.22B shows a sampling of the 214 ESTs in the human kibra cluster. The clone names often help to reveal more about the tissue of origin. For example, BRHIP is from the brain hippocampus, BRALZ is from the brain of an Alzheimer's disease patient, and BRAMY is from the brain amygdala.

Clicking on the accession numbers reveals additional information. For example, clicking on accession number DA228506.1 will take you to the sequence summary (**Figure 2.23**). From here you can go to two additional resources to learn about this clone or this library. Clicking on the GenBank entry hyperlink DA228506.1 will take you to a record containing the sequence. The dbEST_18318 hyperlink will take you to information about the cDNA library, BRAWH3.

The top of the BRAWH3 library's Web page contains fundamental information about the library (**Figure 2.24A**). It is from a normal human brain but no other

Figure 2.22 UniGene sequences.
(A) The sequences that make up a UniGene cluster are divided to separate the longer, often annotated mRNAs from the shorter and largely uncharacterized ESTs. When there are many ESTs present, only some (here 10 of 214) are shown in the main window but all are available for browsing (click on "Show all sequences") or download (see the button in this panel). The key to the other annotation symbols can be found on the UniGene page.
(B) A number of the ESTs found in the human kibra UniGene cluster.

(A)

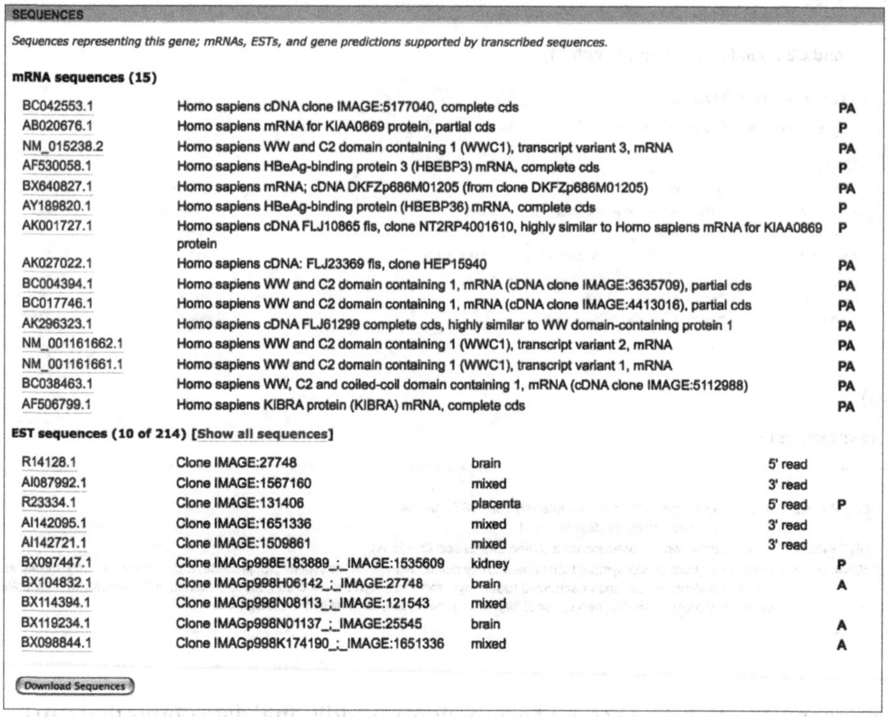

SEQUENCES

Sequences representing this gene; mRNAs, ESTs, and gene predictions supported by transcribed sequences.

mRNA sequences (15)

BC042553.1	Homo sapiens cDNA clone IMAGE:5177040, complete cds		PA
AB020676.1	Homo sapiens mRNA for KIAA0869 protein, partial cds		P
NM_015238.2	Homo sapiens WW and C2 domain containing 1 (WWC1), transcript variant 3, mRNA		PA
AF530058.1	Homo sapiens HBeAg-binding protein 3 (HBEBP3) mRNA, complete cds		P
BX640827.1	Homo sapiens mRNA; cDNA DKFZp686M01205 (from clone DKFZp686M01205)		PA
AY189820.1	Homo sapiens HBeAg-binding protein (HBEBP36) mRNA, complete cds		P
AK001727.1	Homo sapiens cDNA FLJ10865 fis, clone NT2RP4001610, highly similar to Homo sapiens mRNA for KIAA0869 protein		P
AK027022.1	Homo sapiens cDNA: FLJ23369 fis, clone HEP15940		PA
BC004394.1	Homo sapiens WW and C2 domain containing 1, mRNA (cDNA clone IMAGE:3635709), partial cds		PA
BC017746.1	Homo sapiens WW and C2 domain containing 1, mRNA (cDNA clone IMAGE:4413016), partial cds		PA
AK296323.1	Homo sapiens cDNA FLJ61299 complete cds, highly similar to WW domain-containing protein 1		PA
NM_001161662.1	Homo sapiens WW and C2 domain containing 1 (WWC1), transcript variant 2, mRNA		PA
NM_001161661.1	Homo sapiens WW and C2 domain containing 1 (WWC1), transcript variant 1, mRNA		PA
BC038463.1	Homo sapiens WW, C2 and coiled-coil domain containing 1, mRNA (cDNA clone IMAGE:5112988)		PA
AF506799.1	Homo sapiens KIBRA protein (KIBRA) mRNA, complete cds		PA

EST sequences (10 of 214) [Show all sequences]

R14128.1	Clone IMAGE:27748	brain	5' read	
AI087992.1	Clone IMAGE:1567160	mixed	3' read	
R23334.1	Clone IMAGE:131406	placenta	5' read	P
AI142095.1	Clone IMAGE:1651336	mixed	3' read	
AI142721.1	Clone IMAGE:1509861	mixed	3' read	
BX097447.1	Clone IMAGp998E183889_;_IMAGE:1535609	kidney		P
BX104832.1	Clone IMAGp998H06142_;_IMAGE:27748	brain		A
BX114394.1	Clone IMAGp998N08113_;_IMAGE:121543	mixed		
BX119234.1	Clone IMAGp998N01137_;_IMAGE:25545	brain		A
BX098844.1	Clone IMAGp998K174190_;_IMAGE:1651336	mixed		A

[Download Sequences]

(B)

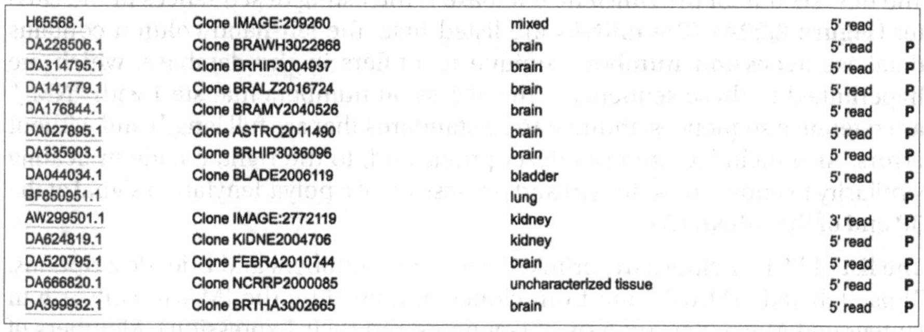

H65568.1	Clone IMAGE:209260	mixed	5' read	
DA228506.1	Clone BRAWH3022868	brain	5' read	P
DA314795.1	Clone BRHIP3004937	brain	5' read	P
DA141779.1	Clone BRALZ2016724	brain	5' read	P
DA176844.1	Clone BRAMY2040526	brain	5' read	P
DA027895.1	Clone ASTRO2011490	brain	5' read	P
DA335903.1	Clone BRHIP3036096	brain	5' read	P
DA044034.1	Clone BLADE2006119	bladder	5' read	P
BF850951.1		lung		P
AW299501.1	Clone IMAGE:2772119	kidney	3' read	P
DA624819.1	Clone KIDNE2004706	kidney	5' read	P
DA520795.1	Clone FEBRA2010744	brain	5' read	P
DA666820.1	Clone NCRRP2000085	uncharacterized tissue	5' read	P
DA399932.1	Clone BRTHA3002665	brain	5' read	P

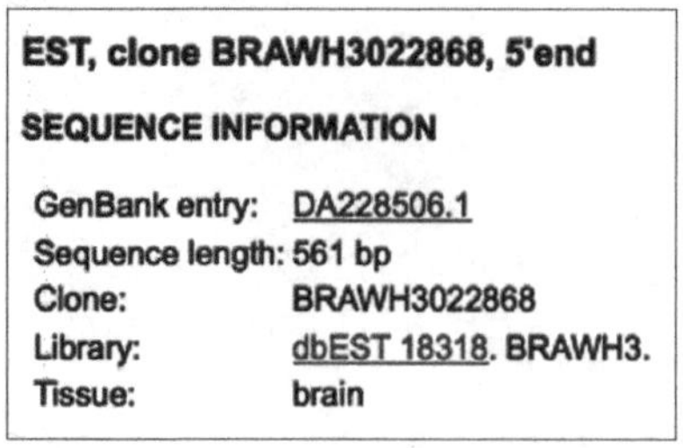

EST, clone BRAWH3022868, 5'end

SEQUENCE INFORMATION

GenBank entry:	DA228506.1
Sequence length:	561 bp
Clone:	BRAWH3022868
Library:	dbEST 18318. BRAWH3.
Tissue:	brain

Figure 2.23 UniGene sequence summary. This includes a link to the full GenBank record (the accession number), the sequence length, clone name, library name, and tissue of origin. ESTs can come from multiple regions on a single clone, for example the 5P and 3P ends.

information is provided as to its origin. Is it derived from the whole brain or just a portion? There are no specifics regarding the exact tissue. Under "Gene Content Analysis" we learn that the library generated over 43,000 ESTs that were collapsed into 7583 UniGene entries, listed here in descending number of ESTs. As can be seen, 1317 ESTs are from the amyloid beta A4 gene. Then there is a big jump to 388 ESTs from a gene called ermin. The EST number is also expressed as TPM (transcripts per million), which is a convenient way of normalizing the EST numbers. Considering that there was a big jump down to the ermin ESTs, you might conclude that amyloid beta A4 is hugely abundant, but the TPM number puts it at 3% of this library. This is still a very large number but the TPM value keeps it in perspective.

Compare and contrast the BRAWH3 library record to that of BRHIP3 (Figure 2.24B). The BRHIP3 library has almost 33,000 ESTs that collapse to 6383 clusters, smaller than the BRAWH3 library but similar in proportions. The BRHIP3 library is from the hippocampus and the most abundant EST is again from the amyloid beta A4 gene. In fact, two of the top five EST clusters are identical between the two libraries. By comparing libraries from different brain regions, gene expression patterns could be recognized which may shed light on the roles and functions of genes.

(A)

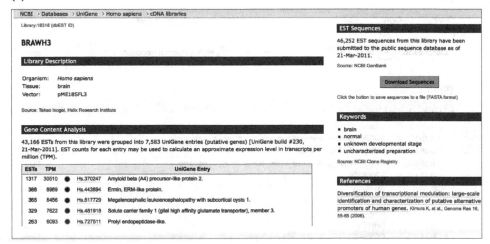

Library:18318 (dbEST ID)

BRAWH3

Library Description

Organism: *Homo sapiens*
Tissue: brain
Vector: pME18SFL3

Source: Takao Isogai, Helix Research Institute

Gene Content Analysis

43,166 ESTs from this library were grouped into 7,583 UniGene entries (putative genes) [UniGene build #230, 21-Mar-2011]. EST counts for each entry may be used to calculate an approximate expression level in transcripts per million (TPM).

ESTs	TPM		UniGene Entry
1317	30510	● Hs.370247	Amyloid beta (A4) precursor-like protein 2.
388	8989	● Hs.443894	Ermin, ERM-like protein.
365	8456	● Hs.517729	Megalencephalic leukoencephalopathy with subcortical cysts 1.
329	7622	● Hs.481918	Solute carrier family 1 (glial high affinity glutamate transporter), member 3.
263	6093	● Hs.727511	Prolyl endopeptidase-like.

EST Sequences

46,252 EST sequences from this library have been submitted to the public sequence database as of 21-Mar-2011.

Source: NCBI GenBank

[Download Sequences]

Click the button to save sequences to a file (FASTA format)

Keywords

- brain
- normal
- unknown developmental stage
- uncharacterized preparation

Source: NCBI Clone Registry

References

Diversification of transcriptional modulation: large-scale identification and characterization of putative alternative promoters of human genes. Kimura K, et al., Genome Res 16, 55-65 (2006).

(B)

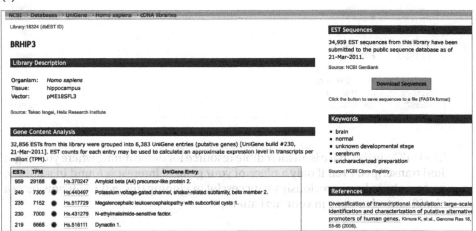

Library:18324 (dbEST ID)

BRHIP3

Library Description

Organism: *Homo sapiens*
Tissue: hippocampus
Vector: pME18SFL3

Source: Takao Isogai, Helix Research Institute

Gene Content Analysis

32,856 ESTs from this library were grouped into 6,383 UniGene entries (putative genes) [UniGene build #230, 21-Mar-2011]. EST counts for each entry may be used to calculate an approximate expression level in transcripts per million (TPM).

ESTs	TPM		UniGene Entry
959	29188	● Hs.370247	Amyloid beta (A4) precursor-like protein 2.
240	7305	● Hs.440497	Potassium voltage-gated channel, shaker-related subfamily, beta member 2.
235	7152	● Hs.517729	Megalencephalic leukoencephalopathy with subcortical cysts 1.
230	7000	● Hs.431279	N-ethylmaleimide-sensitive factor.
219	6665	● Hs.516111	Dynactin 1.

EST Sequences

34,959 EST sequences from this library have been submitted to the public sequence database as of 21-Mar-2011.

Source: NCBI GenBank

[Download Sequences]

Click the button to save sequences to a file (FASTA format)

Keywords

- brain
- normal
- unknown developmental stage
- cerebrum
- uncharacterized preparation

Source: NCBI Clone Registry

References

Diversification of transcriptional modulation: large-scale identification and characterization of putative alternative promoters of human genes. Kimura K, et al., Genome Res 16, 55-65 (2006).

Figure 2.24 UniGene library exploration. (A) The library BRAWH3 can be accessed by clicking on its hyperlink in the UniGene sequence summary shown in Figure 2.23. The tissue is described as "brain" and, under the "Gene Content Analysis" section, the library consists of over 43,000 sequences which collapse to over 7500 loci. The EST clusters from this library are listed, with the most abundant being amyloid beta (over 1300 transcripts in this single library). (B) The details of another library, BRHIP3, derived from the hippocampus region of the brain. The most abundant transcript from this library is also amyloid beta.

2.12 THE UniGene LIBRARY BROWSER

The UniGene Library Browser allows access to all the cDNA libraries represented in the database. You can navigate to this page from the NCBI Site Map (Figure 2.18), or from the UniGene home page to the Library Browser seen in a number of the UniGene pages (for example, Figure 2.19A). Once on the Library Browser page, you are met with a very simple query form (**Figure 2.25A**). Here you pick the species and the minimum library size to view. For example, do you only want to see huge libraries that may show the most rare of transcripts? Or do you want to see as many libraries as possible, especially from rarely seen tissues that yield only a few clones? In this example, the form is set to find human libraries containing 1000 ESTs or more. Using these settings, you are shown many pages of EST and cDNA libraries. There are 977 libraries from brain tissue alone, with "only" 109 libraries above 1000 ESTs. The top of the brain library list is shown in Figure 2.25B.

Some of the information in UniGene is difficult to interpret. For example, why can you find albumin ESTs, a gene normally expressed in the liver, in brain libraries? You can explain blood-specific genes found in the brain because of blood contamination in the brain preparation, but not liver genes. Rather than being a problem, perhaps it points to something interesting: a low level of expression of every gene in many tissues. Alternatively, perhaps many proteins have been re-purposed to serve a function in the brain.

Figure 2.25 The UniGene Library Browser. (A) The simple query form where the request is to show all human libraries with a minimum number of sequences equal to 1000. (B) This query brings up a long list, alphabetically organized by tissue and then library, the top of which is shown here. Notice that BRAWH3 and BRHIP3 libraries are on this list.

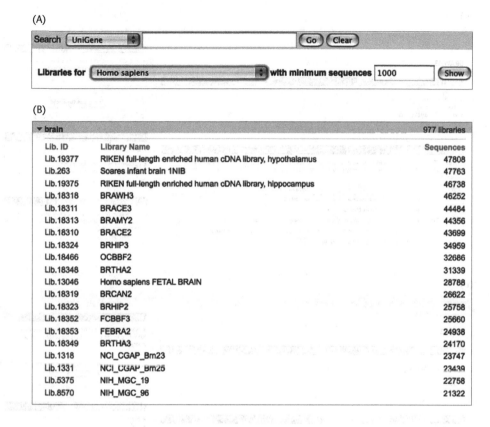

The UniGene database is an incredible resource for determining where you might find transcripts. Even if only a piece of your gene of interest is found in a library, and you have been previously unsuccessful in cloning this gene, you now have a significant advantage in your next attempt.

2.13 SUMMARY

In this chapter, we learned about the databases maintained at the NCBI and elsewhere. At these Websites, we did text-based searches using accession numbers, gene names, authors, and patent numbers. We searched the medical literature and databases of DNA sequences, and we did global searches where a single text query found hits in the nucleotide, protein, literature, and other databases.

EXERCISES

Williams syndrome and oxytocin: research with Internet tools

Oxytocin-neurophysin 1 is a protein expressed in the hypothalamus of the brain. It is cleaved into two different proteins: the 9 amino acid oxytocin and the 94 amino acid neurophysin. Oxytocin causes the smooth muscle of the uterus to contract during labor and has been used for years to induce labor in both humans and animals. Oxytocin has also been shown to have effects on behavior. In fact, the popular press often refers to it as the "trust hormone" and experiments support this claim.

Williams syndrome is a disorder characterized by both physical features and developmental delays. The OMIM database entry includes a description of individuals with Williams syndrome as "empathetic, loquacious, and sociable." The behavior also includes a tendency to trust others, and some have drawn comparisons between Williams syndrome and the effects of oxytocin.

In this series of exercises, the various tools covered in this chapter are to be employed to explore both oxytocin and Williams syndrome.

1. The bond between the domesticated dog and their owner can be quite strong. It is also believed that owning and interacting with a dog can lower blood pressure, increase longevity, and give the owner a feeling of well-being. Could it be that some of these physiological effects are mediated by the dog inducing the production of oxytocin in the owner? Search PubMed for evidence.

2. Oxytocin has been shown to induce trust, even when money is concerned. By administering an oxytocin nasal spray to experimental subjects, oxytocin has been shown to make people more willing to share money with strangers. Search PubMed for articles pertaining to this topic and others related to trust.

3. Find the human oxytocin Gene record at the NCBI. What is the official gene symbol for this gene?

4. Using the UniGene Website, what human tissues are known to express oxytocin?

5. Oxytocinase is an enzyme that degrades oxytocin. Using www.ihop-net.org, find the current official symbol for this human protein. Sorting the reference sentences by date, what are some of the earliest names used for this protein?

6. Using the uspto.gov Website, find a patent where oxytocin is used to induce labor in farm animals. In what year was this patent issued?

7. Using the OMIM Website, can you find any diseases associated with oxytocin?

8. Using the NIH RePORTER Website, are there any grant applications on Williams syndrome?

9. Using the OMIM Website, can you find any records associated with Williams syndrome? Make sure your strategy does not result in a number greater than 50 hits. What is the proper name for this syndrome?

10. There is an article in the *New England Journal of Medicine* on Williams syndrome written by a Dr. Barbara Pober. Construct a specific search of PubMed to find this article.

11. A characteristic of those afflicted by Williams syndrome is excessive trust and acceptance of strangers. Using any tool you were introduced to in this chapter, can you find any connection between Williams syndrome and oxytocin?

FURTHER READING

Ashburner M, Ball CA, Blake JA et al. (2000) Gene ontology: tool for the unification of biology. The Gene Ontology Consortium. *Nat. Genet.* 25, 25–29.

Baxevanis AD (2008) Searching NCBI databases using Entrez. *Curr. Protoc. Bioinformatics* Chapter 1, Unit 1.3.

Grusche FA, Richardson HE & Harvey KF (2010) Upstream regulation of the hippo size control pathway. *Curr. Biol.* 20, R574–R582. Review article on network of gene regulation for kibra.

Hoffmann R & Valencia A (2004) A gene network for navigating the literature. *Nat. Genet.* 36, 664. A very short article on iHOP.

Lee HJ, Macbeth AH, Pagani JH & Young WS 3rd (2009) Oxytocin: the great facilitator of life. *Prog. Neurobiol.* 88, 127–151. A wonderful and thorough review article on the various roles oxytocin plays.

Lennon GG, Auffray C, Polymeropoulos M & Soares MB (1996) The I.M.A.G.E. Consortium: an integrated molecular analysis of genomes and their expression. *Genomics* 33, 151–152.

Lu Z (2011) PubMed and beyond: a survey of web tools for searching biomedical literature. *Database (Oxford)* (DOI: 10.1093/database/baq036). Like any other mountain of data, the medical literature is being mined by sophisticated tools. This article reviews the different approaches used to extract more information out of the medical literature.

Maglott D, Ostell J, Pruitt KD & Tatusova T (2011) Entrez Gene: gene-centered information at NCBI. *Nucleic Acids Res.* 39 (Database issue), D52–D57.

Online Mendelian Inheritance in Man, OMIM®. McKusick-Nathans Institute of Genetic Medicine, Johns Hopkins University (Baltimore, MD), 2011. World Wide Web URL: omim.org

Papassotiropoulos A, Stephan DA, Huentelman MJ et al. (2006) Common Kibra alleles are associated with human memory performance. *Science* 314, 475–478. Ties between kibra and memory are being actively pursued. There were four papers on kibra prior to 2006, and 36 papers in the five years that followed, including papers that suggest that the role of kibra in memory is minor or "complex."

Wheeler DL, Church DM, Federhen S et al. (2003) Database resources of the National Center for Biotechnology. *Nucleic Acids Res.* 31, 28–33. General and brief reference on UniGene and other resources at the NCBI.

CHAPTER 3

Introduction to the BLAST Suite and BLASTN

Key concepts

- Why and how to search a sequence database
- An introduction to nucleotide BLAST (BLASTN)
- Interpreting BLASTN results
- Cross-species searches: paralogs, orthologs, and homologs

3.1 INTRODUCTION

In Chapter 2 we learned how to search databases with text queries. All of these were exact matches—that is, we were expecting to find the exact accession number or exactly spelled words. In this chapter, a much harder database-searching problem is introduced. How do you find matches when your **query** is not a short accession number or a text term, but instead a DNA sequence that is 500 nucleotides long? In addition to finding all the exact matches, can you find those sequences with mismatches, clearly related to the query but not 100% identical? For all the hits that are not exact matches, can calculations generate statistics that help evaluate which hits are significant, and which should be ignored? On top of these challenges, can this search of a database, that contains millions of sequences, show the results in a reasonable time? These and other questions will be answered here. A computer program called BLAST is one of the most commonly used tools in bioinformatics and will be introduced in this chapter. The next three chapters will explore further uses of BLAST.

Why search a database?

Let's assume that you have an unknown sequence and you use it as a query to search a bioinformatics database:

- Is the query identical to something already in the database? Is the query a known gene? If so, you can learn a lot about this gene by looking at the annotation in the sequence records.

- Is your query just a small piece of a much larger gene? If so, you may have just found a way to obtain the rest of the gene. Or based on these results, your laboratory technique may need to be improved upon.

- Is the query similar to something already in the database? Has it already been found in another organism, or is it similar to something in the same organism? Did you just find members of a gene family? These **sequence similarities** may tell you something about the function of your sequence.

- Is the query unique? Never been seen before? Perhaps you have discovered a new gene!

You are asking a handful of questions every time you do a sequence similarity search.

3.2 WHAT IS BLAST?

BLAST, or the Basic Local Alignment Search Tool, was specifically designed to search nucleotide and protein databases. It takes your query (DNA or protein sequence) and searches either DNA or protein databases for levels of **identity** that range from perfect matches to very low similarity. Using statistics, it reports back to you what it finds, in order of decreasing significance, and in the form of graphics, tables, and alignments. There are multiple forms of BLAST, but in this chapter we concentrate on nucleotide BLAST (**BLASTN**, pronounced "blast en"). The query is a DNA sequence and the database you search is populated with DNA sequences, too. **Table 3.1** outlines the query and the subjects of the search.

How does BLAST work?

As mentioned previously, many millions of DNA sequences have been collected in databases and are available for searching at Websites such as those at the NCBI. To conduct sequence-based queries with BLAST, databases must be of a special format to optimize for this type of query. The annotation associated with these individual files must be removed leaving just the sequences. Links to the annotation are still maintained so the identities of these sequences are not lost. Each sequence in the database is then broken into **words** or short sequences for comparison to the query.

When a search is submitted, BLASTN first takes your query DNA sequence and breaks it into words that are quite short (11 nucleotides). It then compares these words to those in the database. As BLAST has to compare many millions of words in this manner, and subsequent steps can be time-consuming, BLAST looks for two adjacent word pairs and, if their similarities and distance between words are acceptable, only those advance to the next set of calculations.

Starting with this local similarity, BLAST then tries to extend the similarity in either direction (**Figure 3.1**). Using the sequence immediately upstream and downstream of the word in the original query, BLAST starts keeping track of the consequences of lengthening the alignment between the query and the sequence in the database. Still matching? The significance score increases. Mismatches encountered? Penalty points accumulate until the cost outweighs the benefit and BLAST stops extending. This approach is finely tuned; if the penalty threshold for extension is too low, BLAST would stop trying to extend similarity very quickly and distant but significant sequence similarities would be missed. If the penalty threshold were too high, BLAST would be given too much freedom to keep extending past real areas of similarity and start collecting many truly insignificant hits.

Finally, all the **alignments** between the query and the database subjects, or "high-scoring subject pairs" (HSPs), are ranked based on length and significance. The best hits are kept and shown to you in the forms of a graphic, a table, and alignments between the query and the hits.

Table 3.1 BLASTN definition

Type	Query	Database
BLASTN	Nucleotide	Nucleotide

Alignment starts with initial word of 11

Figure 3.1 Simple extension example for BLASTN. Starting with an initial match of "words," BLAST extends the alignment between query and hit, keeping track of penalty points against, and increasing significance for, extending the alignment.

```
ACACTGAGTGA
| | | | | | | | | | |
ACACTGAGTGA
```

Extension to the left has no mismatches, no penalty points
Extension to the right has mismatches and penalty points

```
GCACCTTTGCCACACTGAGTGAGCTGCTCTATG
| | | | | | | | | | | | | | | | | | | | | | | | | | |  | | | |  | |  |
GCACCTTTGCCACACTGAGTGACCTGCACTGTA
```

Extension to the left has no penalty points and can continue to grow
Extension to the right accumulates too many mismatch penalty points; extension in this direction stops

```
CAACCTCAAGGGCACCTTTGCCACACTGAGTGAGCTGCTCTATGGTCCTTTGGGG
| | | | | | | | | | | | | | | | | | | | | | | | | | | | | | | | | | | |  | | | |  | |  |        | | | |
CAACCTCAAGGGCACCTTTGCCACACTGAGTGACCTGCACTGTAAAGTTTTGCAT
```

If left side cannot grow any more, the final alignment looks like this:

```
CAACCTCAAGGGCACCTTTGCCACACTGAGTGAGCTGCTCTATG
| | | | | | | | | | | | | | | | | | | | | | | | | | | | | | | | | | | |  | | | |  | |  |
CAACCTCAAGGGCACCTTTGCCACACTGAGTGACCTGCACTGTA
```

3.3 YOUR FIRST BLAST SEARCH

BLAST may be the most widely used sequence analysis program in the world. It is available as a tool from many Websites but is also downloadable as an application that works on your local personal computer or powerful server. It is free, but commercial parties have also created enhanced BLAST applications and charge a fee for these products. Please note that here only the Web form of BLAST will be discussed.

Below is a step-by-step description of your first BLAST search. In future use, these searches can take less than a minute, including your analysis.

Find the query sequence in GenBank

For your first BLASTN search, we'll use a sequence where information is very limited. This may be very typical of what you will face: you have an unknown sequence and you want to determine its identity, if possible.

1. Go to the NCBI Website, ncbi.nlm.nih.gov.

2. Near the top of the home page is the Entrez drop-down menu of database choices. Select "Nucleotide."

3. In the text field next to the menu, enter the GenBank accession number, DD148865. Accession numbers are unique identifiers for sequence records in the GenBank database. By searching with this accession number, you will find just one DNA sequence file.

4. Either press the return key on your keyboard or hit the Search button on the Web page. The Web page refreshes and you will see the nucleotide file.

There are two main sections to this file (**Figure 3.2**): on top, information about this sequence and, below it, the DNA sequence. The information of a sequence record is generally referred to as the annotation. Although the DNA sequence is the key information for all GenBank records, the annotation section can be a rich source of knowledge about the sequence and should not be overlooked. Databases such as GenBank have a specific structure to their annotation: fields

**Figure 3.2
A GenBank file.**

```
LOCUS       DD148865                631 bp    DNA     linear   PAT 04-NOV-2005
DEFINITION  A group of genes which is differentially expressed in peripheral
            blood cells, and diagnostic methods and assay methods using the
            same.
ACCESSION   DD148865
VERSION     DD148865.1  GI:92839210
KEYWORDS    JP 2005102694-A/175.
SOURCE      Homo sapiens (human)
  ORGANISM  Homo sapiens
            Eukaryota; Metazoa; Chordata; Craniata; Vertebrata; Euteleostomi;
            Mammalia; Eutheria; Euarchontoglires; Primates; Haplorrhini;
            Catarrhini; Hominidae; Homo.
REFERENCE   1  (bases 1 to 631)
  AUTHORS   Nojima,H.
  TITLE     A group of genes which is differentially expressed in peripheral
            blood cells, and diagnostic methods and assay methods using the
            same
  JOURNAL   Patent: JP 2005102694-A 175 21-APR-2005;
            Japan Science and Technology Agency,Hiroshi NOJIMA,GeneDesign Inc
COMMENT     OS   Homo sapiens
            PN   JP 2005102694-A/175
            PD   21-APR-2005
            PF   09-SEP-2004 JP 2004263092
            PI   hiroshi nojima
            CC
            FH   Key             Location/Qualifiers
            FT   misc_feature    (17)..(17)
            FT                   /note='n is a, c, g, or t'.
FEATURES             Location/Qualifiers
     source          1..631
                     /organism="Homo sapiens"
                     /mol_type="unassigned DNA"
                     /db_xref="taxon:9606"
ORIGIN
        1 gcaactgtgt tcactancaa cctcaaacag acaccatggt gcatctgact cctgaggaga
       61 agtctgccgt tactgccctg tggggcaagg tgaacgtgga tgaagttggt ggtgabgccc
      121 tgggcaggct gctggtggtc tacccttgga cccagaggtt ctttgagtcc tttggggatc
      181 tgtccactcc tgatgctgtt atgggcaacc ctaaggtgaa ggctcatggc aagaaagtgc
      241 tcggtgcctt tagtgatggc ctggctcacc tggacaacct caagggcacc tttgccacac
      301 tgagtgagct gcactgtgac aagctgcacg tggatcctga gaacttcagg ctcctgggca
      361 acgtgctggt ctgtgtgctg gcccatcact ttggcaaaga attcaccca ccagtgcagg
      421 ctgcctatca gaaagtggtg gctggtgtgg ctaatgccct ggcccacaag tatsactaag
      481 ctcgctttct tgctgtccaa tttctattaa aggttccttt gttccctaag tccaactact
      541 aaactggggg atattatgaa gggccttgag catctggatt ctgcctaata aaaaagcatt
      601 tattttcatt gcaaaaaaaa aaaaaaaaa a
```

of information that are the same in each record. Examples include definition, accession number, and species of origin. We will spend a lot of time examining and utilizing the annotation of files so let's look at this record's annotation closely.

Down the left side of the annotation are section labels, all in uppercase. The LOCUS line contains some basic information: the name, usually synonymous with the accession number (DD148865); the length (631 bp); the type of molecule which was the source of this sequence (DNA); the topology of the source material (linear or circular); the division code (in this case, PAT, which stands for the Patent division of GenBank); and the date when the file was created or underwent revision (04-NOV-2005).

For a full description of the fields in a GenBank file, see the "Release Notes" associated with the database. One way to find these is by using the drop-down menu at the top of the NCBI home page: select "NCBI Web Site" and enter "GenBank release notes" as a query.

The DEFINITION field is usually a brief description of the sequence, its origin, and any additional information that may prove valuable to the reader. The definition line for this record is unusually long, clearly taken from the title of the patent: "A group of genes which is differentially expressed in peripheral blood cells, and diagnostic methods and assay methods using the same."

The ACCESSION and VERSION lines are related. GenBank uses two types of unique identifiers. *Accession numbers* are unique to a sequence record, but should that record be revised, then the accession number is given a new version number. When this sequence was submitted to GenBank, it was given an accession number of DD148865, and the version was DD148865.1. If this record is updated, for example if the annotation is revised by the author, then it will be given a new version number, DD148865.2. Both versions will be kept. Rather than force you to know which version is the latest, GenBank lets you retrieve the latest version by just entering the accession number (DD148865), without the version number extension (.1), as you did when you were asked to retrieve this sequence. *GI numbers* are also assigned as unique identifiers for each specific record, for practical reasons of constructing the GenBank database. Different versions of sequence records could have completely different GI numbers (for example, they won't be sequential or appended with numbers or letters). We will only be working with accession numbers in this book.

KEYWORDS is a field where authors can record any other information they feel might be useful. However, as this is an optional field, you cannot rely on it to comprehensively search a database for related records. In this case, the authors (inventors) put the patent number (JP 2005102694-A/175) in this field so a search of Nucleotide for "JP 2005102694" finds 305 sequences that are associated with this patent. Note that not all terms are indexed for searches. For example, using more of the original patent number (adding "-A/175") does not find these sequences.

The SOURCE and ORGANISM fields are related. The SOURCE is the genus and species according to the author of this record. This is a free text field, and so you might find variations and typographic mistakes; for example, *Homo sapien* instead of the correct *Homo sapiens*. The ORGANISM field is constrained and contains the full taxonomic classification for the source organism.

The REFERENCE field contains details about the sequence origin and can include publications. Also present under FEATURES is information found elsewhere in the annotation (for example, organism) but it may include additional facts or partial analysis results. We'll see this later when we examine sequence records more rich with information.

The COMMENT section may include information such as accession numbers of related sequences or references to other databases. In this patent record, it repeats basic details about the patent.

The last section of this file is the DNA sequence. It is one continuous sequence, broken up into groups of 10 nucleotides (with blank spaces in between), with 60 nucleotides per line. Each line is prefixed with the number of the first base it contains.

Convert the file to another format

The GenBank file described above contains annotation and features that are not used by BLAST and it is necessary to remove these components to run the search. This can be easily accomplished by converting this file into another file format called **FASTA** (pronounced "FAST-AY," the second syllable rhymes with "say").

1. Near the top of the sequence Web page there is a list of formats under "Display Settings" which includes GenBank, FASTA, Graphics, and more. Choose FASTA.

2. When the page refreshes you see the file in FASTA format (**Figure 3.3**).

```
>gi|92839210|dbj|DD148865.1| A group of genes which is differentially expressed in peripheral blood cells,
and diagnostic methods and assay methods using the same
GCAACTGTGTTCACTANCAACCTCAAACAGACACCATGGTGCATCTGACTCCTGAGGAGAAGTCTGCCGT
TACTGCCCTGTGGGGCAAGGTGAACGTGGATGAAGTTGGTGGTGABGCCCTGGGCAGGCTGCTGGTGGTC
TACCCTTGGACCCAGAGGTTCTTTGAGTCCTTTGGGGATCTGTCCACTCCTGATGCTGTTATGGGCAACC
CTAAGGTGAAGGCTCATGGCAAGAAAGTGCTCGGTGCCTTTAGTGATGGCCTGGCTCACCTGGACAACCT
CAAGGGCACCTTTGCCACACTGAGTGAGCTGCACTGTGACAAGCTGCACGTGGATCCTGAGAACTTCAGG
CTCCTGGGCAACGTGCTGGTCTGTGTGCTGGCCCATCACTTTGGCAAAGAATTCACCCCACCAGTGCAGG
CTGCCTATCAGAAAGTGGTGGCTGGTGTGGCTAATGCCCTGGCCCACAAGTATSACTAAGCTCGCTTTCT
TGCTGTCCAATTTCTATTAAAGGTTCCTTTGTTCCCTAAGTCCAACTACTAAACTGGGGGATATTATGAA
GGGCCTTGAGCATCTGGATTCTGCCTAATAAAAAAGCATTTATTTTCATTGCAAAAAAAAAAAAAAAAAA
A
```

Figure 3.3 GenBank file DD148865 in FASTA format.

Almost all the annotation is now missing. What description remains is on the top line which, in all FASTA files, begins with a ">" symbol. Next are fields, separated by vertical bars. The first field indicates the unique identifier for this file record (GI number 92839210). The next field shows the organization that first received this sequence (another sequence database called the DNA Data Bank of Japan or DBJ). This is followed by the accession number, and the definition. The DNA sequence begins on a new line. This signals "everything past this point is just DNA sequence": no numbers, spaces, or anything else among the sequence characters. This is the essence of the FASTA format: minimum annotation followed by sequence. In Figure 3.3 it appears that there are two lines of annotation because of the long definition. In fact, the annotation text has been "wrapped" to the next line due to page width.

This is a good opportunity to look at the sequence with your eyes. There are no numbers and spaces and with the simplicity of the format you might recognize details not visible in GenBank format. Do you see any pattern in the sequence? Any regions rich in As? Do you see any doublets or triplets that seem to be repeated throughout the sequence? Are there any bases that are not A, T, G, or C (**Box 3.1**)? This moment of "low-tech" examination will often reveal details that may help you interpret findings derived by other means. Throughout this book, take time to look at the sequences you encounter and your eyes will become trained at recognizing interesting features without the use of software. An obvious feature of this cDNA is the poly(A) tail.

Performing BLASTN searches

At the NCBI home page, the link to the BLAST forms is usually near the top of the list of Popular Resources. This hyperlink will take you to a page listing different types of BLAST. In this case we are using a DNA sequence query and will be searching a DNA database, so you will be using the "nucleotide blast" or BLASTN form. Navigate to this choice and we'll begin to populate the fields on the form.

- The Query Sequence field. The best way to provide sequence is the plain text found in FASTA format. Some Websites will take your DNA sequence and strip away any numbers, spaces, or otherwise non-DNA characters. This convenience is not found everywhere so it is best to get into the habit of using FASTA format. The NCBI also allows you to enter accession numbers, but in this case, paste the FASTA format of DD148865 into the Query Sequence field, including the annotation line. The BLAST program will grab the definition from this FASTA file and label your results. This will help you stay organized when conducting more than one BLAST search at a time. Depending on the Internet browser you are using, you may see the sequence underlined because the browser will interpret the 631 As, Ts, Gs, and Cs as a spelling error and underline the sequence, but you can ignore that warning.

- The Database field. Next we have to choose the database to search. There are many to choose from the drop-down menu; select the Reference RNA sequences, also known as **RefSeq** mRNA (refseq_rna) (**Figure 3.4**). This is a specialized, nonredundant database of sequences containing the definitive reference sequence for RNAs. Note that RNA sequences are represented as DNA: you will see Ts instead of Us.

- The Organism field. This field allows you to limit your search to a specific organism's DNA. We could search all species, which is the default, but since the query is from *Homo sapiens*, and every human gene has (in theory) been sequenced, why not search a much smaller database and get a direct answer? Your results will also come back faster because of the smaller database size and the NCBI will use fewer computer resources to deliver your results. Enter "Homo sapiens" in this field (Figure 3.4). You'll see that as you begin to type this in the text box, choices will come up and you can finish your entry by clicking on "Homo sapiens (taxid:9606)."

- Program Selection. The default setting uses a version of BLASTN called megaBLAST. This version of BLAST is optimized to find nearly identical hits and is only found at the NCBI Website. We will often be looking for distantly related hits so BLASTN will be used in this book (**Figure 3.5**). Click on the radio button next to this choice.

Box 3.1 Uncertainty codes

As you work with nucleic acid sequence records and read scientific articles, you may come across representations of bases which are not A, T, G, or C. In fact there are several of these in Figure 3.3. These nonstandard letters are actually the uncertainty codes proposed by the International Union of Biochemistry and Molecular Biology (IUBMB) to represent certain groupings of nucleotides. On occasion, software or a scientist is unable to decide if a newly sequenced base is, for example, an A or a G. The three IUBMB symbols most frequently encountered are R for the puRines (A or G), Y for the pYrimidines (C or T), and N for aNy nucleotide. Here are the uncertainty codes:

IUBMB symbol	Definition
R	A or G
Y	C or T
K	G or T
M	A or C
W	A or T
S	C or G
B	C or G or T
D	A or G or T
V	A or C or G
H	A or C or T
N	G or A or T or C

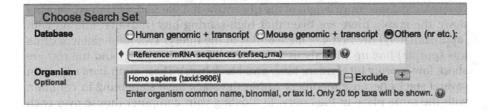

Figure 3.4 BLASTN database choices. The drop-down menu lists the databases and the species can be entered in the "Organism" field.

Figure 3.5 Select BLASTN as the program to use.

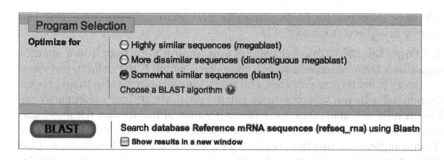

The search is now ready. Click on the "BLAST" button and wait for the results. Times will vary but this incredibly fast-working program will break your 631-nucleotide sequence into small "words," search a database measured in the many thousands of sequences, identify the best hits, try to extend and join the sequences of similarity, then generate the statistics so you can better evaluate the hits. By the time you finish reading this brief description, all of this may be performed and you are now looking at the results.

Depending on your browser settings, the next time you visit this BLASTN form, the defaults may now be changed to what you used in this search. Be sure to make it a habit to review all settings before initiating the search.

3.4 BLAST RESULTS

The results from a BLAST search are divided into three sections: the graphic pane, a results table, and the alignments between the query and the hits. Although some conclusions can be obtained based on interpretations of sections individually, it is best to consider all three sections and draw upon their complementary content during your analysis. These sections are described sequentially below, but when reviewing your results, you should move back and forth between sections as needed.

Graphic

The colorful graphic of the BLAST search results shows the query length across the top (see **Figure 3.6** and color plates). The line goes from 0 to over 600, corresponding to the 5P end and 3P end, respectively. The sequences found by BLAST, the "hits," appear below as horizontal bars in rows. If a hit lined up to every nucleotide of the query, then the bar goes across completely, as is almost the case in the first row. If other hits were similar to only the extreme 3P end of the query, then there would be short bars to the right, as there are in various rows of this graphic. To save space, several hits are often placed in the same row, as space and location allows. For example, the second row has a long red bar, and much shorter black and green bars (see Figure 3.6 and color plates). Bars in the same row do not represent different regions from the same sequence; they are different hits. Different regions of the same sequence are joined by a thin line, but there are none in this example. The color-coding within the graphic is generated by the statistics of each hit. As indicated by the key at the top of the graphic, hits with the highest score are red, the next highest are purple, then green, and so on. In this BLAST search, hits from all ranges of scores are found.

What we see in Figure 3.6 (see also color plates) is a single high-scoring hit which covers almost the entire length of the query (the red bar in the first row). Moving down the graph, there are five other high-scoring hits (red) that line up with (approximately) the first 475 nucleotides of the query. Two moderately scoring hits (purple) line up with about 200 nucleotides of the query, and the rest are short, low-scoring hits (green, blue, and black bars). These short bars appear in several stacks because members within the stack have something in common with the same place in the query. By floating your computer mouse over each

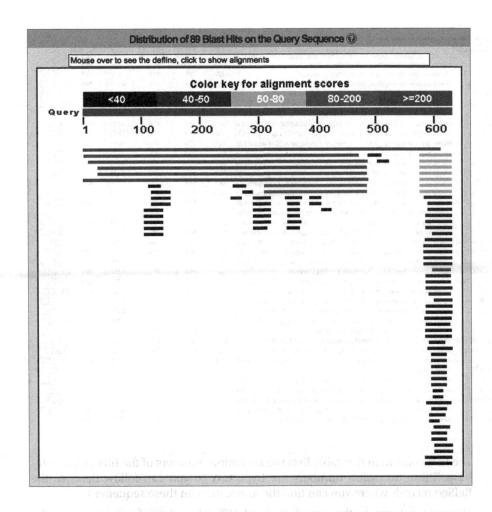

Figure 3.6 The graphic pane of the NCBI BLASTN results.
The query coordinates and length correspond to the numbered scale across the top. Sequences found by BLAST, "hits," are represented as horizontal bars below this scale. These will vary in length, position, and color-coded scoring, shown here in gray shades. See color plates for a color version of this figure.

bar, the sequence definition line appears in the small window above the graphic. By clicking on the individual bars you can navigate to the alignments between the query and the hits.

It is important to realize that these short bars show the length of the similarity between the query and the hit and do not necessarily represent the entire length of the hit. For example, the hit might be one million bases long, but only contain 30 nucleotides in common with the query. This identity will be represented by a bar that is 30 nucleotides long.

Interpretation of the graphic

Even without knowing the identities of the hits shown in the graphic, you can still conclude several things. First, there appears to be only one *Homo sapiens* reference mRNA sequence that aligns with almost the entire query. That is, BLAST was able to align almost every section of the query, in a continuous fashion, across a similar length of sequence within the hit.

Considering the other high-scoring hits, it appears that the 3P end of the query does not align with anything else in the database that is high scoring (ignoring the low-scoring hits for now). The endpoint of the five short red bars is around nucleotide 475, and two other purple bars align between nucleotides 300 and 475. Does this reflect some kind of feature about the query? More information is needed to find the answer and this is discussed later.

Results table

Below the graphic is the BLAST results table that provides basic information about the hits along with the statistics of each hit. This table is long but the top hits are shown in **Figure 3.7**.

Sequences producing significant alignments:

Accession	Description	Max score	Total score	Query coverage	E value	Max ident	Links
NM_000518.4	Homo sapiens hemoglobin, beta (HBB), mRNA	1085	1085	96%	0.0	99%	U E G M
NM_000519.3	Homo sapiens hemoglobin, delta (HBD), mRNA	706	706	74%	0.0	93%	U E G M
NM_005330.3	Homo sapiens hemoglobin, epsilon 1 (HBE1), mRNA	389	389	75%	6e-107	78%	U E G M
NM_000184.2	Homo sapiens hemoglobin, gamma G (HBG2), mRNA	347	347	72%	2e-94	77%	U E G M
NM_000559.2	Homo sapiens hemoglobin, gamma A (HBG1), mRNA	338	338	72%	9e-92	76%	U E G M
NR_001589.1	Homo sapiens hemoglobin, beta pseudogene 1 (HBBP1), non-coding R	232	232	77%	1e-59	71%	E G M
XR_132577.1	PREDICTED: Homo sapiens hypothetical LOC100653006 (LOC100653006	141	141	27%	1e-32	78%	G M
XR_132954.1	PREDICTED: Homo sapiens hypothetical LOC100653319 (LOC100653319	141	141	27%	1e-32	78%	G M
NR_045035.1	Homo sapiens anterior pharynx defective 1 homolog A (C. elegans) (A	53.6	53.6	8%	5e-06	82%	G M
NR_045034.1	Homo sapiens anterior pharynx defective 1 homolog A (C. elegans) (A	53.6	53.6	8%	5e-06	82%	G M
NR_045033.1	Homo sapiens anterior pharynx defective 1 homolog A (C. elegans) (A	53.6	53.6	8%	5e-06	82%	G M
NM_001243772.1	Homo sapiens anterior pharynx defective 1 homolog A (C. elegans) (A	53.6	53.6	8%	5e-06	82%	G M
NM_001243771.1	Homo sapiens anterior pharynx defective 1 homolog A (C. elegans) (A	53.6	53.6	8%	5e-06	82%	G M
NM_016022.3	Homo sapiens anterior pharynx defective 1 homolog A (C. elegans) (A	53.6	53.6	8%	5e-06	82%	G M
NM_001077628.2	Homo sapiens anterior pharynx defective 1 homolog A (C. elegans) (A	53.6	53.6	8%	5e-06	82%	G M
NM_031212.3	Homo sapiens solute carrier family 25, member 28 (SLC25A28), mRNA	48.2	48.2	7%	2e-04	83%	U E G M
NM_197972.1	Homo sapiens non-metastatic cells 7, protein expressed in (nucleoside	46.4	46.4	6%	8e-04	86%	U E G M
NM_013330.3	Homo sapiens non-metastatic cells 7, protein expressed in (nucleoside	46.4	46.4	6%	8e-04	88%	U E G M
NM_001167929.1	Homo sapiens interleukin 1 receptor accessory protein (IL1RAP), trans	44.6	44.6	7%	0.003	79%	U G M
NM_001167928.1	Homo sapiens interleukin 1 receptor accessory protein (IL1RAP), trans	44.6	44.6	7%	0.003	79%	U G M
NM_002182.3	Homo sapiens interleukin 1 receptor accessory protein (IL1RAP), trans	44.6	44.6	7%	0.003	79%	U E G M
NM_001523.2	Homo sapiens hyaluronan synthase 1 (HAS1), mRNA	44.6	44.6	5%	0.003	88%	U E G M
NM_001012993.2	Homo sapiens chromosome 9 open reading frame 152 (C9orf152), mR	44.6	44.6	7%	0.003	83%	U E G M
NM_207034.1	Homo sapiens endothelin 3 (EDN3), transcript variant 4, mRNA	44.6	44.6	7%	0.003	80%	U E G M
NM_000114.2	Homo sapiens endothelin 3 (EDN3), transcript variant 1, mRNA	44.6	44.6	7%	0.003	80%	U E G M
NM_207033.1	Homo sapiens endothelin 3 (EDN3), transcript variant 3, mRNA	44.6	44.6	7%	0.003	80%	U E G M
NM_207032.1	Homo sapiens endothelin 3 (EDN3), transcript variant 2, mRNA	44.6	44.6	7%	0.003	80%	U E G M
NM_153259.2	Homo sapiens mucolipin 2 (MCOLN2), mRNA	44.6	44.6	5%	0.003	89%	U E G M
NM_001198557.1	Homo sapiens lamin B1 (LMNB1), transcript variant 2, mRNA	42.8	42.8	6%	0.010	84%	U G M
NM_005573.3	Homo sapiens lamin B1 (LMNB1), transcript variant 1, mRNA	42.8	42.8	6%	0.010	84%	U G M
NM_001129981.1	Homo sapiens ankyrin repeat domain 2 (stretch responsive muscle) (A	42.8	42.8	7%	0.010	82%	U E G M
NM_021998.4	Homo sapiens zinc finger protein 711 (ZNF711), mRNA	42.8	42.8	6%	0.010	87%	U E G M
NM_002047.2	Homo sapiens proteasome (prosome, macropain) 26S subunit, non-AT	41.0	41.0	5%	0.034	88%	U E G M
NM_001167858.1	Homo sapiens protein phosphatase 1, regulatory subunit 12B (PPP1R1	37.4	37.4	7%	0.42	78%	U G M
NM_001167857.1	Homo sapiens protein phosphatase 1, regulatory subunit 12B (PPP1R1	37.4	37.4	7%	0.42	78%	U G M
NM_203413.1	Homo sapiens chromosome 17 open reading frame 81 (C17orf81), tra	37.4	37.4	7%	0.42	79%	U E G M
NM_199076.2	Homo sapiens cyclin M2 (CNNM2), transcript variant 2, mRNA	35.6	35.6	3%	1.5	95%	

The first column in this table lists the accession numbers of the hits in the database. These accession numbers are hypertext so you can follow these to the RefSeq records where you can find the annotation on these sequences.

The next column is the description. BLAST takes the information from the DEFINITION line in the database record and places it here, but because of space limitations, many of the descriptions will appear truncated.

The next two columns are associated with the statistics of the database search. As mentioned earlier, BLAST uses statistics to sort through all the hits, shows you only the best, and then tells you why (statistically) they are the best. The first of these numbers is called the **Max Score**. Although the average user of BLAST often overlooks it, the change in this score is often important. If you see a sudden drop in the Max Score, expect to see a change in the query–hit alignment length, quality, or both.

The next column, "Total Score," becomes important when BLAST finds multiple, but not joined, sections of similarity between the query and the hit. For each area of similarity, BLAST generates an alignment and a score. If the Max Score is equal to the Total Score, then only a single alignment is present. If the Total Score is larger than the Max Score, then multiple alignments must be present and their individual scores have contributed to the Total Score. For this BLASTN with DD148865, the values in Max Score and Total Score are identical, indicating that only single alignments were generated for this BLASTN search.

"Query Coverage" is the next column. The original query, DD148865, is 631 nucleotides long. If BLAST can align all 631 nucleotides of this query against a hit, then that would be 100% coverage. Remember, Query Coverage does not take into account the length of the hit, only the percentage of the query that aligns with the hit.

Next is the **E value** or **Expect value**, which represents the number of hits you would expect to find by chance given the quality of the alignment and the size of the database. If a database of only As, Ts, Gs, and Cs gets sufficiently large, you start finding sequence similarities by chance, particularly with short queries. The E value in BLAST takes into account both the length and composition of the alignment along with the percentage identity found. A number close to zero means that the hit has to be significant and not due to chance. BLAST results tables are sorted by E value, the most significant hits appearing at the top. When there are two or more identical E values, the Max Score is then used to sort the hits.

The next column is called "Maximum Identity." BLAST calculates the percentage identity between the query and the hit in a nucleotide-to-nucleotide alignment. If there are multiple alignments with a single hit, then only the highest percent identity is shown. The last column, "Links," contains links to databases for that hit and these will not be discussed here.

Interpretation of the table

The top two hits are very significant, both having E values of 0.0. Even the eighth hit has a very small E value (1e–32) so this search has found a number of significant hits.

Based on the first line of the table, it appears that the unknown query used in this BLASTN is nearly identical to the human beta hemoglobin mRNA sequence. Over 96% of the query's length is 99% identical to the first hit. Just concentrating on the Maximum Identity column, the other top hits, although strong, show 93% or less identity when considering large portions of the query length. Note that the 6% drop in Maximum Identity and 22% drop in Query Coverage translated to a significant drop in the Max Score between the first and second hit. The descriptions of the next four hits reveal that the query has found other members of the hemoglobin family: delta, epsilon 1, gamma G, and gamma A. The next three hits, starting with accession number NR_001589, are a hemoglobin pseudogene and two predicted genes. The pseudogene may be both divergent and missing sequence found in the functional family members. Note that the query coverage is higher than the hit above it, but the percent identity is lower and the E value is greater. Gene predictions are often generated automatically, without human supervision, and can be based on incomplete experimental evidence. These may be missing portions, such as exons, of the real gene. The Maximum Identity of these two predictions to the query is identical to that seen with the epsilon 1 transcript (the third hit), but the Query Coverage is significantly lower. It would be interesting to analyze these sequences in detail to support or refute these predictions.

After the two predicted genes, there is a dramatic drop in the Query Coverage, and the E value makes a huge jump. Although there are multiple hits that have high Maximum Identity, the large E value (approaching or greater than 1) indicates that the identities seen can be due to chance. Exploring the annotation of these hits would find that they have significantly different functions from hemoglobin. Not knowing anything else about the identities that are seen for these low-scoring hits, it is easy to conclude that these are not significant.

The alignments

Below the table are the alignments (also called high-scoring subject pairs, HSPs) between the query and the hits. The statistics seen in the BLAST table are repeated here, along with additional important numbers. Within the alignments, the E value is called "Expect." The description line is now shown in its entirety and the length of the hit is shown. The alignments clearly show the relationships that BLAST has found between your query and the hits.

Examine the data in **Figure 3.8** above the alignment between the query (DD148865) and the first hit, NM_000518. The identity is 99%, with 607 nucleotides out of 611 nucleotides aligned. In the graphic above, this hit was shown as a solid red line almost all the way across the length of the query. Here, BLAST shows this by displaying horizontal pairs of sequence; the query is shown above the hit, now labeled the **Subject (Sbjct)**. There are vertical lines between the two sequences wherever they have the same nucleotide. Continuous alignments are

Figure 3.8 BLASTN alignment between Query DD148865 and Sbjct, NM_000518. The Query and Sbjct lines are labeled and aligned bases have a vertical bar "|" between identical bases. Notice that nucleotide 2 of DD148865 aligns with nucleotide 17 of NM_000518, and nucleotide 61 aligns with nucleotide 76.

```
>ref|NM_000518.4| Homo sapiens hemoglobin, beta (HBB), mRNA
Length=626

 Score = 1085 bits (1202),  Expect = 0.0
 Identities = 607/611 (99%), Gaps = 1/611 (0%)
 Strand=Plus/Plus

Query  2    CAACTGTGTTCACTANCAACCTCAAACAGACACCATGGTGCATCTGACTCCTGAGGAGAA  61
            ||||||||||||||| ||||||||||||||||||||||||||||||||||||||||||||
Sbjct  17   CAACTGTGTTCACTAGCAACCTCAAACAGACACCATGGTGCATCTGACTCCTGAGGAGAA  76

Query  62   GTCTGCCGTTACTGCCCTGTGGGGCAAGGTGAACGTGGATGAAGTTGGTGGTGABGCCCT  121
            |||||||||||||||||||||||||||||||||||||||||||||||||||||| |||||
Sbjct  77   GTCTGCCGTTACTGCCCTGTGGGGCAAGGTGAACGTGGATGAAGTTGGTGGTGAGGCCCT  136

Query  122  GGGCAGGCTGCTGGTGGTCTACCCTTGGACCCAGAGGTTCTTTGAGTCCTTTGGGGATCT  181
            ||||||||||||||||||||||||||||||||||||||||||||||||||||||||||||
Sbjct  137  GGGCAGGCTGCTGGTGGTCTACCCTTGGACCCAGAGGTTCTTTGAGTCCTTTGGGGATCT  196

Query  182  GTCCACTCCTGATGCTGTTATGGGCAACCCTAAGGTGAAGGCTCATGGCAAGAAAGTGCT  241
            ||||||||||||||||||||||||||||||||||||||||||||||||||||||||||||
Sbjct  197  GTCCACTCCTGATGCTGTTATGGGCAACCCTAAGGTGAAGGCTCATGGCAAGAAAGTGCT  256

Query  242  CGGTGCCTTTAGTGATGGCCTGGCTCACCTGGACAACCTCAAGGGCACCTTTGCCACACT  301
            ||||||||||||||||||||||||||||||||||||||||||||||||||||||||||||
Sbjct  257  CGGTGCCTTTAGTGATGGCCTGGCTCACCTGGACAACCTCAAGGGCACCTTTGCCACACT  316

Query  302  GAGTGAGCTGCACTGTGACAAGCTGCACGTGGATCCTGAGAACTTCAGGCTCCTGGGCAA  361
            ||||||||||||||||||||||||||||||||||||||||||||||||||||||||||||
Sbjct  317  GAGTGAGCTGCACTGTGACAAGCTGCACGTGGATCCTGAGAACTTCAGGCTCCTGGGCAA  376

Query  362  CGTGCTGGTCTGTGTGCTGGCCCATCACTTTGGCAAAGAATTCACCCCACCAGTGCAGGC  421
            ||||||||||||||||||||||||||||||||||||||||||||||||||||||||||||
Sbjct  377  CGTGCTGGTCTGTGTGCTGGCCCATCACTTTGGCAAAGAATTCACCCCACCAGTGCAGGC  436

Query  422  TGCCTATCAGAAAGTGGTGGCTGGTGTGGCTAATGCCCTGGCCCACAAGTATSACTAAGC  481
            ||||||||||||||||||||||||||||||||||||||||||||||||||||| |||||||
Sbjct  437  TGCCTATCAGAAAGTGGTGGCTGGTGTGGCTAATGCCCTGGCCCACAAGTATCACTAAGC  496

Query  482  TCGCTTTCTTGCTGTCCAATTTCTATTAAAGGTTCCTTTGTTCCCTAAGTCCAACTACTA  541
            ||||||||||||||||||||||||||||||||||||||||||||||||||||||||||||
Sbjct  497  TCGCTTTCTTGCTGTCCAATTTCTATTAAAGGTTCCTTTGTTCCCTAAGTCCAACTACTA  556

Query  542  AACTGGGGGATATTATGAAGGGCCTTGAGCATCTGGATTCTGCCTAATAAAAAAGCATTT  601
            ||||||||||||||||||||||||||||||||||||||||||||||||||||||| |||||
Sbjct  557  AACTGGGGGATATTATGAAGGGCCTTGAGCATCTGGATTCTGCCTAATAAAAAA-CATTT  615

Query  602  ATTTTCATTGC  612
            |||||||||||
Sbjct  616  ATTTTCATTGC  626
```

displayed in lines of 60 nucleotides, with both query and subject wrapping to new lines until the alignment ends. Each line is numbered at the beginning and the end, allowing you to see where the alignment begins and ends.

Looking at this first alignment you can easily conclude that your "unknown" query, DD148865, is *Homo sapiens* beta hemoglobin. The query is nearly fully aligned to a reference sequence for human beta hemoglobin, NM_000518. What do the statistics tell you? This first hit is not a chance finding. In fact, the chance that you would encounter this randomly is so low that the calculated E value is 0.0: there is a zero chance you would find this by accident. The unknown must be *Homo sapiens* beta hemoglobin mRNA. Look back at Figure 3.2 and read the title of the patent associated with this sequence: "A group of genes which is differentially expressed in peripheral blood cells, and diagnostic methods and assay methods using the same." Our conclusion is consistent with the subject matter of the patent. Would it have been easier to just read the patent? It is important to remember that many sequences are published in patents before their identities are known, and so reading the patent may not reveal the identity. In this case, a simple BLAST search gave us the identity.

The graphic (see Figure 3.6 and color plates) and the alignment (Figure 3.8) might suggest that DD148865 is a complete copy of beta hemoglobin, at least compared to this reference sequence, but it is not so. Notice the numbering of the lines at the beginning, or the 5P end, of the sequence. These coordinates indicate that the second nucleotide of the query aligns to nucleotide 17 of the reference sequence. At the 5P end, DD148865 is missing 16 nucleotides possessed by the reference sequence, NM_000518. In DD148865, the first base is a G while the 16th base of NM_000518 is an A. Rather than start the alignment with a mismatch, BLAST just trimmed off the first base of DD148865. At the other end of this first line, see that nucleotide 61 of the Query lines up with nucleotide 76 of the Sbjct.

The statistics just above this alignment indicate that there are 607 nucleotides aligned out of 611 in this alignment. Where are the four mismatches? The one discussed above at the beginning of the query is not included in these numbers because it is not shown. The first is in the first row of the alignment at coordinate 17 of the query. The authors knew there was a base here but could not tell which base. So the IUBMB symbol N for "any base" was used. The second mismatch is in line two of the alignment, at coordinate 116 of the query. The authors were not sure which nucleotide was here but they knew it was a C, G, or T. Thus, the IUBMB symbol for these possibilities (B) appears in this position. In the reference sequence NM_000518, the nucleotide in this position is a G, consistent with the nucleotide code (B) in DD148865. The third mismatch is at query coordinate 474 where there is an S (C or G) in the query and a C in the reference position, again consistent with the correct base.

The final mismatch seen at coordinate 596 of the query is different (Figure 3.8). Here, the query has an extra base, a G, so a gap (–) is inserted in the reference sequence at this position. This maintains the alignment for the remainder of both sequences. Which is correct? The reference sequence should definitely be given serious consideration; however, this single base may be key to the author's reason for submitting this sequence. Is this an interesting mutation, either natural or synthetic? Or is the sequence just incorrect at this position? Further reading and analysis would be required to answer this question. For example, can you find any other human beta globin mRNA sequences with this extra base?

Look at the last lines of the alignment between this query and the hit: nucleotide 612 of DD148865 aligns with nucleotide 626 of NM_000518. What is the length of the entire query sequence? Looking back at the annotation for this record, the DD148865 sequence is 631 nucleotides long (top line of Figure 3.2). The alignment stops short of the poly(A) stretch, visible earlier at the end of the DD148865 sequence file (see Figure 3.3). Notice that the length of the reference sequence NM_000518 is given in Figure 3.8 as "Length=626." BLAST took the alignment as far as it could go and then stopped when it ran out of reference sequence at the 3P end.

An added complication to the alignment interpretation is the poly(A) tail in DD148865. The poly(A) tails of many sequences are trimmed before the sequence is submitted to GenBank and other databases. Others are not trimmed, as seen in this example. It is possible that the original poly(A) tail for NM_000518 started in the same location as DD148865. But the poly(A) tail for sequences is variable in both length and location. Furthermore, RefSeq sequences are compiled from one or more sequences and the curators of RefSeq may have trimmed this sequence here for other reasons. Regardless, we can't be sure what's going on at the 3P end, but the rest of the alignment is excellent and requires little consideration.

As you work through this book, it is highly recommended that you generate simple drawings (pencil and paper are fine!) to understand the relationships that BLAST finds for you. **Figure 3.9** is an example "sketch" which shows the relationship

Figure 3.9 A rough "sketch" outlining the relationship between the query and the hit. Coordinates are taken directly from the BLAST alignment. Although this is an "electronic" sketch, pencil and paper work fine.

```
                       1                                612 631
Query         5P |--------------------------------|--| 3P
Sbjct 5P |--------|--------------------------------| 3P
          1        16                               626
```

between the sequences described above, based on the coordinates. It shows that although the query is missing the 5P end, it has more nucleotides at the 3P end, although they are a stretch of As.

Does this mean that DD148865 is only a fragment? This is a strong possibility, but it will take more analysis to confirm this. Remember that cDNA synthesis is an imperfect process and truncations, particularly at the 5P end of cDNAs, are very common. The poly(A) tail is an obvious difference but database records are inconsistent about this feature. There are many sequences in RefSeq that have poly(A) tails, but NM_000518 does not. Remember that a poly(A) tail is added to transcripts after transcription takes place, so if you look at the genomic DNA, you would not see a stretch of As at the end of the last exon.

Other BLASTN hits from this query

It is a good habit to look at additional alignments to confirm any conclusions you make from the first alignment and to learn more about your query. The second hit, NM_000519, is delta hemoglobin (**Figure 3.10**). If you knew nothing about the beta hemoglobin gene before, you now know that it is a member of a gene family and it is quite similar to another member, delta. The identity between the query and this second hit is 93%, 438 nucleotides out of 470.

Looking at the alignment closely, you can now see nucleotides that do not align. For example, in the second line you can see that seven nucleotides are mismatched and missing the vertical line of identity between the query and reference sequences. Still, 93% identity is high and visually the alignment still appears strong. If these nucleotide differences fell within the coding region of the delta hemoglobin transcript, these could translate into a different amino acid sequence, but delta and beta hemoglobin would look quite similar in protein sequence nevertheless.

The alignment length is shorter than that seen with the first hit. Only 438 nucleotides out of 470 nucleotides from the query align with delta hemoglobin (remember, the total length of the query is 631 nucleotides). The decrease in this length came from the 3P end. This can be seen with the 3P coordinates of the query (472 versus 612, above). However, the 5P end and 3P end of delta hemoglobin mRNA appear to be longer than that of beta hemoglobin. Draw a picture to see the difference.

The annotation for the delta hemoglobin sequence indicates that the coding region is from nucleotides 196 to 639 of NM_000519. These numbers are in good (but not exact) agreement with the alignment. That is, the coding region between these two globins is relatively conserved and aligns well with BLAST, but the noncoding regions have diverged enough to not align. It appears that sequence immediately upstream of the coding regions (approximately nucleotides 163–195) is still quite conserved, reflecting the importance of this sequence.

Comparing the first two hits, the E value stays the same (0.0), but the Score (Max Score in the results table) drops from 1085 to 706. The drop in score reflects the decrease in alignment length between the query and the hit, from 611 to 470 nucleotides, as well as the drop in percent identity. This shows how the Max Score can sometimes tell you more than the E value.

Let's now skip to the seventh alignment between the query and XR_132577 (**Figure 3.11**). This is a predicted mRNA. It has never been isolated in the laboratory, but instead it was stitched together by a gene prediction tool that took cDNAs

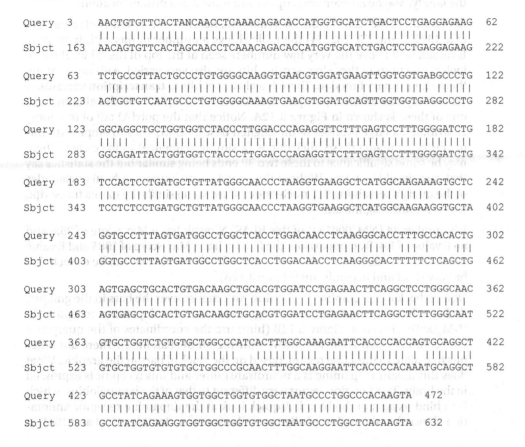

>ref|NM_000519.3 Homo sapiens hemoglobin, delta (HBD), mRNA
Length=774

```
 Score =  706 bits (782),  Expect = 0.0
 Identities = 438/470 (93%), Gaps = 0/470 (0%)
 Strand=Plus/Plus

Query  3    AACTGTGTTCACTANCAACCTCAAACAGACACCATGGTGCATCTGACTCCTGAGGAGAAG  62
            |||  ||||||||||  |||||||||||||||||||||||||||||||||||||||||||
Sbjct  163  AACAGTGTTCACTAGCAACCTCAAACAGACACCATGGTGCATCTGACTCCTGAGGAGAAG  222

Query  63   TCTGCCGTTACTGCCCTGTGGGGCAAGGTGAACGTGGATGAAGTTGGTGGTGABGCCCTG  122
            ||||  ||   |||||||||||||||||| ||||||||||||| |||||||||||  ||||
Sbjct  223  ACTGCTGTCAATGCCCTGTGGGGCAAAGTGAACGTGGATGCAGTTGGTGGTGAGGCCCTG  282

Query  123  GGCAGGCTGCTGGTGGTCTACCCTTGGACCCAGAGGTTCTTTGAGTCCTTTGGGGATCTG  182
            |||||  |  |||||||||||||||||||||||||||||||||||||||||||||||||||
Sbjct  283  GGCAGATTACTGGTGGTCTACCCTTGGACCCAGAGGTTCTTTGAGTCCTTTGGGGATCTG  342

Query  183  TCCACTCCTGATGCTGTTATGGGCAACCCTAAGGTGAAGGCTCATGGCAAGAAAGTGCTC  242
            |||  |||||||||||||||||||||||||||||||||||||||||||||||||  |||||
Sbjct  343  TCCTCTCCTGATGCTGTTATGGGCAACCCTAAGGTGAAGGCTCATGGCAAGAAGGTGCTA  402

Query  243  GGTGCCTTTAGTGATGGCCTGGCTCACCTGGACAACCTCAAGGGCACCTTTGCCACACTG  302
            ||||||||||||||||||||||||||||||||||||||||||||||||||||   |||  |||
Sbjct  403  GGTGCCTTTAGTGATGGCCTGGCTCACCTGGACAACCTCAAGGGCACTTTTTCTCAGCTG  462

Query  303  AGTGAGCTGCACTGTGACAAGCTGCACGTGGATCCTGAGAACTTCAGGCTCCTGGGCAAC  362
            ||||||||||||||||||||||||||||||||||||||||||||||||||||||  |||||||
Sbjct  463  AGTGAGCTGCACTGTGACAAGCTGCACGTGGATCCTGAGAACTTCAGGCTCTTGGGCAAT  522

Query  363  GTGCTGGTCTGTGTGCTGGCCCATCACTTTGGCAAAGAATTCACCCCACCAGTGCAGGCT  422
            |||||||| |||||||||||||||||  |||||||||||||||||||||  ||||||||||
Sbjct  523  GTGCTGGTGTGTGTGCTGGCCCGCAACTTTGGCAAGGAATTCACCCCACAAATGCAGGCT  582

Query  423  GCCTATCAGAAAGTGGTGGCTGGTGTGGCTAATGCCCTGGCCCACAAGTA  472
            ||||||||||||  |||||||||||||||||||||||||||||||  ||||||||
Sbjct  583  GCCTATCAGAAGGTGGTGGCTGGTGTGGCTAATGCCCTGGCTCACAAGTA  632
```

Figure 3.10 Alignment between query DD148865 and the second BLAST hit, delta hemoglobin NM_000519. Nonidentical nucleotides lack the vertical bar "|" seen between identical nucleotides.

>ref|XR_132577.1| PREDICTED: Homo sapiens hypothetical LOC100663006
(LOC100663006), miscRNA
Length=255

```
 Score =  141 bits (156),  Expect = 1e-32
 Identities = 136/175 (78%), Gaps = 0/175 (0%)
 Strand=Plus/Minus

Query  312  CACTGTGACAAGCTGCACGTGGATCCTGAGAACTTCAGGCTCCTGGGCAACGTGCTGGTC  371
            |||||||||||||||||| |||||||||||||||||| | |||||||| || ||||||||
Sbjct  255  CACTGTGACAAGCTGCATGTGGATCCTGAGAACTTAAAGCTCCTGGGAAATGTGCTGGTG  196

Query  372  TGTGTGCTGGCCCATCACTTTGGCAAAGAATTCACCCCACCAGTGCAGGCTGCCTATCAG  431
            || ||||     || || |||||||||||||||||||| |||||||| ||| |||
Sbjct  195  ACCGTTTTGGCAATCCATTTCGGCAAAGAATTCACCCCTGAGGTGCAGGCTTCCTGGCAG  136

Query  432  AAAGTGGTGGCTGGTGTGGCTAATGCCCTGGCCCACAAGTATSACTAAGCTCGCT  486
            ||  |||||| |||| ||||| |  |||||||  ||  ||   || ||||| ||
Sbjct  135  AAGATGGTGACTGGAGTGGCCAGTGCCCTGTCCTCCAGATACCACTGAGCTCACT  81
```

Figure 3.11 The seventh hit (XR_132577) from the BLASTN results with query DD148865. This gene prediction is very divergent compared to the top hits.

originating from this genomic region as evidence for its existence. This predicted transcript shows enough similarity to our query to appear on the list of hits with a very significant E value: 1e–32. There is 78% identity over a 175-nucleotide alignment, much smaller than the values seen with beta and delta hemoglobin. The sequence seems to be related to the query, but is clearly not a close member of the family. Maybe an exon was copied and moved to a distant location.

Let's return to the results table for a moment. Below the two predicted sequences (XR_132577 and XR_132954) the table shows a number of hits with increasing E values, well above the very low numbers seen at the top of the table. The biological functions of these low-scoring hits are also very varied, with no common theme for primary function. There is a cluster of hits to **transcription variants** of "Homo sapiens anterior pharynx defective 1 homolog A," and the alignment of one of these is shown in **Figure 3.12A**. Notice that the poly(A) tail of the query contributed a large number of the aligned bases. BLAST tries to emphasize this to you by changing these nucleotides (a simple sequence) to lowercase. There may be some significance to these two 3P ends being similar but the statistics say that care should be taken. Without the poly(A) contribution, only 36 nucleotides of the query's remaining 612 nucleotides show some similarity to this transcript, which is hardly significant.

Hit number 44 (NM_000794) of this BLASTN result has a Max Score of 35.6 and an E value of 1.4. By comparison, the first hit has a Max Score of 1085 and E value of 0.0. Alignments with numbers like those for hit number 44 can be expected to be very short and insignificant (Figure 3.12B).

Now go back to the graphic. By floating your mouse over the bars in the graphic, find the bar that represents the alignment with the dopamine receptor D1 (NM_000794) seen in Figure 3.12B (hint: use the coordinates of the query and the score-expected color to narrow your region of searching). There are other hits that align to the same region, based on the stack of bars in the graphic. What does this mean? Dopamine is a neurotransmitter and this receptor is expressed in the brain. This function seems very different to that of the hemoglobins, which is to bind oxygen in the blood. A quick look at the dopamine receptor annotation shows that the region of alignment, between nucleotides 1053 and 1081, is

Figure 3.12 Low-scoring alignments from BLASTN with query DD148865. Compared to the top hit, these alignments have a very low Score, a high Expect value, and a very short alignment length. (A) Hit number 9, NR_045035. (B) Hit number 44, NM_000794.

(A)
```
>ref|NR_045035.1| Homo sapiens anterior pharynx defective 1 homolog A (C. elegans)
(APH1A), transcript variant 7, non-coding RNA
Length=2216

 Score = 53.6 bits (58),  Expect = 5e-06
 Identities = 45/55 (82%), Gaps = 3/55 (5%)
 Strand=Plus/Plus

Query  577   GATTCTGCCTAATAAAAAAGCATTTATTTTCATTGCaaaaaaaaaaaaaaaaaaa  631
             |||| || |||||||||||| ||  || ||| ||||||||||||||||||||
Sbjct  2159  GATTTTGACTAATAAAAAAGAAT---TTGTAATTGTGAAAAAAAAAAAAAAAAAA  2210
```

(B)
```
>ref|NM_000794.3| Homo sapiens dopamine receptor D1 (DRD1), mRNA
Length=3373

 Score = 35.6 bits (38),  Expect = 1.4
 Identities = 25/29 (86%), Gaps = 0/29 (0%)
 Strand=Plus/Plus

Query  347   CAGGCTCCTGGGCAACGTGCTGGTCTGTG  375
             || |||||||||| |||  |||||||||||
Sbjct  1053  CACGCTCCTGGGGAACACGCTGGTCTGTG  1081
```

in the coding region of the receptor. Remembering that codons are in groups of three nucleotides, a quick calculation says that this small alignment of 29 nucleotides encodes about 10 amino acids which appear to be similar between beta hemoglobin and the dopamine D1 receptor. Considering that these two proteins are 147 and 477 amino acids long, respectively, such a short similarity does not suggest similarity in protein function. But, nevertheless, it would be fun to investigate this small stretch of amino acids, especially since many other proteins share something in common with hemoglobin. You will need additional skills to study this, and you will receive them later on in this book.

Simultaneous review of the graphic, table, and alignments

Although we reviewed the results sections independently, we did move back and forth between the sections to get a better view of the independent pieces of data. When getting your first look at BLAST results, it is often helpful to do a quick review of all the sections to get a general understanding of the results, and then examine the sections individually and more slowly for details. While looking at the table, you can visualize the hits across the query because you saw them in the graphics panel. When looking at the alignments, you can visualize the trend in the scores because you saw them in the table. And when you look at the graphics, you can visualize the identities because you have scanned the alignments. Now that you understand the components of the BLAST output, let's look at the results again more quickly and introduce some additional results. Below is a fast-paced narrative to better demonstrate the quick review. In each case, look back at the section figure and follow along.

- The graphic shows a single high-scoring hit which stretches from end to end of the query. Five other high-scoring hits appear to end around query coordinate 475. Another pair of hits aligns from approximately nucleotide 300 to 475. Finally, there are multiple groups of low-scoring hits that are quite short in length. Some of these are at the 3P end of the query—maybe the poly(A) stretch is responsible for most of these hits. Floating the mouse over the top red bar shows (in the small text window above the graphics pane) that the best hit is beta hemoglobin. The next five red bars represent the four members of the hemoglobin family (delta, epsilon 1, gamma G, and gamma A) and a pseudogene. Their bars are shorter than that of beta hemoglobin, consistent with the drop in Query Coverage seen in the table. The black bars are short and correspond to the low-scoring hits in the table. There is some vertical stacking of the black bars so this will have to be analyzed further by looking at the alignments.

- The table descriptions show multiple hits to members of the hemoglobin family. The Max Scores reflect a drop in quality and length of alignment (Query Coverage) after the best hit, beta hemoglobin. Four family members and a pseudogene show good alignment length and E values, but none are as good as the first hit, so it is safe to say that our query is closest to beta hemoglobin. Two predicted genes appear high on the list, but have shorter query coverage than the other family members. There are many other hits listed but their scores, query coverage, and E values are all very poor. Their descriptions are also varied, with names that look very different than the globins that bind oxygen.

- The alignments reflect what we have seen in the graphic and table. The best hit is an almost base-for-base alignment between the query and human beta hemoglobin. There are some missing bases at the ends of the alignments, but these should have little impact on deciding the identity of our unknown query. The next few alignments show that the sequences of the hemoglobin family are quite similar but some significant differences at the 3P ends prevent full alignment. Except for some predictions, the rest of the hits appear to be insignificant.

3.5 BLASTN ACROSS SPECIES

Now let's perform another BLASTN search. Rather than try to identify an unknown, the goal of this search is to find hemoglobin genes in other animals.

BLASTN of the reference sequence for human beta hemoglobin against nonhuman transcripts

In the first search of this chapter, we identified the reference mRNA sequence for human beta hemoglobin. Go back to those results and, using the accession number, retrieve this sequence from the database in FASTA format. Keep this beta hemoglobin browser window open for later reference.

In another window, navigate to the NCBI BLASTN form and paste in the FASTA format for the reference sequence for human beta hemoglobin. When trying to identify our unknown, above, we restricted the BLASTN search to just reference sequences annotated as coming from *Homo sapiens*. This time, enter "vertebrates" in the Organism box, broadening the search but still restricting it to organisms that are likely to have hemoglobin (**Figure 3.13**). Again, select BLASTN (not megaBLAST) and launch the BLAST search.

When the screen refreshes and the results appear, perform your quick review of the results and then look at the details. Let's first look at the table (**Figure 3.14**).

The first hit is the reference human beta hemoglobin cDNA: as expected, this BLAST query found itself in the database. Below the human sequence are hits from a variety of vertebrates. Should these Latin names look unfamiliar to you, use the NCBI Taxonomy database to look up the common name for these species.

The leading hits are from primates, for example chimpanzee (*Pan troglodytes*), gibbon (*Nomascus leucogenys*), orangutan (*Pongo abelii*), marmoset (*Callithrix*

Figure 3.13 Configuring the BLASTN form to search reference mRNAs from other species.

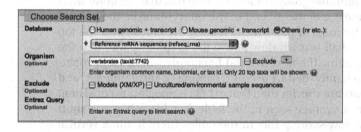

Figure 3.14 Human beta hemoglobin BLASTN results table, showing hits across many species.

Sequences producing significant alignments:

Accession	Description	Max score	Total score	Query coverage	E value	Max ident	Links
NM_000518.4	Homo sapiens hemoglobin, beta (HBB), mRNA	1130	1130	100%	0.0	100%	U E G M
XM_508242.3	PREDICTED: Pan troglodytes hemoglobin, beta, transcript variant 2 (H	1124	1124	100%	0.0	99%	G M
XM_003312881.1	PREDICTED: Pan troglodytes hemoglobin, beta, transcript variant 1 (H	1068	1068	95%	0.0	99%	G M
XR_120944.1	PREDICTED: Nomascus leucogenys hemoglobin subunit beta-like (LOC	1067	1067	100%	0.0	98%	G M
XM_002822127.1	PREDICTED: Pongo abelii hemoglobin subunit beta-like (LOC10043852	1029	1029	95%	0.0	98%	G M
XM_002754891.1	PREDICTED: Callithrix jacchus hemoglobin subunit beta-like (LOC100	886	886	92%	0.0	94%	G M
NM_001164428.1	Macaca mulatta globin, beta (HBB), mRNA >gb|GQ205391.1| Macaca	764	764	75%	0.0	96%	U G M
XM_003254823.1	PREDICTED: Nomascus leucogenys hemoglobin subunit delta-like, tra	749	749	79%	0.0	93%	G M
XM_003254822.1	PREDICTED: Nomascus leucogenys hemoglobin subunit delta-like, tra	749	749	79%	0.0	93%	G M
NM_000519.3	Homo sapiens hemoglobin, delta (HBD), mRNA	731	731	79%	0.0	93%	U E G M
XM_003312882.1	PREDICTED: Pan troglodytes hemoglobin, beta, transcript variant 2 (H	728	728	79%	0.0	92%	G M
XM_001162045.2	PREDICTED: Pan troglodytes hemoglobin, beta, transcript variant 1 (H	728	728	79%	0.0	92%	G M
XM_002822129.1	PREDICTED: Pongo abelii hemoglobin subunit delta-like (LOC1004392	722	722	79%	0.0	92%	G M
NM_001168847.1	Papio anubis hemoglobin, beta (HBB), mRNA	706	706	70%	0.0	95%	U G M
XR_013983.2	PREDICTED: Macaca mulatta hemoglobin subunit delta-like (HBD), mi	679	679	77%	0.0	91%	G M
NM_173917.2	Bos taurus hemoglobin, beta (HBB), mRNA	652	652	100%	0.0	83%	U G M
XM_003584649.1	PREDICTED: Bos taurus hemoglobin subunit beta-like (LOC100850059	648	648	99%	0.0	83%	G
NM_001082260.2	Oryctolagus cuniculus hemoglobin, beta (HBB2), mRNA >emb|V00087	634	707	90%	1e-179	88%	U G M
XM_003433019.1	PREDICTED: Canis lupus familiaris hemoglobin subunit beta-like, tran	632	632	79%	3e-179	88%	G M
XM_537902.3	PREDICTED: Canis lupus familiaris hemoglobin subunit beta-like, tran	632	632	79%	3e-179	88%	G M
XM_002754892.1	PREDICTED: Callithrix jacchus hemoglobin subunit delta-like (LOC100	627	627	70%	1e-177	92%	G M
XM_857349.2	PREDICTED: Canis lupus familiaris hemoglobin subunit beta-like, tran	587	587	74%	1e-165	88%	G M
XM_001250141.4	PREDICTED: Bos taurus hemoglobin fetal subunit beta-like (LOC78167	583	583	100%	2e-164	81%	G
NM_001014902.2	Bos taurus hemoglobin, gamma (HBG), mRNA	583	583	100%	2e-164	81%	U G M
NM_001110509.1	Bos taurus hemoglobin, gamma 2 (LOC788610), mRNA	583	583	100%	2e-164	81%	U G M

jacchus), and Rhesus monkey (*Macaca mulatta*). Note that for many of these and other species, the mRNA is annotated as "Predicted" reflecting, in most cases, that the sequence is derived from genomic sequencing, not sequencing of cDNA. This indicates the amount of attention these other animals have received in studying their genes. Once you get outside of mainstream model organisms, relatively little mRNA/cDNA cloning of globins and many other genes has taken place. These genes were predicted based on very strong similarity to known mRNA sequences from human, mouse, rat, and other well-studied organisms.

Figure 3.14 shows that the E value for many of the top hits is 0.0. For these sequences, there is almost a twofold drop in Max Scores. The Query Coverage drops throughout the list until it gets to the cow (*Bos taurus*) where it is 100%. However, the percent identity between the human and cow sequence is only 81%, which pushed it down the list of hits.

Looking at the alignment to the predicted chimpanzee beta hemoglobin it is easy to see strong similarity between the human and chimpanzee sequences. The identities measurement indicates that the human and chimpanzee sequences are 99% identical. The alignment of 625 out of 626 nucleotides is nearly 100%, but this field uses whole numbers and rounds this value down to 99%. With this and many other human–chimpanzee sequence alignments, you can clearly see that at the DNA level, chimpanzees are our closest relatives.

Moving down the table or the alignments, certain trends are seen. Unlike the first BLASTN, which found all the members of the human hemoglobin family among the top hits, this search shows that most of the other human hemoglobin family members are not seen before many beta hemoglobin sequences in other species are found. *Homo sapiens* delta hemoglobin is the next human hit in the table, but no other human sequences are seen until much further down on the list. This indicates that, at least at the mRNA sequence level, the beta hemoglobins in these top species are more similar than human hemoglobin family members are to each other.

At the bottom of the table in Figure 3.14, among all the beta hemoglobin hits, notice that the cow (*Bos taurus*) hits include "gamma" and "gamma 2," which is a family member not seen in humans. As you explore genes from other species, you will see variation in names that will often reflect differences in physiology between organisms, differences in naming conventions, history, and even mistakes.

Looking at the graphic display for this BLASTN search (**Figure 3.15**), notice that many hits do not align with the 5P or the 3P ends of the query. Look at the alignment between the query and the *Equus caballus* (horse) beta hemoglobin (**Figure 3.16**). The alignment with the query starts at nucleotide 51 and ends with nucleotide 492. Why are so many sequences failing to align from end to end? The answer can be gathered from the alignment and the annotation of the sequence records.

Go to the GenBank record (NM_000518) and look at the annotation for the human beta hemoglobin and you see that the coding region sequence, abbreviated as **CDS**, starts at nucleotide 51, the A of the ATG start codon. The horse sequence, NM_001164018, starts at this exact base (horse nucleotide number 1 aligns with human nucleotide 51). The 5P untranslated region (UTR) is not even present in this horse reference sequence. Based on this BLAST search, it appears that the 5P untranslated regions of many vertebrate mRNAs are not present in the sequence records or cannot align well with the human untranslated region; hence the common start site for many of the alignments is in the vicinity of human nucleotide 51.

The 3P ends of many of these alignments terminate around query nucleotide 494 (seen in the graphic, Figure 3.15), which is the 3P boundary of the human beta hemoglobin coding region. Certainly, divergence of sequence in the 3P UTR explains some of these terminations. In the case of the horse sequence, there is

Figure 3.15 The BLASTN graphic of human beta hemoglobin mRNA against many species.
Note the truncations at the 5P (left) end of the graphic as well as the distinct boundary around nucleotide 500 at the 3P end.

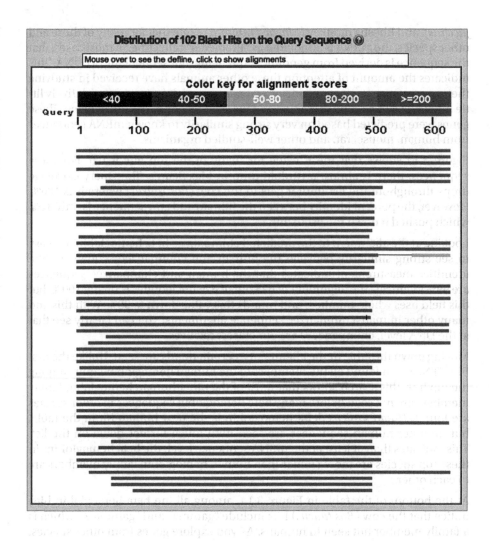

a simple reason to explain the sudden stop of alignment at horse nucleotide 442: the horse sequence comes to an end. Look at the description line of this alignment and it says "Length=444." Scroll through the alignments and you'll notice that many sequences do not include the 3P UTR of the mRNA. Many genomes are annotated in an automated fashion and genes are predicted based on similarity to known, well-annotated genes from other organisms. Like the 5P UTR, the 3P UTRs of many predicted transcripts are underrepresented in the database.

In general, gene coding regions are more conserved than the noncoding regions. A single nucleotide change in the coding region can change the amino acid sequence, possibly alter the structure and function of the protein, or introduce a stop codon and truncate the translation product. There are fewer constraints for sequence and function on untranslated regions. However, there are very important regulatory elements at work in untranslated regions. As long as regulatory elements, if present, are not disrupted, many nucleotide substitutions, insertions, and deletions are tolerated and lead to sequence differences in untranslated regions so extensive that they fail to align using BLAST.

Paralogs, orthologs, and homologs

In the first BLASTN search of this chapter, we were able to identify members of the human hemoglobin family: in order, beta (the unknown found the reference sequence for itself), delta, epsilon 1, gamma G, and gamma A hemoglobins. Based on the identities between our beta hemoglobin query and these hits, it is clear the family members are still closely related to each other, the lowest identity

```
>ref|NM_001164018.1|  Equus caballus hemoglobin, beta (HBB), mRNA
Length=444


Score =  522 bits (578),  Expect = 3e-146
 Identities = 381/442 (86%), Gaps = 0/442 (0%)
 Strand=Plus/Plus

Query  51   ATGGTGCATCTGACTCCTGAGGAGAAGTCTGCCGTTACTGCCCTGTGGGGCAAGGTGAAC  110
            ||||||||| |||| | ||| |||||| | || || |||||||||| |||||||||
Sbjct  1    ATGGTGCAACTGAGTGGTGAAGAGAAGGCAGCTGTCTTGGCCCTGTGGGACAAGGTGAAT  60

Query  111  GTGGATGAAGTTGGTGGTGAGGCCCTGGGCAGGCTGCTGGTGGTCTACCCTTGGACCCAG  170
            | ||| |||||||||||||||| |||||||||||||||||||| |||||| ||||||||||
Sbjct  61   GAGGAAGAAGTTGGTGGTGAAGCCCTGGGCAGGCTGCTGGTTGTCTACCCATGGACTCAG  120

Query  171  AGGTTCTTTGAGTCCTTTGGGGATCTGTCCACTCCTGATGCTGTTATGGGCAACCCTAAG  230
            |||||||||||| ||||||||||||||||||||  ||| ||||||| ||||||||||| |||
Sbjct  121  AGGTTCTTTGACTCCTTTGGGGATCTGTCCAATCCTGGTGCTGTGATGGGCAACCCCAAG  180

Query  231  GTGAAGGCTCATGGCAAGAAAGTGCTCGGTGCCTTTAGTGATGGCCTGGCTCACCTGGAC  290
            ||||||| || ||||||||||||||| ||||   |||| |||| || ||  ||| || |||
Sbjct  181  GTGAAGGCCCACGGCAAGAAAGTGCTACACTCCTTTGGTGAGGGCGTGCATCATCTTGAC  240

Query  291  AACCTCAAGGGCACCTTTGCCACACTGAGTGAGCTGCACTGTGACAGCTGCACGTGGAT  350
            |||||||||||||||||||||   | ||||||||||||||||||||||| ||||||||||
Sbjct  241  AACCTCAAGGGCACCTTTGCTGCGCTGAGTGAGCTGCACTGTGACAAGCTGCACGTGGAT  300

Query  351  CCTGAGAACTTCAGGCTCCTGGGCAACGTGCTGGTCTGTGTGCTGGCCCATCACTTTGGC  410
            |||||||||||||||||||||||||||||||||||| ||||  |||||||| | ||||||||
Sbjct  301  CCTGAGAACTTCAGGCTCCTGGGCAACGTGCTGGTTGTTGTGCTGGCTCGCCACTTTGGC  360

Query  411  AAAGAATTCACCCCACCAGTGCAGGCTGCCTATCAGAAAGTGGTGGCTGGTGTGGCTAAT  470
            || || |||||||||   |||||||| ||||||| || ||||||||||||||||||| |||
Sbjct  361  AAGGATTTCACCCCAGAGTTGCAGGCTTCCTATCAAAAGGTGGTGGCTGGTGTGGCCAAT  420

Query  471  GCCCTGGCCCACAAGTATCACT  492
            || |||||||||||| ||| |||
Sbjct  421  GCACTGGCCCACAAATACCACT  442
```

Figure 3.16 BLASTN alignment between human (NM_000518) and horse beta hemoglobin (NM_001164018). Note that the alignment starts at the "ATG" of the coding region (underlined), base number one of the horse sequence. There is no horse 5P untranslated region.

being 76% between beta and gamma A. The simplest explanation for these very strong identities is that they all share a common ancestor. Some time in the distant past, there was a first hemoglobin gene. Then, through gene duplication events, other members arose, diverged in sequence, and became specialized in function. We can find them today as we did in the above BLASTN search. This stands as a model of what has happened throughout evolution; one gene gave rise to other family members.

Gene family members within the same organism are referred to as **paralogs**. Paralogs share a common ancestor and reside in the same genome. They are clearly related to each other but are usually specialized and have different functions. Beta hemoglobin is expressed in adults, while gamma A hemoglobin protein is only found in the fetus. This specialization reflects the distinctly different oxygen-binding needs between an air-breathing adult and a fetus growing in a womb and getting oxygen from its mother's blood. This is further illustrated by the human diseases called thalassemias, where a globin gene does not function and the other globin family members cannot adequately substitute for the lost function.

As seen in the last BLASTN search, many other animals also have hemoglobins. Genes that perform identical functions in different organisms are called **orthologs**. The human beta hemoglobin gene is orthologous to the horse beta hemoglobin gene. Both are expressed in adult animals. Did they evolve independently? No. The simplest explanation is that there was a common animal ancestor who had the first beta hemoglobin gene, and the evolutionary descendants of that ancestor inherited this gene.

Homolog is a term that describes both paralogs and orthologs. When comparing genes between organisms, and it is not clear if they are orthologous, then the genes are described as homologs. When describing genes that show some identity but it is not clear if they are family members, then it is safer to describe them as sharing homology. The human alpha hemoglobin and mouse beta hemoglobin clearly have a common ancestor but perform different functions. They are homologs, not orthologs, and certainly not paralogs.

3.6 BLAST OUTPUT FORMAT

The output of BLAST described above is the HTML or Web format of the results. This format allows easy navigation between the graphic, table, and alignments, as well as instant access to sequence files through hypertext. The NCBI and other Websites provide this to you because of these obvious advantages. However, you may encounter Websites or an instance of BLAST you run from a command-line interface like UNIX where the output is raw text. In this case, the results may look like **Figure 3.17**. The advantage of this format is the simplicity; copying from this output, or parsing using a simple programming script, is uncomplicated by hidden formatting. The NCBI gives you the option to output your results as "plain text" by clicking on "Formatting options" near the top of the results page. In fact, if you are copying your own BLAST results and pasting them into reports, you may wish to use this format. This book will often show you the raw text format for simplicity. Note that in this raw form, the only columns after the description are the Score and E Value.

3.7 SUMMARY

In this chapter, the focus was BLASTN, a Web application that allows you to search nucleotide databases with nucleotide sequence queries. You learned how to paste a sequence into the query window, select a database to search, sometimes narrow your search to sequences of a certain species, and then launch the search. When the results came back you were able to quickly review the search by looking at the graphic and the table. You could look at the graphic to get a visual idea of how your query lined up to the hits, and sometimes saw that the query found smaller sequences. In the BLASTN results table, you saw the hits along with the statistics relating to them. Besides obvious criteria such as percent identity, there was the E value that showed you the probability that hits were found by chance. An E value of 0.0 meant that these hits were very real and could not be explained by random occurrences. Finally, you looked at the alignments where you could see, base by base, how your query lined up with the hits.

EXERCISES

Exercise 1: Biofilm analysis

Public water supply lines are immersed in water for decades and a community of microorganisms thrives on these wet surfaces. These slippery coatings are referred to as biofilms and the bacterial makeup is generally unknown because scientists are unable to culture and study the vast majority of these organisms in the laboratory. In 2003, Schmeisser and colleagues published a study where they collected and sequenced the DNA from bacteria growing on pipe valves of a drinking water network in Northern Germany. Through sequence similarity, they were able to classify a large number of these organisms as belonging to certain species or groups. In this process they identified many new species. In this

(A)

		Score	E
Sequences producing significant alignments:		(Bits)	Value

```
ref|NM_000518.4|  Homo sapiens hemoglobin, beta (HBB), mRNA        1034    0.0
ref|NM_000519.3|  Homo sapiens hemoglobin, delta (HBD), mRNA        654     0.0
ref|NM_005330.3|  Homo sapiens hemoglobin, epsilon 1 (HBE1), mRNA   381     8e-105
ref|NM_000184.2|  Homo sapiens hemoglobin, gamma G (HBG2), mRNA     343     2e-93
ref|NM_000559.2|  Homo sapiens hemoglobin, gamma A (HBG1), mRNA     334     1e-90
ref|XM_002344540.1|  PREDICTED: Homo sapiens similar to PRO298...   241     2e-62
ref|XM_002347218.1|  PREDICTED: Homo sapiens similar to PRO298...   241     2e-62
ref|XM_002343046.1|  PREDICTED: Homo sapiens similar to PRO298...   241     2e-62
ref|NR_001589.1|  Homo sapiens hemoglobin, beta pseudogene 1 (...   233     2e-60
ref|NM_001128602.1|  Homo sapiens RAS guanyl releasing protein...   35.6    1.3
ref|NM_005739.3|  Homo sapiens RAS guanyl releasing protein 1 ...   35.6    1.3
ref|NM_080723.4|  Homo sapiens neurensin 1 (NRSN1), mRNA            35.6    1.3
ref|NM_016642.2|  Homo sapiens spectrin, beta, non-erythrocyti...   35.6    1.3
ref|NM_000794.3|  Homo sapiens dopamine receptor D1 (DRD1), mRNA    35.6    1.3
ref|NM_199077.1|  Homo sapiens cyclin M2 (CNNM2), transcript v...   35.6    1.3
ref|NM_199076.1|  Homo sapiens cyclin M2 (CNNM2), transcript v...   35.6    1.3
ref|NM_017649.3|  Homo sapiens cyclin M2 (CNNM2), transcript v...   35.6    1.3
ref|NM_144666.2|  Homo sapiens dynein heavy chain domain 1 (DN...   33.7    4.5
ref|NM_021020.2|  Homo sapiens leucine zipper, putative tumor ...   33.7    4.5
ref|NM_000798.4|  Homo sapiens dopamine receptor D5 (DRD5), mRNA    33.7    4.5
ref|NM_015221.2|  Homo sapiens dynamin binding protein (DNMBP)...   33.7    4.5
ref|NM_080539.3|  Homo sapiens collagen-like tail subunit (sin...   33.7    4.5
ref|NM_182515.2|  Homo sapiens zinc finger protein 714 (ZNF714...   33.7    4.5
ref|NM_080538.2|  Homo sapiens collagen-like tail subunit (sin...   33.7    4.5
ref|NM_005677.3|  Homo sapiens collagen-like tail subunit (sin...   33.7    4.5
ref|NM_016315.2|  Homo sapiens GULP, engulfment adaptor PTB do...   33.7    4.5
ref|NM_024686.4|  Homo sapiens tubulin tyrosine ligase-like fa...   33.7    4.5
ref|NM_015540.2|  Homo sapiens RNA polymerase II associated pr...   33.7    4.5
ref|NM_032444.2|  Homo sapiens BTB (POZ) domain containing 12 ...   33.7    4.5
ref|NM_014234.3|  Homo sapiens hydroxysteroid (17-beta) dehydr...   33.7    4.5
ref|NM_148414.1|  Homo sapiens ataxin 2-like (ATXN2L), transcr...   33.7    4.5
ref|NM_007245.2|  Homo sapiens ataxin 2-like (ATXN2L), transcr...   33.7    4.5
ref|NM_145714.1|  Homo sapiens ataxin 2-like (ATXN2L), transcr...   33.7    4.5
ref|NM_148415.1|  Homo sapiens ataxin 2-like (ATXN2L), transcr...   33.7    4.5
ref|NM_148416.1|  Homo sapiens ataxin 2-like (ATXN2L), transcr...   33.7    4.5
ref|NM_016261.2|  Homo sapiens tubulin, delta 1 (TUBD1), mRNA       33.7    4.5
ref|NM_000911.3|  Homo sapiens opioid receptor, delta 1 (OPRD1...   33.7    4.5
ref|NM_007261.2|  Homo sapiens CD300a molecule (CD300A), mRNA       33.7    4.5
ref|NM_020857.2|  Homo sapiens vacuolar protein sorting 18 hom...   33.7    4.5
```

(B)

```
>ref|NM_000518.4| Homo sapiens hemoglobin, beta (HBB), mRNA
Length=626

 Score = 1034 bits (1146),  Expect = 0.0
 Identities = 573/573 (100%), Gaps = 0/573 (0%)
 Strand=Plus/Plus

Query  1    GTGCATCTGACTCCTGAGGAGAAGTCTGCCGTTACTGCCCTGTGGGGCAAGGTGAACGTG   60
            ||||||||||||||||||||||||||||||||||||||||||||||||||||||||||||
Sbjct  54   GTGCATCTGACTCCTGAGGAGAAGTCTGCCGTTACTGCCCTGTGGGGCAAGGTGAACGTG   113

Query  61   GATGAAGTTGGTGGTGAGGCCCTGGGCAGGCTGCTGGTGGTCTACCCTTGGACCCAGAGG   120
            ||||||||||||||||||||||||||||||||||||||||||||||||||||||||||||
Sbjct  114  GATGAAGTTGGTGGTGAGGCCCTGGGCAGGCTGCTGGTGGTCTACCCTTGGACCCAGAGG   173

Query  121  TTCTTTGAGTCCTTTGGGGATCTGTCCACTCCTGATGCTGTTATGGGCAACCCTAAGGTG   180
            ||||||||||||||||||||||||||||||||||||||||||||||||||||||||||||
Sbjct  174  TTCTTTGAGTCCTTTGGGGATCTGTCCACTCCTGATGCTGTTATGGGCAACCCTAAGGTG   233
```

Figure 3.17 The "plain text" format of NCBI BLAST results. (A) The BLASTN results table. (B) The top of a BLASTN alignment.

exercise, you are to use BLASTN to repeat some of their analysis and identify the makeup of these biofilms.

Below is a list of 10 sequence accession numbers from their study. You are to use the NCBI BLASTN Web form to search for sequence similarities to try to identify the bacteria growing within these biofilms.

AY187314

AY187315

AY187316

AY187317

AY187318

AY187325

AY187326

AY187330

AY187332

AY187333

1. Retrieve each sequence from the NCBI GenBank and, based on the annotation of these sequence records, identify what gene was used in their analysis.

2. For each sequence, convert the file format to FASTA using the "Display Settings."

3. Navigate to the NCBI BLASTN Web form and paste the FASTA format of each DNA sequence into the Query window.

4. Choose the "Nucleotide collection (nr/nt)" as the database to be searched.

5. To save lots of time for your searches, restrict your search to "bacteria (taxid:2)" in the Organism field.

6. Pick "Somewhat similar sequences (BLASTN)" as the program to be used in the search.

7. When ready, launch the search by clicking on the "BLAST" button.

8. Open up additional Internet browser windows and launch the other searches.

9. Ten individual windows of results will be returned within a few minutes. Be sure to stay organized and record your conclusions for each accession number.

10. For each BLASTN search, survey the results graphic, table, and alignments to assign each unknown sequence to an organism. You may not find 100% identity between your query and the hits, except for the self-hit. Note that the first hit may also be an unknown so you should examine all the hits before drawing any conclusions as to what kind of bacteria the sequence came from.

11. Using the NCBI PubMed database or other Internet resources, try to find basic information about the genus and/or species; for example, habitats where these bacteria grow, and if they are associated with any diseases or environmental pollutants.

Exercise 2: RuBisCO

It is often said that ribulose bisphosphate carboxylase (RuBisCO) is the most abundant protein on the planet. This enzyme is part of the Calvin cycle and is the key enzyme in the incorporation of carbon from carbon dioxide into living organisms. It is part of an enzyme complex found in plants, terrestrial or aquatic, and most probably played an important role in the development of our atmosphere and life on Earth.

Arabidopsis thaliana, a member of the mustard family, is an important model system for higher plants. It is easily cultivated in the laboratory, undergoes rapid development, and produces a large number of seeds, making it amenable to genetic studies. Although not important agronomically, *Arabidopsis* has provided fundamental knowledge of plant biology and it was the first plant genome to be sequenced (in 2000).

In this exercise, you will use BLASTN to identify members of the RuBisCO gene family in *Arabidopsis*.

1. Retrieve the reference mRNA for the *Arabidopsis* RuBisCO small chain subunit 1b, NM_123204, at the NCBI Website.

2. Change the format to FASTA and paste the sequence into the NCBI BLASTN Web form Query window.

3. Set the database to "Reference RNA sequences (refseq_rna)" and restrict the organism to "Arabidopsis thaliana (taxid:3702)."

4. Set the program selection to "Somewhat similar sequences (BLASTN)" and click on the "BLAST" button to launch the search.

5. When the results are returned, you should now utilize the graphic, table, and alignments to identify the family members.

6. The Reference RNA database should not have any redundancy but two family members have alternatively spliced mRNAs. Compare the alignments carefully and examine the annotation (especially the coordinates of the coding regions) of all the relevant sequence records to describe and understand the major differences between these family member transcripts.

7. Create a table with a listing of the names of family member transcripts and their accession numbers, their mRNA length, the coordinates of the coding regions (CDS), and a brief description of what is observed in the alignments.

FURTHER READING

Altschul SF, Gish W, Miller W et al. (1990) Basic local alignment search tool. *J. Mol. Biol.* 215, 403–410. The first BLAST paper, a classic in the field of bioinformatics.

Altschul SF, Madden TL, Schäffer AA et al. (1997) Gapped BLAST and PSI-BLAST: a new generation of protein database search programs. *Nucleic Acids Res.* 25, 3389–3402. This is another foundation paper by the creators of the BLAST suite of programs, describing significant improvements to the algorithm.

Schmeisser C, Stöckigt C, Raasch C et al. (2003) Metagenome survey of biofilms in drinking-water networks. *Appl. Environ. Microbiol.* 69, 7298–7309. This paper is relevant to the biofilms exercise.

Internet resources

To learn more about *Arabidopsis* and the *Arabidopsis* genome sequencing project, go to www.arabidopsis.org and click on their Education and Outreach portal. This is an extensive Website with many resources ranging from fundamental learning about *Arabidopsis* to detailed workings of plant research and how the genome was sequenced.

To learn more about the globin family of genes, go to the NCBI Bookshelf of electronic books, choose "Human Molecular Genetics" from the extensive list of textbooks, and in the search window for this text, enter "globin." This gene was discussed or illustrated 74 times in this book, within a variety of topics. For example, there is a figure showing the "Evolution of the globin superfamily." Due to their history, biology, biochemistry, genetics, diseases, size, and genomic structure, globin genes are frequent subjects in textbooks.

ACAAGGGACTAGAGAAACCAAAA
AGAAACCAAAACGAAAGGTGCAGAA
AACGAAAGGTGCAGAAGGGGAAACAGATGCAGA
GAAGGGGAAACAGATGCAGAAAGCATC
AGAAAGCATC
ACAAGGGACTAGAGAAACCAAAACGAAAGGTGCAGAAGGGGAAACAGATGCAGAAAGCATC
AGAAACCAAAACGAAAGGTGCAGAA
AACGAAAGGTGCAGAAGGGG
GAAGGGG

CHAPTER 4

Protein BLAST: BLASTP

Key concepts

- Protein BLAST (BLASTP) at the NCBI and ExPASy Websites
- The genetic code
- Amino acids and their overlapping properties
- The BLASTP scoring matrix

4.1 INTRODUCTION

In the last chapter, the focus was on nucleotide BLAST (BLASTN), a Web application that allows you to search nucleotide databases with nucleotide sequence queries. BLASTN gives you powerful search capabilities over simple text-based searches. You searched with unknowns and determined their identities based on similarities or identities to reference sequences. Now, if you were asked to find the insulin gene for the pig, you could start with the human version of insulin and search a DNA database, perhaps limiting your search to DNA sequences from *Sus scrofa*. There are other types of BLAST, giving you additional power and flexibility of approaches, and in this chapter we will cover protein BLAST (BLASTP, **Table 4.1**).

4.2 CODONS AND THE GENETIC CODE

With just four possible nucleotides (A, T, G, or C) in each position, a small gene such as insulin has a coding region that looks deceptively simple (**Figure 4.1**).

Historically, it must have been a surprise when the chemical makeup of DNA was determined and it consisted of just four nucleotides. How could something so simple encode 20 amino acids and the phenomenal diversity of life in the world? The answer was equally simple; DNA sequence is read in "threes," three nucleotides at a time, during the translation into protein sequence. These groups of

Table 4.1 Comparison of BLAST definition

Type	Query	Database
BLASTN	Nucleotide	Nucleotide
BLASTP	Protein	Protein

The names, BLASTN and BLASTP, are easy to remember: you use a Nucleotide query to search a nucleotide database, and a Protein query to search a protein database, respectively.

Figure 4.1 DNA sequence of the human insulin mRNA.

```
>NM_000207 Homo sapiens insulin (INS), mRNA, coding region only
ATGGCCCTGTGGATGCGCCTCCTGCCCCTGCTGGCGCTGCTGGCCCTCTGGGGACCTGACCCAGCCGCAG
CCTTTGTGAACCAACACCTGTGCGGCTCACACCTGGTGGAAGCTCTCTACCTAGTGTGCGGGGAACGAGG
CTTCTTCTACACACCCAAGACCCGCCGGGAGGCAGAGGACCTGCAGGTGGGGCAGGTGGAGCTGGGCGGG
GGCCCTGGTGCAGGCAGCCTGCAGCCCTTGGCCCTGGAGGGGTCCCTGCAGAAGCGTGGCATTGTGGAAC
AATGCTGTACCAGCATCTGCTCCCTCTACCAGCTGGAGAACTACTGCAACTAG
```

three are referred to as **codons** and, collectively, they make up the genetic code. If you consider all the possible combinations of three that can be made with four nucleotides, there are 64 possible triplets of the four nucleotides. It can be calculated as follows:

```
(4 possible bases in the first position)
(4 possible bases in the second position)
(4 possible bases in the third position)
          4 × 4 × 4 = 64 codons
```

Twenty codons are adequate for the 20 amino acids. But all possible codons are utilized, with some redundancy, usually referred to as **degeneracy**. In addition, there are three codons that signal the termination of mRNA translation, and these are called terminators. There are several ways to portray the genetic code but a commonly used chart appears in **Figure 4.2**.

As you examine the genetic code, you should notice a number of features. First, the degeneracy described above: there are two codons for phenylalanine (TTT, TTC), four codons for valine (GTT, GTC, GTA, GTG), and six codons for leucine (TTA, TTG, CTT, CTC, CTA, CTG). Codon assignments are not always multiples of two: methionine (ATG) and tryptophan (TGG) each have a single codon while isoleucine has three (ATT, ATC, ATA). Note that it is common to refer to a group of codons by using an "N" to represent "any base." For example, the four codons for valine can be written as GTN.

Figure 4.2 The genetic code. All 64 possible triplets of the four nucleotides create a code for translating mRNA into protein sequence. To find the amino acid assigned to a given triplet or codon, start with the first nucleotide on the far left. Then find the column of the second nucleotide. Finally, find the narrow row of the third nucleotide on the far right. The intersection of all three vectors (Row × Column × Row) arrives at the amino acid assignment for a nucleotide triplet. For example, "GGG" is found in the intersection of the bottom large row, the far right column, and the last row. The three-letter and single-letter abbreviations for the amino acids appear in this figure.

```
                              Second Nucleotide
        |      T       |       C       |       A       |       G       |
    ----+--------------+---------------+---------------+---------------+----
        | TTT  Phe (F) | TCT  Ser (S)  | TAT  Tyr (Y)  | TGT  Cys (C)  | T
 F   T  | TTC  Phe     | TCC  Ser      | TAC  Tyr      | TGC  Cys      | C  T
 i      | TTA  Leu (L) | TCA  Ser      | TAA  Ter (*)  | TGA  Ter (*)  | A  h
 r      | TTG  Leu     | TCG  Ser      | TAG  Ter (*)  | TGG  Trp (W)  | G  i
 s   ---+--------------+---------------+---------------+---------------+--  r
 t      | CTT  Leu (L) | CCT  Pro (P)  | CAT  His (H)  | CGT  Arg (R)  | T  d
     C  | CTC  Leu     | CCC  Pro      | CAC  His      | CGC  Arg      | C
 N      | CTA  Leu     | CCA  Pro      | CAA  Gln (Q)  | CGA  Arg      | A  N
 u      | CTG  Leu     | CCG  Pro      | CAG  Gln      | CGG  Arg      | G  u
 c   ---+--------------+---------------+---------------+---------------+--  c
 l      | ATT  Ile (I) | ACT  Thr (T)  | AAT  Asn (N)  | AGT  Ser (S)  | T  l
 e   A  | ATC  Ile     | ACC  Thr      | AAC  Asn      | AGC  Ser      | C  e
 o      | ATA  Ile     | ACA  Thr      | AAA  Lys (K)  | AGA  Arg (R)  | A  o
 t      | ATG  Met (M) | ACG  Thr      | AAG  Lys      | AGG  Arg      | G  t
 i   ---+--------------+---------------+---------------+---------------+--  i
 d      | GTT  Val (V) | GCT  Ala (A)  | GAT  Asp (D)  | GGT  Gly (G)  | T  d
 e   G  | GTC  Val     | GCC  Ala      | GAC  Asp      | GGC  Gly      | C  e
        | GTA  Val     | GCA  Ala      | GAA  Glu (E)  | GGA  Gly      | A
        | GTG  Val     | GCG  Ala      | GAG  Glu      | GGG  Gly      | G
    ----+--------------+---------------+---------------+---------------+----
```

(*) Termination codons are labeled as "Ter" and "*".

For the most part, the genetic code is orderly. Related codons encode the same amino acid. For example, all proline codons begin with CC, all glycine codons begin with GG, and all valine codons begin with GT. Leucine codons can start with either a T or C, but all leucine codons have a T in the second position. With only four possibilities, having one nucleotide in common isn't necessarily significant.

Looking elsewhere in the code, terminator codons are unique in that they do not encode an amino acid but, instead, signal the termination of translation. There are other groups of two or four codons for certain amino acids. A notable set of codons is for serine: there is one group of four (TCN) but another pair which seems to have little in common with the other four (AGT, AGC).

There is a large advantage for having a genetic code with redundancy. The homologous DNA sequences that were found with BLASTN in Chapter 3 showed good levels of conservation of DNA sequence, but there were definitely differences, especially across species. There is evolutionary pressure to maintain the same amino acid sequence for a protein. But when a change in DNA sequence occurs, the redundancy of codons means that there is not necessarily a change of the amino acid. In many cases, the third nucleotide of the codon can be modified with reduced or no consequences on the encoded amino acid.

In fact, we can easily find instances where the conservation of amino acid sequence leaves a specific signature involving the third base. Here is another case where your eyes can easily recognize a pattern if you take the time to look and you know what you are looking for. Recalling the Chapter 3 BLASTN search of human beta hemoglobin (NM_000518) against RefSeq, included on the list of hits was the rat epsilon hemoglobin mRNA (NM_001008890). One pair of aligned sequences from rat epsilon hemoglobin shows a pattern where there are pairs of identical bases and then the third is different (**Figure 4.3**). Despite there being 12 nonidentities in this alignment of 60 nucleotides, there is only one that results in an amino acid difference for this region. Ten of those nucleotide differences occurred in the third position resulting in no change in amino acid (for example, the codon labeled "2" in Figure 4.3B). One third-position change (codon "1" in Figure 4.3B) resulted in a D (GAT) to E (GAG) substitution but these amino acids are biochemically similar so this difference is probably less significant than many other possible substitutions. The final difference in this aligned stretch is in the first position of the codon labeled "3" (CTG vs TTG) but both codons are for leucine.

It is important to remember that an organism doesn't choose where the changes in nucleotide sequence will occur. What we see today are changes that were tolerated after they happened. Mutations are allowed in positions where there is little or no impact on the structure or function of a particular sequence. For example, should the third base of an alanine codon change from GCG to GCA, GCC, or GCT, there is no consequence on the protein sequence since all four codons are for alanine. Importantly, mutations that lead to an advantage for the organism are positively selected. If an enzyme that breaks down sugar has a change in protein sequence and can now digest sugar more efficiently, a bacterium that

> **GCG and the Alanines**
> There once was a company that sold a large suite of sequence analysis programs and although they are no longer in business, many institutions still use this old software. The company's name was Genetics Computer Group, but everyone called it "GCG." They had a company softball team called the "Alanines," named after the codon GCG.

Figure 4.3 Conservation of protein sequence signature. (A) BLASTN alignment between a coding region of human beta hemoglobin (NM_000518) and rat epsilon hemoglobin (NM_001008890) mRNAs. (B) The same alignment as in (A) with every other codon underlined for easier viewing and three codons marked with 1, 2, and 3. (C) Translations of this same coding region with one-letter abbreviations for the amino acids. The single amino acid change for this sequence is bold and underlined.

(A)
```
human beta hemoglobin    GTGGATGAAGTTGGTGGTGAGGCCCTGGGCAGGCTGCTGGTGGTCTACCCTTGGACCCAG
                         || || ||  |||||||||||| ||| |||| || || || || |||| ||||||||||
rat epsilon hemoglobin   GTTGAGGAGGTTGGTGGTGAAGCCTTGGGAAGACTTCTCGTTGTGTACCCATGGACCCAG
```

(B)
```
                              1                  2     3
human beta hemoglobin    GTGGATGAAGTTGGTGGTGAGGCCCTGGGCAGGCTGCTGGTGGTCTACCCTTGGACCCAG
                         || || ||  |||||||||||| ||| |||| || || || || |||| ||||||||||
rat epsilon hemoglobin   GTTGAGGAGGTTGGTGGTGAAGCCTTGGGAAGACTTCTCGTTGTGTACCCATGGACCCAG
```

(C)
```
human beta hemoglobin     V  D  E  V  G  G  E  A  L  G  R  L  L  V  V  Y  P  W  T  Q
rat epsilon hemoglobin    V  E  E  V  G  G  E  A  L  G  R  L  L  V  V  Y  P  W  T  Q
```

harbors this new protein will grow faster than its neighbors and soon come to dominate the local environment.

Memorizing the genetic code

There are people who have memorized the entire genetic code. This is not essential to perform sequence analysis, but as you work your way through this book, there will be times where you will find it helpful to know some of the genetic code without having to look it up. Perhaps you could start with the beginnings and ends of coding regions.

- Most proteins begin with the codon ATG, which is translated to methionine (look at the first codon of the insulin-coding region in Figure 4.1).

- The translation of proteins ends with one of the codons TAA, TAG, TGA (look at the last codon of the insulin-coding region in Figure 4.1).

4.3 AMINO ACIDS

For convenience, amino acids are abbreviated two different ways, using three-letter and one-letter codes (**Figure 4.4**).

Even with a small protein such as preproinsulin, you see all 20 amino acids represented. **Figure 4.5** shows the 110 amino acid protein sequence encoded by the insulin DNA sequence (see Figure 4.1).

mRNAs are translated from the 5P end to the 3P end. The first amino acid of the protein is the **N-terminus** while the last amino acid is the **C-terminus**. However, these terms are also used to describe the terminal regions, not just individual amino acids. For example, sometimes the first five amino acids are described as the N-terminus, but it is just as appropriate to label the first 30 amino acids the same way. If you find it difficult to remember that the N-terminus is encoded by the 5P end and the C-terminus by the 3P end, just recall that "3" rhymes with "C."

Figure 4.4 Amino acids and their abbreviations. The literature and specialized bioinformatics applications will often use the three-letter abbreviations, most of which are the first three letters of the amino acid's name. But in most books, publications, and bioinformatics applications, the one-letter abbreviations for amino acids are seen. The column on the far right contains common ways to remember the abbreviations. Protein sequence can be derived from the coding sequence of an mRNA, or can be chemically determined. If the amino acid sequence of a protein is uncertain, an "X" is used as a placeholder. It represents any amino acid.

AMINO ACIDS AND THEIR SYMBOLS			CODONS	MEMORY AID
A	Ala	Alanine	GCA GCC GCG GCT	**A**lanine
C	Cys	Cysteine	TGC TGT	**C**ysteine
D	Asp	Aspartic acid	GAC GAT	Aspar**D**ic acid
E	Glu	Glutamic acid	GAA GAG	Glu**E**tamic acid
F	Phe	Phenylalanine	TTC TTT	**F**enylalanine
G	Gly	Glycine	GGA GGC GGG GGT	**G**lycine
H	His	Histidine	CAC CAT	**H**istidine
I	Ile	Isoleucine	ATA ATC ATT	**I**soleucine
K	Lys	Lysine	AAA AAG	**K** is adjacent to **L**ysine in the alphabet
L	Leu	Leucine	TTA TTG CTA CTC CTG CTT	**L**eucine
M	Met	Methionine	ATG	**M**ethionine
N	Asn	Asparagine	AAC AAT	Asparagi**N**e
P	Pro	Proline	CCA CCC CCG CCT	**P**roline
Q	Gln	Glutamine	CAA CAG	**Q**tamine
R	Arg	Arginine	AGA AGG CGA CGC CGG CGT	a**R**ginine
S	Ser	Serine	AGC AGT TCA TCC TCG TCT	**S**erine
T	Thr	Threonine	ACA ACC ACG ACT	**T**hreonine
V	Val	Valine	GTA GTC GTG GTT	**V**aline
W	Trp	Tryptophan	TGG	t**W**iptophan
Y	Tyr	Tyrosine	TAC TAT	t**Y**rosine
X		Any amino acid		

Figure 4.5 Protein sequence of human preproinsulin.

```
>NP_000198  insulin preproprotein [Homo sapiens]
MALWMRLLPLLALLALWGPDPAAAFVNQHLCGSHLVEALYLVCGERGFFYTPKTRREAEDLQVGQVELGG
GPGAGSLQPLALEGSLQKRGIVEQCCTSICSLYQLENYCN
```

Take a moment to look at this preproinsulin protein sequence. What do you notice? Are some amino acids abundant, and others rare? Is it surprising that more than one-third of this protein is made up of three amino acids: 10 alanines (A), 12 glycines (G), and 20 leucines (L)? Do you see adjacent pairs of amino acids (LL, FF, RR)? Even English words appear ("VEAL," "GAG," "LEGS") at random. To test your powers of observation, try and find these three words and others.

Amino acid properties

Let's return to the substitution we saw in the human–rat hemoglobin mRNA alignment above (Figure 4.3). We saw there was only one amino acid substitution, aspartic acid (D) for glutamic acid (E), and these amino acids were referred to as biochemically **similar**. Looking at **Figure 4.6**, these two amino acids are quite similar in appearance, with a common backbone and variable side chains or "R groups." Both side chains are acidic and interact with their environment in a similar fashion. If evolution substitutes a glutamic acid for an aspartic acid, this may not change the shape of the protein very much. Glutamic acid is larger by three atoms (CH_2), which can be a critical feature of this position in the protein. If the amino acid in this position were pointing inward to a reactive pocket, glutamic acid would protrude deeper into that pocket than aspartic acid and that could be crucial to the proper function of the protein. Alternatively, it may be crucial to provide room in a pocket and glutamic acid may be too large. Nature allows certain substitutions in many positions as long as both the structure and the properties of this region of the protein are conserved. Still, there is change, and sometimes abandonment, of one function as another function is acquired.

As we compare similar proteins, you will notice that certain regions tolerate insertions or deletions, which are collectively known as **indels**. If you examine the three-dimensional structure of these proteins, you often see that these variable regions are on the exterior surface of the protein. As long as a change does not alter the hydrophilic nature of this region (which is important since it comes in contact with the aqueous environment) and does not significantly alter the structure or function of the rest of the protein, it can be tolerated. In contrast, place a new hydrophilic or biochemically reactive amino acid in the hydrophobic interior of a protein which is normally folded into a tight ball, and the structure, function, and stability are liable to be affected.

The amino acid side chains that are brought in close proximity may be orchestrating a complex chain of biochemical events that are extremely sensitive to the slightest change. Little or no substitution is tolerated in these regions because the amino acids here must have certain dimensions, charge, and other properties. These critical amino acids are often not adjacent to each other in the polypeptide chain. In **Figure 4.7**, amino acid side chains from distant regions of a hypothetical unfolded protein are brought together to create an active site. This active site may bind to a substrate and catalyze a change, for example. As we examine protein sequence alignments generated by BLASTP, you will notice "islands" of conservation that may indeed represent distant regions brought together for a key protein function.

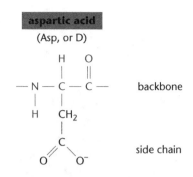

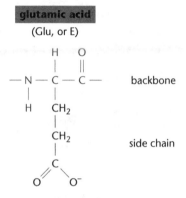

Figure 4.6 The chemical structures of the amino acids aspartic acid and glutamic acid. Both amino acids share a common backbone (N–C–C). The variable lower part of each molecule is known as the side chain. From, Zvelebil M and Baum JO (2008) Understanding Bioinformatics. Garland Science.

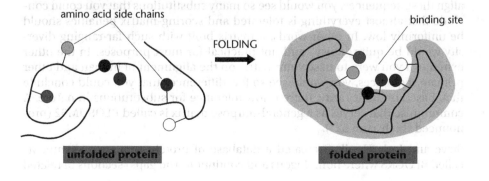

Figure 4.7 A hypothetical protein showing side chains being brought into close proximity by folding of the protein. From, Alberts B, Johnson A, Lewis J et al. (2007) Molecular Biology of the Cell, 5th ed. Garland Science.

4.4 BLASTP AND THE SCORING MATRIX

As described in Chapter 3, a BLAST algorithm matches queries with database entries and, based on matches, tries to extend the alignments until penalty points accumulate past a certain threshold. With BLASTN, you have five possibilities in each position of an alignment between the query and a hit: A, T, G, C, and N (for any nucleotide). As BLASTN tries to find alignments, the scoring is relatively simple: an A should line up with an A, a T with a T, a G with a G, and a C with a C. Any base can align with N. Any other pairs, such as A lining up with a T, is a mismatch and a penalty is applied.

When working with protein sequences, you have 21 possibilities for each position (20 amino acids plus X or any amino acid). If you tried to keep it simple, you could design your search program to just reward matches and penalize mismatches. But as discussed above, there is an added layer of complication when trying to align protein sequences: nature often tolerates conservative differences in amino acid sequences. **Figure 4.8** shows the 20 amino acids grouped by the properties of the side chains. A smaller penalty should be applied if an amino acid substitution maintains the properties important for a protein to function.

Rather than just counting matches and mismatches, BLASTP uses a **scoring matrix** for matches, mismatches, and conservative substitutions. A scoring matrix rewards identities, gives "partial credit" for some mismatches, and penalizes others. The scoring on a matrix is critical to the success of a BLASTP search. If penalties are too high, and rewards too low, sequences distant but related to your query would go undetected by BLASTP. But if penalties are too low and rewards are too high, then sequences unrelated to your query would clutter the results.

Building a matrix

So how do you approach building a matrix? You could focus on physical size and create matrix scores based on the similarities of amino acid dimensions. Certainly, there are critical amino acids filling a cavity in a protein, or maybe not filling that cavity, and any change in size disrupts the important dimensions of, for example, an enzyme's reactive site. Or, you could also place emphasis on the biochemical properties of the amino acid; whether it is acidic, basic, nonpolar, uncharged, hydrophobic, or hydrophilic. Like physical size, there are circumstances where the biochemical properties of a given position on the protein have to be maintained in order to maintain function or structure.

People who study protein structures and functions will tell you that all of these properties are important. For every amino acid of every protein, there are different forces at work that shape the most important property for substitutions at those positions. Rather than try to design a scoring matrix to cover all of these specific issues, a powerful approach would be to build a matrix based on what we see in protein families. For millions of years, nature has been experimenting and natural selection has left us with substitutions that work. Why not just catalog all of these substitutions and let the frequencies at each position guide us?

The flaw in the approach as stated is that "all of these substitutions" are too many. There are bacterial proteins that are related to human proteins and if you were to align these sequences, you would see so many substitutions that you could conclude that almost everything is tolerated and scoring-matrix penalties should be uniformly low. In other words, a matrix built with such far-ranging diversity would be quite "noisy" and not practical for most purposes. In the other extreme, if you were to base your matrix on the alignment of human and other primate sequences, you would see so few differences that you could conclude that, based on their rarity, there is low tolerance for substitutions in nature. A compromise that serves as a general-purpose matrix is called BLOSUM62 (pronounced "blah sum 62").

Steve and Jorja Henikoff created a database of protein sequence alignments called BLOCKS where homologous and continuous (no gaps) sections of related

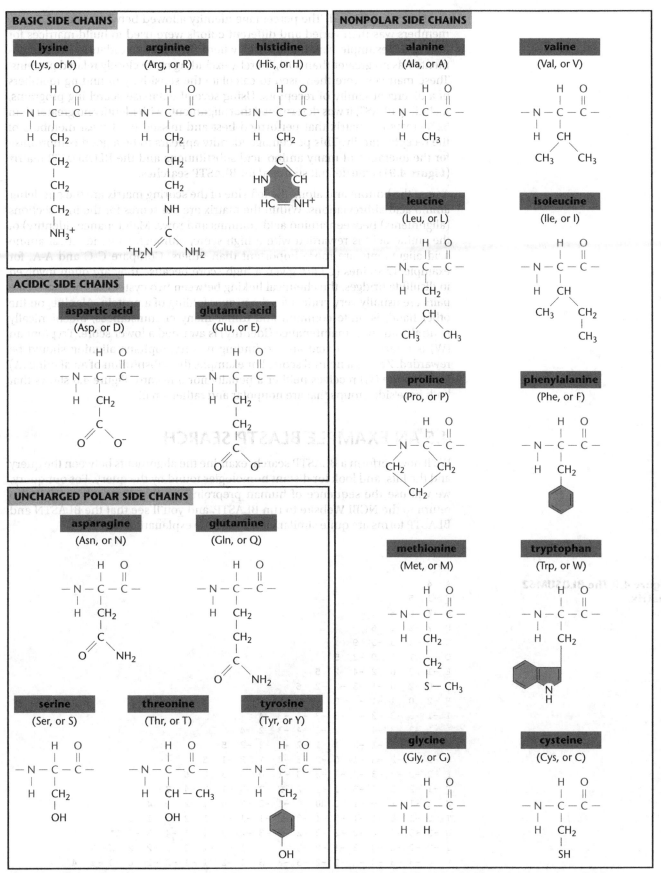

Figure 4.8 Amino acids can be grouped based on the properties of their side chains. It is common to divide them into these four groups: basic, acidic, uncharged, and nonpolar. From, Zvelebil M and Baum JO (2008) Understanding Bioinformatics. Garland Science.

proteins were aligned. The percentage identity allowed between aligned family members was then varied and different cutoffs were used to build matrices for testing. For example, a matrix was built where the percentage identity in the alignments was no greater than 80%, thereby excluding more closely related proteins. These matrices were then used to calculate the sensitivity of finding members of a divergent family of receptors. Using several database-searching programs, including BLAST, it was determined that alignments with identity no greater than 62% created a matrix that performed best and missed the fewest members of the receptor family. This percentage identity appears to be a good compromise for the tolerance of many amino acid substitutions and the BLOSUM62 matrix (**Figure 4.9**) is the default standard for BLASTP searches.

Across the bottom and along the left side of the scoring matrix are the one-letter amino acid abbreviations. Within the matrix are the scores for the intersections (alignments) between amino acid columns and rows. Maintenance (identity) of the amino acid is rewarded with a high score, although some identical amino acid alignments are more important than others. Compare C-C and A-A, for example. Cysteines (C) are given a high score because they are often involved in disulfide bridges, the chemical linking between two cysteines. These cysteine pairs are usually very critical for the proper folding of a protein. Alanine, on the other hand, is quite common and under many circumstances, biochemically neutral, and so its maintenance (identity) is awarded a lower score. Tryptophan (W) is a rare amino acid and so finding two tryptophans aligning should be rewarded. Zero is a neutral score. For example, the substitution of an alanine (A) for a glycine (G) receives neither a penalty nor a reward. Figure 4.8 shows that both have side groups that are nonpolar and rather small.

4.5 AN EXAMPLE BLASTP SEARCH

We'll now perform a BLASTP search, examine the alignments between the query and the hits, and look for distant homologies found by the query. For our query, we will use the sequence of human preproinsulin (see Figure 4.5). We'll also return to the NCBI Website to run BLASTP, and you'll see that the BLASTN and BLASTP forms are quite similar, requiring little explanation here.

Figure 4.9 The BLOSUM62 matrix.

```
A   4
R  -1   5
N  -2   0   6
D  -2  -2   1   6
C   0  -3  -3  -3   9
Q  -1   1   0   0  -3   5
E  -1   0   0   2  -4   2   5
G   0  -2   0  -1  -3  -2  -2   6
H  -2   0   1  -1  -3   0   0  -2   8
I  -1  -3  -3  -3  -1  -3  -3  -4  -3   4
L  -1  -2  -3  -4  -1  -2  -3  -4  -3   2   4
K  -1   2   0  -1  -3   1   1  -2  -1  -3  -2   5
M  -1  -1  -2  -3  -1   0  -2  -3  -2   1   2  -1   5
F  -2  -3  -3  -3  -2  -3  -3  -3  -1   0   0  -3   0   6
P  -1  -2  -2  -1  -3  -1  -1  -2  -2  -3  -3  -1  -2  -4   7
S   1  -1   1   0  -1   0   0   0  -1  -2  -2   0  -1  -2  -1   4
T   0  -1   0  -1  -1  -1  -1  -2  -2  -1  -1  -1  -1  -2  -1   1   5
W  -3  -3  -4  -4  -2  -2  -3  -2  -2  -3  -2  -3  -1   1  -4  -3  -2  11
Y  -2  -2  -2  -3  -2  -1  -2  -3   2  -1  -1  -2  -1   3  -3  -2  -2   2   7
V   0  -3  -3  -3  -1  -2  -2  -3  -3   3   1  -2   1  -1  -2  -2   0  -3  -1   4
    A   R   N   D   C   Q   E   G   H   I   L   K   M   F   P   S   T   W   Y   V
```

Retrieving protein records

Protein records can be found much the same way as nucleotide records. Text queries at the NCBI main page give you access to all their major databases, including one for proteins (**Figure 4.10**).

Most NCBI mRNA or Gene files will give you easy access to the protein sequences encoded by that nucleic acid. For example, in a RefSeq mRNA file, the translation product is displayed in an annotation feature called CDS (coding sequence) and a screenshot of this region of the annotation is in **Figure 4.11**.

Although it is certainly convenient for the protein sequence to be visible from within the nucleotide record, it is not intended for any serious analysis other than a visual examination. Copying and pasting from this field will also yield a format that needs "cleaning up" to make it into FASTA format, for example. However, within the CDS section is a link to the protein record. Notice the line labeled "protein_id" in Figure 4.11 and the hypertext "NP_000198.1" which will take you to the protein RefSeq record. Once there, you will find a link back to the corresponding mRNA record near the top of the file in the line labeled "DBSOURCE" (**Figure 4.12**). When viewing any database record at any Website, a small amount of hunting will often find links to the files you need.

Running BLASTP

1. At the NCBI Website, find the protein RefSeq sequence for human preproinsulin precursor, NP_000198.

2. Open another Web page and go to the NCBI BLAST page and choose Protein BLAST.

3. Paste in the sequence or enter the RefSeq accession number.

4. Change the database to "Reference proteins."

5. Make sure BLASTP is the algorithm and click on the BLAST button.

Figure 4.10 The NCBI drop-down menu. This appears on the home page and allows convenient searching of the Protein database with text queries.

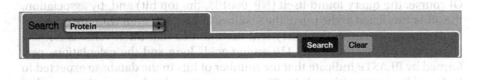

Figure 4.11 The CDS section of an NCBI mRNA record. This contains a translation of the protein encoded by this mRNA.

```
CDS            60..392
               /gene="INS"
               /gene_synonym="IDDM2; ILPR; IRDN; MODY10"
               /note="proinsulin"
               /codon_start=1
               /product="insulin preproprotein"
               /protein_id="NP_000198.1"
               /db_xref="GI:4557671"
               /db_xref="CCDS:CCDS7729.1"
               /db_xref="GeneID:3630"
               /db_xref="HGNC:6081"
               /db_xref="HPRD:01455"
               /db_xref="MIM:176730"
               /translation="MALWMRLLPLLALLALWGPDPAAAFVNQHLCGSHLVEALYLVCG
               ERGFFYTPKTRREAEDLQVGQVELGGGPGAGSLQPLALEGSLQKRGIVEQCCTSICSL
               YQLENYCN"
```

```
LOCUS       NP_000198               110 aa            linear   PRI 03-AUG-2010
DEFINITION  insulin preproprotein [Homo sapiens].
ACCESSION   NP_000198
VERSION     NP_000198.1  GI:4557671
DBSOURCE    REFSEQ: accession NM_000207.2
```

Figure 4.12 A RefSeq protein record. The line "DBSOURCE" has a hyperlink to the DNA record encoding this protein.

The results

The top of the BLASTP search results can often look very different from that seen with BLASTN. In this case, the query description has been greatly expanded and many different protein records are listed, each marked with a ">" sign. This is because all of these sequences (20 are seen in **Figure 4.13**) are identical to each other. As you look closely at this list, you can see that these files come from a number of databases. The gi (gene identification) number appears first, followed by those from the RefSeq (ref), Swiss-Prot (sp), GenBank (gb), EMBL (emb), and Protein Research Foundation (prf) databases. These different databases each have a human preproinsulin protein sequence which is 100% identical to the original query, NP_000198. The NCBI shows them here to spare you from these being the top 20 hits in your BLAST results. Looking closely at the list, you may be surprised to see the gorilla sequence, AAN06935, indicating that it is 100% identical to the human sequence.

Scrolling down, there is a graphic (see **Figure 4.14** and color plates) that is quite similar to the BLASTN figure, with the query length going horizontally across the top and the hits being represented as colored bars. What is different is that the top of the graphic now has an additional section depicting any similarities between your query and the sequences of known protein families. In addition to searching a database of sequences (RefSeq), your query was searched against a database of conserved domains, and the two major hits are depicted here as thick bars. Appropriately, the domains are specific to insulin-like growth factors (ILGF) and a larger family or **"superfamily"** that includes insulin and many related sequences. Based on the scale across the top of this section, the domain that was detected begins near amino acid 25 and goes to the end of the query. Looking down at the rest of the graphic, colored differently because of lower scores, the hits are beginning to lose similarity to the N-terminus but are retaining similarity from around amino acid 25 to the end of the query, the same coordinates seen with the domain search results.

The columns of the results table should also look familiar (**Figure 4.15**). Of course, the query found itself (NP_000198, the top hit) and, by association, the other hits listed at the top of the results window. But even with the expected 100% identity between the query and itself, notice that the E value is not 0.0. Human preproinsulin is only 110 amino acids long and the calculations performed by BLASTP indicate that the number of hits in the database expected to be found is not 0.0, although 1e–73 is quite small and nobody would argue that

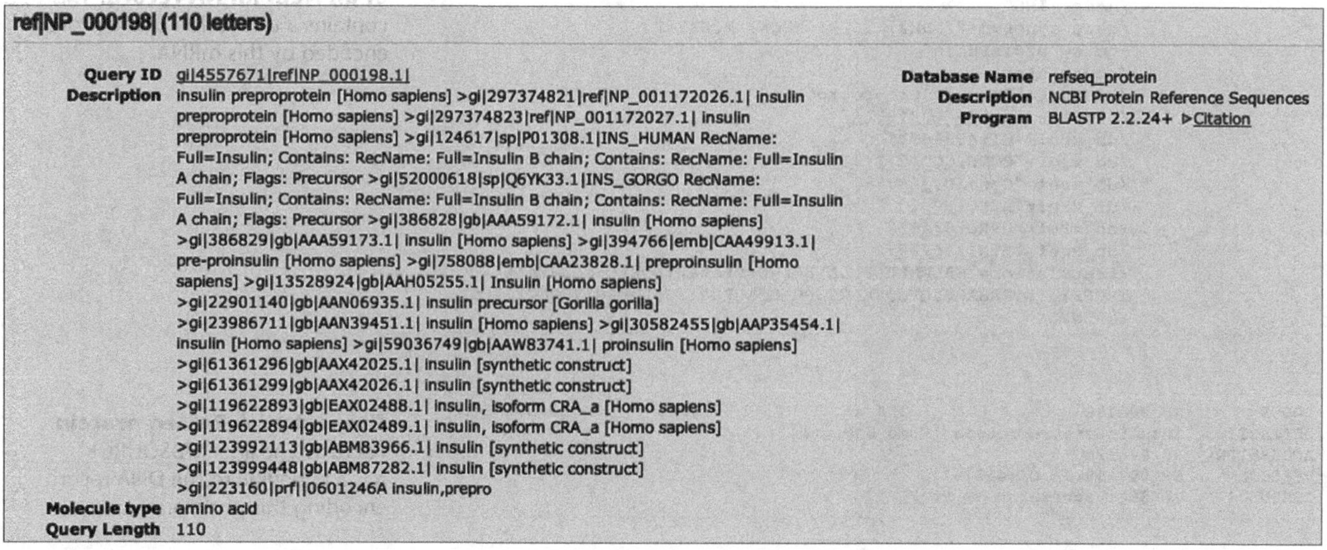

Figure 4.13 The top of the NCBI BLASTP results.

```
ref|NP_000198| (110 letters)
```

Query ID	gi|4557671|ref|NP_000198.1|
Description	insulin preproprotein [Homo sapiens] >gi|297374821|ref|NP_001172026.1| insulin preproprotein [Homo sapiens] >gi|297374823|ref|NP_001172027.1| insulin preproprotein [Homo sapiens] >gi|124617|sp|P01308.1|INS_HUMAN RecName: Full=Insulin; Contains: RecName: Full=Insulin B chain; Contains: RecName: Full=Insulin A chain; Flags: Precursor >gi|52000618|sp|Q6YK33.1|INS_GORGO RecName: Full=Insulin; Contains: RecName: Full=Insulin B chain; Contains: RecName: Full=Insulin A chain; Flags: Precursor >gi|386828|gb|AAA59172.1| insulin [Homo sapiens] >gi|386829|gb|AAA59173.1| insulin [Homo sapiens] >gi|394766|emb|CAA49913.1| pre-proinsulin [Homo sapiens] >gi|758088|emb|CAA23828.1| preproinsulin [Homo sapiens] >gi|13528924|gb|AAH05255.1| Insulin [Homo sapiens] >gi|22901140|gb|AAN06935.1| insulin precursor [Gorilla gorilla] >gi|23986711|gb|AAN39451.1| insulin [Homo sapiens] >gi|30582455|gb|AAP35454.1| insulin [Homo sapiens] >gi|59036749|gb|AAW83741.1| proinsulin [Homo sapiens] >gi|61361296|gb|AAX42025.1| insulin [synthetic construct] >gi|61361299|gb|AAX42026.1| insulin [synthetic construct] >gi|119622893|gb|EAX02488.1| insulin, isoform CRA_a [Homo sapiens] >gi|119622894|gb|EAX02489.1| insulin, isoform CRA_a [Homo sapiens] >gi|123992113|gb|ABM83966.1| insulin [synthetic construct] >gi|123999448|gb|ABM87282.1| insulin [synthetic construct] >gi|223160|prf||0601246A insulin,prepro
Molecule type	amino acid
Query Length	110

Database Name	refseq_protein
Description	NCBI Protein Reference Sequences
Program	BLASTP 2.2.24+ ▷Citation

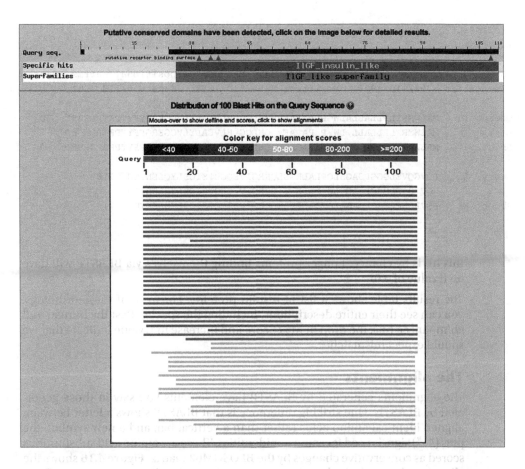

Figure 4.14 The graphic from the NCBI BLASTP results. The query is the human preproinsulin protein sequence NP_000198 and the database is RefSeq protein. See color plates for a color version of this figure.

Sequences producing significant alignments:

Accession	Description	Max score	Total score	Query coverage	E value	Max ident	Links
NP_000198.1	insulin preproprotein [Homo sapiens] >gi|297374821|ref|NP_001172(	223	223	100%	1e-73	100%	G M
NP_001008996.1	insulin preproprotein [Pan troglodytes]	221	221	100%	1e-72	98%	U G M
XP_003281396.1	PREDICTED: insulin-like isoform 1 [Nomascus leucogenys] >ref|XP_0(	219	219	100%	1e-71	98%	G M
XP_003508128.1	PREDICTED: insulin-like isoform 1 [Cricetulus griseus]	175	175	100%	2e-54	86%	G M
NP_062003.1	insulin-2 preproprotein [Rattus norvegicus]	172	172	100%	2e-53	83%	U G M
NP_062002.1	insulin-1 preproprotein [Rattus norvegicus]	171	171	100%	7e-53	83%	U G M
NP_001123565.1	insulin precursor [Canis lupus familiaris]	170	170	100%	2e-52	88%	U G M
NP_032413.1	insulin-2 preproprotein [Mus musculus] >ref|NP_001172012.1| insulir	169	169	100%	3e-52	82%	G M
XP_002755713.1	PREDICTED: insulin-like [Callithrix jacchus]	168	168	100%	1e-51	85%	G M
NP_001009272.1	insulin precursor [Felis catus]	163	163	100%	6e-50	81%	G
XP_002920166.1	PREDICTED: insulin-like [Ailuropoda melanoleuca]	160	160	100%	1e-48	80%	G M
XP_003422420.1	PREDICTED: insulin-like isoform 1 [Loxodonta africana]	157	157	100%	2e-47	79%	G M
NP_001075804.1	insulin precursor [Oryctolagus cuniculus]	154	154	82%	2e-46	90%	U G M
NP_032412.3	insulin-1 preproprotein [Mus musculus]	154	154	100%	2e-46	78%	U G M
XP_001496855.2	PREDICTED: insulin-like isoform 1 [Equus caballus]	152	152	100%	2e-45	88%	U G M
XP_003362685.1	PREDICTED: insulin-like [Equus caballus]	147	147	100%	9e-44	84%	G M
XP_003362686.1	PREDICTED: insulin-like [Equus caballus]	147	147	100%	2e-43	86%	U G M
XP_002198969.1	PREDICTED: insulin [Taeniopygia guttata]	131	131	100%	2e-37	64%	U G M
NP_001103242.1	insulin precursor [Sus scrofa]	131	131	100%	2e-37	85%	U G M
NP_990553.1	insulin precursor [Gallus gallus]	129	129	100%	1e-36	64%	U G M
XP_003206326.1	PREDICTED: insulin-like [Meleagris gallopavo]	129	129	100%	3e-36	63%	U G M
NP_001166362.1	insulin precursor [Cavia porcellus]	127	127	100%	9e-36	69%	G M
NP_001079350.1	insulin-2 precursor [Xenopus laevis]	120	120	100%	3e-33	57%	U G
NP_776351.2	insulin preproprotein [Bos taurus] >ref|NP_001172055.1| insulin prep	119	119	100%	2e-32	76%	U G M
NP_001079351.1	insulin-1 precursor [Xenopus laevis]	119	119	100%	2e-32	57%	U G
XP_003214805.1	PREDICTED: insulin-like [Anolis carolinensis]	117	117	100%	4e-32	55%	U G M
NP_001093706.1	insulin precursor [Xenopus (Silurana) tropicalis]	117	117	100%	7e-32	56%	U G M
NP_001035835.1	insulin- insulin-like growth factor 2 read-through product precursor [H	112	112	56%	5e-29	100%	U G M
XP_003281401.1	PREDICTED: insulin-like isoform 6 [Nomascus leucogenys]	108	108	56%	2e-27	97%	G M
XP_003508129.1	PREDICTED: insulin-like isoform 2 [Cricetulus griseus]	99.0	99.0	56%	9e-24	89%	G M
XP_003422421.1	PREDICTED: insulin-like isoform 2 [Loxodonta africana]	98.2	98.2	57%	2e-23	86%	G M
XP_003458727.1	PREDICTED: insulin-like [Oreochromis niloticus]	92.0	92.0	99%	6e-22	45%	U G M
NP_001034153.1	preproinsulin b precursor [Danio rerio]	88.6	88.6	76%	1e-20	54%	U G M
NP_001118142.1	preproinsulin 2 precursor [Oncorhynchus mykiss]	83.6	83.6	100%	8e-19	49%	U G
NP_001118141.1	preproinsulin 1 precursor [Oncorhynchus mykiss]	82.8	82.8	84%	2e-18	48%	U G
NP_571131.1	insulin preproprotein [Danio rerio]	79.7	79.7	75%	3e-17	49%	U G M

Figure 4.15 The table from the NCBI BLASTP results. The RefSeq protein database was searched with the human insulin protein sequence, NP_000198.

Figure 4.16 The BLASTP alignment between the human preproinsulin protein (Query) and the rat hit (Sbjct).

```
>ref|NP_062002.1| insulin-1 preproprotein [Rattus norvegicus]
Length=110


Score =  171 bits (433),  Expect = 2e-41, Method: Compositional matrix adjust.
Identities = 91/110 (82%), Positives = 95/110 (86%), Gaps = 0/110 (0%)

Query  1    MALWMRLLPLLALLALWGPDPAAAFVNQHLCGSHLVEALYLVCGERGFFYTPKTRREAED  60
            MALWMR LPLLALL LW P PA AFV QHLCG HLVEALYLVCGERGFFYTPK+RRE ED
Sbjct  1    MALWMRFLPLLALLVLWEPKPAQAFVKQHLCGPHLVEALYLVCGERGFFYTPKSRREVED  60


Query  61   LQVGQVELGGGPGAGSLQPLALEGSLQKRGIVEQCCTSICSLYQLENYCN  110
             QV Q+ELGGGP AG LQ LALE + QKRGIV+QCCTSICSLYQLENYCN
Sbjct  61   PQVPQLELGGGPEAGDLQTLALEVARQKRGIVDQCCTSICSLYQLENYCN  110
```

this hit is by chance. Longer proteins finding themselves via BLASTP will have an E value of 0.0.

The results table shows a list of insulin proteins. For most of these orthologs, you can see their entire descriptions, including the species. Past the human and chimpanzee hits, the drop in Max Score and increase in E value indicate that we should expect mismatches.

The alignments

The alignments generated by BLASTP have elements you saw in those generated by BLASTN. One striking difference is that BLASTP shows a letter between aligned identical amino acids rather than a vertical bar, and a new symbol, the plus ("+") sign, is used for some nonidentities. These are amino acid substitutions scored as conservative changes by the BLOSUM62 matrix. **Figure 4.16** shows the alignment between the human query and the rat insulin preproprotein. There are blank spaces for most nonidentities between amino acids, but four pairs where a "+" sign is used. These four conservative amino acid differences have a BLOSUM62 score of "1" or larger. All nonidentities where the BLOSUM62 scores are "0" or negative numbers have a blank space.

Referring back to the BLOSUM62 matrix in Figure 4.9, we see the scores for selected pairs of amino acids in **Figure 4.17**.

Notice that no gaps were needed to generate a strong alignment between these distant species (Figure 4.16). Both proteins are 110 amino acids long. Although there are some regions which have mismatches (for example, the first 27 amino acids), there are stretches where there are none. Of the 19 mismatches, four are marked with a "+."

Pair	BLOSUM62 Score
T + S	1
V + L	1
S + A	1
E + D	2
L F	0
A V	0
G E	-2
L P	-3

Figure 4.17 BLOSUM62 scores of eight selected pairings from the alignment shown in Figure 4.16. The "+" symbol between some pairs indicates conservative substitutions while the other pairs listed here have a blank space as seen in the BLASTP alignments in Figure 4.16.

Like BLASTN, we see some calculations immediately above each alignment, such as the "Score" and "Expect" values. In addition to "Identities," there is now "Positives" which is the sum of the Identities plus any amino acids aligned with a "+" sign. In the case of the human–rat alignment, this would be 91 identities plus 4 conservative changes = 95. Another way to refer to the Positives is **similarities**.

Distant homologies

Scroll through the alignments and you will observe a high degree of conservation. Insulin is critical to the regulation of blood sugar in many animals so it should not be surprising to see the similarities between distant species. Stop at the alignment with preproinsulin 1 of *Oncorhynchus mykiss*, the rainbow trout (**Figure 4.18**). Trout preproinsulin is 105 amino acids long, quite similar to the 110 amino acid human form.

Despite the millions of years since we shared an ancestor with the fish, you see that trout preproinsulin is similar to the human sequence. You can also see the

>ref|NP_001118141.1|preproinsulin 1 [Oncorhynchus mykiss]
Length=105

 Score = 82.8 bits (203),   Expect = 9e-15, Method: Compositional matrix adjust.
 Identities = 46/96 (47%), Positives = 59/96 (61%), Gaps = 12/96 (12%)

Query 18   GPDPAAAFVNQHLCGSHLVEALYLVCGERGFFYTPKTRREAEDLQVGQVELGGGPGAGSL   77
            G D AAA   QHLCGSHLV+ALYLVCGE+GFFY PK R+ + L   +   A
Sbjct 19   GADAAAA---QHLCGSHLVDALYLVCGEKGFFYNPK--RDVDPL----IGFLSPKSAKEN   69

Query 78   QPLALEGSLQ---KRGIVEQCCTSICSLYQLENYCN   110
            +   + ++   KRGIVEQCC   C+++ L+NYCN
Sbjct 70   EEYPFKDQMEMMVKRGIVEQCCHKPCNIFDLQNYCN   105

**Figure 4.18 BLASTP
alignment between
human preproinsulin
(Query) and trout
preproinsulin (Sbjct).**

impact the "Positives" scoring has on the percentages. The human–trout alignment is 47% identical, but 61% positive or similar. A big jump. Prior to amino acid 18, BLASTP could not align the two insulins but there is a solid region of similarity starting around trout amino acid 26 where the sequence conservation is high and probably essential to the function of the protein. As you recall from Figure 4.14, lower-scoring hits viewed with the graphic failed to align with the N-terminus of human preproinsulin but did show similarity starting somewhere around the amino acid 26 coordinate.

After the insulin mRNA is translated, the N-terminus of the preproinsulin is cleaved off during the secretion process. The protein folds and creates three pairs of disulfide bonds, chemically linking distant cysteines. The proinsulin protein is then cleaved into two chains, but they remain linked to each other through these disulfide bonds and attain a structure critical to the function of the insulin protein. There are no mismatches in the important cysteine (C) positions, suggesting that there is no radical change in secondary structure. Other regions are very divergent, only showing a limited number of aligned amino acids or conservative substitutions.

4.6 PAIRWISE BLAST

The NCBI BLAST Web forms have the option of performing a pairwise alignment between two sequences of your choosing. It will not align two sequences from end to end; should only a short sequence of similarity be detected by BLASTP, this will be the only alignment you will see.

To demonstrate the pairwise BLAST function, let's return to the gorilla preproinsulin protein. Earlier in this chapter, BLASTP concatenated the names of all the protein sequences that were identical to the human preproinsulin protein when we used this sequence as a query (see Figure 4.13). By running a pairwise BLASTP alignment we should be able to verify the 100% identity between the gorilla and human preproinsulins.

1. Go to the NCBI BLASTP Web form as before. Notice a check box called "Align two or more sequences."

2. Check that box and the window refreshes, replacing the database choice section with a second window for entering a sequence (see **Figure 4.19**).

3. Obtain the accession numbers for the gorilla and human preproinsulin sequences from Figure 4.13 or by other means.

4. Enter these two accession numbers as indicated in Figure 4.19.

5. Click on the BLAST button as usual.

Figure 4.19 The NCBI pairwise BLASTP Web form. By clicking on the "Align two or more sequences" box, the form switches from normal BLASTP, with a query searching a database, to a form that aligns sequences of your choosing. This check box is found on all NCBI BLAST forms.

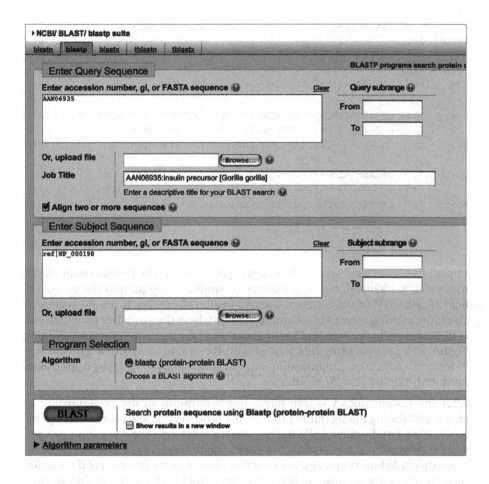

Very quickly, the results will appear and you will see a list of all the identical sequences as before (**Figure 4.20**). But below that is the alignment between the gorilla and human preproinsulin proteins. This alignment confirms the 100% identity between the two preproinsulins. Notice that the "hit" in the results is the human protein. The human preproinsulin, NP_000198, was entered in the lower part of the form (see Figure 4.19) and is therefore treated as the subject while the gorilla sequence entered at the top of the form is the query.

4.7 RUNNING BLASTP AT THE ExPASy WEBSITE

Until now, we have only used sequence analysis tools that are found on the NCBI Website. Although we could use the NCBI tools for almost every chapter of this book, many excellent Websites and bioinformatics applications would be overlooked. ExPASy (rhymes with ecstasy), the Proteomics Server of the Swiss Institute of Bioinformatics, is among the best for tools specifically designed for the study of proteins. The ExPASy version of BLASTP is no exception and we will perform two database searches to demonstrate the features of its interface.

First, let's go to the home page, www.expasy.ch. Take a moment to see what is offered here by clicking on the left sidebar link called "Resources A..Z": there are a number of databases and tools almost entirely focused on proteins. A comprehensive database created here is the UniProtKB database, made up of two divisions: TrEMBL and the Swiss-Prot databases. TrEMBL (pronounced as "tremble") stands for "translated EMBL," and is the non-curated translations of the EMBL DNA database. TrEMBL records are automatically generated by software with little or no human intervention. Many of the entries in Swiss-Prot ("swiss prote," rhymes with wrote) are famously curated by people who take great care in providing accurate and often deep annotation of a sequence based on the

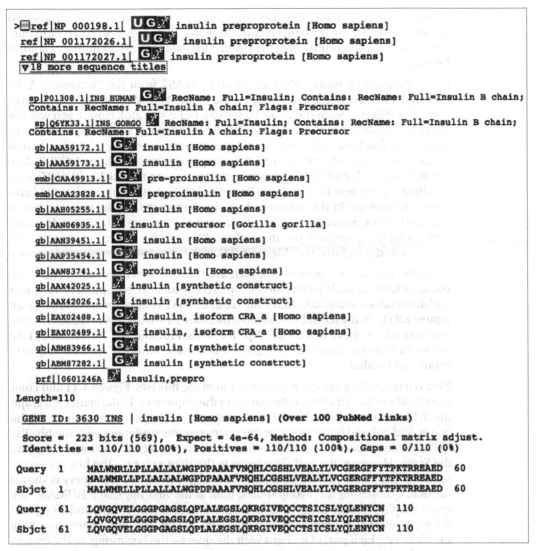

Figure 4.20
Alignment between gorilla (Query) and human (Sbjct) preproinsulins.

review of the literature and consultation with area experts. When given a choice between using an automatically generated file and a file curated by a proteomics expert, go with the curated file.

Searching for pro-opiomelanocortin using a protein sequence fragment

To perform the following two exercises, we'll use protein sequences from the ExPASy Website. Near the top of the home page is a search function with a drop-down menu and text box (**Figure 4.21**). From the menu, select "UniProtKB."

Enter the UniProtKB accession number Q53WY7 and click on the "search" button. Examine this TrEMBL sequence record and you will see that it has little annotation. It is a 30 amino acid fragment of the human pro-opiomelanocortin protein (POMC). With BLASTP we'll find the full-length version of this protein, gain access to a wealth of annotation, and see exactly where the fragment is within the complete protein. Like the NCBI Web form, you can use either the

query | UniProtKB ▾ | Q53WY7 | ✖ | search | help

Figure 4.21 Search function for the ExPASy Website.

protein sequence or the accession number as a query so there is no need to retrieve the sequence from this record. However, this ExPASy BLASTP form only works on accession numbers of the UniProtKB database; NCBI numbers cannot be used.

Go to the BLASTP form, either by selecting BLAST from the "Resources A..Z" link on the ExPASy home page or by typing in the Web address, web.expasy.org/blast. It is different from the NCBI Web form, but if you look closely you will see many of the same features: a window to enter the query sequence; a section to choose the database and apply species restrictions; a section to change search options; and a launch button. Hopefully, your experience with the NCBI BLAST forms for BLASTN and BLASTP has prepared you to use the ExPASy BLASTP form and little explanation is given here. Paste Q53WY7 into the appropriate field and choose Mammalia in the database drop-down menu to restrict the searches to mammals. For reasons that will become apparent soon, the only change from the defaults for this exercise is you must *uncheck* the box for "Filter the sequence for low-complexity regions" (see **Figure 4.22**).

Click the "Run BLAST" button to launch the search. When the window refreshes, you see a familiar basic format for the top of the results: a description of the query and the database searched. Unlike the NCBI BLASTP results, the table is next (see **Figure 4.23**). On the left side of the table there are two-letter abbreviations for the database where these hits reside: "sp" for Swiss-Prot and "tr" for TrEMBL. Like the NCBI Website, there are columns for accession number (AC), Description, Score, and E-value.

Next is a fantastic graphical representation of the hits (see **Figure 4.24** and color plates). Along the left side are the names of the sequences. In the center is a graph much like that seen at the NCBI BLAST page: the query goes across the top of the graph, and colored bars represent the query sequence aligning with the hit. The ExPASy color is based on percent identity, while the NCBI coloring is based on the Score. However, the real showpiece is the right graphic. The hits are represented as gray bars and the position of any alignments with the query is shown in color. Comparing the two graphics, look at the first hit, COLI_HUMAN. The query graphic (center panel) shows a solid bar from left to right indicating that the entire query (30 amino acids) aligns with the hit. You don't know how long the hit is, or which part of it aligns with the query—the beginning, or the end, or somewhere in the middle? On the right graphic the answers appear; the first hit is roughly 200 amino acids long but the area aligning with the query is only a short segment near the middle, depicted in green.

There is so much information here, in a simple graphic. Although you can determine the exact location of similarities based on the coordinates of the alignments, this one graph visualizes a lot of this information. You can see that most of the hits are about the same length and the location of the similarities is often aligned as well (in the first half of the protein sequence). There are exceptions and perhaps these can be grouped in some way by exploring their annotation. Note that the scale used on the right graph is not linear.

Figure 4.22 The options menu for ExPASy BLASTP. Filtering of low-complexity regions is controlled here.

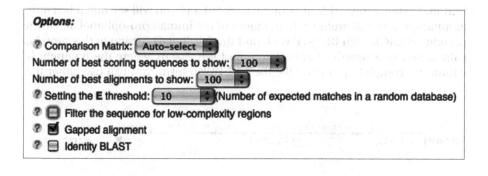

Options:

? Comparison Matrix: Auto-select

Number of best scoring sequences to show: 100

Number of best alignments to show: 100

? Setting the E threshold: 10 (Number of expected matches in a random database)

? ☐ Filter the sequence for low-complexity regions

? ☑ Gapped alignment

? ☐ Identity BLAST

Db	AC	Description	Score	E-value
sp	P01189	COLI_HUMAN Pro-opiomelanocortin precursor (POMC) (Cort...	99	2e-21
tr	Q6FHC8	_HUMAN POMC protein (Fragment) [POMC] [Homo sapiens (Hu...	99	2e-21
tr	Q5TZZ7	_HUMAN Proopiomelanocortin (Adrenocorticotropin/ beta-l...	99	2e-21
tr	Q53WY7	_HUMAN Proopiomelanocortin (Fragment) [proopiomelanocor...	99	2e-21
tr	Q53T23	_HUMAN Proopiomelanocortin (Adrenocorticotropin/ beta-l...	99	2e-21
tr	A6XND7	_HUMAN Proopiomelanocortin preproprotein [Homo sapiens ...	94	7e-20
tr	Q8HZB2	_9PRIM Proopiomelanocortin (Fragment) [Gorilla gorilla ...	92	2e-19
tr	Q8HZB3	_PANTR Proopiomelanocortin (Fragment) [Pan troglodytes ...	91	4e-19
tr	Q8HZB1	_PONPY Proopiomelanocortin (Fragment) [Pongo pygmaeus (...	85	2e-17
sp	P01192	COLI_PIG Pro-opiomelanocortin precursor (POMC) (Cortic...	83	1e-16
tr	A8VWB5	_PIG Proopiomelanocortin protein [POMC] [Sus scrofa (Pig)]	83	1e-16
sp	P01190	COLI_BOVIN Pro-opiomelanocortin precursor (POMC) (Cort...	83	1e-16
tr	Q8HZB0	_9PRIM Proopiomelanocortin (Fragment) [Macaca sp]	83	2e-16
tr	Q8HZA9	_SAGOE Proopiomelanocortin (Fragment) [Saguinus oedipus...	83	2e-16
sp	P01191	COLI_SHEEP Pro-opiomelanocortin precursor (POMC) (Cort...	82	3e-16
tr	Q8MIC5	_SHEEP Proopiomelanocortin [pomc] [Ovis aries (Sheep)]	82	3e-16
tr	B4XH73	_CAPHI Alpha-melanocyte-stimulating hormone (Fragment) ...	82	3e-16

Figure 4.23 The results table for ExPASy BLASTP. In this search, the UniProtKB sequence Q53WY7 was used as a query against the mammalian division of the UniProtKB database.

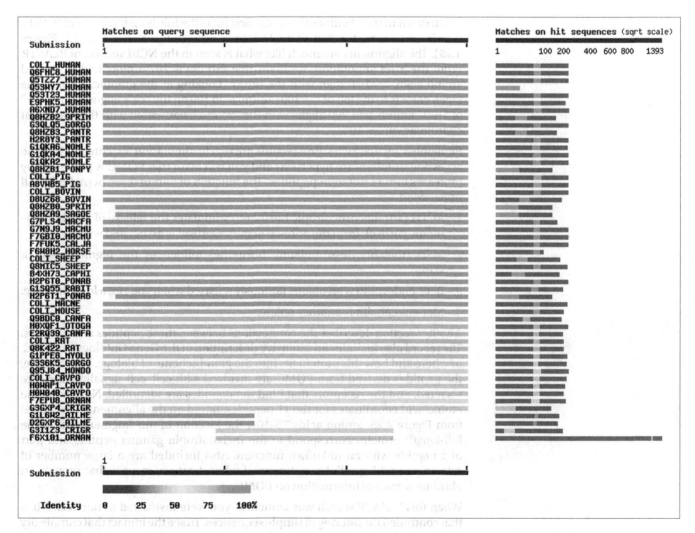

Figure 4.24 The results graphic for ExPASy BLASTP. The POMC fragment, UniProtKB accession number Q53WY7, was used as a query against the mammalian division of the UniProtKB database. On the far left are sequence names for each row. The center panel depicts the query and the best areas of identity with the hits. The right panel shows the hits as gray bars with the sequence identities with the query shaded lighter gray. See color plates for a color version of this figure.

Figure 4.25 The 30 amino acid query has perfect identity with P01189 between amino acids 75 and 104. The UniProtKB sequence Q53WY7 was used as a query against the mammalian division of the UniProtKB database.

```
sp        P01189
COLI_HUMAN        Pro-opiomelanocortin precursor (POMC)
(Corticotropin-lipotropin)
[Contains: NPP; Melanotropin gamma (Gamma-MSH);
Potential peptide; Corticotropin (Adrenocorticotropic
hormone) (ACTH); Melanotropin alpha (Alpha-MSH);
Corticotropin-like intermediary peptide (CLIP);
Lipotropin beta (Beta-LPH); Lipotropin gamma
(Gamma-LPH); Melanotropin beta (Beta-MSH);
Beta-endorphin; Met-enkephalin] [POMC] [Homo sapiens
(Human)]        267 AA


Score = 99.0 bits (226), Expect = 2e-21
Identities = 30/30 (100%), Positives = 30/30 (100%)


Query: 1    RKYVMGHFRWDRFGRRNSSSSGSSGAGQKR 30
            RKYVMGHFRWDRFGRRNSSSSGSSGAGQKR
Sbjct: 75   RKYVMGHFRWDRFGRRNSSSSGSSGAGQKR 104
```

Scroll down to the alignments and the first hit is the full-length version of POMC, a Swiss-Prot record called COLI_HUMAN, accession number P01189 (**Figure 4.25**). The alignments are much like what is seen in the NCBI version of BLASTP. From the brief annotation, we learn that P01189 is 267 amino acids long and amino acids 75–104 align with the query. Looking at the graphic, most of the green color is to the left of the 100 (amino acid 100 on the scale aligns with the "1" of "100," amino acid 200 aligns with the "2," etc.), consistent with the alignment coordinates of 75–104.

Remembering that Swiss-Prot records are rich in annotation, explore P01189 (the accession numbers are hypertext) and you will learn that POMC is processed by proteases into 11 different peptides. The function of four of these peptides is well understood.

1. ACTH (adrenocorticotropic hormone) stimulates the release of another hormone, cortisol, from the adrenal glands. Cortisol is associated with stress.

2. MSH (melanocyte-stimulating hormone) influences the pigmentation of skin.

3. Beta-endorphin is an opiate, famously released when you run a long distance.

4. Met-enkephalin is another opiate.

Another section describes the connections between these peptides and diseases. The record also includes an extensive description of the Gene Ontology of POMC, or where and how this protein fits into the grand scheme of biology. For example, the peptides derived from POMC are involved with cell–cell signaling, they are secreted, and the receptors that bind the peptides are identified. Next come the amino acid coordinates for the 11 peptides. Based on the alignment coordinates from Figure 4.25, amino acids 75–104, the location of our fragment within the full-length protein, correspond to the melanotropin gamma peptide and part of a peptide with an unknown function. Also included are a large number of references and hyperlinks to dozens of other database records that provide an absolute wealth of information on POMC.

When this BLASTP search was launched, you were instructed to uncheck a box that controlled the filtering of simple sequences. To see the impact that complexity filtering has on the BLASTP result, repeat this BLASTP search with Q53WY7 and, this time, check the box under Options for "Filter the sequence for low-complexity regions." Be sure to keep the current BLASTP results window, too, so you may easily compare them. You will see (**Figure 4.26**) that when this filtering

```
Score = 75.3 bits (170), Expect = 3e-14
 Identities = 22/30 (73%), Positives = 22/30 (73%)


Query: 1    RKYVMGHFRWDRFGRRNXXXXXXXXAGQKR 30
            RKYVMGHFRWDRFGRRN        AGQKR
Sbjct: 75   RKYVMGHFRWDRFGRRNSSSSGSSGAGQKR 104
```

Figure 4.26 With low-complexity filtering turned on, eight amino acids of Q53WY7 are turned to "X." UniProtKB sequence Q53WY7 was used as a query against the mammalian division of the UniProtKB database. Unlike the previous search, the low-complexity filtering parameter was checked. The percent identity has dropped from 100% to 73%.

is applied, the serine- and glycine-rich region of our query, SSSSGSSG, is turned into XXXXXXXX. This filtering is applied by default to shield you from nonspecific hits caused by regions of low complexity. Also notice that the calculations now indicate lower percent identity (that is, 73% versus 100%), a lower Score, and a less significant Expect value. In this case, turning off the filtering allows a clearer alignment and a more accurate representation for our very short query.

Searching for repeated domains in alpha-1 collagen

Now let's perform a search with a more complicated result so you can see the additional power of the graphics. On the ExPASy home page, retrieve the record for UniProtKB accession number Q9UMA6. This record is a 27 amino acid fragment of the human alpha-1 procollagen protein. Like the POMC example, above, we'll find the full-length version of this protein, gain access to the extensive annotation on this protein, and see the relationship between the query and the hits.

Go back to the ExPASy BLASTP form and enter Q9UMA6 as the query. From the database choice drop-down menu, select *Homo sapiens* so the search is restricted to the human sequences. Launch the search.

When the window refreshes, look at the BLASTP results table (**Figure 4.27**). The original table is quite long and, in this shortened version, rows were deleted (indicated by dots) to bring some at the bottom into view.

Going down the first column, we see a new label for a number of hits, "sp_vs." This indicates that they are Swiss-Prot records where there is variable splicing of the transcript, resulting in different protein products or isoforms from the same gene.

Figure 4.27 The BLASTP results of query Q9UMA6. The human division of the UniProtKB database was searched. Many rows of the results table were deleted, indicated by dots, to shorten the table.

```
Db     AC          Description                                        Score E-value

sp     P02452      CO1A1_HUMAN Collagen alpha-1(I) chain precursor (Alpha... 84 2e-17
tr     Q9UMA6      _HUMAN Type I collagen alpha 1 chain (Fragment) [COL1A1... 84 2e-17
tr     Q6LAN8      _HUMAN Collagen type I alpha 1 (Fragment) [COL1A1] [Hom... 84 2e-17
sp     P05997      CO5A2_HUMAN Collagen alpha-2(V) chain precursor [COL5A... 57 3e-09
tr     B4DNJ0      _HUMAN cDNA FLJ60734, highly similar to Collagen alpha-... 57 3e-09
sp     P13942      COBA2_HUMAN Collagen alpha-2(XI) chain precursor [COL1... 45 1e-05
tr     Q5STP6      _HUMAN Uncharacterized protein [COL11A2] [Homo sapiens ... 45 1e-05
tr     H0YIS1      _HUMAN Collagen alpha-2(XI) chain (Collagen, type XI, a... 45 1e-05
tr     H0YHY3      _HUMAN Collagen alpha-2(XI) chain [COL11A2] [Homo sapie... 45 1e-05
tr     H0YHN2      _HUMAN Collagen alpha-2(XI) chain [COL11A2] [Homo sapie... 45 1e-05
tr     H0YHM9      _HUMAN Collagen alpha-2(XI) chain [COL11A2] [Homo sapie... 45 1e-05
.
.

.
sp_vs  P13942-2 Isoform 2 of Collagen alpha-2(XI) chain OS=Homo sapi...     45 1e-05
sp_vs  P13942-3 Isoform 3 of Collagen alpha-2(XI) chain OS=Homo sapi...     45 1e-05
sp_vs  P13942-4 Isoform 4 of Collagen alpha-2(XI) chain OS=Homo sapi...     45 1e-05
sp_vs  P13942-5 Isoform 5 of Collagen alpha-2(XI) chain OS=Homo sapi...     45 1e-05
sp_vs  P13942-6 Isoform 6 of Collagen alpha-2(XI) chain OS=Homo sapi...     45 1e-05
sp_vs  P13942-7 Isoform 7 of Collagen alpha-2(XI) chain OS=Homo sapi...     45 1e-05
sp_vs  P13942-8 Isoform 8 of Collagen alpha-2(XI) chain OS=Homo sapi...     45 1e-05
```

The first hit, Swiss-Prot record P02452, is the full-length version of the alpha-1 procollagen protein. Visit the Swiss-Prot record for this protein and you will be taken to a richly annotated file, as we saw for POMC. Collagen is the most abundant protein in your body and this is a great opportunity to learn more about it.

Scroll down to the graphic of the BLASTP search (see **Figure 4.28** and color plates) and you will now see something that is perhaps puzzling. The center panel graphic mostly indicates single colored bars, for example the top bars are green. But the right side of the graph shows a single green bar and multiple yellow bars on the gray background of the hits; the query aligns to multiple places on the hit. There is one area of strong similarity (green) but there are other locations with weaker similarity (shades of yellow). This can't be shown on the left graphic but is easily seen on the right graphic.

This may be better understood while looking at the alignments. The top six alignments of the first hit are shown in **Figure 4.29**. Notice that the first alignment is 100% identical and the coordinates are amino acids 353–379 (this is the green bar on the graphic). However, there are other places on the hit where the query can be aligned and they can be seen on both the right graph and the alignments. There are many other hits that can align in multiple places with the query. They are all clearly related, based on the similar pattern of "stripes" this query has left on them in the graphic seen in the right side of Figure 4.28 (see also color plates). Exploring the annotation of these hits will shed light on their relationships.

Let's return to the table of hits for the query, a collagen fragment (Figure 4.27). The first hit is the full-length collagen alpha-1 that contains the fragment sequence, and the second hit is the fragment itself (self-hit). Despite the fact that the full-length sequence is 1464 amino acids long and the fragment is 27 amino acids long, the values for the Score and the E value are identical for the two hits. This demonstrates that these values are dependent on the length and quality of the alignment, and independent of the length of the Subject.

Figure 4.28 The results graphic from BLASTP query Q9UMA6. The human division of UniProtKB was searched. Note the multiple regions of similarity between the query and the hits in the right panel. See color plates for a color version of this figure.

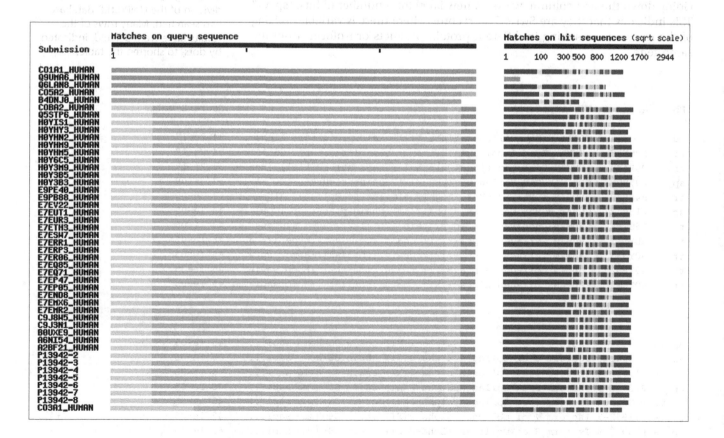

```
sp      P02452
CO1A1_HUMAN    Collagen alpha-1(I) chain precursor (Alpha-1 type I collagen)
[COL1A1] [Homo sapiens (Human)]        1464 AA

 Score = 84.2 bits (191), Expect = 6e-17
 Identities = 27/27 (100%), Positives = 27/27 (100%)

Query: 1    GEAGPQGPRGSEGPQGVRGEPGPPGPA 27
            GEAGPQGPRGSEGPQGVRGEPGPPGPA
Sbjct: 353  GEAGPQGPRGSEGPQGVRGEPGPPGPA 379


 Score = 44.3 bits (97), Expect = 6e-05
 Identities = 20/36 (55%), Positives = 20/36 (55%), Gaps = 9/36 (25%)

Query: 1    GEAGPQGPRGSEGPQGVRGEPGP---------PGPA 27
            GEAG QGP G  GP G RGE GP         PGPA
Sbjct: 614  GEAGAQGPPGPAGPAGERGEQGPAGSPGFQGLPGPA 649


 Score = 43.1 bits (94), Expect = 1e-04
 Identities = 19/30 (63%), Positives = 19/30 (63%), Gaps = 3/30 (10%)

Query: 1    GEAGPQGPRGSEGPQGV---RGEPGPPGPA 27
            GE GP GP G  G  G    RGEPGPPGPA
Sbjct: 782  GESGPSGPAGPTGARGAPGDRGEPGPPGPA 811


 Score = 40.5 bits (88), Expect = 8e-04
 Identities = 20/35 (57%), Positives = 20/35 (57%), Gaps = 9/35 (25%)

Query: 1    GEAGP---QGPRGS---EGPQGVRGEPGP---PGP 26
            GE GP   QGP G    EG G RGEPGP   PGP
Sbjct: 449  GEPGPVGVQGPPGPAGEEGKRGARGEPGPTGLPGP 483


 Score = 38.8 bits (84), Expect = 0.003
 Identities = 18/36 (50%), Positives = 19/36 (52%), Gaps = 11/36 (30%)

Query: 1    GEAGPQGPRGSEGPQGVR----------GEPGPPGP 26
            G+ GP GPRG  GP G R          G PGPPGP
Sbjct: 115  GDTGPRGPRGPAGPPG-RDGIPGQPGLPGPPGPPGP 149


 Score = 36.7 bits (79), Expect = 0.012
 Identities = 20/42 (47%), Positives = 20/42 (47%), Gaps = 15/42 (35%)

Query: 1    GEAGPQGPRGS---EGPQGVRGE-----------PGPPGPA 27
            G AGP GP G    EG  G RGE           PGPPGPA
Sbjct: 890  GNAGPPGPPGPAGKEGGKGPRGETGPAGRPGEVGPPGPPGPA 931
```

Finally, there are seven **isoforms**, or alternative splice forms, listed for collagen alpha-2 (XI) in the results table. As indicated earlier, they are labeled "sp_vs" on the left column. They also appear in the graphic as P13942-2 to -8. Looking very carefully at the right graph, you can see that these sequences vary slightly in length and the yellow areas of similarity with the query move to the left or right, depending on the regions that are spliced out. This can be gathered from the alignments as well, with the added benefit of getting exact coordinates from the alignments. But the graphic captures the complex details of all these alignments in an easy-to-understand view.

But what do all these "stripes" on the right graphic, and these multiple regions of similarity to our query, mean? What is witnessed here is the history of this protein. Early in evolution there was a protein with one domain similar to the

Figure 4.29 Six of the regions of similarity between the query Q9UMA6 and hit P02452. The top alignment of 100% identity is represented as the green bar in Figure 4.28 (see color plates) while the other alignments are shown as shades of yellow.

27 amino acid query. In time, the gene with this single domain had duplication events that generated multiple copies of this domain within the same protein. Over time, these domains diverged from each other, perhaps taking on new functions, and what we now see are those multiple domains scattered on the length of full-length alpha-1 procollagen. This domain is seen on other proteins as well, indicating that a single gene with multiple domains probably gave rise to other genes which have since diverged in sequence and function, but nevertheless maintain similarity in these domains.

4.8 SUMMARY

In this chapter, another member of the BLAST suite of database searching tools was introduced; BLASTP uses protein queries to search a protein database. Amino acids were described, and their overlapping properties allow many biochemically conservative substitutions in nature. In BLASTP alignments, these non-identical pairings are scored according to a substitutions matrix, allowing distantly related proteins to be identified and compared. BLASTP searches were run both at the NCBI and the ExPASy Websites. Interesting properties of proteins were introduced, including the processing of proteins into smaller peptides and the repetition of related domains within a single protein.

EXERCISES

Exercise 1: Typing contest

This contest is actually educational and many students of sequence analysis find it fun. The E value and the other statistics generated by BLASTP are a function of the length and the quality of the alignment. Before getting to more serious exercises, we'll have a typing contest which can be run in a number of ways: (a) type for a fixed time (for example, 2 minutes); or (b) have a race with others where everyone stops typing when the winner finishes first. Even fast and accurate typists will find that they are slower and less accurate when typing sequences rather than English words. With no spaces and all the characters in uppercase, it is easy to lose track of where you are.

1. Type the sequence (**Figure 4.30**) directly into an NCBI BLASTP query window.

2. When the typing has ended, run a BLASTP search against RefSeq proteins (no species defined).

3. Look at the Scores, E values, and percent identities. The sequence is from a real protein (can you identify it?) so you can achieve 100% identity to the first hit. If you were a fast and accurate typist, your E value will be lower (better) than other attempts where fewer amino acids were typed or more errors were made.

4. Record your query length, Score, percent identity, and E value.

5. Repeat the contest and compare these values to your first run.

6. Look at the alignments and notice if you accidentally typed in any conservative substitutions.

7. Based on the differences between repeat runs, determine why your E value changed.

Figure 4.30 Typing contest sequence.

DGQRGGGGGATGSVGGGKGSGVGISTGGWVGGSYFTDSYVITKNTRQFLVK
IQNNHQYKTELISPSTSQGKSQRCVSTPWSYFNFNQYSSHFSPQDWQRLTN
EYKRFRPKGMHVKIYNLQIKQILSNGADTTYNNDLTAGVHIFCDGEHAYPN
ATHPWDEDVMPELPYQTWYLFQYGYIPVIHELAEMEDSNAVEKAICLQIPF
FMLENSDHEVLRTGESTEFTFNFDCEWINNER

Exercise 2: How mammoths adapted to cold

Mammals have adapted to cold climates through both physical and physiological changes that allow them to survive and thrive in colder temperatures. Compared to their relatives in warmer climates, a heavier coat of hair and more abundant fat keep the animal warm. Mammals can also limit heat loss by allowing their limbs to be colder than their bodies. However, these colder legs present a problem for oxygen transport. Globin proteins within red blood cells bind oxygen in the lungs and carry it elsewhere to be released. However, when blood cells are cold, they do not release oxygen very readily and cold limbs could quickly become starved of oxygen.

Changes to the globin protein sequence are necessary to facilitate oxygen release at lower temperatures. This was dramatically demonstrated when the globin genes of an extinct species, the woolly mammoth, were identified. Mammoths lived in an arctic climate and died off approximately 10,000 years ago, but well-preserved remains are frequently found in the present-day tundra. Globin DNA sequence was recovered from these remains and, through genetic engineering, translated and tested for oxygen binding at different temperatures. Campbell and colleagues were able to demonstrate that, compared to the Asiatic elephant (a close relative), the mammoth globin sequence had amino acid substitutions that allowed it to release oxygen at low temperatures.

In this exercise you will identify these changes by comparing the mammoth protein sequence to that of the Asiatic elephant. Specifically, a globin protein called beta/delta hybrid will be compared. This will be accomplished using pairwise BLASTP at the NCBI Website. You will later compare the nucleotide sequences and identify the changes at the codon level.

Pairwise comparison of mammoth and elephant globin proteins

1. Go to the NCBI BLASTP Web form and click the check box to turn it into a pairwise comparison tool.

2. In the two sequence fields, enter the accession number for the mammoth protein, ACV41408, and the Asiatic elephant protein, ACV41395, and click on the BLAST button.

3. When the screen refreshes, examine the alignment and identify the amino acid substitutions in the globin protein.

4. Using the BLOSUM62 matrix (Figure 4.9), what are the matrix scores for these substitutions?

5. Keep this screen open for later reference.

Nucleotide changes in the woolly mammoth beta/delta hybrid gene

For this part of the analysis, you will need to retrieve the coding sequences of the two globin mRNAs. That is, we want to only compare the nucleotides that encode the amino acids and ignore the nucleotides of the untranslated regions of the mRNAs. The NCBI makes this quite easy with just a few clicks.

When looking at a nucleotide record for each animal, there is an annotation "Feature" section called CDS (coding sequence). Click on the hypertext "CDS" and the screen will refresh, and highlighted are the regions of the genomic DNA sequence that correspond to the predicted mRNA. In the bottom-right corner of the screen you will see "Display: FASTA GenBank Help." Click on the "GenBank" link and a new page is created where you just have the coding sequence. You can verify this by looking at the first three nucleotides (usually the codon for methionine, ATG) and the last three nucleotides (a termination codon, either TAA, TAG, or TGA). The length of the sequence will now be much shorter. Convert this to FASTA format by clicking on either the FASTA link on the bottom of the page, or the other FASTA link at the upper left of the page.

1. Retrieve the NCBI files for the woolly mammoth (FJ716094) and Asiatic elephant (FJ716083) beta/delta hybrid globin mRNAs.

2. For each record, retrieve only the coding region as instructed above.

3. For each record, display the FASTA formatted file and paste these sequences separately into the two text fields of the NCBI BLASTN Web form for a pairwise alignment.

4. Click on the BLAST button. When the screen refreshes, you will see the pairwise alignment between the mammoth and Asiatic elephant globin coding sequences. In the protein sequence comparison, above, you identified three amino acid substitutions. You can now see the nucleotide changes responsible for these amino acid substitutions. Normally you would need coordinates to verify that you are working with the correct nucleotide changes. But in this case, there are only three nucleotide differences between these related species.

5. Use the genetic code (Figure 4.2) to determine the nucleotide changes that occurred to create these three amino acid substitutions. Manually translate the nucleotides into amino acids in the vicinity of each nucleotide mismatch to determine if the change was in the first, second, or third position of the affected codons. Since you know the amino acid sequence at each substitution for each animal, translating different groups of three nucleotides will allow you to answer this question.

Exercise 3: Longevity genes?

It has been known for many years that a calorie-restricted diet can lead to the extension of life expectancy. Early experiments in yeast, roundworm, and fly have shown that by limiting food, life expectancy can be extended by 30–40% with no apparent ill effects. Similar experiments with mice, rats, and nonhuman primates have shown that this effect has been conserved throughout evolution. In a spectacular demonstration of the effects of calorie restriction on the aging process, Colman and colleagues placed Rhesus macaque monkeys on a calorie-restricted diet and studied them for 20 years. Be sure to look at pictures of these monkeys, and their matched controls with no calorie restrictions, in the reference provided at the end of this chapter.

Many people are interested in living longer yet balk at the idea of reducing calorie consumption by one-third, the restriction necessary to attain benefit in other mammals. However, a factor in red wine called resveratrol has also been shown to modulate aging factors in flies. Could this be the answer to living longer?

Research has led to the discovery of a protein family called sirtuins that are regulated through calorie restriction and appear to be players in the pathways responsible for extending life span. Manipulation of these genes has defined some of their functions and they include effects on gene regulation, reactions to stress, respiration, and metabolism. Interestingly, the wine compound resveratrol has also been shown to modulate sirtuin activity, thus joining the two manipulations, calorie restriction and resveratrol treatment, at the molecular level.

In this exercise, you are going to find orthologs and paralogs of human *SIR2*, a member of the sirtuin family. Since the original experiments were done in yeast (*Saccharomyces cerevisiae*), you'll start with the yeast Sir2p protein as a query and search for protein sequences using BLASTP. Finding the SIR2 ortholog in a higher organism, you will then use that organism's protein sequence to find paralogs.

You will need to stay organized. There will be multiple queries and multiple database restrictions. You should always be aware of what you are doing, and why you are doing it.

1. Go to the NCBI Website and use Sir2p and *cerevisiae* as your query terms for the Entrez interface. When you have a list of protein sequences, find the hypertext filter for the results (upper right of the results page) and click on "RefSeq." From the shorter list of results, identify the yeast reference sequence for Sir2p. Most RefSeq proteins begin with "NP_."

2. Using this yeast Sir2p reference protein sequence as a query, run a BLASTP search against RefSeq proteins, with the database restricted to *Saccharomyces cerevisiae.*

3. You should find your query in the list of hits. In addition, did you find any paralogs? Do they have names similar to Sir2p?

4. Exploring the RefSeq annotation for these top hits (via the hypertext RefSeq accession numbers in your BLASTP results), do these suspected paralogs have similar function? Make a list of their names and probable functions.

5. Using the yeast Sir2p reference protein sequence as a query, run another BLASTP search against RefSeq proteins, restricted to proteins of *Drosophila melanogaster.*

6. Can you identify the *Drosophila* ortholog to yeast Sir2p? What is its name?

7. Can you identify any *Drosophila* paralogs to Sir2? Make a list of their names and leave this window open.

8. Using the *Drosophila* Sir2 reference protein as a query, run another BLASTP search against *Drosophila* RefSeq proteins.

9. Compare the hits from the yeast query (step 5) to the hits with the *Drosophila* query (step 8). How do they compare?

10. How does the number of hits compare between the searches performed in step 5 and step 8? Explain why the numbers in the lists are different.

11. Using the yeast Sir2p reference protein as a query, perform a BLASTP search against RefSeq proteins, restricted to proteins of *Homo sapiens.*

12. Can you identify the *Homo sapiens* ortholog to yeast Sir2p? Is it an easy decision?

13. How many human SIR2 paralogs can you identify? Do not count the different isoforms. Keep this window open.

14. Using the human SIR2 reference protein as a query, run another BLASTP search against the human reference proteins.

15. How does the list of hits from step 11 compare to the list of hits from step 14?

16. Looking back at the graphics when the yeast Sir2p protein was used against the *Drosophila* (step 5) and human (step 11) databases, what were the coordinates of the domain with the highest similarity to the best hits? Using the Conserved Domains graphic, what does this domain correspond to?

17. Using the human SIR2 reference protein sequence, find the *Macaca* SIR2 ortholog using BLASTP, with species restriction to *Macaca mulatta.* How similar are these two orthologs?

FURTHER READING

Alberts B, Johnson A, Lewis J et al. (2008) Molecular Biology of the Cell, 5th ed. New York: Garland Science. A textbook for general knowledge of amino acids and proteins.

Campbell KL, Roberts JE, Watson LN et al. (2010) Substitutions in woolly mammoth hemoglobin confer biochemical properties adaptive for cold tolerance. *Nat. Genet.* 42, 536–540. This paper describes the work on the mammoth adaption to cold weather.

Colman RJ, Anderson RM, Johnson SC et al. (2009) Caloric restriction delays disease onset and mortality in rhesus monkeys. *Science* 325, 201–204. This paper is mentioned in the exercise on longevity genes. Be sure to see the pictures of the monkeys in this article. There are additional pictures on the Internet.

Donmez G & Guarente L (2010) Aging and disease: connections to sirtuins. *Aging Cell* 9, 285–290. This paper describes the work mentioned in the longevity genes exercise.

Eddy S (2004) Where did the BLOSUM62 alignment score matrix come from? *Nat. Biotechnol.* 22, 1035–1036. The short article is a description of the BLOSUM62 matrix, widely used in bioinformatics applications.

Gasteiger E, Gattiker A, Hoogland C et al. (2003) ExPASy: the proteomics server for in-depth protein knowledge and analysis. *Nucleic Acids Res.* 31, 3784–3788. A general article on the ExPASy Website.

Henikoff S & Henikoff JG (1992) Amino acid substitution matrices from protein blocks. *Proc. Natl Acad. Sci. USA* 89, 10915–10919. This article describes the construction of the blocks database and the derived matrices.

Pritchard LE & White A (2007) Neuropeptide processing and its impact on melanocortin pathways. *Endocrinology* 148, 4201–4207. This article discusses the proteolytic processing of POMC and the biology of the peptides generated by this cleavage.

CAAGGGACTAGAGAAACCAAAA
AGAAACCAAAACGAAAGGTGCAGAA
AACGAAAGGTGCAGAAGGGGAAACAGATGCAGA
GAAGGGAAACAGATGCAGAAAGCATC
AGAAAGCATC
CAAGGGACTAGAGAAACCAAAACGAAAGGTGCAGAAGGGGAAACAGATGCAGAAAGCATC
AGAAACCAAAACGAAAGGTGCAGAA
AACGAAAGGTGCAGAAGGGG
GAAGGGG

CHAPTER 5

Cross-Molecular Searches: BLASTX and TBLASTN

Key concepts

- mRNA structure and cDNA synthesis
- BLASTX uses a nucleotide query to search a protein database
- TBLASTN does the opposite: it uses a protein query to search a nucleotide database
- Using BLASTX and TBLASTN to analyze cDNA sequences and probe huge cDNA libraries for sequences of interest

5.1 INTRODUCTION

So far, we have studied versions of BLAST which look for sequence identities and similarities within the same molecule type: BLASTN uses a nucleotide query to search a nucleotide database, and BLASTP uses a protein query to search a protein database. There are two other versions of BLAST which both cross molecular boundaries. **BLASTX** uses a nucleotide query to search a protein database, and **TBLASTN** uses a protein query to search a nucleotide database (**Table 5.1**). BLASTX (pronounced "BLAST-ex") and TBLASTN (pronounced "tea-BLAST-en") are particularly useful when studying cDNA, so this chapter will begin by revisiting mRNA structure and cDNA synthesis, providing details that will make interpreting BLAST output easier.

Table 5.1 Four of the BLAST applications

Type	Query	Database
BLASTN	Nucleotide	Nucleotide
BLASTP	Protein	Protein
BLASTX	Nucleotide	Protein
TBLASTN	Protein	Nucleotide

BLASTN was covered in Chapter 3 and BLASTP was covered in Chapter 4.

5.2 MESSENGER RNA STRUCTURE

Eukaryotic protein-encoding genes are transcribed by a protein complex containing RNA polymerase II. In an exquisitely evolved process, this complex will faithfully transcribe genomic DNA, often for thousands of nucleotides, and generate an RNA copy of the gene. Processing by other enzymes removes the transcribed introns and enzymes perform other post-transcriptional processing, resulting in the mature RNA known as messenger RNA (mRNA).

There are three major regions of mRNA that are important for function (**Figure 5.1**). As the name suggests, the **coding region** is translated into protein sequence. The coding region is flanked by untranslated regions (UTR, pronounced "you-tea-are") that are referred to as the 5P untranslated and 3P untranslated regions, or the 5P UTR and 3P UTR, respectively.

The 5P UTR is usually less than several hundred nucleotides in length, and may be encoded by multiple exons. Shortly after transcription starts, the end of the nascent transcript is modified with a 5P methyl "cap." Features within the 5P UTR are recognized by the ribosome and associated co-factors to initiate translation. Although not the principal translation products of the mRNA, there are often ATG codons in the 5P UTR that mark the beginning of short open reading frames.

The coding region starts with ATG (the codon for methionine) and ends with one of the three terminator codons: TAA, TGA, or TAG. The size of the coding region ranges from several hundred to thousands of nucleotides in length. For example, titin is the largest human gene, with over 300 exons and a coding region of over 80,000 nucleotides. Even though many amino acids can be specified by several different codons (see Figure 4.2), there is a bias to use certain codons and the differences between organisms can be quite striking. For example, in humans, the leucine codon CTG is used almost fivefold more often than CTA (40% versus 7% of the time). In *Arabidopsis* (thale cress), CTG is only used 5% of the time and the most favored codon for leucine is CTC (45%). It is thought that the observed codon bias reflects the need to either optimize or regulate translation rates and that codon choice is tied to the abundance of corresponding transfer RNAs for the codons.

The length of the 3P UTR ranges from less than one hundred nucleotides to thousands of bases long. It is not unusual for the 3P UTR to be longer than the 5P UTR and coding region, combined. Shortly after a gene is completely transcribed, a "tail" of A nucleotides is added to the 3P end of the mRNA. Unlike the rest of the mRNA, the poly(A) tail is synthesized in the absence of a template: there is no genomic sequence that corresponds to the poly(A) tail at the 3P end of the last exon. This tail is quite long, often 200 nucleotides in length, and is thought to delay the degradation of mRNA. The 3P UTR was widely ignored until it was recognized as containing important regulatory elements for translation and mRNA stability. In fact, you will find many sequence records in which only the coding region of mRNA is present and the UTR sequence is absent. In recent years, these elements in the 3P UTR have been studied intensely, and we'll be learning about them in Chapter 10.

From an evolutionary viewpoint, there are many constraints on the coding region. As we learned in Chapter 4, if a protein sequence changes too much the protein may no longer fold properly and/or function. The 5P and 3P untranslated regions, however, have fewer constraints and evolve at a much faster rate. As long as the small but important regulatory elements found in each remain functional, the UTRs can diverge to the point where orthologous UTRs may no longer look similar yet the protein-coding regions can be easily aligned.

Figure 5.1 Messenger RNA (mRNA) structure. Eukaryotic mRNA has a coding region flanked by untranslated regions (5P UTR and 3P UTR). The 3P end is polyadenylated after transcription of the genomic DNA is completed.

5P	5P UTR	Coding region	3P UTR	AAAAA-3P

5.3 cDNA

Many of the details of mRNA and gene structure have been determined by the isolation and analysis of DNA that is complementary to mRNA, called cDNA. In this section, cDNA will be described extensively, including its synthesis, the different forms, and the bioinformatics databases where cDNAs are found.

Synthesis

For cloning and sequencing purposes, mRNA is usually synthetically converted to complementary DNA (cDNA). Techniques vary but a common approach is to first anneal a short run of Ts (oligo(dT) primer) to the 3P poly(A) tail of mRNA. Reverse transcriptase, the enzyme used to convert mRNA to cDNA, is then added, which binds to the 3P end of this oligo(dT) primer. The enzyme then starts moving toward the 5P end of the mRNA, synthesizing cDNA based on the mRNA sequence that it is "reading."

Once the first strand of cDNA is made, how do you make the cDNA double-stranded, usually a prerequisite for cloning? Again, techniques vary, but what often happens is the 3P end of the first strand folds back onto itself and acts as a priming spot for an enzyme to begin synthesis of the second strand. Synthesis of the second strand then proceeds using the first strand cDNA as a template.

As a result of this self-priming, some of the original 5P end of the mRNA will not be represented. Today there are new approaches that target the 5P end of the mRNA during cDNA synthesis. For example, a "tail" of known sequence can be added to the 3P end of the first strand cDNA, and then a primer with the complement of this sequence can be used to begin synthesis of the second strand. There are many different methods to reach the 5P cap site so you will encounter sequences and databases where the annotation specifically states this goal.

Synthesizing cDNA can be technically difficult (**Figure 5.2**). Early in the history of genetic engineering, scientists struggled to clone and sequence full-length copies of cDNA. Even today, there are still significant limitations, not the least of which is that mRNA is very susceptible to degradation. If it is not handled carefully, mRNA can easily break down into small pieces. Subsequent steps to synthesize cDNA from this partially degraded mRNA would generate even smaller pieces (Figure 5.2C). Another concern is that mRNA can form complex secondary structures. Since mRNA is single-stranded and composed of only four bases, regions can fold back onto each other and base-pair. These structures can then impede the movement of reverse transcriptase along the molecule and cause the enzyme to

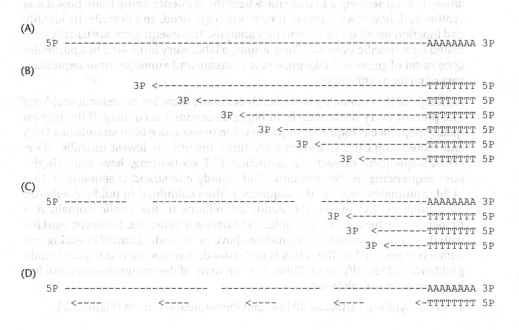

Figure 5.2 mRNA and cDNA synthesis. (A) An mRNA with a poly(A) tail. (B) cDNA synthesis primed with an oligo(dT) primer commences at the extreme 3P end of mRNA. Due to limitations of the *in vitro* reaction, reverse transcriptase will often "fall off" the mRNA resulting in a population of cDNA that is biased toward the 3P end. (C) The impact of partial degradation of mRNA. Starting at the primer, cDNA synthesis can only go as far as the break in mRNA. (D) Randomly primed cDNA synthesis will reduce the bias toward the 3P end of mRNAs, and allow partially degraded mRNA to generate cDNA from all regions.

disengage with the mRNA. So even if the mRNA is full length, reverse transcriptase can literally "fall off" the mRNA before reaching the 5P end (Figure 5.2B).

If cDNA synthesis starts at the extreme 3P end of the mRNA, and there is some level of degradation or impediment of enzyme synthesis, the cDNA generated will be biased toward the 3P end of the mRNA. In fact, many oligo(dT)-primed cDNAs do not extend into the coding region and are entirely 3P UTR sequence. One approach to solve this problem is to randomly prime mRNA. Rather than use oligo(dT) priming at the extreme 3P end, random oligomers are added to the reaction which are complementary to various locations along the length of the mRNA (Figure 5.2D). Of course, you have replaced one problem with another: the cDNA you synthesize will not represent full-length copies of the mRNA. However, bioinformatics tools can nicely identify overlapping pieces and assemble them into a long, perhaps full length, cDNA.

cDNA in databases

In the above discussion, the steps and problems of cDNA synthesis were described for a single mRNA transcript. When cDNA is synthesized in a laboratory setting, transcripts from many different genes are simultaneously reverse transcribed into cDNA. From even small amounts of mRNA isolated from a single tissue, many hundreds to thousands of different cDNAs can be synthesized and then cloned into bacteria, and this is referred to as a **cDNA library**. Library descriptions usually contain the origin of the mRNA (for example, liver) and the methods used to generate the cDNA (for example, oligo(dT) primed). The cDNAs are then sequenced and, very often, deposited into a database such as GenBank. In light of the issues regarding synthesizing cDNA, explained above, considerable variability of data quality exists. The way scientists submit their sequences to GenBank varies as well: some will submit every nucleotide they have, while others will submit only the coding region or the region of their interest (for example, an exon). Although almost every mRNA has a poly(A) tail, the reporting of this sequence is inconsistent and it is often not submitted with the rest of the cDNA sequence. Of course, it does not contain a lot of information but it does indicate that you are looking at the 3P end of the mRNA.

At one extreme of analysis, a scientist may compare each sequence to genomic DNA or other overlapping cDNA to insure that the sequence is correct. They may verify that artifacts were not introduced while cloning the cDNA. The sequence from both strands, perhaps generated multiple times, or with different protocols, may be aligned to definitively identify each base. Finally, they may carefully annotate each sequence to describe where the sequence came from, how it was synthesized, how it was cloned, how it was sequenced, and describe its identity and function based on bioinformatics analysis. These sequences are usually generated for a specific purpose. For example, a laboratory interested in a particular gene or set of genes will take great care to create and annotate these sequences, very often for a publication.

Although DNA sequencing is extremely accurate, there are occasional problems that prevent every nucleotide from being determined accurately. If the highest quality sequence is required, steps are taken to sequence both strands of a DNA and sequencing each strand multiple times insures the fewest miscalls. Many cDNA sequencing projects, in particular EST sequencing, have only **single-pass sequencing** as the standard. That is, only one strand is sequenced. In a highly automated process, the sequence is then submitted to public databases. Submitting cDNAs by the thousands or millions to the public domain is a tremendous achievement, valuable, and extremely generous. However, working with these huge numbers, compromises have to be made. Manual checking and curation is impossible. But if this is understood, then you can anticipate certain problems and identify them. Below is a summary of the common issues, but it is by no means a complete list.

- cDNA synthesis artifacts. 3P bias and incomplete synthesis (Figure 5.2).

- Cloning artifacts. cDNA may be rearranged, or chimeric. Cloning often results in the annealing of two cDNAs into the same cloning vector.

- Sequence artifacts. Errors introduced in "reading" the sequence. In first-generation sequencing, reading can jump from one lane to an adjacent one, artificially joining two sequences.

- Sequence quality errors. "Single-pass" sequencing is often not checked for errors.

- Submission or annotation artifacts. Computer/software errors.

- The cDNA is actually genomic DNA. Even a minor DNA contamination of an mRNA sample can easily be cloned. However, as about 1% of the human genome codes for proteins, the chances are that the genomic sequence will be noncoding.

- Human error. When dealing with thousands of repetitive tasks, errors may occur.

Thus, cDNA is far from perfect and there is often both technical and human variability. Much has improved since the early days of cloning and sequencing so, in general, recently generated data will have fewer artifacts than those of the past. Also, it is technically less challenging to make cDNA for small mRNAs than it is for larger mRNAs. Finally, there are many records annotated as being "full length" or tissue-specific. Don't believe it.

ESTs

The sequence content of a cDNA library provides a view of gene expression in the cells or tissues from which it was derived. It is not a perfect relationship, but cDNA as a proxy for mRNA should roughly reflect the genes that are expressed in cells. If you want to know what proteins are expressed in a given cell type, you can sequence the cDNA derived from mRNA from those cells. In general, RNAs that are abundant in a particular tissue will be well represented in a cDNA library. Certain tissues (for example, nervous tissue and testes) tend to express many different genes in an apparent nonspecific manner. Other cDNAs are extremely difficult to find, even though the proteins of these mRNAs can be detected with antibodies.

Scientists have discovered that high-throughput, single-pass, often non-optimized synthesis of cDNA is a fast and inexpensive way to look into cells and ask, "what is expressed in here?" With cDNAs only hundreds of nucleotides long, representing a fraction of the mRNA length, there is often enough sequence to answer that question. These small cDNAs, first introduced in Chapter 1, are known as expressed sequence tags or ESTs (pronounced "ee-es-teas").

If you are exploring expression in a well-studied organism, an unknown EST can be identified by looking for perfect alignment with a reference sequence. A BLASTN search can show an alignment of your EST query with (hopefully) a well-annotated sequence. The quality and length of the alignment will give you confidence that you have found a piece of a gene. If you are working with an organism not well represented in GenBank, DNA identity may not always help you. Should the EST sequence extend into the coding region, the similarity to homologous and well-known genes may be strong enough to positively identify the EST. However, should the EST be from the 3P UTR, you might see little or no homology to known UTRs.

EST libraries are often huge, some numbering in the tens of thousands. So why sequence so many ESTs? Why not just sequence 100 cDNAs and determine the top 100 genes that are expressed? This approach will not work because genes are not transcribed in equal numbers. There are transcripts which dominate the mRNA population for a cell type and would, in turn, dominate a cDNA library. For example, globin is an abundant mRNA in red blood cells, albumin in liver, and actin and myosin in muscle. These types of transcripts could add up to several percent of all cDNAs from a cell.

Other classes of genes, although not dominating the mRNA population, are less interesting if you are looking for rare, tissue-specific transcripts. There are roughly 21,000 protein-encoding genes in the human genome. A large number of these genes are considered **housekeeping genes**, encoding proteins found in every human cell. These can make up a large portion of an EST library and include proteins necessary for basic cellular functions such as transcription, translation, and energy metabolism.

Since housekeeping genes will be transcribed in every cell, and other transcripts will dominate a cDNA library, sequencing a low number of cDNAs will reveal relatively few (new) genes of interest. You would only scratch the surface of the pathways and intricate regulation going on in your cell of interest. By sequencing many thousands of ESTs you will encounter the more rare mRNAs, getting past the genes you already know about.

There are currently over eight million human ESTs in GenBank (**Box 5.1**). They are from almost every tissue in the body, and from different developmental stages. EST libraries have been made from different health states, from cells grown in culture, and from cells treated with many agents. For example, there are over 2300 ESTs for human beta globin alone.

Normalized cDNA libraries

Rather than sequence large numbers of ESTs to obtain rare transcripts, another approach is to optimize the construction of the cDNA library. A "normalized" cDNA library is constructed by depleting abundant transcripts or cDNAs from the reaction. An illustration of this approach is shown in **Figure 5.3**. Normalization can be accomplished by synthesizing double-stranded cDNA and then heating it to break the bonds between the strands (known as denaturing). The single strands of cDNAs are then cooled which allows complementary strands to come together (re-anneal) to form double strands.

Consider the chances of two different mRNAs re-annealing: one is very abundant in cells (for example, 10% of the mRNAs are from a single gene) and the other is very rare (for example, 0.1% of the mRNAs are from another gene). Simple rules of kinetics say that the abundant single-stranded cDNAs will find their complementary sequences much faster in solution, while the rare cDNA will likely not encounter a complementary sequence. So, after a short period of re-annealing,

Figure 5.3 Normalization of a cDNA library. (A) An example cDNA library construction where the ratio of one cDNA, "abundant," and another cDNA, "rare," is 9:1. The rare cDNA is in bold. (B) The double-stranded cDNA is heated to denature it to single strands. (C) The mixture is cooled for a short period of time, long enough to allow many of the abundant single-stranded cDNAs to base-pair, but not for the rare cDNA strands to find each other in the mixture. (D) Double-stranded cDNAs are removed from the library, leaving behind single-stranded cDNA. The ratio between the abundant and rare cDNAs is now 2:1. Subsequent annealing and cloning result in a normalized cDNA library.

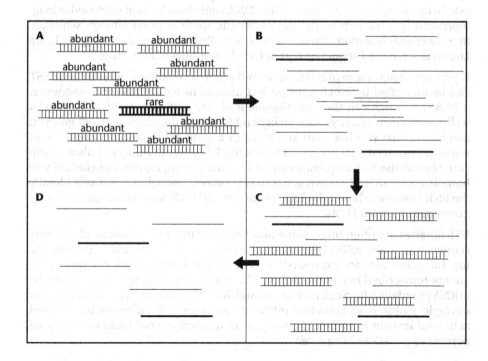

Box 5.1 How many EST records?

In Chapter 2, the NCBI UniGene database was extensively described and Figure 2.19 illustrates the number of ESTs from individual species. There are other ways to obtain this information, independent of the UniGene database. The NCBI EST database can be searched directly through the Entrez drop-down menu as shown in **Figure 1**. Here, *Homo sapiens* was the query and over 8.4 million ESTs were found. Also listed is the total number of nucleotide records, including ESTs (21 million).

Are they all human sequences? No. They were found using the query *Homo sapiens* but this found many records containing these two words outside of the Source or Organism fields, such as *Mus musculus* sequences with a definition line containing "similar to Homo sapiens Chromosome 16" or *Plasmodium falciparum* sequences with "Lab host=Homo sapiens" in the Features section. Looking on the right side of Figure 1, notice the Top Organisms panel listing over 8.3 million ESTs for human and thousands of ESTs from other organisms. Clicking on the hypertext numbers next to these species will take you to these records.

In another approach, the number of ESTs from an organism can be found using the NCBI Taxonomy database. Again, use the Entrez drop-down menu (Taxonomy database selected) to directly search with the query *Homo sapiens*. The next screen requires you to click on a link for *Homo sapiens* and that takes you to a full biological classification of humans as seen in **Figure 2**.

Near the center of Figure 2 is a bold link "Homo sapiens (human)." Clicking on this hypertext brings you an extensive page of links associated with *Homo sapiens* but near the top is a listing of the number of Entrez records for *Homo sapiens*. Although we were pursuing the number of human ESTs, going this route shows the numbers of proteins, structures, SNPs, and many other classes of biomedical data captured in the Entrez database. This is shown in **Figure 3**. When there are sub-classifications for related species, these are totaled and referred to as Subtree links. Direct links are for the original query. For example, looking at the bottom of Figure 3, there are two Taxonomy records for *Homo sapiens* and one Taxonomy record for *Homo sapiens neanderthalensis*, both found with the query *Homo sapiens*. For a thorough explanation of the Taxonomy database, see the NCBI Handbook, available through the NCBI Bookshelf or the Site Map (A-Z).

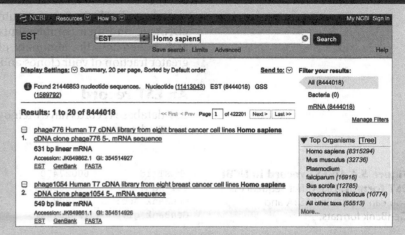

Figure 1 Determining EST numbers using the NCBI Entrez interface. By putting the species into the Entrez query field, the number of ESTs originating from that species can be determined.

Lineage (full): root; cellular organisms; Eukaryota; Opisthokonta; Metazoa; Eumetazoa; Bilateria; Coelomata; Deuterostomia; Chordata; Craniata; Vertebrata; Gnathostomata; Teleostomi; Euteleostomi; Sarcopterygii; Tetrapoda; Amniota; Mammalia; Theria; Eutheria; Euarchontoglires; Primates; Haplorrhini; Simiiformes; Catarrhini; Hominoidea; Hominidae; Homininae; Homo

o **Homo sapiens** (human) *Click on organism name to get more information.*

   ▪ **Homo sapiens neanderthalensis** (Neandertal)

Figure 2 The NCBI Taxonomy database listing of a species biological classification. Pictured here is the full taxonomic description of *Homo sapiens*. Clicking on the bold "Homo sapiens (human)" link near the center will display what is shown in Figure 3.

Entrez records		
Database name	Subtree links	Direct links
Nucleotide	9,892,226	9,892,201
Nucleotide EST	8,315,294	8,315,294
Nucleotide GSS	1,581,050	1,579,724
Protein	592,489	592,393
Structure	18,991	18,991
Genome Sequences	51	50
Popset	22,240	22,240
SNP	60,481,170	60,481,170
Domains	10	10
GEO Datasets	339,865	339,865
GEO Expressions	27,034,750	27,034,750
UniGene	122,727	122,727
UniSTS	328,581	328,581
PubMed Central	11,011	11,006
Gene	42,110	42,073
HomoloGene	18,431	18,431
SRA Experiments	65,923	65,921
Genome Projects	140	138
Probe	9,032,879	9,032,879
Taxonomy	2	1

Figure 3 Biomedical record numbers as shown in the NCBI Taxonomy database. The numbers are hypertext and will take you to Entrez collections of those records.

only the most abundant cDNAs become double-stranded. You can then use physical or biochemical methods to remove the double-stranded fraction (the abundant cDNA). You then resume the annealing conditions, allowing the rare cDNAs to find their complements and become double-stranded. After cloning and sequencing, the abundant cDNAs have now been greatly reduced and a greater fraction of your clones and sequences are from the rare transcripts.

An EST record

EST database records have their own format (**Figure 5.4**), which is different from GenBank format.

Figure 5.4 An EST record in NCBI EST format. Like other sequences, you can view it in FASTA and GenBank formats.

```
dbEST Id:       60027470
EST name:       MCBL05H04.F
GenBank Acc:    FL590802
GenBank gi:     196121421

CLONE INFO
Clone Id:       (5')
Source:         Centro de Biotecnologia, Instituto Butantan
DNA type:       cDNA

PRIMERS
Sequencing:     M13F
PolyA Tail:     Unknown

SEQUENCE

                AGAGAAGATTTCACGATGAAAGCTCTGCTGCTGACCTTGGTGGTGGTGACAATGT
                GTGCCTGGACTTAGGATACACCCGAAAATGTTACGAGGGAGAGGGTACTAGGAAA
                TCTGTGACTTGTCCAAAGGGGGAGAAAGTATGCTATACCATATTTCTTGTGGGAC
                CTAGTCATCCTGCAAAGGTACTCAAATGGGGATGTGCTGCTTCTTGCCCTAAAGT
                AGGACTCGGTGCGCGTATTTCCTGTTGCTCAAAAGACAATTGCAACTCACATCGT
                TGAAAATTTAAGGTTCTGGCTTTTACCTCCAAATGGTGATTCATTCCCTCTCAAC
                CCTGCTGTCTTTGACACCTCAACATCTTTCCCTTTTCTCTTGTTCTGTAAATTTC
                CTTCTGCTAGTTCTGTAGTTTGAGAATCGAATAAACCTCAGCATTCAAA

Entry Created:  Aug 14 2008
Last Updated:   Aug 14 2008

LIBRARY
Lib Name:       Coral snake Micrurus corallinus cDNA library
Organism:       Micrurus corallinus
Organ:          venom glands
Tissue type:    venom glands
Vector:         Lambda GT11
R. Site 1:      EcoRI
R. Site 2:      NotI

SUBMITTER
Name:           Inacio de L. M. Junqueira-de-Azevedo
Lab:            Centro de Biotecnologia
Institution:    Instituto Butantan
Address:        Av. Vital Brazil, 1500, Sao Paulo SP, BRAZIL, 05503-900.
Tel:            55 11 37 26 7222 ext. 2244
Fax:            55 11 37 26 1505
E-mail:         ijuncaze@butantan.gov.br

CITATIONS
Title:          Transcriptomic basis for an antiserum against Micrurus
                corallinus (coral snake) venom
Authors:        Leao,L.I., Ho,P.L., Junqueira-de-Azevedo,I.L.M.
Year:           2008
Status:         Unpublished
```

There is a minimal amount of annotation associated with this record: the gene identity is unknown, you can't tell if it is full length, and, even though it was derived from mRNA, you don't know if it encodes part or all of a protein. To view the length (435 nucleotides), you have to convert it to GenBank format. Is it the sense strand that is translated, or the complementary strand? The record doesn't tell you. Since the cDNA synthesis may have been started at the 3P end, this sequence may be from the 3P UTR and may not even reach the protein-coding sequence. But it is interesting: you can see from this record that someone was interested in the genes that are expressed in the venom gland of a coral snake. A program like BLASTX can help you determine more about this sequence.

5.4 BLASTX

BLASTX crosses boundaries between molecules. It takes a DNA query and then searches a protein database. How is this accomplished? BLASTX takes your nucleotide query, translates it into protein sequence, and then searches a protein database much like BLASTP. You don't have to tell BLASTX where to start translating your DNA query. It generates all the possible translations for you.

Reading frames in nucleic acids

To understand how BLASTX works, we have to first review how DNA is translated into protein sequence. Within a cell, mRNA is read by the ribosome, three nucleotides at a time. If an mRNA is full length, enough information will be present to direct the translation to start at the ATG at the 5P end of the coding region, and the proper **reading frame** would be read starting from this first group of three. In the example in **Figure 5.5A**, there is a piece of cDNA that contains coding sequence. The complementary second strand is also shown.

There are six potential reading frames. For the first three, translations could start at position one, two, or three and the nucleotides are read in groups of three at a time. Figure 5.5B shows the three forward frames from this unknown cDNA (find them in Figure 5.5A too).

The three reverse frames are taken from the complementary DNA strand (in Figure 5.5A), going in the reverse direction. Although many cDNA libraries are cloned in a manner where the orientation of the cDNA is planned (for example, the 3P end will be next to a particular sequence in the cloning vector), things do not always go as planned. The result is a piece of double-stranded cDNA and the true orientation is not known. So probing the translations of the complementary strand is required. These reading frames on the reverse strand are depicted in Figure 5.5C.

Any guesses as to which is the real reading frame? Frame 1 looks promising because it contains no terminator codons from end to end and is referred to as "open" or an **open reading frame** (**ORF**). If codons appeared randomly, you would expect a terminator to be present, approximately, every 22 codons (there are 3 terminators in the 64 codons of the genetic code). There are 20 codons in Frame 1 so it could be "open" by chance. Another possibility is that Frame 3 may be the end of the real reading frame with a terminator after the tyrosine (Y). Frame 2 has two terminators so, assuming the DNA sequence is without errors, this frame does not seem adequate to encode a protein. The two terminators are only separated by 15 amino acids, and there are no ATG codons.

Since EST sequencing is single pass and the large number of sequences makes it impossible to manually inspect them, there are frequent sequencing errors. An error can create a termination codon where there is none, or insert or delete one or more bases that put the sequence "out of frame." Is "Frame 1" correct, or are there errors that mask the correct frame from being recognized? BLASTX may tell us.

Figure 5.5 The reading frames of a DNA sequence. (A) A section of an unknown EST with coding sequence. Both strands are shown. (B) The three forward reading frames of the sequence in (A). The codons are separated by spaces and the translations of these codons appear as the one-letter amino acid code below the second base of the codon; a "–" indicates the position of a terminator codon. For Frames 1, 2, and 3, only the starting point of the codons is shifting. (C) The three reverse reading frames of the sequence in (A), originating from the bottom strand. To remind you that the frames are from the opposite strand, the frame numbers are preceded by a minus sign, "–." For Frames –4, –5, and –6, only the starting point of the codons is shifting.

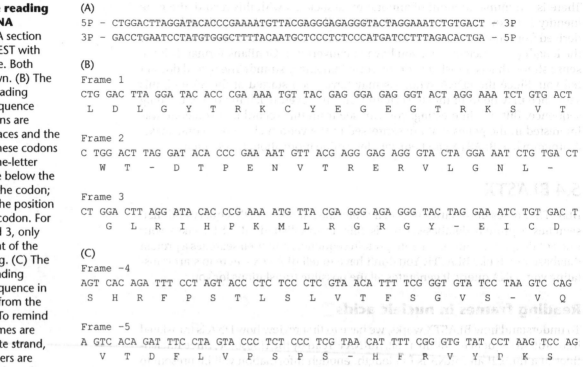

(A)
```
5P - CTGGACTTAGGATACACCCGAAAATGTTACGAGGGAGAGGGTACTAGGAAATCTGTGACT - 3P
3P - GACCTGAATCCTATGTGGGCTTTTACAATGCTCCCTCTCCCATGATCCTTTAGACACTGA - 5P
```

(B)
```
Frame 1

CTG GAC TTA GGA TAC ACC CGA AAA TGT TAC GAG GGA GAG GGT ACT AGG AAA TCT GTG ACT
 L   D   L   G   Y   T   R   K   C   Y   E   G   E   G   T   R   K   S   V   T

Frame 2

C TGG ACT TAG GAT ACA CCC GAA AAT GTT ACG AGG GAG AGG GTA CTA GGA AAT CTG TGA CT
   W   T   -   D   T   P   E   N   V   T   R   E   R   V   L   G   N   L   -

Frame 3

CT GGA CTT AGG ATA CAC CCG AAA ATG TTA CGA GGG AGA GGG TAC TAG GAA ATC TGT GAC T
    G   L   R   I   H   P   K   M   L   R   G   R   G   Y   -   E   I   C   D
```

(C)
```
Frame -4

AGT CAC AGA TTT CCT AGT ACC CTC TCC CTC GTA ACA TTT TCG GGT GTA TCC TAA GTC CAG
 S   H   R   F   P   S   T   L   S   L   V   T   F   S   G   V   S   -   V   Q

Frame -5

A GTC ACA GAT TTC CTA GTA CCC TCT CCC TCG TAA CAT TTT CGG GTG TAT CCT AAG TCC AG
   V   T   D   F   L   V   P   S   P   S   -   H   F   R   V   Y   P   K   S

Frame -6

AG TCA CAG ATT TCC TAG TAC CCT CTC CCT CGT AAC ATT TTC GGG TGT ATC CTA AGT CCA G
    S   Q   I   S   -   Y   P   L   P   R   N   I   F   G   C   I   L   S   P
```

BLASTX generates six protein sequences from the six reading frames to use as queries against the protein database. When there is a hit, BLASTX shows you the alignment between the translation and the protein sequence hit. Therefore BLASTX is particularly useful when you have a DNA sequence that may be encoding a protein sequence but little else is known.

A simple BLASTX search

By crossing the boundaries between nucleotides and translation products, BLASTX can be used to define the coding and noncoding regions of an mRNA. Remember that a cDNA can consist of either translated or untranslated regions, or both. If the untranslated regions are never translated, then there should be no protein records for these regions in the database. If you take a RefSeq cDNA record, for example human renin (NM_000537), and perform a BLASTX against RefSeq proteins, you get the alignment from the best hit seen in **Figure 5.6**.

The Subject of the alignment is the renin protein sequence (NP_000528). The protein sequence of the query is derived from the RefSeq cDNA record for human renin so it is no surprise that the identity is 100%. Position 45 is the 45th nucleotide of the query. Starting at this position, the first three nucleotides are ATG, which codes for methionine (M), so you see a methionine at this position in the alignment.

The "DNA to protein" nature of BLASTX is reflected in the coordinates of BLASTX alignments. In Figure 5.6, the protein coordinates of the first line of the alignment are 1–60, or 60 amino acids. But the query coordinates reflect a number three times that length: 45–224, or 180. This is because the query is translated nucleotides which are read in groups of three.

This entire alignment shows that the protein sequence for human renin is derived from nucleotides 45–1262 of the query (mRNA or cDNA) and this aligns to amino

```
>ref|NP_000528.1| renin preproprotein [Homo sapiens]
Length=406


Score =  830 bits (2143),  Expect = 0.0
Identities = 406/406 (100%), Positives = 406/406 (100%), Gaps = 0/406 (0%)
Frame = +3


Query  45    MDGWRRMPRWGLLLLLWGSCTFGLPTDTTTFKRIFLKRMPSIRESLKERGVDMARLGPEW  224
             MDGWRRMPRWGLLLLLWGSCTFGLPTDTTTFKRIFLKRMPSIRESLKERGVDMARLGPEW
Sbjct  1     MDGWRRMPRWGLLLLLWGSCTFGLPTDTTTFKRIFLKRMPSIRESLKERGVDMARLGPEW  60


Query  225   SQPMKRLTLGNTTSSVILTNYMDTQYYGEIGIGTPPQTFKVVFDTGSSNVWVPSSKCSRL  404
             SQPMKRLTLGNTTSSVILTNYMDTQYYGEIGIGTPPQTFKVVFDTGSSNVWVPSSKCSRL
Sbjct  61    SQPMKRLTLGNTTSSVILTNYMDTQYYGEIGIGTPPQTFKVVFDTGSSNVWVPSSKCSRL  120


Query  405   YTACVYHKLFDASDSSSYKHNGTELTLRYSTGTVSGFLSQDIITVGGITVTQMFGEVTEM  584
             YTACVYHKLFDASDSSSYKHNGTELTLRYSTGTVSGFLSQDIITVGGITVTQMFGEVTEM
Sbjct  121   YTACVYHKLFDASDSSSYKHNGTELTLRYSTGTVSGFLSQDIITVGGITVTQMFGEVTEM  180


Query  585   PALPFMLAEFDGVVGMGFIEQAIGRVTPIFDNIISQGVLKEDVFSFYYNRDSENSQSLGG  764
             PALPFMLAEFDGVVGMGFIEQAIGRVTPIFDNIISQGVLKEDVFSFYYNRDSENSQSLGG
Sbjct  181   PALPFMLAEFDGVVGMGFIEQAIGRVTPIFDNIISQGVLKEDVFSFYYNRDSENSQSLGG  240


Query  765   QIVLGGSDPQHYEGNFHYINLIKTGVWQIQMKGVSVGSSTLLCEDGCLALVDTGASYISG  944
             QIVLGGSDPQHYEGNFHYINLIKTGVWQIQMKGVSVGSSTLLCEDGCLALVDTGASYISG
Sbjct  241   QIVLGGSDPQHYEGNFHYINLIKTGVWQIQMKGVSVGSSTLLCEDGCLALVDTGASYISG  300


Query  945   STSSIEKLMEALGAKKRLFDYVVKCNEGPTLPDISFHLGGKEYTLTSADYVFQESYSSKK  1124
             STSSIEKLMEALGAKKRLFDYVVKCNEGPTLPDISFHLGGKEYTLTSADYVFQESYSSKK
Sbjct  301   STSSIEKLMEALGAKKRLFDYVVKCNEGPTLPDISFHLGGKEYTLTSADYVFQESYSSKK  360


Query  1125  LCTLAIHAMDIPPPTGPTWALGATFIRKFYTEFDRRNNRIGFALAR  1262
             LCTLAIHAMDIPPPTGPTWALGATFIRKFYTEFDRRNNRIGFALAR
Sbjct  361   LCTLAIHAMDIPPPTGPTWALGATFIRKFYTEFDRRNNRIGFALAR  406
```

Figure 5.6 A BLASTX alignment between Query (NM_000537) and Subject (Sbjct; NP_000528). Note that the reading frame is indicated (+3) in the last line before the alignment.

acids 1–406, the entire protein sequence. In fact, if you look at the annotation of the NM_000537 record you see the coding sequence (CDS) is defined as 45–1265.

Notice the discrepancy between the last number in BLASTX (1262) and the last number in the CDS (1265). The NCBI CDS annotation includes the termination codon for this gene, TGA, which is nucleotides 1263–1265 in the Nucleotide record. But this last codon is not translated into any amino acid; that is, there is no amino acid 407.

A more complex BLASTX

The above example was easy to understand. We used a full-length cDNA that found and cleanly aligned to a full-length protein. In addition, the identity between query and subject was 100% so it required little interpretation. Next, we will perform a more challenging BLASTX search with an unknown query and the hit will not be 100%. Confusion over the interpretation of BLASTX coordinates is quite common so we will revisit the topic in this second BLASTX search.

For this example, the coral snake EST seen earlier in Figure 5.4 will be used. Go to the NCBI BLASTX form and enter FL590802 into the query field. For the database, choose "UniProtKB/Swiss-Prot(swissprot)" and then click on the BLAST button to launch the search.

When the page refreshes, you are presented with a graphical representation of the hits (**Figure 5.7**). So what is going on here? All the hits are lined up, slightly off-center, with approximately the same coordinates. Remember that the query

Figure 5.7 BLASTX results of FL590802 versus UniProtKB/ Swiss-Prot(swissprot).
The graphic demonstrates a fairly consistent pattern of similarity between a central region of the query and many proteins.

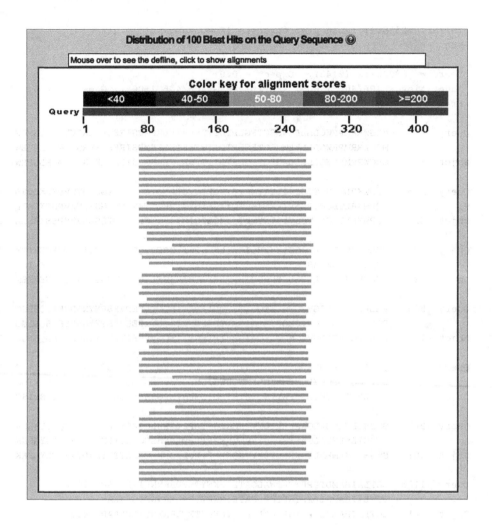

is a cDNA, which potentially could have both 5P and 3P untranslated regions. Since we are looking for similarities to proteins (the target database is a protein database), you would not expect the UTRs to find any significant similarity to any protein because they are not translated. Although it may not be apparent from the grayscale figure, these top hits are the same color, signifying that they all share roughly the same score range (50–80).

Another explanation for this graphic pattern is that the cDNA encodes a protein with a domain that is conserved across many proteins, and this domain has distinct boundaries. Other interpretations are possible but let's move on to the table to see what was found and look for evidence for the two models.

The results in the table are also quite uniform (**Figure 5.8**). The top hits are all neurotoxins and the rest of the hits (not shown) continue the listing of this interesting type of protein. The hits that are shown in this figure have very similar scores, query coverage, E values, and maximum identities. None of the hits have 100% identity so the EST encodes an uncharacterized protein.

Scrolling down to the alignments, the first hit and many of the subsequent hits look similar to that shown in **Figure 5.9**. First, notice that the hit is a neurotoxin taken from another species of snake, *Laticauda semifasciata*, the broad-banded blue sea krait. The Expect value of the alignment is not spectacular but there are a number of identities and similarities over a 70 amino acid stretch. Notice that BLASTX has a "Frame =" field to assist you. It says that the translation from the first frame (+1) of the query was used in this alignment. This BLASTX alignment is more complex than the one we looked at in Figure 5.6, as it includes mismatches, + signs, and gaps.

Sequences producing significant alignments:		Max score	Total score	Query coverage	E value	Max Ident
Accession	Description					
P01379.2	RecName: Full=Long neurotoxin Ls3; AltName: Full=LsIII; AltName: F	67.0	67.0	46%	2e-14	44%
A8N285.1	RecName: Full=Long neurotoxin LNTX-2 homolog; Flags: Precursor	64.7	64.7	45%	2e-13	51%
Q53B53.1	RecName: Full=Long neurotoxin OH-34; Flags: Precursor	62.0	62.0	45%	2e-12	48%
Q2VBP8.1	RecName: Full=Long neurotoxin LNTX1; Flags: Precursor	62.0	62.0	46%	2e-12	44%
Q2VBP3.1	RecName: Full=Long neurotoxin LNTX37; Flags: Precursor	62.0	62.0	46%	2e-12	44%
A6MFK4.1	RecName: Full=Long neurotoxin LNTX-1; Flags: Precursor	61.6	61.6	46%	3e-12	44%
Q53B57.1	RecName: Full=Long neurotoxin OH-56; Flags: Precursor	61.6	61.6	45%	3e-12	47%
Q53B58.1	RecName: Full=Long neurotoxin OH-55; AltName: Full=CM-11; Flags:	61.6	61.6	45%	3e-12	47%
Q2VBP6.1	RecName: Full=Long neurotoxin LNTX8; Flags: Precursor	61.6	61.6	46%	3e-12	44%
P01395.1	RecName: Full=Long neurotoxin 2; AltName: Full=Toxin I/V	60.8	60.8	43%	4e-12	47%
Q7T2I3.1	RecName: Full=Long neurotoxin LiLong; Flags: Precursor	61.2	61.2	46%	4e-12	44%
Q2VBP5.1	RecName: Full=Long neurotoxin LNTX22; Flags: Precursor	61.2	61.2	45%	5e-12	50%
A6MFK5.1	RecName: Full=Long neurotoxin LNTX-2; Flags: Precursor	60.8	60.8	46%	6e-12	44%
Q53B56.1	RecName: Full=Long neurotoxin OH-57; Flags: Precursor	60.8	60.8	46%	6e-12	43%
P01383.1	RecName: Full=Long neurotoxin 1; AltName: Full=Neurotoxin 3.9.4	60.1	60.1	44%	8e-12	44%
P01393.1	RecName: Full=Long neurotoxin 1; AltName: Full=Toxin V-III-N1	60.1	60.1	43%	8e-12	47%
P01397.2	RecName: Full=Long neurotoxin 2; AltName: Full=Neurotoxin delta; A	60.1	60.1	38%	8e-12	46%
P82662.2	RecName: Full=Long neurotoxin OH-6A/OH-6B; AltName: Full=Alpha-	60.5	60.5	46%	9e-12	45%
A1IVR8.1	RecName: Full=Alpha-delta-bungarotoxin-1; Flags: Precursor	59.3	59.3	46%	2e-11	40%
Q53B54.1	RecName: Full=Long neurotoxin OH-17	58.9	58.9	42%	2e-11	50%
P25667.1	RecName: Full=Long neurotoxin 3; AltName: Full=Toxin VN2	58.9	58.9	36%	2e-11	49%
A1IVR7.1	RecName: Full=Alpha-delta-bungarotoxin-3; Flags: Precursor	58.9	58.9	46%	2e-11	40%
Q2VBP4.1	RecName: Full=Long neurotoxin LNTX28; Flags: Precursor	59.3	59.3	46%	3e-11	44%
A1IVR9.1	RecName: Full=Alpha-delta-bungarotoxin-2; Flags: Precursor	57.8	57.8	46%	7e-11	40%
Q53B59.1	RecName: Full=Long neurotoxin OH-37; Flags: Precursor	58.2	58.2	46%	7e-11	44%
P01394.1	RecName: Full=Long neurotoxin 1; AltName: Full=Neurotoxin 4.7.3/4	57.4	57.4	43%	9e-11	46%
Q9YGI8.1	RecName: Full=Short neurotoxin homolog NTL4; Flags: Precursor	57.0	57.0	45%	2e-10	41%

The alignment required the introduction of gaps in two places to optimize the lining up of identical amino acids. Without those gaps, many random alignments would be made and the scoring would have been a lot lower. Notably, a number of cysteines are aligned between the query and the hit, although there are two in the region with the gaps that do not align. Cysteines play a major role in holding folded proteins in place because they form disulfide bonds, covalently bonding distant regions to each other. The + signs indicate conservative substitutions between these two protein sequences, so the similarity is good.

Our query is most probably a neurotoxin: from the millions of protein sequences in the database, the similarity is to a protein from the same kind of animal and a protein we would expect to find encoded by an EST from the venom gland. The E value is not spectacularly low, but the simplest interpretation is that you are on the right track for identifying this EST.

Looking at Figure 5.9, similarity between the query and the hit started at nucleotide 70 and ended at 273. The total length of the query is 435 nucleotides, as can

Figure 5.8 The BLASTX results table of FL590802 versus UniProtKB/ Swiss-Prot(swissprot).

```
>sp|P01379.2|NXL1_LATSE  RecName: Full=Long neurotoxin Ls3; AltName: Full=LsIII;
AltName: Full=Neurotoxin 1; AltName: Full=Neurotoxin alpha; Flags:
Precursor
Length=87

 Score = 67.0 bits (162),  Expect = 2e-14
 Identities = 31/70 (44%), Positives = 41/70 (59%), Gaps = 4/70 (6%)
 Frame = +1

Query  70   GYTRKCYEGEGTRKSVTCPKGEKVCYTIFLVGP--SHPAKVLKWGCAASCPKVGLGARIS  243
            GYTR+CY        + TCP G+++CY      S   KVL++GCAA+CP V  G  I
Sbjct  19   GYTRECYLNP--HDTQTCPSGQEICYVKSWCNAWCSSRGKVLEFGCAATCPSVNTGTEIK  76

Query  244  CCSKDNCNSH  273
            CCS D CN++
Sbjct  77   CCSADKCNTY  86
```

Figure 5.9 The BLASTX alignment between query FL590802 and the hit, Long neurotoxin Ls3. There are two pairs of gaps that optimize this alignment, along with many + signs signifying conservative substitutions. Note that many of the cysteines align, suggesting a similarity in tertiary structure.

Figure 5.10 A "sketch" of the BLASTX alignment between EST FL590802 and protein P01379. This crude representation of the alignment will often help with the interpretation of the query–hit relationship.

```
                            1          70                    273                    435
EST FL590802            5P----------------------------------------------------------3P
                                        ||||||||||||||||||||
P01379                                  ---------------------------
                            1          19                    86
```

be seen either at the top of the BLAST results page or in the GenBank format of the nucleotide. So this leaves 5P and 3P regions with no apparent similarity to proteins. The hit, the snake neurotoxin, has a length of 87 amino acids. The coordinates of the alignment show that the first 18 amino acids do not align with the EST translation but the alignment then continues to amino acid 86, almost the last amino acid of the protein. This alignment is "sketched" in **Figure 5.10**.

Notice, again, that the numbering of the query and subject lines is quite different. The query is DNA sequence that was translated into amino acids but the numbering of the query line corresponds to the DNA sequence coordinates. There are over 170 nucleotides represented in the first line in Figure 5.9 but roughly one-third this number of amino acids in the lower subject line. Each gap in the query line corresponds to one codon or three nucleotides, while a gap in the subject line corresponds to a single amino acid. It is important to do the math correctly (**Box 5.2**).

The sketch in Figure 5.10 indicates that the EST most probably includes the 3P UTR of the neurotoxin transcript. This assumes that the neurotoxin encoded by the query is like the hit and is approximately 87 amino acids long. If this is true, the remaining nucleotides after coordinate 273 do not encode any more protein sequence. Indeed, the rest of the neurotoxin hits in the BLASTX results range in length from 68 to 94 amino acids, so there appears to be adequate DNA in our query to encode a longer protein and still have plenty of 3P untranslated DNA sequence remaining. Is there a terminator codon after the query's predicted coding region? Look at the DNA record and determine this.

The sketch also indicates that our query's translation of the 5P end does not align with the first 18 amino acids of the hit. In fact, it fails to align to any of the hits in the results and all of these alignments are missing approximately the same length of this N-terminus. This is clearly seen in the graphics (Figure 5.7). What is going on here? To explain, we have to go off topic for a few moments. The database searched was UniProtKB/Swiss-Prot(swissprot), from the Swiss Institute of Bioinformatics. This database was introduced earlier (Chapter 4) when a BLASTP search was run at the ExPASy Website. When you launched the recent BLASTX search, the NCBI BLASTX form did not reach across the Internet and search the database at the ExPASy Website. The NCBI has a copy of the UniProtKB database and that is what was searched. The NCBI also formatted the UniProtKB annotation to the NCBI style. By clicking on the results hypertext for the first hit, you are brought to the NCBI version of the UniProtKB record, a part of which is seen in **Figure 5.11A**. At the end of the "DBSOURCE" line is the hyperlink to the original UniProtKB record. Click on that hypertext and you are taken to the ExPASy Website; a portion of the UniProtKB record is shown in Figure 5.11B.

What is seen in the description line of this BLASTX hit (Figure 5.9) are some of the lines from the NCBI's Swiss-Prot record. "RecName" and "AltName" stand for "Recommended name" and "Alternative name," respectively, and you can see these in the ExPASy record in Figure 5.11B. In Figure 5.9, the NCBI also introduces a field, "Flags," to bring to your attention any particular features of this database entry that are important. In this case, the flag "Precursor" indicates that the protein is "further processed into a mature form" using the exact words from the ExPASy record.

This is a key fact in understanding the failure to align the EST 5P end and the N-terminus of the hit. By examining the UniProtKB record, it is seen that the N-terminus of this hit, and many other snake venom neurotoxins, is a signal

Box 5.2 Do the math (correctly!)

When calculating the lengths of the lines in BLAST alignments, it is important to consider the coordinates as molecular coordinates, not simply numbers which can be subtracted from each other. For example:

```
Query  1     MDGWRRMPRWGLLLLLWGSCTFGLPTDTTTFKRIFLKRMPSIRESLKERGVDMARLGPEW  60
```

There are 60 amino acids in this line, the standard for BLASTP. If asked how long is this line, you would *not* do the following: 60 − 1 = 59. Instead, you would remember to add one: 60 − 1 (+ 1) = 60.

So consider this line, below, taken from a BLASTP alignment:

```
Sbjct  8     SAVTALWGKVNVDEVGGEALGRLLVVYPWTQRFFESFGDLSTPDAVMGNPKVKAHGKKVL  67
```

How long is this sequence? To calculate the correct length of an alignment, you must add one: 67 − 8 (+ 1) = 60, the correct length.

With BLASTX, the query line represents translated nucleotides, and the nucleotide coordinates are used. The subject line is a protein sequence, and the amino acid coordinates are used. Consider the following alignment from BLASTX:

```
Query  51    MVHLTPEEKSAVTALWGKVNVDEVGGEALGRLLVVYPWTQRFFESFGDLSTPDAVMGNPK  230
             MVHLTPEEKSAVTALWGKVNVDEVGGEALGRLLVVYPWTQRFFESFGDLSTPDAVMGNPK
Sbjct  1     MVHLTPEEKSAVTALWGKVNVDEVGGEALGRLLVVYPWTQRFFESFGDLSTPDAVMGNPK  60
```

Since groups of three nucleotides (codons) are translated into single-letter amino acid abbreviations, the query coordinates will reflect a sequence of 180 nucleotides. In the example, above, the first amino acid (M) is encoded by nucleotides 51, 52, and 53. The last amino acid (K) is encoded by nucleotides 228, 229, and 230. The subject protein is 60 amino acids long and the coordinates reflect this. How many nucleotides are translated in the query line, above? Remembering to add one: 230 − 51 (+ 1) = 180.

Things get a little more complicated when there are gaps. If the line has *protein* coordinates (for example, in BLASTP, or the subject line in BLASTX) then each gap (−) counts as one. The example, below, has 60 *positions* but four of them are gaps. Thus, 307 − 252 (+1) = 56. Then 56 + 4 gaps = 60 *positions*.

```
Query  252   YVSISKPGSWQIRMKGVSV--RSTTLLCEE--GCMVVVDTGASYISGPTSSLRLLMEALG  307
```

Why isn't the last coordinate, above, 311 instead of 307, to reflect that there are four gaps? It is because these are coordinates within the *protein*, not the BLAST search results. Even if there were 15 gaps in the above line, the 307th amino acid of this protein would still be a glycine (G) at coordinate 307.

Now consider the alignment of a BLASTX result (below). The query line is a nucleotide sequence that is translated into amino acids. The subject line is the amino acid sequence of a protein. Gaps are necessary to align the translation with a protein sequence, each gap representing three nucleotides.

```
Query  51    MVHLTPEEKSAVTALWGKVNVDEVGGEALGR----LLVVYPWTQRFFESFGDLSTPDAVM  218
             MV L+ EEK+A+ ALWGKVN +E+GGEALGR    LL+VYPWTQRFF+SFG L+T  AV
Sbjct  1     MVQLSGEEKTAILALWGKVNEEIGGEALGRVVSRLLIVYPWTQRFFDSFGHLTTSAAVT  60
```

Do the math: 218 − 51 (+ 1) = 168. Four gaps × three nucleotides = 12. Then 168 + 12 = 180 *positions*.

peptide (see **Figure 5.12**). This protein sequence directs the newly translated protein to be secreted outside the cell. Again, this makes sense biologically. Venom is a liquid that is stored in a venom gland, and for neurotoxin to be present here, it must be secreted from cells into the liquid. During this secretion process, the signal sequence is cleaved off, hence the description "processed." A characteristic of secretion signals is that the sequence is not conserved over evolution, only the biochemical properties. So secretion signals do not align well between related proteins. It is possible that the secretion signal of our query is encoded by the EST. According to the sketch in Figure 5.10, there are 69 nucleotides upstream of the BLASTX alignment, which is an adequate length to encode a signal sequence of at least 18 amino acids, the length of the signal sequence in the hit. Although

Figure 5.11 The top portion of the P01379 UniProtKB record.
(A) The NCBI version of the Swiss-Prot record for the sea krait neurotoxin. At the end of the DBSOURCE line is the hypertext link (underlined) to the record at the ExPASy Website. (B) A portion of the ExPASy version of the same record. Note the difference in common names between the two records.

(A)
```
LOCUS       NXL1_LATSE                    87 aa        linear   VRT 27-JUL-2011
DEFINITION  RecName: Full=Long neurotoxin Ls3; AltName: Full=LsIII; AltName:
            Full=Neurotoxin 1; AltName: Full=Neurotoxin alpha; Flags:
            Precursor.
ACCESSION   P01379
VERSION     P01379.2  GI:25108918
DBSOURCE    UniProtKB: locus NXL1_LATSE, accession P01379;
            class: standard.
            extra accessions:O93496,Q7T2I4
            created: Jul 21, 1986.
            sequence updated: Nov 15, 2002.
            annotation updated: Jul 27, 2011.
            xrefs: AB015512.1, BAA32991.1, AB015513.1, BAA32992.1, AB098531.1,
            BAC78204.1, 1LSI_A
            xrefs (non-sequence databases): PDBsum:1LSI,
            ProteinModelPortal:P01379, SMR:P01379, HOVERGEN:HBG006553,
            GO:0035792, GO:0030550, GO:0009405, InterPro:IPR003571,
            InterPro:IPR018354, Pfam:PF00087, PROSITE:PS00272
KEYWORDS    3D-structure; Acetylcholine receptor inhibitor; Direct protein
            sequencing; Disulfide bond; Neurotoxin; Postsynaptic neurotoxin;
            Secreted; Signal; Toxin.
SOURCE      Laticauda semifasciata (broad-banded blue sea krait)
```

(B)

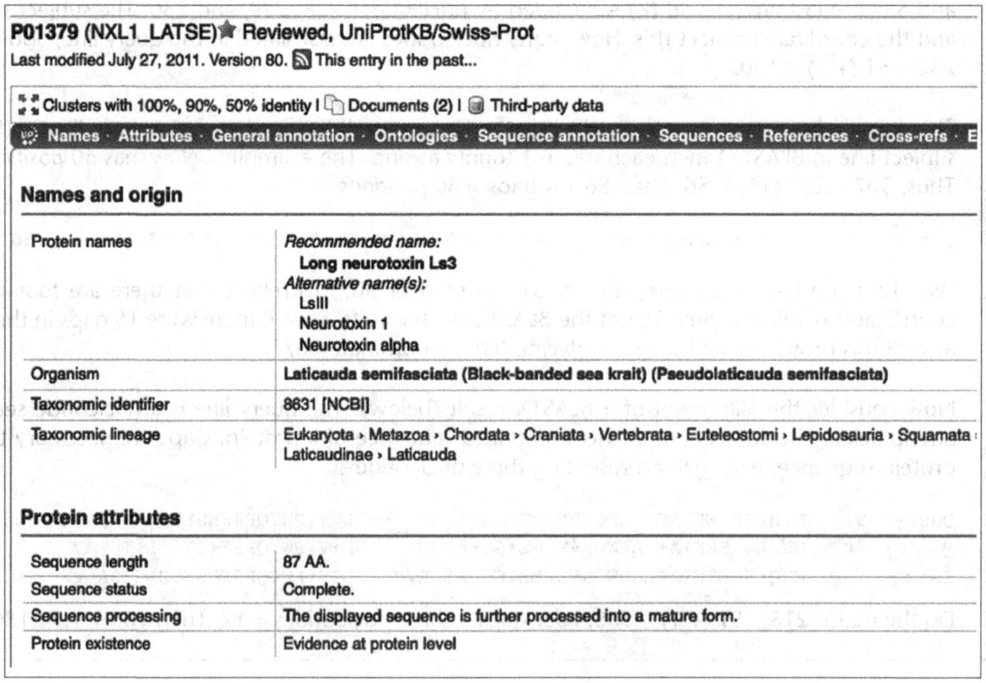

P01379 (NXL1_LATSE) ★ Reviewed, UniProtKB/Swiss-Prot

Last modified July 27, 2011. Version 80. 🔊 This entry in the past...

⚏ Clusters with 100%, 90%, 50% identity I ▢ Documents (2) I ▤ Third-party data

Names · Attributes · General annotation · Ontologies · Sequence annotation · Sequences · References · Cross-refs · E

Names and origin

Protein names	*Recommended name:* **Long neurotoxin Ls3** *Alternative name(s):* LsIII Neurotoxin 1 Neurotoxin alpha
Organism	**Laticauda semifasciata (Black-banded sea krait) (Pseudolaticauda semifasciata)**
Taxonomic identifier	8631 [NCBI]
Taxonomic lineage	Eukaryota › Metazoa › Chordata › Craniata › Vertebrata › Euteleostomi › Lepidosauria › Squamata › Laticaudinae › Laticauda

Protein attributes

Sequence length	87 AA.
Sequence status	Complete.
Sequence processing	The displayed sequence is further processed into a mature form.
Protein existence	Evidence at protein level

intended to be only a sample of what is transcribed in the venom gland, this EST might be long enough to encode the entire length of this relatively short protein.

The annotation in the UniProtKB record shows the secretion signal both in coordinates and graphically (Figure 5.12). Here, the signal peptide is defined as amino acids 1–21. Below this are the mapped disulfide bonds of the cysteines, showing the coordinates and small vertical lines on the horizontal lines of the graphic. Under "Amino acid modifications" in this figure, notice that the distant amino acids (for example, the cysteines at positions 24 and 41) are linked by double-headed arrows. Together with the other pairs, the cysteines hold this short protein sequence in a tight structure for proper function. It is interesting

Sequence annotation (Features)

	Feature key	Position(s)	Length	Description	Graphical view	Feature identifier
Molecule processing						
▢	Signal peptide	1 – 21	21	(Ref.3)		
▢	Chain	22 – 87	66	Long neurotoxin Ls3		PRO_0000035430
Amino acid modifications						
▢	Disulfide bond	24 ↔ 41		(Ref.5)		
▢	Disulfide bond	34 ↔ 62		(Ref.5)		
▢	Disulfide bond	47 ↔ 51		(Ref.5)		
▢	Disulfide bond	66 ↔ 77		(Ref.5)		
▢	Disulfide bond	78 ↔ 83		(Ref.5)		

Figure 5.12 A portion of the UniProtKB annotation for P01379. The signal sequence information along with the disulfide bonds appear in this part of the record.

that the protein encoded by the EST is missing two of these cysteines (mentioned above), suggesting that the three-dimensional structure may be different than the hit.

Using the annotation of sequence records

Let's do another BLASTX search using an EST from *Apis mellifera*, the honeybee. Examine the annotation for sequence record DB773907. The annotation does not identify the gene that this transcript is from, which is the case for most ESTs. If you are viewing DB773907 in the EST format, change it to GenBank format so you can see the length: 547 nucleotides. The cDNA library was made from the head of a honeybee. Run the BLASTX with the default settings, searching the "Nonredundant protein sequences (nr)" database (**Box 5.3**).

The BLASTX graphic tells you something immediately about your query (**Figure 5.13**). It appears that only the first 150 nucleotides, when translated to amino acids, are finding anything in the database. Notice, too, that the query length on this graphic is the nucleotide length, not the translated length: approximately $547/3 = 182$ amino acids. Without looking at the alignments or the annotations of the hits, there are several possible interpretations: (a) although the 5P end of the EST aligns with known proteins, the remaining bases encode something novel; (b) the 5P end encodes something real but there was a cloning or sequencing artifact and the rest of the sequence is from the vector, for example; or (c) this EST bridges the junction between coding (first 150 nucleotides) and noncoding (3P UTR) regions.

The information from the first alignment and associated annotation tells the story. Based on the 100% identity, this EST sequence is from a member of the royal jelly family of proteins. Identity to the royal jelly protein sequence stops at

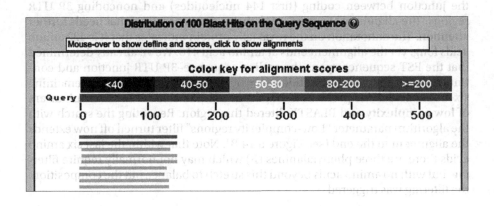

Figure 5.13 BLASTX search results of DB773907 against the NCBI nr protein database. Only the first 150 nucleotides of the query hit anything in the database.

Box 5.3 Reference, non-redundant, and special collection databases

The NCBI RefSeq databases are intended to provide a reference sequence, representing the final word on the sequence of an mRNA, protein, or gene. Through both automated and manual efforts, the RefSeq databases provide a place to go where, with good confidence, you may find a complete and error-free sequence. These databases are also free of duplications. Even though human beta globin has been sequenced and deposited in GenBank hundreds of times, there is only one human beta globin mRNA, protein, and gene sequence in RefSeq. But each alternatively spliced mRNA (variant) and corresponding protein (isoform) has a separate file and accession number at the NCBI.

At the ExPASy Website, the UniProtKB database has two divisions: (1) Swiss-Prot records, which are both automatically and manually reviewed protein sequence records; and (2) TrEMBL records, or automatically derived translation records of the EMBL DNA database. Swiss-Prot records should be thought of as the final word, much like RefSeq sequences, with no redundancy and no errors. But TrEMBL records may have protein fragments and errors. Swiss-Prot handles isoforms by having one record for a protein, for example accession number Q9UKP4 for human ADAMTS7, but suffixes for the different isoforms. For example, Q9UKP4-1 is thought of as the "canonical" representation of the protein, while Q9UKP4-2 is an isoform.

Non-redundant (nr) databases are collections with a stated goal of providing only one copy of a sequence, particularly when found by database-searching tools such as BLAST. Non-redundant databases are automatically generated. When two or more sequences are found to have the exact same sequence, the accession numbers and brief annotations of the records appear in the same record so you can see the origins of the individual sequences. But the sequence of these multiple records appears only once in an nr database. However, should two sequences differ

by a single nucleotide or amino acid, they are thought of as unique and are not combined. For an example, there are dozens of human beta globin sequences in the NCBI nr protein database, differing by one or more amino acids. But there are over 70 human beta globin records which have identical sequence and have been combined to only one record.

Population studies often generate many records showing the variation of sequence in a population. Of the human beta globins, 36 came from a study of sequence variation in Melanesia. You would not want these variations to disappear into one file, even if they only differed by one amino acid. In addition, there are mistakes in the databases. A change in sequence can come about via technology or human error. So having all the sequences available to you is a great advantage when you find something unusual.

Importantly, the nr database is the collection you should search if you can't find a sequence elsewhere. A BLASTP search of RefSeq can yield no hits while the same query will find hundreds of hits in an nr database. If a sequence originates from a model organism, there is a good chance that extra efforts to curate and finalize this sequence will take place and it may end up in a reference collection. However, newly discovered sequences or sequences from organisms not considered "models" will often not get the same attention and will stay in the nr database.

There are *special databases* with sequences isolated from the rest by virtue of their sheer numbers. ESTs and Metagenomics DNA sequences are two databases which house the absolutely enormous output from these exploratory efforts. You should not find these sequences in the nr database. However, should an EST cDNA be re-sequenced and studied, the new and improved sequence may appear in the nr database or even RefSeq.

Royal jelly

Royal jelly is a substance secreted by bees and fed to all bee larvae during their first three days of development. Should larvae be fed royal jelly for the remainder of their development, they become queens, while those who stopped receiving royal jelly at day three are destined to become sterile worker bees. An interesting protein family!

nucleotide 144 of the query, aligned with amino acid 416, which is very close to the end of the protein sequence (see **Figure 5.14A**). Remember that the query is 547 nucleotides long and only the first 144 are needed (translated) to align with the (near) end of the protein. So interpretation (c) is correct: this EST bridges the junction between coding (first 144 nucleotides) and noncoding 3P UTR regions. The alignment (Figure 5.14A) shows one more thing that needs further attention. The annotation on the royal jelly protein indicates that it is 422 amino acids long, yet the alignment ends at amino acid 416. We've already determined that the EST sequence spans the coding sequence–3P UTR junction and continues for another 400 nucleotides, so why didn't BLASTX show the remaining six amino acids? The problem is that the last amino acids at the C-terminus are of **low complexity** and BLASTX filtered this region. Repeating the search with the algorithm parameter "Low-complexity regions" filter turned off now extends the alignment to the end (see Figure 5.14 B). Note that within the last six amino acids there are three phenylalanines (F) which may not normally require filtering, but with no amino acids beyond this stretch to balance out the composition, the filtering was triggered.

(A)

```
>gb|ACD84800.1| major royal jelly protein 9 [Apis mellifera]
Length=422

 Score =  104 bits (260),  Expect = 4e-21
 Identities = 48/48 (100%),  Positives = 48/48 (100%),  Gaps = 0/48 (0%)
 Frame = +1

Query  1    NEYIWIVSNKYQKIANGDLNFNEVNFRILNAPVNQLIRYTRCENPKTN  144
            NEYIWIVSNKYQKIANGDLNFNEVNFRILNAPVNQLIRYTRCENPKTN
Sbjct  369  NEYIWIVSNKYQKIANGDLNFNEVNFRILNAPVNQLIRYTRCENPKTN  416
```

(B)

```
Query  1    NEYIWIVSNKYQKIANGDLNFNEVNFRILNAPVNQLIRYTRCENPKTNFFSIFL  162
            NEYIWIVSNKYQKIANGDLNFNEVNFRILNAPVNQLIRYTRCENPKTNFFSIFL
Sbjct  369  NEYIWIVSNKYQKIANGDLNFNEVNFRILNAPVNQLIRYTRCENPKTNFFSIFL  422
```

Figure 5.14 BLASTX alignment between EST DB773907 and royal jelly protein 9. (A) This honeybee protein is 422 amino acids long (indicated in the hit annotation, "Length=422") yet the BLASTX alignment does not reach the end of the reading frame. (B) Repeating the same BLASTX search with the "Low complexity regions" filter turned off now shows the alignment to the end of the reading frame (Query) and the C-terminus (Sbjct). The six amino acids that were filtered are underlined.

In this case, using the annotation was critical to understanding how to interpret the fine details of the alignment and to knowing that the BLASTX needed repeating. The length of the royal jelly protein gave a clue that filtering was hiding some of the results.

BLASTX alignments with the reverse strand

If the reverse (or negative) strand of a BLASTX DNA query is translated into a virtual protein and used to align with the subject (protein), then the query line goes in reverse. BLASTX indicates which frame is used in the translation of the Query sequence in the "Frame =" field, just above the first lines of the alignment. For example, below is a BLASTX alignment and the frame is "–1" (see **Figure 5.15**). This says that a translation of the negative strand is used to align to the protein hit. Since the translation is from the negative strand, the numbering decreases, going from 692 to 513. The numbering of the protein (the BLASTX Sbjct) is *never* in reverse.

5.5 TBLASTN

A database has tens of thousands of EST sequences, and one of these ESTs may be your sequence of interest. Now what? If both your sequence of interest and the EST database are from a well-characterized organism that has had many, if not all, of its genes sequenced, then a BLASTN search could probably be used to identify ESTs through sequence identity. But if you want to analyze sequences from an organism which has not been studied extensively, BLASTN may not be able to help you. Particularly if the evolutionary distance between your query and the target EST is significant.

Well-equipped computer laboratories could contemplate taking ESTs from a newly characterized organism and running BLASTX searches, one for each EST. Computer programs would automatically run these searches, identify the best hits in a protein database, and then annotate the EST records. These

Figure 5.15 An example of a BLASTX alignment with a translation from Frame –1. The coordinate numbering of the nucleotide query decreases, from left to right, 692–513, while the protein coordinates (Sbjct) increase from left to right, as normal.

```
Frame = -1

Query  692  NIELHLEQDLDHGKLGRGCWRHIAEQRVMEYCAEMEAPDGSFLVRESGTYVGDYVLSIWW  513
            N ELH ++  HGKL G R  A+Q + Y     DG+FLVRES T+VGDY LS W
Sbjct  620  NDELHFGENWFHGKLEGG--RQEADQLLETY---KHLGDGTFLVRESATFVGDYSLSFWR  674
```

"high-throughput" approaches are most often performed by the laboratory generating these EST sequences by the thousands. Many laboratories have their own extensive computing power and copy the public databases and programs to their servers. These laboratories have people skilled in bioinformatics programming, database design, and analysis. They can then set up thousands of BLAST searches or other analysis programs to run in a matter of hours to minutes.

What about translating these EST sequences? After all, protein sequence is often more conserved than DNA sequence. Historically, protein databases began with sequences which were determined by sequencing proteins directly. When DNA sequences became inexpensive to generate (in money and in effort), cDNA was translated computationally and these protein sequences started to dominate the databases. In time, it became easier to predict the coding regions of genes (for example, RefSeq mRNA sequence accession numbers usually begin with "XM_"), so predicted protein sequences (for example, RefSeq sequence accession numbers usually begin with "XP_") now appear in databases as well.

But translating EST sequences is difficult. As described earlier, the sequence quality is not always the best, the orientation can be uncertain, and the reading frame is not defined and may shift with indels, so placing all these translations into a permanent database is just not a good idea. Yet, ESTs can be a gold mine of genes, waiting to be discovered. How can you use a protein query to probe this large amount of information? TBLASTN solves this problem in an incredible way: it translates an entire DNA database, in all six reading frames, to create a temporary protein database to search with your protein query. As you can imagine, this is computationally intensive compared to BLASTN or BLASTP. It is as if you are searching each EST with six BLAST searches.

What about all the poor-quality concerns and uncertainty of translations described in the last paragraph? Some potential misalignments due to sequence problems or laboratory-introduced artifact can be compensated for, if allowed by the algorithm. Upon encountering a termination codon, TBLASTN will skip over this codon and resume translation in that frame until it runs out of DNA sequence. A "*" then appears in the subject line of the alignment where the terminator was encountered. Like other BLAST programs, TBLASTN will introduce a gap in the protein sequence or translated DNA in order to maintain the alignment between query and hit.

A TBLASTN search

LIM homeobox transcription factor 1-beta is a protein that contributes to proper limb development in humans and is widely conserved in nature. In fact, if you were to take a human reference sequence for this protein, NP_001167617, and run a BLASTP search, you can easily detect homologous proteins in insects. But what about a phylum that is more distant in evolution, and with no apparent homologous structure like a limb?

Let's use TBLASTN to examine the ESTs found in sponges. According to www.timetree.org, it has been over a billion years since we last shared a common ancestor with the sponge. Sponges don't have limbs but perhaps we could detect some distant homolog with a human protein. The genes of sponges have not been well characterized, but according to the NCBI Taxonomy Browser, there are over 110,000 ESTs from this phylum (*Porifera*).

Go to the NCBI Web form for TBLASTN and enter NP_001167617 into the query window. For the database choice, pick "Expressed sequence tags." Since we are only interested in sponge ESTs, narrow the search considerably by entering "Porifera" in the organism field. Even though TBLASTN is computationally intensive, there are not that many *Porifera* ESTs to translate so the search results arrive quickly. Many of the hits from the search are from *Amphimedon queenslandica*. The annotation from these records indicates that they are from a

```
>gb|GW174989.1|  CAYI2890.b1 CAYI Amphimedon queenslandica competent larva (one
day old) (L) Amphimedon queenslandica cDNA clone CAYI2890 5', mRNA sequence.
Length=711

 Score =  115 bits (287),  Expect = 9e-26, Method: Compositional matrix adjust.
 Identities = 49/118 (41%), Positives = 75/118 (63%), Gaps = 1/118 (0%)
 Frame = +2

Query  53    PAVCEGCQRPISDRFLMRVNESSWHEECLQCAACQQALTTSCYFRDRKLYCKQDYQQLFA  112
             P +C GC  PI +RFLM+V + SWH +C++C+ CQ  L+  C+ RD KLYC+ D+ + +
Sbjct  47    PFICAGCSEPIMERFLMKVLDKSWHVQCVKCSDCQCLLSEKCFSRDNKLYCRSDFFRQYG  226

Query  113   AKCSGCMEKIAPTEFVMRALECVYHLGCFCCCVCERQLRKGDEFVLKEGQ-LLCKGDY    169
             +C+ C E + P + V R +  +YH+ CF C VC+RQL  G++  L +G+  LC   Y
Sbjct  227   TQCASCKEGLCPEDLVRRGVNKIYHVQCFKCSVCQRQLNTGEQLYLVQGEKFLCDSCY    400
```

Department of Energy "Joint Genome Institute *Amphimedon queenslandica* EST project." This species of sponge is from the Australian Great Barrier Reef and the genome was recently sequenced. **Figure 5.16** shows the alignment with a strong hit from this search.

Let's look closely at this alignment. The Query, NP_001167617, is 406 amino acids long and this alignment shows amino acids 53 through 169, so the alignment is toward the N-terminus of the Query. The Subject line is the translated EST so the coordinates are from the nucleotide sequence of the EST. There is sufficient DNA sequence in the EST to encode more of the protein; the EST length is 711 nucleotides but the similarity stops at nucleotide 400. This alignment is from Frame +2. No other alignments from this hit are present in TBLASTN (for example, Frame +1 or +3) so it is unlikely that this level of similarity continues after a frameshift in the EST. Perhaps the sponge protein-coding region ends here and the rest is 3P UTR, although the DNA sequence of the hit at this spot does not show a stop codon (**Figure 5.17**).

Figure 5.16 shows that there is 41% identity between the query and the hit, but the similarity is 63%, a big jump. The alignment scoring was strengthened by all the aligned cysteines, which were described earlier as important for tertiary structure and proper folding. There are only a few cysteines in each sequence that are unaligned.

Don't forget to look at the annotation of both the query and this hit. As mentioned earlier, the LIM homeobox transcription factor is highly conserved in nature and has been the subject of many studies. The homeobox protein family is large and its members are responsible for a number of developmental events, so the likelihood of this hit being the ortholog is low. However, it does look similar enough to characterize it as a possible family member.

Finally, let's use another feature of the NCBI BLAST forms. We will now concentrate on a species of sponge other than *Amphimedon queenslandica*. Look next to the field where you typed in "Porifera." There is a "+" symbol, and clicking it will create an additional organism field. Adjacent to this field is a box labeled "Exclude" (see **Figure 5.18**). Check this box and fill in *Amphimedon queenslandica* then resubmit this TBLASTN search. When the results return, you will see there are no *Amphimedon queenslandica* hits listed. This feature allows easier access to lower-scoring hits that may be overpowered by sequences from a highly sequenced species.

Figure 5.16 TBLASTN alignment of NP_001167617 with the sponge EST GW174989. The last three amino acids for the Subject (Sbjct) are underlined for reference to Figure 5.17.

Figure 5.17 Isolated region of the sponge EST. Nucleotides 361–420 of the sponge EST, GW174989, obtained from the EST record. Shown here is the region from the last part of the TBLASTN alignment, translated in Figure 5.16, with the serine, cysteine, and tyrosine underlined. The alignment between the sponge EST and the human transcription factor ends at nucleotide 400. The Frame +2 translation is shown below the nucleotides. Alternating triplets are underlined for easier viewing. No stop codons are seen in any forward frame immediately after this translation.

```
361 cctagttcaaggagagaagttcctatgtgacagttgctatccagctcccgccccatcaca 420
      L  V  Q  G  E  K  F  L  C  D  S  C  Y  P  A  P  A  P  S
```

Figure 5.18 Excluding organisms in BLAST. Shown here is the Organism selection area of the NCBI BLAST Web form. By including a species or taxonomic group, you narrow your search to what is entered. By checking the "Exclude" box you block sequences from the entered species or group from appearing in your BLAST results.

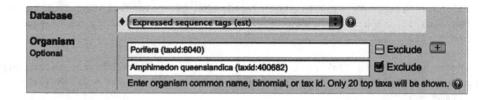

Metagenomics and TBLASTN

To further illustrate the use of TBLASTN, as well as demonstrate a possible workflow for a project, consider the following scenario. You are a scientist and believe that novel cellulase enzymes can be harnessed to break down material from crops that is normally discarded (for example, roots, stems, leaves). Once the cellulose is broken down, conversion to biofuels such as ethanol can proceed efficiently, thereby generating a large supply of energy.

First, to survey the legal landscape around this approach, you want to search for the bacterial cellulases that have already been patented. You start with your cellulase protein sequence of choice, YP_001376471, from the bacteria *Bacillus cytotoxicus*. You paste this protein accession number into the NCBI BLASTP Web form and select "Patented Protein Sequences (pat)" from the database dropdown menu. Performing this search, you get a list of hits (see **Figure 5.19**) with very little annotation. However, the annotation includes the patent numbers, allowing you to examine these patents to determine why these sequences were patented. Based on the number of high-quality hits, you fear that there might be patents covering the cellulases of *B. cytotoxicus* already.

Figure 5.19 BLASTP search results after searching the patent division of GenBank. Note the sparse sequence descriptions.

Now you want to go looking for cellulases quite far from the common proteins that have been already identified and patented. These novel proteins might have additional properties allowing easier transition to industrial production. To accomplish this, you turn to the DNA sequences generated from environmental sequencing, or **metagenomics**.

Sequences producing significant alignments:						
Accession	**Description**	**Max score**	**Total score**	**Query coverage**	**E value**	**Links**
ACP62620.1	Sequence 7114 from patent US 7494798	612	612	100%	6e-175	
AAR59171.1	Sequence 5705 from patent US 6617156	462	462	96%	1e-129	
ADA17908.1	Sequence 5302 from patent US 7608276	397	397	99%	4e-110	
ABH89371.1	Sequence 4744 from patent US 7060458 >gb|ABP09460.1| Sequence 4744 from patent US 71	382	382	97%	1e-105	
ACP62722.1	Sequence 7216 from patent US 7494798	368	368	97%	2e-101	
AAR59318.1	Sequence 5852 from patent US 6617156	293	293	98%	1e-78	
AAE60668.1	Sequence 1 from patent US 6194190	261	261	94%	2e-69	
AAE60675.1	Sequence 10 from patent US 6194190	261	261	94%	3e-69	
CAT69568.1	unnamed protein product [Thermococcus kodakarensis KOD1]	247	247	94%	8e-65	
AAT15610.1	Sequence 2980 from patent US 6699703 >gb|ABI02941.1| Sequence 2980 from patent US 707	240	240	96%	7e-63	
ADC19454.1	Sequence 157 from patent US 7638136	237	237	96%	7e-62	
ABH90529.1	Sequence 5902 from patent US 7060458 >gb|ABP10618.1| Sequence 5902 from patent US 71	236	236	98%	9e-62	
AAW09068.1	Sequence 4464 from patent US 6800744	235	235	96%	2e-61	
ADA15701.1	Sequence 888 from patent US 7608276	223	223	98%	1e-57	
AAR58625.1	Sequence 5159 from patent US 6617156	220	220	98%	8e-57	
AAQ45075.1	Sequence 5633 from patent US 6583275	216	216	45%	1e-55	
ABH79717.1	Sequence 9560 from patent US 7041814	186	186	96%	1e-46	
CAT68961.1	unnamed protein product [Thermococcus kodakarensis KOD1]	185	185	95%	4e-46	G
ACK26594.1	Sequence 102 from patent US 7455992	177	177	97%	7e-44	
AAQ43253.1	Sequence 3811 from patent US 6583275	86.7	86.7	19%	2e-16	
AAQ44830.1	Sequence 5388 from patent US 6583275	86.3	86.3	24%	2e-16	
CAT69562.1	unnamed protein product [Thermococcus kodakarensis KOD1]	80.5	80.5	83%	1e-14	G
ADA17390.1	Sequence 4266 from patent US 7608276	76.3	76.3	83%	2e-13	
ACP60722.1	Sequence 5216 from patent US 7494798	74.7	74.7	95%	6e-13	
ACG81759.1	Sequence 259 from patent US 7384775	73.9	73.9	83%	1e-12	
ACQ32266.1	Sequence 24780 from patent US 7517684 >gb|ADA94297.1| Sequence 24780 from patent US	63.9	63.9	61%	1e-09	
AAT17920.1	Sequence 5290 from patent US 6699703 >gb|ABI05251.1| Sequence 5290 from patent US 707	46.6	46.6	53%	2e-04	
ACQ24948.1	Sequence 17462 from patent US 7517684 >gb|ADA86979.1| Sequence 17462 from patent US	46.2	46.2	23%	2e-04	
CAT72166.1	unnamed protein product [Thermococcus kodakarensis KOD1]	36.2	36.2	19%	0.29	
CAI54195.1	unnamed protein product [Ammonifex degensii]	33.9	33.9	20%	1.3	
ABT78527.1	Sequence 165997 from patent US 7214786	33.5	33.5	14%	1.4	
ACC31159.1	Sequence 32 from patent US 7348420 >gb|ACS07389.1| Sequence 32 from patent US 753487	32.7	32.7	21%	2.9	
CAY08024.1	unnamed protein product [Candida glabrata CBS 138] >emb|CBN65509.1| unnamed protein pr	32.0	32.0	34%	4.3	G
CBM35783.1	unnamed protein product [Oryza sativa]	31.6	31.6	8%	6.8	
ADF22110.1	Sequence 4 from patent US 7674959	31.2	31.2	12%	8.2	
CAS91887.1	unnamed protein product [Leishmania mexicana] >emb|CAS92099.1| unnamed protein produc	31.2	31.2	26%	8.3	
ADF22115.1	Sequence 9 from patent US 7674959	31.2	31.2	21%	8.5	
ABH99819.1	Sequence 31 from patent US 7071375 >gb|ABZ25933.1| Sequence 31 from patent US 731497	31.2	31.2	19%	8.8	

Metagenomics is a fascinating field where DNA is collected from an environment and sequenced, with little prior knowledge as to what lives there. This approach solves a very large problem for studying microbes: only a tiny fraction of the world's microorganisms can be grown in culture and studied. Scientists were left to guess what lived in the world around us, frustrated that they could literally see that life was present but could not grow it in a test tube. This is a severe limitation. Many of our analysis methods depend on pure and abundant quantities of biological materials, for example flasks of cells where every cell is genetically identical. But due to tremendous advances in DNA sequencing and bioinformatics, DNA randomly isolated from the environment can be sequenced and often assigned to unclassified organisms. Homologies to well-studied sequences, found with bioinformatics analysis, have revealed literally thousands of novel proteins and microorganisms, revealing a biochemically diverse and exciting world, heretofore invisible. One of the exercises at the end of this chapter focuses on metagenomics.

Now, let's run a TBLASTN search at the NCBI Website using the cellulase protein from the above workflow. In the TBLASTN Web form, select "Whole-genome shotgun contigs (wgs)" from the database menu and under organism, restrict the search to "metagenome (taxid:408169)". The majority of these DNAs are not ESTs, but genomic DNA. Unlike the eukaryotic genome where only a small percentage of the genome encodes proteins, the majority of prokaryotic genomes codes for proteins: almost everything you sequence from a microbe is a gene. Running a TBLASTN search with the sequence YP_001376471, a long list of interesting hits is observed (note that the results may take a long time to appear). For biofuel production, there is a distinct biochemical advantage to operate at high temperatures, so one alignment, very far down the list, catches your eye: a hit isolated from a hot spring in Wyoming (**Figure 5.20**).

The query is 361 amino acids long and this alignment covers roughly 80% of the *B. cytotoxicus* sequence. Remember that environmental samples are normally genomic DNA, not cDNA, so the length of the hit (in this case 1821 nucleotides) is not important. It is a section of genomic DNA, not terminated by the beginning or end of a gene. The length is dependent on the sequencing technology used to

> **Metagenomics aboard the Sorcerer II**
>
> Metagenomics has been applied to a wide range of settings. Historically, an early and large study was started during a voyage of the Sorcerer II sailing ship, captained by Dr. Craig Venter. Collecting many samples from the Atlantic and Pacific oceans, Dr. Venter and collaborators identified literally hundreds of new species and thousands of novel genes and proteins. Using metagenomics, scientists have explored microbes that live in many other environments around the world such as hot springs, deep caves, glaciers, deserts, jungles, and high-altitude lakes.

```
>gb|ADKH01000114.1|  Hot springs metagenome ctg_1106445186637, whole genome
shotgun sequence
Length=1821


 Score =  218 bits (556),  Expect = 1e-55, Method: Compositional matrix adjust.
 Identities = 118/284 (41%), Positives = 175/284 (61%), Gaps = 8/284 (2%)
 Frame = +2

Query  8     LTMLKELTDARGIAGNEREPREVMKKYIEPFADELSTDNLGSLVAKKVGEENGPKIMVAG    67
             +  +LK+ +++A G +G E E R+ +   + P+ DEL  D  G+++  K G++   K MVA
Sbjct  977   IQLLKKFSEAFGPSGFEDEVRDFVIDELNPYVDELFIDRWGNVIGIKYGKQRDLKAMVAA   1156

Query  68    HLDEVGFMITQIDNKGFLRFQTVGGWWSQVMLAQRVTIITCKG-DITGVIGSKPPHILSA   126
             H+DE+G +I   ID  GFLRF+ +GGW    ++ QRV I    G  + GVIGS+PPH+
Sbjct  1157  HMDEIGLLIDNIDKNGFLRFRAIGGWNEVTLVGQRVIIKASNGKKVKGVIGSRPPHVTPP   1336

Query  127   EARKKPVEIKDMFIDIGASSKEEAMEWGVRPGDQVVPYFEFQVMKNEKLLLAKAWDNRIG   186
                  ++ E+KD+FID+GASS EE + G+ + G  V   EF+V+ NE ++ KA+D+R+G
Sbjct  1337  GKEREAPELKDLFIDVGASSDEEVRKLGINVGSVAVLDREFEVL-NENVVTGKAFDDRVG   1513

Query  187   CAITIDVLRQLKDEKHPNIVYGVGTVQEEVGLRGAKTSANYIKPDIAFAVDVGIAGDTPG   246
              A+ +  RQL++ +       Y V TVQEEVGLRGA+ +A+ + PD+A A+D  IA D PG
Sbjct  1514  LAVMLWAARQLREAEVTT--YFVATVQEEVGLRGAQVAADRVYPDLAIALDTTIAADVPG   1687

Query  247   VTEKEAQSKMGDGPQIILYDAS----VIGHTGLRNFVVDVADEL    286
             V E+E  +++G GP I +   D         I H L  F+V VA+EL
Sbjct  1688  VGEREYVTRIGKGPAIKVMDGGRGNVFIAHPKLTGFLVSVAEEL   1819
```

Figure 5.20 TBLASTN alignment. The Query, YP_001376471, is aligned with ADKH01000114, an unknown DNA sequence from a hot springs metagenomics collection.

Box 5.4 A convenient way to BLAST

The NCBI makes it very easy to launch BLAST searches. For every sequence record, viewed in GenBank format, there is a "Run BLAST" link in the upper right of the screen (Figure 1).

Clicking on this link will take you directly to the BLAST Web form, with the accession number already entered in the query field. By clicking on the various tabs you can switch between the forms of BLAST and the query accession number remains in place. But beware: you can pick any BLAST, even if it doesn't make sense. For example, you can launch a BLASTP search with a nucleotide accession number.

Analyze this sequence
Run BLAST

Figure 1 The convenient "Run BLAST" link found on every sequence record.

---

determine the sequence. Very few gaps are needed to maintain the alignment, and the gaps are short. Identity is modest (41%) but similarity is 61%. You are encouraged by this alignment but there are two cysteines (C) and both are unmatched in the alignment. A study of known cellulases should reveal details that could focus further analysis. For example, are there many cellulases without these cysteines? Or have you discovered the first cellulase without these cysteines?

For this project, the search for novel cellulase sequences, there is one piece of unfinished business. Is the hit ADKH01000114 a novel sequence? All you have shown is that it is not identical to your query, a *B. cytotoxicus* sequence. But ADKH01000114 could be 100% identical to another known bacterial gene found in RefSeq. ADKH01000114 is not annotated, as such, but ESTs and metagenomic samples are usually not annotated. Let's run one more test with BLASTN. If you haven't discovered this feature of the Website already, the NCBI tries to make launching a BLAST as convenient as possible, so run this BLASTN as directed here (**Box 5.4**). A BLASTN search of ADKH01000114 against the NCBI nr nucleotide database shows only 79% identity to a sequence, accession number CP002529, from a bacterium called *Vulcanisaeta moutnovskia*. This bacterium is found in a Russian hot spring so it may be related to your thermophile isolated from the Wyoming hot spring. But based on the BLASTN result, ADKH01000114 is *novel*.

Other uses for TBLASTN, such as identifying exons in genomic DNA, will be covered in Chapter 6.

5.6 SUMMARY

In this chapter, you learned about mRNA structure through descriptions of the 5P and 3P untranslated regions, along with the coding sequence. To better understand cDNA and EST sequences, frequent objects of sequence analysis, their synthesis and technical limitations were extensively covered. BLASTX, a program that searches a protein database with a DNA query, was introduced. Interpreting the sequence coordinates in BLASTX alignments was covered in several examples of both known and unknown queries. TBLASTN was also introduced using several examples to demonstrate how a protein query can be used to search a DNA database.

EXERCISES

Exercise 1: Analyzing an unknown sequence

Identify this unknown EST, NCBI accession number AA617657.

1. Using BLASTN, what is its identity?

2. Using BLASTX, what is its identity?

3. Explain your results, making sure you discuss the data from both part (1) and (2).

Exercise 2: Snake venom proteins

Many animals, including reptiles, insects, arachnids, mollusks, fishes, and even mammals, produce toxic liquids known as venoms. Venoms are produced for the purposes of defense or predation, and delivered through a bite or sting. They have evolved to match the lifestyle of the venomous animal and the desired result. For example, insect stings often result in severe pain, discouraging present and future attacks. Spiders use venom to digest their prey, providing easy consumption. Venoms of carnivorous snakes can be neurotoxic, hemotoxic, and powerful enough to kill a large animal through uncontrollable bleeding or paralysis.

Venoms are often complex mixtures of many biochemically different proteins. These proteins can interfere with the blood-clotting cascade, proteolytically digest tissues, interfere with neurotransmission, or severely lower blood pressure. Because of this range of biochemical reactions, venoms have been studied quite extensively and hold much promise in the discovery of novel biochemical pathways and therapeutic agents.

To help define the complex nature of venoms, scientists have created cDNA libraries from venom glands, sequenced the cDNA, and used sequence analysis to discover the proteins responsible for the toxic reactions. Jia et al. (2008) created such a library from the venom gland of the Western cottonmouth snake (*Agkistrodon piscivorus leucostoma*) and made these EST sequences available for study through public databases.

EST identification using BLASTX

In this exercise, try to assign a function and role of the encoded proteins from the following ESTs originating from the venom gland library described above.

EV854943

EV854936

EV854934

EV854921

EV854889

EV854885

EV854883

EV854846

EV854843

EV854840

EV854836

Suggested approach:

- Use BLASTX.

- Search against the non-redundant protein database.

- Enter all of these accession numbers in the query field (as a column) at once. When the results are returned, you will notice a drop-down menu at the top of the Web page so you can examine the results for each.

- You will not need to examine many hits to determine the probable function of the encoded protein. So limit the number of alignments to view to 10. Considering the time each BLASTX search will take, setting this parameter will save considerable time and computer resources at the NCBI.

- Hits that are hypothetical or predictions may not be as trustworthy. When possible, choose hits from well-studied organisms first as they are more likely to have good annotation.

- Use the annotation of these proteins to guess the function of the venom component.

Predicting gene function using BLASTX

Here are ESTs from the venom glands of three other organisms: *Echis pyramidum*, the Egyptian carpet viper; *Microctonus hyperodae*, a parasitic wasp that is used by man to control other insect infestations; and *Rhabdophis tigrinus*, also known as the Asian Tiger Keelback. Using a similar approach to the above exercise, try to find the function of the proteins encoded by these ESTs, and a possible mode of toxicity for these venoms. Although each sequence was annotated as coming from one of these organisms, at least one of these sequences appears to be identical to a gene from a different organism.

Echis pyramidum
GR951114
GR951108
GR951106
GR951090

Microctonus hyperodae
EY187078
EY187061
EY187050
EY187423

Rhabdophis tigrinus
FS611067
FS611066
FS611019
FS610998
FS610884
FS610868

Exercise 3: Metagenomics

Environmental sequencing and analysis, also known as metagenomics, has opened up completely new avenues of research. The approach is quite simple and can be illustrated using an example from an aqueous environment. First, the sample is partially purified by filtering the water to capture and concentrate the desired organisms. The DNA is then isolated and sequenced. The data generated by this random sequencing are analyzed by software and pieced together based on overlapping identity. This is also known as "shotgun" sequencing and is the approach used for most genomic sequencing projects. The large twist in the metagenomics approach is that the DNA originates not from one organism, but instead from hundreds or maybe thousands of species, many unknown to science. Only powerful computers and equally powerful bioinformatics software can sort through these millions of sequences and assemble them into genes. An "electronic" filter can also be applied where the DNA sequence from known organisms is discarded.

Metagenomics can be used to study who is living on and in our bodies. Our skin is covered with bacteria and our gut harbors many microbial species. In fact, in terms of *number* of cells, we are 10% human, and 90% bacteria! Metagenomics studies have shown that the bacterial populations differ depending on the area of the body they are collected from. The life on our scalp differs from what lives on our feet. Bacteria on the left hand may be different than that collected from the right. What lives on us varies with our health, age, diet, gender, the season, who we interact with, and where we live.

In this exercise, we are going to use TBLASTN to identify new enzymes from environmental samples. The enzyme alkane 1-monooxygenase is part of a pathway of proteins that can break down petroleum products. An enzyme such as this would be of interest to groups studying pollution from oil. You will use the NCBI TBLASTN Web form to search the environmental samples using the alkane 1-monooxygenase from the bacteria *Pseudomonas mendocina*.

```
  1 mlekhrvlds apeyvdkkky lwilstlwpa tpmigiwlan etgwgifygl vllvwygalp
 61 lldamfgedf nnppeevvpk lekeryyrvl tyltvpmhya alivsawwvg tqpmswleig
121 alalslgivn glalntghel ghkketfdrw makivlavvg yghffiehnk ghhrdvatpm
181 dpatsrmges iykfsireip gafirawgle eqrlsrrgqs vwsfdneilq pmiitvilya
241 vllalfgpkm lvflpiqmaf gwwqltsany iehygllrqk medgryehqk phhswnsnhi
301 vsnlvlfhlq rhsdhhahpt rsyqslrdfp glpalptgyp gaflmamipq wfrsvmdpkv
361 vdwaggdlnk iqiddsmret ylkkfgtssa ghssstsava s
```

Figure 5.21 Alkane 1-monooxygenase of *Pseudomonas mendocina*. The three histidine motifs, a signature for alkane hydroxylase family members, are shown in bold.

Once you have the results, you will study the alignments, paying particular attention to three sequence markers for the alkane hydroxylase family. Members of this family all possess three histidine "motifs" or patterns. These histidines are thought to bind iron and are essential for the enzyme activity. They are:

HXXXH

HXXXHH

HXXHH

These three motifs must be present to qualify as a possible alkane hydroxylase, and they should match up in approximately the same place and order on the query. Remember, the "X" means any amino acid. To place these in perspective, **Figure 5.21** shows the sequence for our query, *Pseudomonas mendocina* alkane 1-monooxygenase, with the three histidine motifs shown in bold:

1. Using the query, RefSeq accession number YP_001185946, and TBLASTN, search against the "Whole-genome shotgun contigs (wgs)" from the database menu and under organism, restrict the search to "metagenome (taxid:408169)".

2. When the results appear, scan through the hits and find five candidates that have good similarity to the query and, importantly, all three histidine motifs. Rather than concentrate near the top of the results list where you will find the hits most similar to your query, move further down the alignments to find more distant similarities. Based on the annotation of the hits, identify bacteria from diverse environments.

3. For each hit that has the three histidine motifs, use BLASTN against nr/nt to test if they are novel; that is, not showing 100% identity to any sequence other than itself.

FURTHER READING

Camacho C, Coulouris G, Avagyan V et al. (2009) BLAST+: architecture and applications. *BMC Bioinformatics* 10, 421–429. This paper describes recent developments and algorithm details of the BLAST programs. Includes many references.

Drapeau MD, Albert S, Kucharski R et al. (2006) Evolution of the Yellow/Major Royal Jelly Protein family and the emergence of social behavior in honey bees. *Genome Res.* 16, 1385–1394. A cDNA for a honeybee royal jelly protein gene was used to demonstrate BLASTX. This paper describes the family in great detail with wonderful figures illustrating gene structures and evolution.

Fullerton SM, Harding RM, Boyce AJ & Clegg JB (1994) Molecular and population genetic analysis of allelic sequence diversity at the human beta-globin locus. *Proc. Natl Acad Sci. USA* 91, 1805–1809. This paper describes a study, mentioned in this chapter, of the beta-globin locus in 36 Melanesians.

Jia Y, Cantu BA, Sánchez EE & Pérez JC (2008) Complementary DNA sequencing and identification of mRNAs from the venomous gland of *Agkistrodon piscivorus leucostoma*. *Toxicon* 51, 1457–1466. This is the paper that describes the venom EST library in one of the exercises.

Nakamura Y, Gojobori T & Ikemura T (2000) Codon usage tabulated from international DNA sequence databases: status for the year 2000. *Nucleic Acids Res.* 28, 292. As described early in this chapter, organisms demonstrate a preference for the codons used to encode proteins. This database has codon usage tables for almost 36,000 organisms. Codon Usage Database: www.kazusa.or.jp/codon

She X, Rohl CA, Castle JC et al. (2009) Definition, conservation and epigenetics of house-keeping and tissue-enriched genes. *BMC Genomics* 10, 269–280. A recent paper describing two major classes of genes: "housekeeping" and "tissue-enriched." These genes were described in the EST section of this chapter.

Srivastava M, Simakov O, Chapman J et al. (2010) The *Amphimedon queenslandica* genome and the evolution of animal complexity. *Nature* 466, 720–726. This paper describes the genomic sequence of the sponge, *Amphimedon queenslandica*, and has an extensive description of developmental structures and events in early evolution.

Wooley JC, Godzik A & Friedberg I (2010) A primer on metagenomics. *PLoS Comput. Biol.* 6, e1000667 (DOI: 10.1371/journal.pcbi.1000667). A recent article covering all the major aspects of metagenomics.

Yooseph S, Sutton G, Rusch DB et al. (2007) The *Sorcerer II* Global Ocean Sampling Expedition: expanding the universe of protein families. *PLoS Biol.* 5, 432–466. This is one of the earliest metagenomics studies. Metagenomics data were used to demonstrate TBLASTN in the chapter.

ACAAGGGACTAGAGAAACCAAAA
AGAAACCAAAACGAAAGGTGCAGAA
AACGAAAGGTGCAGAAGGGGAAACAGATGCAGA
GAAGGGGAAACAGATGCAGAAAGCATC
AGAAAGCATC
ACAAGGGACTAGAGAAACCAAAACGAAAGGTGCAGAAGGGGAAACAGATGCAGAAAGCATC
AGAAACCAAAACGAAAGGTGCAGAA
AACGAAAGGTGCAGAAGGGG
GAAGGGG

CHAPTER 6

Advanced Topics in BLAST

- Reciprocal BLAST, a simple test to help verify the identity of hits
- Changing the default BLAST settings
- Detecting exons with either BLASTN or TBLASTN
- Repetitive DNA
- Guidelines for evaluating BLAST hits

6.1 INTRODUCTION

BLASTN, BLASTP, BLASTX, and TBLASTN have been introduced in the last three chapters and you now have experience to address a variety of sequence analysis tasks and problems. This chapter will extend this experience by addressing additional problems of gene identity and structure. Importantly, BLAST parameters will be varied to help identify sequences missed by the default settings. As they are frequently encountered during sequence analysis, several families of repetitive DNA will be described to better prepare you for the unexpected. Finally, a summary of guidelines is presented to assist when making decisions about distant relationships based on sequence similarity.

6.2 RECIPROCAL BLAST: CONFIRMING IDENTITIES

When doing any kind of cross-species database search based on similarity (for example, BLAST), the top hit does not necessarily represent the orthologous sequence. Imagine the following scenario.

There are 5000 protein sequences of the tiger deposited in a database. As tigers and humans are both mammals, the sequence of blood proteins of both species is expected to be quite similar. To identify the tiger beta globin protein, you perform a BLASTP search of the tiger database using human beta globin as a query and get a single hit that is quite good. Is this hit the tiger beta globin protein? Maybe not. As we saw in earlier chapters, the various globin proteins (for example, alpha, beta, epsilon) are quite similar to each other and appear as a set of excellent hits in a BLAST search with a single query. So when you did your tiger BLASTP search, you may have found the alpha globin protein sequence and the beta globin may not even be in the database! As illustrated in **Figure 6.1**, when sequences are missing, excellent hits may still be possible. One clue is that you only got a single hit and you were expecting several. Another big hint that you may have missed is that there are only 5000 tiger proteins in the database; most mammals have approximately 21,000 genes that encode proteins, so this database contains less than a quarter of the possible proteins.

Figure 6.1 A hypothetical situation addressed with a reciprocal BLAST search. A BLAST search of tiger proteins with human beta globin would find the tiger alpha globin because of strong similarity, but would not find any other tiger proteins because they are missing from the database. A reciprocal BLAST using the tiger sequence would find all the human globin proteins because of the strong sequence similarity, but the strongest identity would be with the alpha protein.

Human globin proteins:	alpha (present)	beta (present)	epsilon (present)
Tiger globin proteins:	alpha (present)	beta (absent)	epsilon (absent)

When looking for orthologs, it is important to know the state of the sequencing and annotation of the target genome. Do a little reading and determine how many genes are predicted in the genome and how many appear in the database. You can determine the latter by looking at the NCBI Taxonomy database. For the Bengal tiger, *Panthera tigris tigris*, there are 124 proteins in the database. A quick look at the proteins shows that there is some redundancy, for example there are five files of NADH dehydrogenase subunit 4. So not only are there very few Bengal tiger proteins in the database, there may be multiple entries of the same protein. In addition, read about your query: is it part of a family or a singleton? If there is little or no information, you can search a well-annotated and finished genome to get an idea of what you can expect. It would be a guess since all organisms are different, but it is evidence that you can add to your body of information.

When trying to identify orthologous sequences, a case must be built which supports your conclusions. As powerful as sequence analysis is, it has its limitations and the above example illustrates how you can easily come to an early and probably incorrect decision. One of the most common mistakes made with BLAST searches is assuming that the best hit in the search results is the sequence you seek. As a scientist, you must perform additional analysis to increase confidence in your decision.

An approach to verify that a hit is the ortholog of your query is to perform a reciprocal BLAST search. In this case, you can take the top tiger hit and use *it* as a BLASTP query against the more complete human protein database. If you search against human proteins, is human beta globin the best hit? Or is it alpha? If it is alpha, you may have originally identified a tiger paralog and might conclude that the tiger beta globin sequence is missing from the database. Another possibility is that tigers do not have a beta globin protein. If you are not sure because the similarity between beta and alpha proteins is strong, then all you can say is that you found a homolog. Perhaps more reading on the globin protein family will identify key amino acids that distinguish family members and these can be used to narrow down the possibilities.

Note that reciprocal is just an adjective. There is no program called "reciprocal BLAST." Other words may be used to describe the concept of using a hit to search against a database containing the original query.

Demonstration of a reciprocal BLASTP

Here is a more detailed example of where a reciprocal BLAST can help verify search results, particularly with closely related family members.

Opsins are light-sensitive proteins found in animals and are key components in vision. They absorb light of different wavelengths (for example, red, blue, green); mutations in the genes that encode them result in various forms of color blindness. There are a number of opsin family members with very similar names: in humans these include opsins 1–5, rhodopsin, and peropsin. Another complication is that names within families are not necessarily the same in other species, and so care must be taken in identifying orthologs.

Let's say you were interested in identifying an orthologous sequence to human opsin-5 in the shark, *Scyliorhinus canicula*. You would start with the human query, NP_859528, and perform a BLASTP search against non-redundant (nr) protein sequences, restricting the species to *Scyliorhinus canicula*. **Figure 6.2** shows the results table.

Based on the description, the first hit is a member of the opsin family, rhodopsin (accession number O93459). After that, there are sequences with a rapid drop in

```
                                               Score      E
Sequences producing significant alignments:    (Bits)   Value
sp|O93459.1|OPSD_SCYCA  RecName: Full=Rhodopsin >emb|CAA76797....   125    9e-36
emb|CAF02299.1|  angiotensin receptor [Scyliorhinus canicula]     35.8    2e-05
gb|AAY83386.1|  chemokine receptor 4 [Scyliorhinus canicula]      32.3    2e-04
gb|AAY68227.1|  chemokine receptor, partial [Scyliorhinus cani... 26.6    0.009
gb|AAR11490.1|  chemokine receptor [Scyliorhinus canicula]        25.0    0.025
gb|AAN73357.1|  ribosomal protein L5 [Scyliorhinus canicula]      25.8    0.033
gb|AAY68226.1|  chemokine receptor [Scyliorhinus canicula]        22.7    0.17
gb|AAB31092.1|  glucagon G-33 [Scyliorhinus canicula=European ... 20.4    0.35
gb|AAB31091.1|  glucagon G-29 [Scyliorhinus canicula=European ... 20.4    0.42
```

Figure 6.2 BLASTP search results with human opsin-5 as the query against the shark, *Scyliorhinus canicula*. NP_859528 was the query and these hits are from the NCBI nr database. This is the complete list of hits.

Score and increase in E Value, and names that suggest completely different families. **Figure 6.3** is the alignment of the first hit.

The hit (UniProtKB protein record O93459) and the DNA link (CAA76797) contained within that protein record indicate that this sequence was generated as part of a study of photoreceptors in this shark. By score and percent identity, this is only a good hit but considering the evolutionary distance between the human and shark (over 500 million years), perhaps this is not surprising. It is called rhodopsin, a clear member of the opsin family. Although the percent identity is low (27%), the Positives score is nearly double that (51%), showing conservation of related amino acids. The length of the query (354 amino acids) and this hit (354 amino acids) is identical, and the alignment covers the majority of the query.

Could this be the shark opsin-5 protein? Here are three possible alternative answers:

1. Yes: this is the best hit, but the shark protein has been given a different name.

2. No: although this is the best hit, the *Scyliorhinus canicula* genes have not been extensively studied, cloned, and sequenced. But when all the *Scyliorhinus canicula* genes have been sequenced, a better hit will be revealed.

3. No: there is no orthologous opsin-5 protein in this shark.

```
>sp|O93459.1|OPSD_SCYCA RecName: Full=Rhodopsin
 emb|CAA76797.1| opsin [Scyliorhinus canicula]
Length=354
```

Figure 6.3 The BLASTP alignment between human opsin-5 (NP_859528) and shark rhodopsin (Sbjct).

```
 Score = 125 bits (315),  Expect = 9e-36, Method: Compositional matrix adjust.
 Identities = 81/295 (27%), Positives = 149/295 (51%), Gaps = 21/295 (7%)

Query  30   WEADLVAG--FYLTIIGILSTFGNGYVLYMSSRRKKKLRPAEIMTINLAVCDLGISVVGK  87
            W+  ++A   F+L I G   F      LY++ + KK  +P     + +NLAV DL +   G
Sbjct  35   WKFSVLAAYMFFLIIAGFPVNF---LTLYVTIQHKKLRQPLNYILLNLAVADLFMIFGGF  91

Query  88   PFTIISCFCHRWVFGWIGCRWYGWAGFFFGCGSLITMTAVSLDRYLKICYLSYGVWLKRK  147
            P T+I+      +VFG  GC + G+      G   L ++  ++++RY+ +C       +
Sbjct  92   PSTMITSMNGYFVFGPSGCNFEGFFATLGGEIGLWSLVVLAIERYVVVCKPMSNFRFGSQ  151

Query  148  HAYICLAAIWAYASFWTTMPLVGLGDYVPEPFGTSCTLDWWLAQASVGGQVFILNILFFC  207
            HA++ +   W A   PLVG  Y+PE    SC +D++  + V  + F++ +
Sbjct  152  HAFMGVGLTWIMAMACAFPPLVGWSRYIPEGMQCSCGIDYYTLKPEVNNESFVIYMFVVH  211

Query  208  LLLPTAVIVFSYVKIIAKVKSSSKEVAHFDSRIHSSHVLEMKLTKVAMLICAGFLIAWIP  267
              +P  +I F Y +++  VK ++ +          ++  E ++T++ +++    FLI W+P
Sbjct  212  FSIPLTIIFFCYGRLVCTVKEAAAQ----QQESETTQRAEREVTRMVIIMVIAFLICWLP  267

Query  268  YAVVSVW------SAFGRPDSIPIQLSVVPTLLAKSAAMYNPIIYQVIDYKFACC  316
            YA V+ +      S FG    PI ++ +P   AK+A++YNP+IY +++ +F  C
Sbjct  268  YASVAFFIFCNQGSEFG-----PIFMT-IPAFFAKAASLYNPLIYILMNKQFRNC  316
```

Figure 6.4 Reciprocal BLAST with shark rhodopsin. O93459 was the query against human RefSeq proteins. Only the top nine hits are shown.

Sequences producing significant alignments:	Score (Bits)	E Value
ref|NP_000530.1| rhodopsin [Homo sapiens]	593	0.0
ref|NP_001699.1| short-wave-sensitive opsin 1 [Homo sapiens]	327	3e-111
ref|NP_064445.1| long-wave-sensitive opsin 1 [Homo sapiens]	285	2e-94
ref|NP_000504.1| medium-wave-sensitive opsin 1 [Homo sapiens]...	284	4e-94
ref|NP_006574.1| visual pigment-like receptor peropsin [Homo ...	152	8e-44
ref|NP_055137.2| opsin-3 [Homo sapiens]	149	3e-42
ref|NP_150598.1| melanopsin isoform 1 [Homo sapiens]	147	2e-41
ref|NP_001025186.1| melanopsin isoform 2 [Homo sapiens]	146	6e-41
ref|NP_859528.1| opsin-5 [Homo sapiens]	126	2e-34

To help identify the correct answer, you need to run a reciprocal BLASTP search. The top shark hit, O93459, is now the query and should be run against the human RefSeq proteins. The results are shown in **Figure 6.4**.

This reciprocal BLAST shows that the shark sequence identified with the first BLASTP search is not opsin-5, but probably rhodopsin. If the shark sequence were the ortholog to human opsin-5, we would have expected human opsin-5 to be the top hit of the reciprocal BLAST. Instead, the human opsin-5 protein is number nine on the list of hits, significantly weaker than the first hit, rhodopsin.

The alignment between the shark sequence and this first human hit is striking (**Figure 6.5**). The two proteins are 81% identical, with an extremely strong E value. Contrast this to the alignment in Figure 6.3. The easiest conclusions to reach are that the shark opsin-5 sequence has not appeared in the database yet, or does not exist.

Considering the evolutionary distance between human and shark, we might have expected low similarity between orthologs, much like that seen in Figure 6.3. The

Figure 6.5 BLASTP alignment between the shark protein (Query) and human rhodopsin (Sbjct). O93459 was the query.

```
>NP_000530.1 rhodopsin [Homo sapiens]
Length=348

 Score =  593 bits (1528),  Expect = 0.0, Method: Compositional matrix adjust.
 Identities = 288/354 (81%), Positives = 322/354 (91%), Gaps = 6/354 (2%)

Query  1    MNGTEGENFYIPMSNKTGVVRSPFDYPQYYLAEPWKFSVLAAYMFFLIIAGFPVNFLTLY  60
            MNGTEG NFY+P SN TGVVRSPF+YPQYYLAEPW+FS+LAAYMF LI+ GFP+NFLTLY
Sbjct  1    MNGTEGPNFYVPFSNATGVVRSPFEYPQYYLAEPWQFSMLAAYMFLLIVLGFPINFLTLY  60

Query  61   VTIQHKKLRQPLNYILLNLAVADLFMIFGGFPSTMITSMNGYFVFGPSGCNFEGFFATLG  120
            VT+QHKKLR PLNYILLNLAVADLFM+ GGF ST+ TS++GYFVFGP+GCN EGFFATLG
Sbjct  61   VTVQHKKLRTPLNYILLNLAVADLFMVLGGFTSTLYTSLHGYFVFGPTGCNLEGFFATLG  120

Query  121  GEIGLWSLVVLAIERYVVVCKPMSNFRFGSQHAFMGVGLTWIMAMACAFPPLVGWSRYIP  180
            GEI LWSLVVLAIERYVVVCKPMSNFRFG  HA MGV TW+MA+ACA PPL GWSRYIP
Sbjct  121  GEIALWSLVVLAIERYVVVCKPMSNFRFGENHAIMGVAFTWVMALACAAPPLAGWSRYIP  180

Query  181  EGMQCSCGIDYYTLKPEVNNESFVIYMFVVHFSIPLTIIFFCYGRLVCTVKEAAAQQQES  240
            EG+QCSCGIDYYTLKPEVNNESFVIYMFVVHF+IP+ IIFFCYG+LV TVKEAAAQQQES
Sbjct  181  EGLQCSCGIDYYTLKPEVNNESFVIYMFVVHFTIPMIIFFCYGQLVFTVKEAAAQQQES  240

Query  241  ETTQRAEREVTRMVIIMVIAFLICWLPYASVAFFIFCNQGSEFGPIFMTIPAFFAKAASL  300
             TTQ+AE+EVTRMVIIMVIAFLICW+PYASVAF+IF +QGS FGPIFMTIPAFFAK+A++
Sbjct  241  ATTQKAEKEVTRMVIIMVIAFLICWVPYASVAFYIFTHQGSNFGPIFMTIPAFFAKSAAI  300

Query  301  YNPLIYILMNKQFRNCMITTICCGKNPFEEEESTSASASKTEASSVSSSQVAPA  354
            YNP+IYI+MNKQFRNCM+TTICCGKNP  ++E+ SA+ SKTE     +SQVAPA
Sbjct  301  YNPVIYIMMNKQFRNCMLTTICCGKNPLGDDEA-SATVSKTE-----TSQVAPA  348
```

identical length between query and hit, 354 amino acids, in the first BLASTP search looked supportive but was misleading. However, Figure 6.5 shows that rhodopsin is extremely conserved between species and it appears that the chances that this shark sequence is opsin-5 are quite slim.

6.3 ADJUSTING BLAST PARAMETERS

BLAST is very good at finding sequences of interest, but you should always look at the alignment with a critical eye and ask if anything could be improved. There may be times where you should step in and either manually correct the alignment (that is, paste it into a word processor document and edit) or adjust the BLAST parameters, which appear on the bottom of the NCBI BLAST forms. For example, in Chapter 4 the low complexity filtering parameter was explored to see the impact on the alignment display, and in Chapter 5, turning off the filtering extended the BLASTX alignment. Departures from the default settings will often not improve things and may reduce the findings or alignment quality. But parameters should be explored to arrive at the optimum settings for your circumstances. Trial and error might be helpful in learning about sequence analysis, and you may discover a better way to achieve your goals. It is important to remember that there is no "best way" to perform BLAST searches.

Gap cost

As discussed in Chapters 3 and 4, BLAST keeps a score as it attempts to lengthen alignments. Contributing to the score is the cost of gaps in the alignment. Evolutionarily distant sequences will often require many gaps to maintain alignment but too many gaps will strain the belief that the finished alignment makes sense. The simple creation of the gap and the extension of the gap length each have a separate scoring, with gap creation having a much larger default penalty.

Often, misalignments appear when your query has a repeated domain or a simple composition that provides multiple possibilities for the algorithm to choose. To explore the "gap cost" parameter, use the human ADAMTS2 protein as a query in BLASTP to find an opossum homolog. ADAMTS2 is a metalloproteinase that has four thrombospondin domains (~50–60 amino acids, each), three of which are found between amino acids 863 and 1029. Thrombospondin domains have been used many times in evolution and so these sequences will find other thrombospondin domains in other proteins. Matters can get even more complicated if the number of repeated domains differs between homologs. What is seen are multiple ways to align these repeated domains. The results can be confusing but, should you get overlapping alignments, a preliminary conclusion should be that some domains are repeated. Should your proteins be unannotated, drawing and redrawing a figure is a good approach to make sense of your results.

Using this human ADAMTS2 protein query (NP_055059) in BLASTP, a search of the RefSeq database (restricted to *Monodelphis domestica*) finds an opossum ortholog for ADAMTS2 among the top hits, XP_001379113, with an E value of 0.0. The default setting for the NCBI BLASTP form was used: a high cost of eleven to introduce the gap and a low cost of one to lengthen the gap. But our interest is in another sequence further down the list of hits, XP_001377024, so we can see how repeated sequences are handled by BLAST. Remember that RefSeq proteins beginning with "XP_" are predictions, not based on actual cDNA sequences.

XP_001377024 is quite similar to the query with an E value of 4e–149 and is identified as ADAMTS7, another member of the ADAMTS family. Since the thrombospondin domains of the two sequences have diverged, and there are repeated thrombospondin domains, it is expected that BLASTP will show us several variations of alignments as they are all possible solutions to aligning these two proteins. For example, human domain one can align with opossum domain one, or human domain one can align with opossum domain two, and so on.

BLASTP found one large alignment (almost 1000 amino acids long) and then aligned the human thrombospondin domains to the opossum thrombospondin domains in overlapping ranges (perform this pairwise alignment yourself to see all the individual alignments).

A "sketch" of the annotated thrombospondin domains of the human protein and the opossum hit (**Figure 6.6A**) will assist in the interpretation of the alignments. The annotation on this opossum is incomplete and/or the sequence is undergoing revision. This is seen in the coordinates of the thrombospondin domains: the > and < signs indicate that the sequence may extend in their respective directions. In human and other organisms, the thrombospondin domains are approximately 50 amino acids long. In opossum, the domain lengths with the > or < signs are much less than that length and will probably be revised upward. In Figure 6.6B, the second alignment of the hit is shown using the default gap costs of 11 (for creating the gap) and 1 (for extending it). Notice that the alignment extends from

Figure 6.6 Adjusting gap cost in BLAST searches. (A) A sketch of the thrombospondin domains on the human and opossum homologs. (B) The second most significant alignment from the default gap-cost settings of 11,1 (gap creation and extension, respectively). (C) The second alignment with a modified gap cost of 7,2.

```
(A)
Human     ----(564-616)---(863-913)-(917-976)-(982-1029)

Opossum   ---(<599-635)---(867-922)---(988->1013)---(1454->1482)--(1561-1620)

(B)
Gap cost: 11,1
Score = 84.3 bits (207),  Expect = 1e-17, Method: Compositional matrix adjust.
 Identities = 73/246 (30%), Positives = 99/246 (40%), Gaps = 52/246 (21%)

Query   858  WALKKWSPCSKPCGGGSQFTKYGCRRRLDHKMVHRGFCA-ALSKPKAIRRACNPQECSQP   916
             W + WS CS+ CGGG +    C   DH+++   C   L KP  + +C    C
Sbjct   1457 WRVGNWSKCSRNCGGGFRVRDVHCVDTRDHRLLRPFHCQPGLFKPLG-QLSCRVDPCLD-  1514

Query   917  VWVTGEWEPCSQTCGRTGMQVRSVRCIQPLHDNTTRSVHAKHCNDA-RPESRRACSRELC   975
             W T W  CS++CG  G Q R V C +        HC++A RP + R C+   C
Sbjct   1515 -WYTSSWRECSESCG-GGEQERLVTCPE-----------FGHCDEASRPNTTRPCNTHPC  1561

Query   976  PGRWRAGPWSQCSVTCGNGTQERPVLC------RTADDSFGICQEERPETARTCRLGPCP  1029
             +W  GPW QC+ TCG G Q R V C       + +DS    E  PE ++ C    CP
Sbjct   1562 -TKWVVGPWGQCTATCGGGIQRRLVKCVNTKTGQPEEDSNLCDHEVWPENSQKCSAQDCP  1620

Query   1030 RNISDPSKKSYVVQWLSRPDPDSPIRKISSKGHCQGDKSIFCRMEVLSRY--CSIPGYNK  1087
             +            D+ I      C D+ F   E L     C +P
Sbjct   1621 TS-------------------DTGIA-------CDRDRLTFSFCETLHLLGRCHLPTVRA  1654

Query   1088 LCCKSC  1093
             CCK+C
Sbjct   1655 QCCKTC  1660

(C)
Gap cost: 7,2
Score = 80.4 bits (218),  Expect = 1e-16, Method: Compositional matrix adjust.
 Identities = 60/180 (33%), Positives = 83/180 (46%), Gaps = 24/180 (13%)

Query   858  WALKKWSPCSKPCGGGSQFTKYGCRRRLDHKMVHRGFCA-ALSKPKAIRRACNPQECSQP   916
             W + WS CS+ CGGG +    C   DH+++   C   L KP  + +C    C
Sbjct   1457 WRVGNWSKCSRNCGGGFRVRDVHCVDTRDHRLLRPFHCQPGLFKPLG-QLSCRVDPCLD-  1514

Query   917  VWVTGEWEPCSQTCGRTGMQVRSVRCIQPLHDNTTRSVHAKHCNDA-RPESRRACSRELC   975
             W T W  CS++CG  G Q R V C +        HC++A RP + R C+   C
Sbjct   1515 -WYTSSWRECSESCG-GGEQERLVTCPE-----------FGHCDEASRPNTTRPCNTHPC  1561

Query   976  PGRWRAGPWSQCSVTCGNGTQERPVLC---RTA--DDSFGICQEE-RPETARTCRLGPCP  1029
             +W  GPW QC+ TCG G Q R V C   +T  ++   +C E PE ++ C    CP
Sbjct   1562 -TKWVVGPWGQCTATCGGGIQRRLVKCVNTKTGQPEEDSNLCDHEVWPENSQKCSAQDCP  1620
```

the second human thrombospondin domain to a region past the last domain (~858–1093). There are numerous gaps to accomplish this, with only 30% identity.

Lowering the cost of introducing a gap to 7 but doubling the cost (to 2) for extending the gap changes the alignment (Figure 6.6C). Notice that there are half the number of gaps (24 versus 52) and the length of the alignment is now shorter (180 versus 246 amino acids). During the first BLAST calculation, it was "expensive" to introduce the gap with the default settings (11), but once it was introduced, it was "inexpensive" (1) to keep extending it. With the 7,2 setting, the alignment was not extended past amino acid coordinates 1029 (Query) and 1620 (Sbjct) because the cost of extending the gap (2) was too high.

For the 11,1 setting, there are three large gaps, while for the 7,2 setting, two of the gaps are now small and placed differently. The 7,2 setting aligned all the cysteines, but the default 11,1 setting did not. Of course, you should not judge similarity based on cysteines alone: other amino acids are also important. Gone is the large gap, seen in the default setting, and both the percent identity and Positives score increase as well. Overall, the second alignment is arguably better for aligning these domains.

The coordinates of the N and C termini (858 and 1029, respectively) for the human protein correspond almost exactly to the annotated coordinates of the three thrombospondin domain boundaries (see Figure 6.6A). The opossum sequence annotation indicates that the aligned range (1457–1620) contains only two thrombospondin domains. Perhaps this will be revised to three as the sequence length, 163 amino acids, is long enough to accommodate three thrombospondin domains.

Compositional adjustments

The amino acid composition varies between proteins and even between the domains of a single protein. These differences can have an effect on alignments and the NCBI BLASTP and TBLASTN forms give you the ability to change how this variability is addressed. In the example below, the parameter for adjusting the compositional matrix is discussed.

Cysteines form disulfide bridges between nonadjacent amino acids and are critical for folding a protein into a specific shape. In **Figure 6.7A**, protein sequences from human and a small marine animal, the lancelet, are aligned with BLASTP. Because of the numerous cysteines, glycines, and prolines, BLASTP can actually find eight different (overlapping) ways to align these two proteins but only the longest alignment is shown in this figure. The numerous gaps suggest that the alignment is based on similar composition and not on sequence similarity. But enough of the cysteines align to suggest also that these proteins are related. These two proteins may have very similar structures (and maybe function) even though their percent identity is only 32%. Notice that 2 out of 72 cysteines do not align (one from each protein; they are bold and underlined in Figure 6.7A) but these are adjacent to gaps, raising the possibility that BLASTP did not align them well in this region. Indeed, manual editing to remove one gap, combined with shifting and reducing the other, brings all the cysteines into alignment (Figure 6.7B).

In this case, adjusting the gap penalties did not maximize the cysteine alignment, as seen in the hand editing. But BLASTP automatically made an adjustment to the scoring matrix based on the composition of the Query and the Subject (notice the wording to the right of the Expect value in Figure 6.7A: "Method: Compositional matrix adjust."). Changing the compositional adjustments from the default "Conditional compositional score matrix adjustment" to "No adjustment" generated what is seen in Figure 6.7C (notice the "Method" is now missing from the area above the alignment). Comparing Figures 6.7A and 6.7C, the gaps have decreased from 45 to 35, the E value improved from 8e–28 to 1e–39, and all the cysteines are now in alignment. The percent identity and Positives count are approximately the same.

Figure 6.7 Compositional matrix adjustment. The human cysteine-rich motor neuron 1 protein, NP_057525, aligned with a hypothetical protein from the lancelet, XP_002600139. (A) With the compositional matrix adjustment, note that 2 cysteines (bold and underlined) out of 72 are unpaired. (B) Hand editing brings these two cysteines into alignment. (C) Removing the compositional matrix adjustment improves the alignment.

(A)

```
Score =  110 bits (274),  Expect = 8e-28, Method: Compositional matrix adjust.
 Identities = 83/261 (32%), Positives = 111/261 (43%), Gaps = 45/261 (17%)

Query  567  CPELSCSKICPLGFQQDSHGCLICKCREASASAGPPILSGTCLTVDGHHHKNEESWH---  623
            C + C   C GF+ D  GC IC+C      A G     G+C   DG  +N + W
Sbjct  962  CKPVRCKMFCKYGFRTDKRGCEICEC----AKEG----EGSCTDADGITRQNGDKWRPSF  1013

Query  624  --DGCRECYCLNGREMCALITCPVPACGN--PTIHPGQCCPSCADDFVVQKPELSTPSIC  679
              D C  C NG+E C  + C  P CGN  P    GQCCP C D  +KP         C
Sbjct  1014 DEDWCNSCTCENGKEACMAVVCARPDCGNEVPVKKEGQCCPVCPGDEKPEKPGDIKEPTC  1073

Query  680  HAPGGEYFVEGETWNIDSCTQCTCHSGRVLCETEVCPPLLCQNPS--------RTQDSCC  731
             G+ + +GE W   CT+C C  G   VC  +LC+ P        +  CC
Sbjct  1074 STKEGKTYKDGEMWEEADCTKCMCMEGEA-----VCTSVLCEVPDCGEGVEAITKEGDCC  1128

Query  732  PQCTDQPFRPSLSRNNSVPNYCKNDEGDIFLAAESWKPD-VCTSCICIDSVISCFSESCP  790
            P C    P + +     C +D+G ++  E+W+ D CT+C C    C + C
Sbjct  1129 PTC------PEVEK-------CTSDDGKEYVTGETWEQDGSCTTCRCEAGKPLCMTMMCD  1175

Query  791  SVSCE---RPVLRKGQCCPYC  808
             CE     PV +G+CCP C
Sbjct  1176 WPECEDGVEPVTEEGECCPSC  1196
```

(B)

```
Query  680  HAPGGEYFVEGETWNIDSCTQCTCHSGRVLCETEVCPPLLCQNPS---RTQDSCCPQCTD  736
             G+ + +GE W   CT+C C  G   C    +C     C        +  CCP C +
Sbjct  1074 STKEGKTYKDGEMWEEADCTKCMCMEGEAVCTSVLCEVPDCGEGVEAITKEGDCCPTCPE  1133
```

(C)

```
Score =  149 bits (376),  Expect = 1e-39
 Identities = 79/256 (31%), Positives = 109/256 (43%), Gaps = 35/256 (13%)

Query  567  CPELSCSKICPLGFQQDSHGCLICKCREASASAGPPILSGTCLTVDGHHHKNEESWH---  623
            C + C   C GF+ D  GC IC+C +            G+C   DG  +N + W
Sbjct  962  CKPVRCKMFCKYGFRTDKRGCEICECAKEG--------EGSCTDADGITRQNGDKWRPSF  1013

Query  624  --DGCRECYCLNGREMCALITCPVPACGN--PTIHPGQCCPSCADDFVVQKPELSTPSIC  679
              D C  C NG+E C  + C  P CGN  P    GQCCP C D  +KP         C
Sbjct  1014 DEDWCNSCTCENGKEACMAVVCARPDCGNEVPVKKEGQCCPVCPGDEKPEKPGDIKEPTC  1073

Query  680  HAPGGEYFVEGETWNIDSCTQCTCHSGRVLCETEVCPPLLCQNPSRT---QDSCCPQCTD  736
             G+ + +GE W   CT+C C  G +C + +C     C        +  CCP C
Sbjct  1074 STKEGKTYKDGEMWEEADCTKCMCMEGEAVCTSVLCEVPDCGEGVEAITKEGDCCPTC--  1131

Query  737  QPFRPSLSRNNSVPNYCKNDEGDIFLAAESWKPD-VCTSCICIDSVISCFSESCPSVSCE  795
             P + +     C +D+G ++  E+W+ D CT+C C    C + C   CE
Sbjct  1132 ----PEVEK-------CTSDDGKEYVTGETWEQDGSCTTCRCEAGKPLCMTMMCDWPECE  1180

Query  796  ---RPVLRKGQCCPYC  808
               PV +G+CCP C
Sbjct  1181 DGVEPVTEEGECCPSC  1196
```

6.4 EXON DETECTION

The majority of eukaryotic protein-coding genes have transcripts with two or more exons that are spliced together to form the mature messenger RNA. These exons can be part of the coding region of the mRNA, part of the noncoding 5P UTR or 3P UTR, or span the junction between a UTR and coding region. With an mRNA or protein sequence as a query, you can detect the exons in genomic DNA using BLAST. With an mRNA query, BLASTN will find most or all of the individual exons. Using TBLASTN and a protein query, only the coding exons can be detected. Here are two examples demonstrating these approaches.

Exon detection with BLASTN

BLASTN is quite sensitive and can easily detect almost all of the smallest exons. The human gamma-aminobutyric acid (GABA) A receptor gamma 2 gene has one transcript (RefSeq mRNA NM_198904) with exons that range in length from over 2400 to 24 nucleotides. Running a BLASTN search against the human RefSeq genomic sequence finds all 10 exons, samples of which are shown in **Figure 6.8A**. When reviewing the alignments of BLASTN, it is important to verify that all aligned sequences are on the same strand (plus or minus) and in the correct exon order based on the genomic coordinates. BLASTN will order the alignments by Score but by looking at the coordinates of the subject lines, the order along the genomic DNA can be verified. For example, the first alignment is from exon six while the second alignment corresponds to exon five (Figure 6.8A). The subject coordinates indicate that exon five (38589) precedes exon six (41248).

The percentage identity for each exon is very important. If you are using an mRNA to find its gene (not ortholog or paralog), then the identity should be 100%. The possible exceptions are the extreme 5P and 3P ends of the alignments. For example, in the second alignment in Figure 6.8A, the alignment extends toward the 5P end until nucleotide 902, with one mismatch at nucleotide 904. The actual end of the exon is at 907 (see Figure 6.8B). BLASTN does not know the true boundaries of exons; it will continue showing you identity beyond these borders if it can extend the alignment and as long as there are not too many mismatches. Figure 6.8B shows that BLASTN found bases upstream and downstream of the actual junctions. How you retrieve precise genomic sequences is explored in **Box 6.1**.

If the identity is not 100% and this value is expected, there are several possible explanations. First, the quality of the query and genomic sequence should be considered. If either the mRNA or the genomic sequence is not a reference sequence, then mismatches due to sequencing error are possible. Nucleotide polymorphisms are present in populations, and the genomic sequence in the database may not exactly match your query sequence. If both are from reference sequences, it is possible that another nearly identical sequence has been found. There are multiple copies of large sections of genomes, a feature known as "segmental duplication." During the evolution of the genome, large stretches of nucleotides were duplicated and inserted elsewhere. These duplicated regions

Box 6.1 How do you retrieve precise genomic nucleotides?

If you've identified an exon in genomic DNA, how do you then recover this exact sequence? You could copy the subject sequence onto your clipboard, but this is only efficient if the sequence is very short, perhaps one line of sequence. If the subject sequence spans multiple lines, perhaps with gaps, then this method is not only inefficient but can be error-prone. A good way to obtain the sequence is by setting a range of the subject sequence to be shown. This can be done from BLAST within an NCBI sequence file.

Within BLASTN search results, clicking on the Sbjct accession number hypertext will take you to the sequence record for the genomic DNA. Many sequence files for genomic DNA actually don't contain sequence, but only the annotation! For an example, see NW_001794284. The sequence is just too long to be included on a Web page unless you request it. To display the sequence, go to the upper-right corner of the

Web page and click on "Change region shown" (**Figure 1**). There you can enter beginning and ending coordinates that you can take from the alignment.

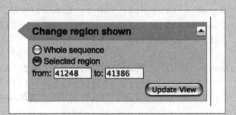

Figure 1 Defining the region to be shown in an NCBI sequence file. Should you suspect that there is more exon upstream or downstream of your BLAST alignment, adjust your coordinates appropriately. After clicking on "Update View," only the sequence you requested is shown.

(A)

```
>ref|NG_009290.1| Homo sapiens gamma-aminobutyric acid (GABA) A receptor, gamma 2 (GABRG2),
RefSeqGene on chromosome 5
Length=94898


Score =  251 bits (278),  Expect = 1e-62
 Identities = 139/139 (100%), Gaps = 0/139 (0%)
 Strand=Plus/Plus

Query  990    ATGGCTATCCACGTGAAGAAATTGTTTATCAATGGAAGCGAAGTTCTGTTGAAGTGGGCG  1049
              ||||||||||||||||||||||||||||||||||||||||||||||||||||||||||||
Sbjct  41248  ATGGCTATCCACGTGAAGAAATTGTTTATCAATGGAAGCGAAGTTCTGTTGAAGTGGGCG  41307

Query  1050   ACACAAGATCCTGGAGGCTTTATCAATTCTCATTTGTTGGTCTAAGAAATACCACCGAAG  1109
              ||||||||||||||||||||||||||||||||||||||||||||||||||||||||||||
Sbjct  41308  ACACAAGATCCTGGAGGCTTTATCAATTCTCATTTGTTGGTCTAAGAAATACCACCGAAG  41367

Query  1110   TAGTGAAGACAACTTCCGG  1128
              |||||||||||||||||||
Sbjct  41368  TAGTGAAGACAACTTCCGG  41386


Score =  154 bits (170),  Expect = 2e-33
 Identities = 87/88 (99%), Gaps = 0/88 (0%)
 Strand=Plus/Plus

Query  902    CTAAGGTTGACAATTGATGCTGAGTGCCAATTACAATTGCACAACTTTCCAATGGATGAA  961
              || |||||||||||||||||||||||||||||||||||||||||||||||||||||||||
Sbjct  38589  CTTAGGTTGACAATTGATGCTGAGTGCCAATTACAATTGCACAACTTTCCAATGGATGAA  38648

Query  962    CACTCCTGCCCCTTGGAGTTCTCCAGTT  989
              ||||||||||||||||||||||||||||
Sbjct  38649  CACTCCTGCCCCTTGGAGTTCTCCAGTT  38676


Score =  129 bits (142),  Expect = 8e-26
 Identities = 71/71 (100%), Gaps = 0/71 (0%)
 Strand=Plus/Plus

Query  616    AGTGAAGCCAACGTTAATTCACACAGACATGTATGTGAATAGCATTGGTCCAGTGAACGC  675
              ||||||||||||||||||||||||||||||||||||||||||||||||||||||||||||
Sbjct  32852  AGTGAAGCCAACGTTAATTCACACAGACATGTATGTGAATAGCATTGGTCCAGTGAACGC  32911

Query  676    TATCAATATGG  686
              |||||||||||
Sbjct  32912  TATCAATATGG  32922


Score =  46.4 bits (50),  Expect = 0.78
 Identities = 25/25 (100%), Gaps = 0/25 (0%)
 Strand=Plus/Plus

Query  1487   CTTCTTCGGATGTTTTCCTTCAAGG  1511
              |||||||||||||||||||||||||
Sbjct  89089  CTTCTTCGGATGTTTTCCTTCAAGG  89113
```

(B)

BLAST exon	RefSeq Annotation
902–989	907–989
616–686	618–685
1487–1511	1487–1510

Figure 6.8 NM_198904 exons found with BLASTN. (A) The four smallest exons detected with the default settings for the NCBI BLASTN form. (B) Exon borders found by BLASTN are compared to the annotated exon coordinates. BLAST will often find sequence identity beyond the true exon borders.

are often almost identical to each other. If the location of the original gene is known, verify that the subject sequence is from the correct chromosomal location. You can do this by looking at the annotation of the RefSeq mRNA record, for example. Finally, it is common for sequences resembling a particular exon to be found nearby in the same genomic region. These may be exons that were duplicated and never used, an exon from an adjacent family member, or perhaps a new exon not yet annotated for this gene.

If you are using BLASTN to find orthologous or paralogous exons, then the identity is not expected to be 100% and will require more careful evaluation and a comparison to what is found with TBLASTN. If exons cannot be found in this fashion, you have to consider several possibilities: (a) the DNA sequence was not conserved in evolution, and probing with protein sequence may be more successful; (b) this organism may have a different gene structure (more or fewer exons); (c) the genomic DNA is incomplete in this region; or (d) BLAST may not be the right tool for this task.

Using BLASTN to search the zebrafish genome with the human query NM_198904 fails to detect several exons, and the 3P UTR is largely missing from the BLAST results. As you recall, noncoding exons are under less selective pressure and often diverge quickly. Considering the evolutionary distance between human and fish, missing exons with BLASTN should not be surprising. One exon is represented several times in the genomic region, as shown in **Figure 6.9**, requiring careful analysis of the coordinates to sort out which sequence is probably the exon. The two exon candidates in Figure 6.9 are over 130,000 nucleotides apart (look at the Subject coordinates) and belong to two different genes.

```
Score =  145 bits (160),  Expect = 2e-31
Identities = 157/208 (75%), Gaps = 0/208 (0%)
Strand=Plus/Plus

Query  1279   AGGTATCACCACTGTCCTGACAATGACCACCCTCAGCACCATTGCCCGGAAATCGCTCCC  1338
              ||| |||||| || || || ||  |||||||| | || |||||||||| |||| || | ||
Sbjct  188627  AGGAATCACAACGGTGCTTACCATGACCACTTTGAGTACCATTGCCAGGAAGTCTTTACC  188686

Query  1339   CAAGGTCTCCTATGTCACAGCGATGGATCTCTTTGTATCTGTTTGTTTCATCTTTGTCTT  1398
              |||||| ||  ||||||||| || |||  ||||| || ||||| ||||| ||||||||||| ||
Sbjct  188687  GAAAGTGTCGTATGTCACGGCTATGGACCTCTTCGTGTCTGTGTGCTTCATCTTTGTTTT  188746

Query  1399   CTCTGCTCTGGTGGAGTATGGCACCTTGCATTATTTTGTCAGCAACCGGAAACCAAGCAA  1458
              || |||||| | ||||| || || || | || |||| ||||| ||||||| || || ||| |
Sbjct  188747  CGCCGCTCTGATCGAGTACGGAACGCTCCACTACTTTGTGAGCAACCGCAAGCCCAGCAA  188806

Query  1459   GGACAAAGATAAAAAGAAGAAAAACCCT  1486
              |    || || ||||||||||||| |||
Sbjct  188807  AAAATCCGACAAGAAGAAGAAAAATCCT  188834
```

```
Score =  82.4 bits (90),  Expect = 2e-12
Identities = 96/130 (74%), Gaps = 0/130 (0%)
Strand=Plus/Plus

Query  1286   ACCACTGTCCTGACAATGACCACCCTCAGCACCATTGCCCGGAAATCGCTCCCCAAGGTC  1345
              ||||| |||||| || |||||||||||||| ||  | |  ||  || || || |||||||||
Sbjct  50214   ACCACCGTCCTCACTATGACCACCCTAAGTATCAGCGCAAGAAACTCTCTCCCCAAGGTG  50273

Query  1346   TCCTATGTCACAGCGATGGATCTCTTTGTATCTGTTTGTTTCATCTTTGTCTTCTCTGCT  1405
              |||||| |||||| ||||| |  || | || |  | || ||| || |||||||| |||
Sbjct  50274   GCCTATGCCACAGCCATGGACTGGTTCATTGCCGTCTGTTACGCCTTTGTCTTCTCGGCT  50333

Query  1406   CTGGTGGAGT  1415
              || | |||||
Sbjct  50334   CTCATTGAGT  50343
```

Figure 6.9 The human query NM_198904 finds at least two *Danio rerio* regions which align with exon 8. Although the percent identity is quite similar, the subject coordinates indicate that the regions belong to two different genes. The Sbjct is accession number NW_001878243.

Look at the coordinates

Gene family members can be scattered about on different chromosomes. Similarity searching tools like BLAST will often find many, if not all, of these different family members, presenting a problem when hunting for exons. Which exons should you splice together? Some exons are more conserved than others so you will often see variability of percent identity. It is tempting to go down the list of hits and look for the best alignment for each exon, but this is not the best approach.

You usually want to stay within a single genomic record when trying to identify the exons of each gene. So as you scroll down your results, be sure not to pass the end of your subject's alignments and start evaluating the alignments of another hit. However, very long genes will span two or more genomic records. If a genomic sequence is only a draft, far from being finished, then the genome can exist in many thousands of pieces. It is entirely possible that the gene you are studying can span one or more of these gaps.

Should you be hunting for exons, be sure to notice the coordinates and length of your subject sequences. If you are near the end of your genomic fragment and there are still more exons upstream or downstream, the additional exons you seek may be on another genomic record, quite possibly another hit in your BLAST results.

Another complication is when family members are adjacent to each other. For example, the mouse hemoglobin cluster has seven family members within 55,000 nucleotides. In cases like this, and many other times, it is best to go "low-tech" and draw a diagram on a piece of paper as you sort things out.

Exon detection with TBLASTN

As seen in Chapter 5, TBLASTN can be quite powerful for finding homologous sequences in EST or metagenomics databases. It can also be used to find exons. There are times when you are looking for orthologous sequences and the genome you are searching is unannotated or underannotated. That is, many genes and their associated exons have not been mapped and you have to find them yourself. If the evolutionary distance between query and subject is too great, the more precise BLASTN search may not work well due to the divergence between DNA sequences. Should you have protein sequence as a query, TBLASTN will often help you identify most, or perhaps all, of the protein-coding exons. The untranslated exons will, of course, not be found with this approach.

For the query, we will use the human protein sequence (NP_944494) for the mRNA used above (GABA receptor subunit gamma-2 isoform 1) and search the reference human genomic DNA. As described earlier, TBLASTN searches are computationally demanding so you should be prepared for a long wait (at least several minutes for this search). You can do yourself a favor, and be a good citizen of a shared resource, by changing the algorithm parameters and only asking for the top 10 hits rather than the default 100 target sequences. We are only expecting one gene to have 100% identity to this protein sequence so a limit of 10 hits is adequate. There are other strategies which can save even more time (see **Box 6.2**).

The GABA receptor gene has multiple exons, so the protein sequence will align with a corresponding number of regions within the genomic sequence. These alignments are arranged by E value (lowest to highest value: best to worst) so they will not appear in order from the N to C termini of the protein. Like BLASTN, TBLASTN does not know where exon boundaries are so it will continue to align outside of these boundaries as the algorithm scoring allows. As a result, the alignments will not be precise representations of exons and some interpretation will be required.

Looking at **Figure 6.10**, the protein sequence for the GABA receptor has been divided by exon distribution. These are the 10 sequences that we are expecting in the TBLASTN alignments. Certain amino acids are encoded by triplets that span

```
  1-36   MSSPNIWSTGSSVYSTPVFSQKMTVWILLLLSLYPG
 37-87   FTSQKSDDDYEDYASNKTWVLTPKVPEGDVTVILNNLLEGYDNKLRPDIGV
 88-109  KPTLIHTDMYVNSIGPVNAINM
110-183  EYTIDIFFAQTWYDRRLKFNSTIKVLRLNSNMVGKIWIPDTFFRNSKKADAHWITTPNRMLRIWNDGRVLYTLR
184-211  LTIDAECQLQLHNFPMDEHSCPLEFSSY
212-257  GYPREEIVYQWKRSSVEVGDTRSWRLYQFSFVGLRNTTEVVKTTSG
258-308  DYVVMSVYFDLSRRMGYFTIQTYIPCTLIVVLSWVSFWINKDAVPARTSLG
309-376  ITTVLTMTTLSTIARKSLPKVSYVTAMDLFVSVCFIFVFSALVEYGTLHYFVSNRKPSKDKDKKKKNP
377-384  LLRMFSFK
385-475  APTIDIRPRSATIQMNNATHLQERDEEYGYECLDGKDCASFFCCFEDCRTGAWRHGRIHIRIAKMDSYARIFFPTAFCLFNLVYWVSYLYL
```

Figure 6.10 The distribution of gamma-aminobutyric acid receptor protein sequence (NP_944494) by exon. The amino acid numbering for exons 1–10 appears on the left.

the junction between two exons. For example, the C-terminal glycine in the first exon (amino acids 1–36) has the codon GGC, and the GG is at the end of exon 1 and the C is at the beginning of exon 2. This can also contribute to the "fuzziness" of the ends in determining the exons' coordinates, with some exons being short or long by an amino acid or more.

Figure 6.11 shows some of the alignments generated by TBLASTN. The default settings were used except the filtering of low-complexity regions was turned off. With this protein sequence, having the filtering on will miss some exons. This cannot always be guessed before doing the TBLASTN search but it can easily be tested.

In evaluating the results, the genomic coordinates and the strand ("+" or "–" frames) should be consistent, with these sequences appearing sequentially on the same strand of genomic DNA. All the alignments should either use positive or negative reading frames, but not both. It is not expected that distant exons, sometimes thousands of nucleotides apart, would maintain the same reading frame, so three different frames can be easily found.

Looking at adjacent exons, the genomic coordinates should show a series, in the correct order, and be non-overlapping. For example, exon 1 encodes amino acids 1–36. The genomic coordinates in Figure 6.11A are from 5359 to 5466. The second exon, amino acid coordinates 37–86, has genomic coordinates 31188–31337. The third exon, amino acids 87–109, has genomic coordinates 32853–32921. Notice that the genomic coordinates are increasing on this positive strand of genomic DNA. Notice, too, that the distance between exons varies and can be many thousands of nucleotides.

TBLASTN does quite well in this search as almost all the protein sequence is accounted for. For brevity, a number of exons are not shown in Figure 6.11A, although nine of the ten were found in this search. As expected, the exon

Box 6.2 Saving computer time (and your time!)

TBLASTN searches of genomic DNA can take a long time to run. Some of the TBLASTN searches in this book took more than 20 minutes to complete! Should you want to repeat the search with a parameter changed, you can save a tremendous amount of time by performing a pairwise TBLASTN (see Figure 4.19). The first TBLASTN search identifies the genomic fragment (for example, NG_009290 in Figure 6.11) so just align the protein sequence to this fragment instead of searching the entire genomic database. With this approach, the results can arrive *in seconds*. Also, at the top of the NCBI BLAST pages are tabs and one is labeled "Recent Results." Click there and you can see the results and BLAST searches you have currently running. Click on the red "x" and you can delete unwanted jobs.

(A)

```
>ref|NG_009290.1
Length=94898


Score = 205 bits (522), Expect = 2e-55, Method: Compositional matrix adjust.
 Identities = 96/128 (75%), Positives = 107/128 (84%), Gaps = 2/128 (1%)
 Frame = +2


Query  350   LVEYGTLHYFVSN--RKPSKDKDKKKKNPLLRMFSFKAPTIDIRPRSATIQMNNATHLQE  407
             L + +H+ + +    K   ++ +P  + S +APTIDIRPRSATIQMNNATHLQE
Sbjct  90341 LSTHFQIHHHIGDIVEKQPRISREQLTDPSPSLPSSQAPTIDIRPRSATIQMNNATHLQE  90520

Query  408   RDEEYGYECLDGKDCASFFCCFEDCRTGAWRHGRIHIRIAKMDSYARIFFPTAFCLFNLV  467
             RDEEYGYECLDGKDCASFFCCFEDCRTGAWRHGRIHIRIAKMDSYARIFFPTAFCLFNLV
Sbjct  90521 RDEEYGYECLDGKDCASFFCCFEDCRTGAWRHGRIHIRIAKMDSYARIFFPTAFCLFNLV  90700


Query  468   YWVSYLYL  475
             YWVSYLYL
Sbjct  90701 YWVSYLYL  90724


 Score = 105 bits (263), Expect = 2e-25, Method: Compositional matrix adjust.
 Identities = 50/50 (100%), Positives = 50/50 (100%), Gaps = 0/50 (0%)
 Frame = +3


Query  37    FTSQKSDDDYEDYASNKTWVLTPKVPEGDVTVILNNLLEGYDNKLRPDIG  86
             FTSQKSDDDYEDYASNKTWVLTPKVPEGDVTVILNNLLEGYDNKLRPDIG
Sbjct  31188 FTSQKSDDDYEDYASNKTWVLTPKVPEGDVTVILNNLLEGYDNKLRPDIG  31337


 Score = 74.7 bits (182), Expect = 5e-16, Method: Compositional matrix adjust.
 Identities = 36/36 (100%), Positives = 36/36 (100%), Gaps = 0/36 (0%)
 Frame = +1


Query  1     MSSPNIWSTGSSVYSTPVFSQKMTVWILLLLSLYPG  36
             MSSPNIWSTGSSVYSTPVFSQKMTVWILLLLSLYPG
Sbjct  5359  MSSPNIWSTGSSVYSTPVFSQKMTVWILLLLSLYPG  5466


 Score = 52.0 bits (123), Expect = 3e-09, Method: Compositional matrix adjust.
 Identities = 23/23 (100%), Positives = 23/23 (100%), Gaps = 0/23 (0%)
 Frame = +3


Query  87    VKPTLIHTDMYVNSIGPVNAINM  109
             VKPTLIHTDMYVNSIGPVNAINM
Sbjct  32853 VKPTLIHTDMYVNSIGPVNAINM  32921
```

(B)
```
Score = 21.6 bits (44), Expect = 4.7
 Identities = 28/83 (34%), Positives = 39/83 (47%), Gaps = 6/83 (7%)
 Frame = +1


Query  307   LGITTVLTMTTLSTIARKSLP--KVSY-VTAMDLFVSVCFIFVFSA-LVEYGTLHYFVSN  362
             L I+ ,V+T  L T+++ +P ++S+ V    F +CF VF  LV   H+F
Sbjct  88867 LKISLVITKLHL-TLSKYFIPSSELSFCVYNFSSFHLLCFQNVFLIYLVSLFFFHFFSPF  89043

Query  363   RKPSKDKDKKKKNP-LLRMFSFK  384
                          NP LLRMFSFK
Sbjct  89044 Y*KQMQFSFLSTNPKLLRMFSFK  89112
```

Figure 6.11 TBLASTN alignments between NP_944494 and genomic DNA. (A) Nine of the ten exons were found with the default settings in TBLASTN. Only four are shown here. (B) Removing the compositional adjustment and low complexity filtering, and raising the E value to 1000, finds the missing exon, an eight amino acid stretch at the end of this alignment.

boundaries were not always precise but the 100% identity segments could be seen in the larger alignments. For example, amino acids 385–475 are part of the alignment with query coordinates 350–475 (Figure 6.11A). The boundaries of the first exon, amino acid coordinates 1–36, were exact and others were close to the amino acid coordinates listed in Figure 6.10.

The only section missing has amino acid coordinates 377–384. Since it is eight amino acids long, a sequence this short will often not be found with the default settings and may be difficult to identify. There is also a distinct possibility that an eight amino acid sequence could be found elsewhere so the genomic coordinates (Sbjct line) have to be examined carefully. Varying the E value, word size, or compositional adjustments will often reveal exons, although it is not always predictable or certain. In this case, changing the E value to 1000, removing the low complexity filtering, and removing the compositional adjustments ("No adjustment") reveals additional alignments, one of which is seen in Figure 6.11B. Although most of the alignment is poor quality (34%), the end has a 100% match to the missing sequence. The genomic coordinates of this sequence, 89,088–89,112, place this sequence between the two adjacent exons. This was also verified by taking these nucleotides of NG_009290 and doing a BLASTN against the cDNA for this gene, NM_198903, and it was in the correct location and an exact match.

Since it was available, the reference genomic sequence was searched in the above example. For organisms where the genomic DNA is not completely sequenced or not sequenced at a depth where the confidence of the nucleotide sequence is high, you would expect some level of inaccuracy in the genomic sequence. Very often, the length of a genomic sequence fragment is known and the flanking sequence of the fragment is obtained. But if the intervening sequence cannot be generated (because of cloning or sequencing limitations), Ns are introduced to fill in the missing sequence. If you are unable to find an exon, it may be because the genomic sequence is missing and you should look at the sequence by eye. For example, **Figure 6.12** is a section of the cow genome.

Orthologous exon searching with TBLASTN

In the above example where we used TBLASTN to identify exons, we had a number of advantages. First, we used a human protein sequence to search the human genome, so 100% identity was expected. Both the query and the genomic sequence were of high quality, so very few mismatches or missing sequences were going to interfere with our success. In addition, we had high-quality cDNA to help verify the last exon. Looking for orthologous exons with TBLASTN is more challenging.

In this analysis, we will use the human amelotin protein (NP_997722) to find the mouse exons in the reference genome for this species. Amelotin is a key protein required for the formation of tooth dentin and enamel, and the sequence is expected to be conserved in the mouse. However, this will be challenging because the exons are small. Using the default parameters of the NCBI TBLASTN

```
GCTGTGCAGACTTCTCCCCGAAGGTCATAAGCTTTGGTCCTGATATTGGGAAACATGTGTTTCTCGTCCT
CGAAAAGATTGTTCTTGGCTTTAAATCCCAGCACTTAGCACTTGTAAAATTGAGATAAAGCGCATTGGC
AGCTTTGGGACTTGGGAGCGCCAATAGCTGAGACTGGACTTTGGAACAACTAAGCTGCTTCATCCATTCT
GGAGCTGCATTTGGTCTTTTCTNNNNNNNNNNNNNNNNNNNNNNNNNNNNNNNNNNNNNNNNNNNNNNNNN
NNNNNNNNNNNNNNNNNNNNNNNNNNNNNNNNNNNNNNNNNNNNNNNNNNNNNNNNNNNNNNNNNNNNNNN
NNNNNNNNNNNNNNNNNNNNNNNNNNNNNNNNNNNNNNNNNNNNNNNNNNNNNNNNNNNNNNNNNNNNNNN
NNNNNNNNNNNNNNNNNNNNNNNNNNNNNNNNNNNNNNNNNNNNNNNNNNNNNNNNNNNNNNAAACAGT
CCACAGTGCTCACCTTGGGTTTCTAAAGCATCTCAGAGAAGCAGCCCGTCCATTCTGCAGCCCTGCCATG
ATGTTTACCTCCCTCCGACATGTTCACAGCAGACTGAGATGGTTCTGACCCTGAAAGCAGGCTGACCTCA
AACATGCCGGAAGACCCAGGCCTGGGCCCGGAACCAGAACCTGGCCTAAGGCTCTGGGGCCACCCAGCTC
TGCGTCCAGGCTTGGAGCACCCACAACCCCATGTACGTGCTTCCAGGGGAAGGAGGCTTGCCCCTATATG
TCTAACACACTCTGTGATCCTGAATTTAGGCTCTTTGCTTAGAAATGGTGTGAGGTTCTAAAAAACATCT
```

Figure 6.12 A region of the cow genome. When sequence cannot be obtained, but the length is known, the missing sequence is filled in with Ns.

Figure 6.13 A false hit which was found with E = 1000.
This TBLASTN alignment was found with the human amelotin protein, NP_997722, used to search against the mouse reference genome. Although 85% identical, the Frame (−1) and location, based on Sbjct coordinates compared to adjacent exons, indicate that this was a false hit.

```
Score = 25.4 bits (54),  Expect =    127, Method: Compositional matrix
adjust.
 Identities = 11/13 (85%), Positives = 11/13 (85%), Gaps = 0/13 (0%)
 Frame = -1

Query  9         CLLGSTRSLPQLK  21
                 CLLG TR LPQLK
Sbjct  149192760  CLLGWTRLLPQLK  149192722
```

Web form, only two exons are found. Turning off the low complexity filtering does not reveal any additional sequence.

The default E value for many BLAST search forms is 10.0. However, if sequences are short, the E values of nearly perfect matches can approach 10 or be much greater. So excluding large E values, although a good idea for an initial BLAST search, may prevent the discovery of some exons. It can also be beneficial to turn compositional adjustments off ("No adjustment") as we did earlier in this chapter.

By raising the E value to 1000 and changing the compositional adjustment to "No adjustment," additional mouse exons can be found with human amelotin. These settings will also generate many alignments, so this approach can be challenging. **Figure 6.13** shows one alignment that could be mistaken for an exon. This region possesses high identity to the protein but the location on the genomic DNA and the frame are incorrect.

Another approach is to use the individual protein sequences (**Figure 6.14A**) for each exon (see **Box 6.3**). As can be seen in Figure 6.14B, almost all the exons can be found in this fashion by only changing the E value to 1000. One protein sequence was not found, the exon encoding amino acids 111–119. Despite a number of

(A)
```
NP_997722 Human amelotin
    1-18   MRSTILLFCLLGSTRSLP
   19-46   QLKPALGLPPTKLAPDQGTLPNQQQSNQ
   47-68   VFPSLSLIPLTQMLTLGPDLHL
   69-98   LNPAAGMTPGTQTHPLTLGGLNVQQQLHPH
   99-110  VLPIFVTQLGAQ
  111-119  GTILSSEEL
  120-207  PQIFTSLIIHSLFPGGILPTSQAGANPDVQDGSLPAGGAGVNPATQGTPAGRLPTPSGTDDDFAVTTPAGIQRST
           HAIEEATTESANG
  208-209  IQ
```

(B)
```
>ref|NC_000071.5| Mus musculus strain C57BL/6J chromosome 5, MGSCv37 C57BL/6J
Length=152537259

Score = 28.1 bits (61),  Expect = 3.5, Method: Compositional matrix adjust.
 Identities = 12/18 (67%), Positives = 15/18 (84%), Gaps = 0/18 (0%)
 Frame = +1

Query  1         MRSTILLFCLLGSTRSLP  18
                 M++ ILL CLLGS +SLP
Sbjct  88805329  MKTMILLLCLLGSAQSLP  88805382
```

Figure 6.14 Using TBLASTN to find the orthologous mouse amelotin exons with the human protein. (A) Amino acid sequence distributed by exon for human amelotin, NP_997722. (B) TBLASTN alignments generated by the individual peptides from the human query aligned with NC_000071. Note that the query coordinates have been changed to match those in (A). In addition, the alignments were ordered to reflect the exon order within the gene. (C) A multi-FASTA file created from sequences in (A). This was pasted directly into the NCBI TBLASTN form.

(B) continued

Score = 23.9 bits (50),   Expect =   95, Method: Compositional matrix adjust.
 Identities = 16/28 (58%), Positives = 17/28 (61%), Gaps = 0/28 (0%)
 Frame = +2

```
Query  19         QLKPALGLPPTKLAPDQGTLPNQQQSNQ   46
                  QL PA G+P TK  P Q T   QQQ NQ
Sbjct  88807061   QLNPASGVPATKPTPGQVTPLPQQQPNQ   88807144
```

Score = 35.4 bits (80),   Expect = 0.026, Method: Compositional matrix adjust.
 Identities = 18/22 (82%), Positives = 20/22 (91%), Gaps = 0/22 (0%)
 Frame = +2

```
Query  47         VFPSLSLIPLTQMLTLGPDLHL   68
                  VFPS+SLIPLTQ+LTLG DL L
Sbjct  88807880   VFPSISLIPLTQLLTLGSDLPL   88807945
```

Score = 28.1 bits (61),   Expect = 5.2, Method: Compositional matrix adjust.
 Identities = 18/29 (63%), Positives = 18/29 (63%), Gaps = 1/29 (3%)
 Frame = +2

```
Query  69         LNPAAGMTPGTQTHPLTLGGLNVQQQLHP   98
                   NPAAG   G T P TLG LN QQQL P
Sbjct  88809302   FNPAAGPH-GAHTLPFTLGPLNGQQQLQP   88809385
```

Score = 20.8 bits (42),   Expect =   351, Method: Compositional matrix adjust.
 Identities = 9/12 (75%), Positives = 10/12 (84%), Gaps = 0/12 (0%)
 Frame = +3

```
Query  99         VLPIFVTQLGAQ   110
                  +LPI V QLGAQ
Sbjct  88810626   MLPIIVAQLGAQ   88810661
```

Query 111-119 not found

Score = 84.0 bits (206),   Expect = 8e-17, Method: Compositional matrix adjust.
 Identities = 45/87 (52%), Positives = 56/87 (65%), Gaps = 1/87 (1%)
 Frame = +2

```
Query  121        QIFTSLIIHSLFPGGILPTSQAGANPDVQDGSLPAGGAGVNPATQGTPAGRLPTPSGT-D   180
                  QIFT L+IH LFPG I P+ QAG  PDVQ+G LP  AG      QGT  G + TP T D
Sbjct  88813922   QIFTGLLIHPLFPGAIPPSGQAGTKPDVQNGVLPTRQAGAKAVNQGTTPGHVTTPGVTDD   88814101
```

```
Query  181        DDFAVTTPAGIQRSTHAIEEATTESAN   206
                  DD+ ++TPAG++R+TH  E  T + N
Sbjct  88814102   DDYEMSTPAGLRRATHTTEGTTIDPPN   88814182
```

Query 208-209 not attempted

(C)
>1-18
MRSTILLFCLLGSTRSLP
>19-46
QLKPALGLPPTKLAPDQGTLPNQQQSNQ
>47-68
VFPSLSLIPLTQMLTLGPDLHL
>69-98
LNPAAGMTPGTQTHPLTLGGLNVQQQLHPH
>99-110
VLPIFVTQLGAQ
>111-119
GTILSSEEL
>120-207
PQIFTSLIIHSLFPGGILPTSQAGANPDVQDGSLPAGGAGVNPATQGTPAGRLPTPSGTDDDFAVTTPAGIQRSTHAIEEATTESANG

Box 6.3 Where can I find a protein sequence distributed by exons?

The distribution of amino acid sequence by coding exons (see Figures 6.10 and 6.14A) can be obtained in a variety of ways, but a convenient route is to look for the CCDS link in the annotation within an NCBI protein record (**Figure 1**). CCDS database entries exist for many model organisms.

```
CDS              1..209
                 /gene="AMTN"
                 /gene_synonym="MGC148132; MGC148133; UNQ689"
                 /coded_by="NM_212557.2:90..719"
                 /db_xref="CCDS:CCDS3542.1"
                 /db_xref="GeneID:401138"
                 /db_xref="HGNC:33188"
                 /db_xref="HPRD:18270"
                 /db_xref="MIM:610912"
```

Figure 1 CCDS link in the NCBI protein record. In bold is the link to the CCDS database.

This "CCDS" link goes to the Consensus CDS Protein Set database where you will find additional annotation of the coding DNA and protein sequence. It includes a colorized version of the sequences with a shift in color at the exon boundaries. In **Figure 2**, alternating exons are underlined for clarity.

```
Nucleotide Sequence (630 nt):
ATGAGGAGTACGATTCTACTGTTTTGTCTTCTAGGATCAACTCGGTCATTACCACAGCTCAAACCTGCTT
TGGGACTCCCTCCCACAAAACTGGCTCCGGATCAGGGAACACTACCAAACCAACAGCAGTCAAATCAGGT
CTTTCCTTCTTTAAGTCTGATACCATTAACACAGATGCTCACACTGGGGCCAGATCTGCATCTGTTAAAT
CCTGCTGCAGGAATGACACCTGGTACCCAGACCCACCCATTGACCCTGGGAGGGTTGAATGTACAACAGC
AACTGCACCCACATGTGTTACCAATTTTTGTCACACAACTTGGAGCCCAGGGCACTATCCTAAGCTCAGA
GGAATTGCCACAAATCTTCACGAGCCTCATCATCCATTCCTTGTTCCCGGGAGGCATCCTGCCCACCAGT
CAGGCAGGGGCTAATCCAGATGTCCAGGATGGAAGCCTTCCAGCAGGAGGAGCAGGTGTAAATCCTGCCA
CCCAGGGAACCCCAGCAGGCCGCCTCCCAACTCCCAGTGGCACAGATGACGACTTTGCAGTGACCACCCC
TGCAGGCATCCAAAGGAGCACACATGCCATCGAGGAAGCCACCACAGAATCAGCAAATGGAATTCAGTAA
```

Figure 2 The Consensus CDS Protein Set database. This portion of the database file includes nucleotide and protein sequences colored by exon.

```
Translation (209 aa):
MRSTILLFCLLGSTRSLPQLKPALGLPPTKLAPDQGTLPNQQQSNQVFPSLSLIPLTQMLTLGPDLHLLN
PAAGMTPGTQTHPLTLGGLNVQQQLHPHVLPIFVTQLGAQGTILSSEELPQIFTSLIIHSLFPGGILPTS
QAGANPDVQDGSLPAGGAGVNPATQGTPAGRLPTPSGTDDDFAVTTPAGIQRSTHAIEEATTESANGIQ
```

You can break the translation into "exons" by copying this sequence, pasting into a word processor document, and introducing paragraph marks. The coordinates (for example, 1–18, 19–54, 55–98, etc.) can be obtained by looking at the protein sequence record.

permutations in parameters, no alignments were generated with BLAST (the NCBI message is "No significant similarity found.") Since we are working with well-annotated genes, proteins, and genomes, it was verified that the mouse genomic sequence did, indeed, have the missing exon. This was checked by performing a BLASTN search using the mouse amelotin cDNA as a query and mouse genomic DNA as the database to search.

The disadvantage of this approach would be the numerous queries that have to be individually run. But again, the NCBI makes this easy by allowing a multi-FASTA file to be used in the Query window (Figure 6.14C).

6.5 REPETITIVE DNA

A large fraction of eukaryotic genomes, ranging from 35 to 50%, is repetitive in nature. The **repeats** are arranged in **tandem arrays**, clustered in groups, or scattered as singletons throughout the genome. Repetitive elements can be as short as two nucleotides, or as long as thousands of bases, and are repeated hundreds to many thousands of times. Many of these elements are considered "junk," leftovers from millions of years of unchecked propagation and change. Others perform a vital function, some of it accidental, and have been major players in the structure and evolution of genomes. Because of their abundance, there are times when repetitive DNA interferes with sequence analysis, particularly similarity searching, so understanding these elements is crucial so you can compensate or alter your methods. Below is a summary of some of the major kinds of repetitive DNA elements. They are studied extensively and have been subdivided into many classes, but only the major classes are described here, organized by size.

Simple sequences

As the name suggests, **simple sequences** consist of very small "repeats" of only two or more nucleotides. Examples include polypyrimidines (for example, CTCTCTCTCTCTCTCT) and single nucleotide stretches (for example, AAAAAAAAAAA, CCCCCCCCCCC, etc.). The functions of these stretches are not known, yet they are often conserved in evolution. BLAST and other database-searching tools have filters that can spare you from the many thousands of hits when simple sequences are present in your query. For example, in BLAST there is a utility called DUST that identifies stretches where your query sequence is simple in composition, and then masks them for the search.

Satellite DNA

Telomeres are specialized structures that cap the ends of chromosomes. Centromeres are located within chromosomes and contain attachment points that bind spindle fibers, which pull apart newly replicated chromosomes. These ends and centers of chromosomes have specialized DNA elements collectively referred to as satellite DNA. These regions contain repeating elements, such as $(GGGTTA)_n$, $(GGGGTT)_n$, and $(GGGGTTTT)_n$, the "n" referring to the number of repeats, which can range in the many thousands. Telomeres and centromeres are almost always condensed in structure and do not harbor any genes. Despite their relative abundance in the genome, you are not likely to encounter them during sequence analysis because these sequences are avoided or are unable to be sequenced in a high-throughput manner.

Mini-satellites

These tandemly arrayed sequences consist of repeated elements of 10 or more nucleotides, and the sequence can vary considerably. Their variability provides genetic markers for identifying the DNA of individuals.

LINEs and SINEs

Long interspersed nuclear elements (**LINEs**) are dispersed repetitive elements that range in size from a few hundred to several thousand bases. They are thought to be parasitic elements that propagate themselves like retroviruses, and they contain similar structural elements. They are transcribed into RNA, then reverse transcribed into DNA and inserted into the genome. In fact, a poly(A) stretch is seen at the end of these elements in the genomic DNA. Much like the synthesis of cDNA, the natural reverse transcription step is thought of as inefficient and leads to the vast majority of the elements being truncated versions of the full-length "parents" (see **Figure 6.15**). In mammalian genomes, upward of several hundred thousand of these elements are found but most are from the 3P end of the parents because of this defective reverse transcription. Looking closely at the structure of the relatively few full-length copies, two overlapping reading frames

```
Truncated                                        -----AAAAAAAAA
versions                                ----------AAAAAAAAA
                              -------------------AAAAAAAAA
                     --------------------------AAAAAAAAA
Full length  -------------------------------------------AAAAAAAAA
```

Figure 6.15 LINE-1 elements.
Reverse transcription of full-length LINE-1 mRNAs is thought to start at the 3P end. After the reverse transcriptase stalls or "falls off" the mRNA, the truncated cDNA becomes double-stranded and is inserted into the genome. Most copies of LINE-1 elements are truncated at the 5P end.

are seen, with the large frame encoding a protein that has distant homology to reverse transcriptase. Since selective pressure is usually not present to maintain the sequence of the truncated versions, they rapidly evolve away from the original sequence until they are barely recognizable as LINE sequence.

There are tens of thousands of copies of **short interspersed nuclear elements (SINEs)** scattered about the genome. Examples include the Alu elements in humans, and the B1 elements in mice. Alu and B1 elements are originally derived from the *7SL* RNA gene. Like the LINE-1 elements, SINE transcripts get reverse transcribed and inserted into the genome. They are approximately 300 nucleotides in length but old elements have rapidly evolved and may be shorter and barely recognizable.

SINEs and LINEs often end up in the 3P UTRs of mRNAs and are frequently adjacent to exons. To give you an idea of how frequently you might encounter Alu repeats, perform a pairwise BLASTN between an 8.45 million nucleotide genomic fragment, CH471135, and U14574, a short sequence containing an Alu element. The alignments will demonstrate the range in identity between Alu elements but the BLASTN graphic (**Figure 6.16**) shows the high frequency in genomic DNA, with each tiny vertical mark on the horizontal line corresponding to an Alu element. The relative abundance of these repetitive sequences was crucial in finding the human genes that cause cancer (see **Box 6.4**).

If you perform a BLASTN search and your query hits many different genes and chromosomes, immediately suspect a repetitive element may be present in your query. Based on the coordinates within your query, you can remove this sequence for further analysis or filter your query using a BLASTN parameter (**Figure 6.17**).

Tandemly arrayed genes

Out of necessity, nature has created gene families that are organized as large arrays of repeated sequences. Rather than scattered about the genome, these members of a gene family are organized into clusters in a very orderly arrangement. These large arrays meet the synthesis demands of the cell or have a structure that allows unique forms for gene regulation. Here are examples you may encounter.

Ribosomal RNAs (rRNAs) are the "scaffolds" of the ribosomes and the genome has evolved to contain many copies of these genes. In humans, there are approximately 5000 copies of 5S rRNA genes, and several hundred copies of 28S, 5.8S, and 18S rRNA genes. They are tandemly arrayed (**Figure 6.18**).

Histone genes are found in high copy number in some organisms out of necessity. The developing sea urchin embryo is a frequently used subject of biology classes due to the high number of cell divisions within the span of a few hours (the length of a laboratory class). As the cells divide, their DNA is replicated at an equally very

Figure 6.16 Alu elements in human genomic DNA. A BLASTN pairwise alignment was generated using the genomic fragment CH471135 as a "Query" and an example Alu element U14574 as the subject. The vertical marks on the horizontal line represent Alu elements distributed on the 8.45 million nucleotide genomic DNA fragment.

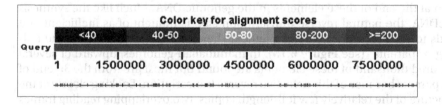

Figure 6.17 A screenshot of the filter parameter in the NCBI BLAST form.

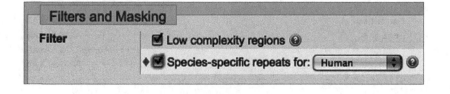

Box 6.4 Repetitive elements and human oncogene discovery

Oncogenes, genes that can cause cancer, were discovered with the help of repetitive DNA. It had been established that you could transfer the DNA from human tumors into cultured mouse cells using a technique called transfection. In this procedure, purified DNA is sheared into small pieces and then mixed with calcium phosphate to form a precipitate. Cultured cells take up this precipitate–DNA mixture and a very small percentage of these cells then stably incorporate this human DNA into their genomes. Mouse cells that took up human genes responsible for cancer-like growth changed shape and growth characteristics, forming piles of cells called foci. These cells could then be harvested from the culture dish. Somewhere in the mouse genome in these cells was a human gene responsible for this transformation, but could it be identified?

In the 1980s, relatively few genes had been sequenced and deposited into GenBank. So you couldn't identify many human genes from mouse genes based on sequence. But the sequence of human and mouse repetitive DNA had been published and so this was used for oncogene identification.

A genomic library was constructed from these cancerous mouse cells. In most clones in this library, all the DNA was mouse genomic DNA. But in a rare number of the clones, human genomic DNA was identified using radioactive probes for human repetitive elements such as Alu. The logic was that when the human tumor genomic DNA was sheared for transfection, pieces included genes plus flanking repetitive elements. By identifying mouse library DNA that contained these human repetitive elements, the adjacent gene must be the human oncogene that transformed the normal mouse cells into cancer cells. This was verified by transfecting normal mouse cells with these suspected, and now purified, human oncogenes. Mouse cells that took up this DNA were transformed into cancer cells.

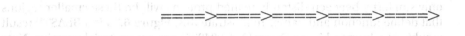

high rate. Once synthesized, genomic DNA must wrap around histone proteins to form chromatin. The generation of these essential proteins keeps pace with DNA synthesis by the transcription of tandemly arrayed clusters of histone genes. In the sea urchin *Echinus esculentus*, each of the early embryonic histone genes (H4, H2B, H3, H2A, and H1) are found in hundreds of repeating units adjacent to each other (**Figure 6.19**).

Figure 6.18 A tandem array. Depicted here is a "head-to-tail" arrangement of repeated elements, with little or no other sequence between elements.

6.6 INTERPRETING DISTANT RELATIONSHIPS

This chapter and the previous three have all focused on the BLAST suite and associated topics. Although BLAST is a very useful tool, if it was replaced tomorrow, you should still understand the makeup of the databases, important areas of consideration, and approaches to detecting and identifying sequences through similarity. This section is an attempt to provide a set of guidelines to consider when judging distant homologies. This list is not comprehensive and should not be used as a "checklist" or an absolute test. In more ways than one, you play detective in sequence analysis, and have to gather evidence through various routes.

Name of the protein

If a hit has the same or similar name to your query, it should be looked at closely, but names can be misleading. As seen earlier in this chapter with the opsin family (Section 6.2), comparisons across species need careful consideration. Many high throughput cDNA sequencing efforts are combined with automated annotation of these cDNAs, and no human being has examined the validity of the names given. For example, something sequenced from a kidney cDNA library might be (incorrectly) called "kidney-specific." Also, if an automated BLAST search found a small region of similarity between the query and the hit, it may be labeled "similar to … ." You will need to manually assess the validity of sequences labeled in this way. Names derived from laboratory experiments and actual human thought are much more likely to be accurate.

Figure 6.19 Tandemly arrayed histone genes in a sea urchin. The early embryonic histone genes of the sea urchin *Echinus esculentus* are arranged in repeating units, each containing one copy of each of the five histone genes. The arrows depict the boundaries between adjacent elements.

```
H4-H2B-H3-H2A-H1→H4-H2B-H3-H2A-H1→H4-H2B-H3-H2A-H1→H4-H2B-H3-H2A-H1→...
```

Percentage identity

Distance in evolution is often tied to percentage identity. Percent identity between orthologs can be very high between closely related species (for example, 95% or greater for human and other *primates*), moderate between related species (for example, 80% or greater for human and other *mammals*), lower for more distant relationships (for example, 40–80% for human and other *vertebrates*), and even lower between kingdoms (for example, less than 40% for human and other *animal cells*). You can sometimes see homology between proteins with the same function (for example, paralogs or orthologs) down to about 25% identity. Short stretches of high identity can be misleading, so consider the length of the identity and the query length. Some proteins are very highly conserved showing little change over evolution (see rhodopsin in Section 6.2), while others diverge rapidly.

Alignment length and length similarity between query and hit

In the strongest situation, your query should align to a hit over its entire length, demonstrating that the similarity goes from end to end. But what if the length of similarity is much shorter than your query length? If your query is 1000 amino acids long but the similarity is limited to 200 amino acids, the size difference should tell you that these might not be related proteins. But to make things complicated, some proteins have conserved and nonconserved domains so the only similarity between distantly related proteins will be these smaller regions that define function and name. To illustrate this, **Figure 6.20** is a BLASTP result graphic of a human kinase Query (NP_004825) against bacterial proteins. Note

Figure 6.20 BLASTP results with a human kinase (NP_004825) as the Query, searching a database of bacterial proteins. Only the kinase domain is conserved.

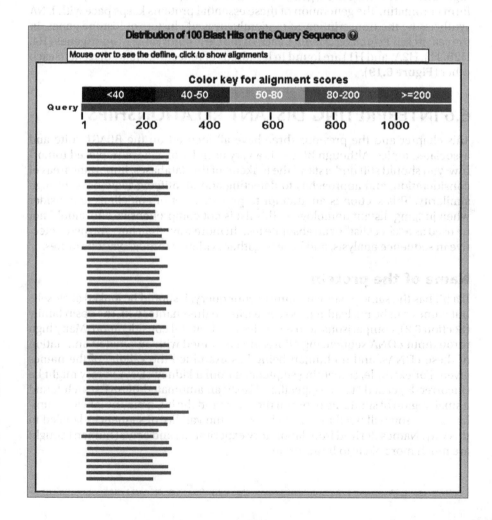

that the Query is over 1100 amino acids long but the 265 amino acid catalytic domain of the kinase is finding the catalytic domains of many bacterial kinases. The rest of the query sequence is less conserved between human and bacteria, and generates no alignments. Nevertheless, these distantly related kinases share similarity in a very important domain, establishing their link.

But, nature also frequently preserves the approximate length (size) of related proteins. For example, human insulin is 110 amino acids long and most real hits are 94–110 amino acids long (Section 4.5). Size—similarity or difference—should be considered carefully.

E value

Are two sequences related to each other? BLAST statistics can assist in the decision-making process but cannot always provide a definitive answer. As first introduced in Chapter 3, and explored further in Chapter 4, and even seen in a simple typing contest (Chapter 4, Exercise 1), E values are influenced by length and the quality of the alignment. Remember that short queries will give large E values, even in the best of alignments. If E values are very small numbers (close to zero) decisions are easier. If the E value is close to, or even above 1.0, be careful with your interpretation.

Hits with very large E values can have description lines that show that distant relatives were found with BLAST. In the example in **Figure 6.21**, you see that hits with E values of 99 will still require consideration. Note that the best alignment (**Figure 6.22A**) and one of the last alignments (Figure 6.22B) are not remarkably different at first glance.

Gaps

BLAST inserts gaps to lengthen alignments but keeps score to make sure it is worth the cost (first seen in Chapter 3). In **Figure 6.23**, a large gap was introduced but it resulted in an alignment at the C-terminus which is comparable to what is seen at the N-terminus. Gap creation and extension parameters can be varied in BLAST, but the results should look convincing. Introducing a large gap to provide a very small extension of the larger alignment is often not appropriate. **Box 6.5** shows an example of where a large gap was appropriate.

```
                                                         Score       E
Sequences producing significant alignments:             (Bits)    Value

ref|YP_593309.1|   serine/threonine protein kinase [Candidatus ...    114    4e-24
ref|YP_001377321.1|   protein kinase [Anaeromyxobacter sp. Fw10...    108    2e-22
ref|YP_002493996.1|   serine/threonine protein kinase with WD40...    105    3e-21
ref|YP_002135874.1|   serine/threonine protein kinase with WD40...    103    8e-21
ref|YP_002431860.1|   serine/threonine protein kinase [Desulfat...    103    1e-20
.
.
many hits not shown...
.
.
ref|YP_739489.1|   serine/threonine protein kinase [Shewanella ...   30.4      96
dbj|BAI92550.1|   hypothetical protein [Arthrospira platensis N...   30.4      99
ref|YP_001500449.1|   serine/threonine protein kinase [Shewanel...   30.4      99
ref|YP_257659.1|   lipopolysaccharide core biosynthesis protein...   30.4      99
ref|ZP_05552205.1|   conserved hypothetical protein [Fusobacter...   30.4     100
```

Figure 6.21 Hits with large E values can still be significant. The catalytic domain (265 amino acids) of human kinase NP_004825 was used in a BLASTP search with the following parameters: the E value maximum was set to 100 and the number of maximum target sequences set to 10,000. The non-redundant protein sequences (nr) database was searched, restricted to bacterial sequences. The top five and the bottom five hits from the results table are listed.

(A)
```
>ref|YP_593309.1| serine/threonine protein kinase [Candidatus Koribacter versatilis
Ellin345] gb|ABF43235.1| serine/threonine protein kinase [Candidatus Koribacter versatilis Ellin345]
Length=381


 Score =  114 bits (286),  Expect = 4e-24, Method: Compositional matrix adjust.
 Identities = 74/221 (34%), Positives = 125/221 (57%), Gaps = 30/221 (13%)

Query   23   GIFELVEVVGNGTYGQVYKGRHVKTGQLAAIKVM-------DVTEDEEEEIKLEINMLK   74
             G +E+V  +GNG  G+VYK RH  + +  A+KV+        +VT+    EI++  N+
Sbjct   10   GAYEIVGPIGNGGMGEVYKVRHTISQRTEAMKVLLSGAARRPEVTDRFVREIRVLANL--   67

Query   75   KYSHHRNIATYYGAFIKKSPPGHDDQLWLVMEFCGAGSITDLVKNTKGNTLKEDWIAYIS   134
                +H NIA + AF       H+DQL +VMEF  ++++++  + G  L+ D +AYI
Sbjct   68   ---NHPNIAALHTAF------HHEDQLIMVMEFIEGKNLSEML--STGMVLR-DSVAYI-   114

Query   135  REILRGLAHLHIHHVIHRDIKGQNVLLTENAEVKLVDFGVS--AQLDRTVGRRNTFIGTP   192
             R+ +  LA+ H   VIHRDIK  N+++   +VKL+DFG++  +  D  +   + +G+
Sbjct   115  RQAVTALAYAHSQGVIHRDIKPSNIMINSAGQVKLLDFGLALMSTPDPRLTSSGSLLGSV   174

Query   193  YWMAPEVIACDENPDATYDYRSDLWSCGITAIEMAEGAPPL   233
             ++++PE I +       T D RSDL++ G+T E+  G P+
Sbjct   175  HYISPEQIRGE-----TMDARSDLYAVGVTLFEVITGRLPI   210
```

(B)
```
>ref|YP_001500449.1| serine/threonine protein kinase [Shewanella pealeana ATCC 700345]
 gb|ABV85914.1| serine/threonine protein kinase [Shewanella pealeana ATCC 700345]
Length=609


 Score = 30.4 bits (67),  Expect =    99, Method: Compositional matrix adjust.
 Identities = 43/169 (26%), Positives = 73/169 (44%), Gaps = 32/169 (18%)

Query   25   FELVEVVGNGTYGQVYKGRHVKTGQLAAIKVMDVT--EDEEEEIKLEINMLKKYSHHRNI   82
             ++ +E+VG G YG V+ G + GQ    K  +T + ++ ++ E ML +  H N+
Sbjct   53   YQELELVGKGAYGFVFAGVN-KLGQAHVFKFSRLTLPQHIQDRLEEEAFMLSQVI-HPNV   110

Query   83   --------ATYYGAFIKKSPPGHDDQLWLVMEFCGAGSITDLVKNTKGNTLKEDWIAYIS   134
                     G +    PG D    + +C          +   L   + I+
Sbjct   111  PPVIKFEHVGKQGILVMARAPGED-----LEQLC-----------IRVGALPVATVMNIA   154

Query   135  REILRGLAHLHIHH-VIHRDIKGQNVLLTENAE-VKLVDFG--VSAQLD   179
             R++   L +LH   +IH DIK  N++   N + + L+D+G  V AQ D
Sbjct   155  RQLAAILQYLHNGRPLIHGDIKPSNLVYDVNTQHLSLIDWGSAVFAQRD   203
```

Figure 6.22 Hits with large E values can still be significant. The catalytic domain (265 amino acids) of human kinase NP_004825 was used in a BLASTP search with the following parameters: the E value maximum was set to 100 and the number of maximum target sequences set to 10,000. The non-redundant protein sequences (nr) database was searched, restricted to bacterial sequences. (A) The alignment from the top hit. (B) The alignment from a bottom hit.

Conserved amino acids

As sequences are aligned to your query, you may notice key amino acids that are conserved throughout evolution—a pattern of cysteines (C), a leucine (L) every seven amino acids, or some other signature that looks either unusual or interesting. This recognition should come with practice. Identifying specific protein motifs with software will be covered in Chapter 8. The NCBI BLAST programs, and other sites, check your query for signatures during the search and will display these domains as graphics (**Figure 6.24**). Should a domain be identified, click on these graphics to reach more information. You may learn that key amino acids

```
>emb|CAG08854.1| unnamed protein product [Tetraodon nigroviridis]
Length=291


 Score =  153 bits (387),  Expect = 1e-37, Method: Compositional matrix adjust.
 Identities = 90/288 (32%), Positives = 156/288 (55%), Gaps = 23/288 (7%)


Query  40   LTIIGILSTFGNGYVLYMSSRRKKKLRPAEIMTINLAVCDLGISVVGKPFTIISCFCHRW  99
            L I +L  GN VL + SR  + L P  ++ +N++   D+ +SV G P ++ +   RW
Sbjct  5    LGFILVLGFLGNFLVLLVFSRFPRLLTPVNLLLVNISASDMLVSVFGTPLSLAASVRGRW  64


Query  100  VFGWIGCRWYGWAGFFFGCGSLITMTAVSLDRYLKICYLSYGVWLKRKHAYICLAAIWAY  159
            + G  GCRWYG++   FG  SL++ + +SL+RY ++ +      + + A I +AA W Y
Sbjct  65   LTGASGCRWYGFSNALFGVVSLVSYSLLSLERYAEVLWDPQTSASRYQRAKIAVAASWFY  124


Query  160  ASFWTTMPLVGLGDYVPEPFGTSCTLDWWLAQASVGGQVFILNILFFCLLLPTAVIVFSY  219
            + FWT  PL G   Y PE  GT+C++  W   Q +   + +I+ +  FCLLLP  V++F Y
Sbjct  125  SLFWTLPPLFGWSSYGPEGLGTTCSVQW--HQRTASSRSYIICLFIFCLLLPLLVMIFCY  182


Query  220  VKIIAKVKSSSKEVAHFDSRIHSSHV--------------------LEMKLTKVAMLIC  258
            +++  +++ S V+    +R  S                        E  + ++ +
Sbjct  183  GRMLLALRAWSLRVSAAGTRSRPSAAGGGSDCCTVCVLQVGGAAGERREALVLQMVLCMV  242


Query  259  AGFLIAWIPYAVVSVWSAFGRPDSIPIQLSVVPTLLAKSAAMYNPIIY  306
            AG+L+ W+PY V++ ++FG P +P  S++P+LLAK++ + NP+IY
Sbjct  243  AGYLLCWMPYGAVAMLASFGPPGVVPPTASLIPSLLAKTSTVLNPVIY  290
```

Figure 6.23 BLASTP alignment between human opsin-5 and CAG08854, from the pufferfish. Note the large gap.

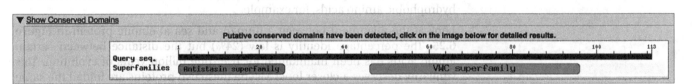

▼ Show Conserved Domains

Putative conserved domains have been detected, click on the image below for detailed results.

Query seq.
Superfamilies   Antistasin superfamily   VWC superfamily

Figure 6.24 Conserved domains identified by the NCBI BLAST program. Near the top of the BLASTP results is a graphic that will display conserved domains present in the query. Clicking on these objects will navigate to the Conserved Domain database where additional information can be found.

Box 6.5 What is going on with this alignment?

NP_005753 was used as a TBLASTN query against human genomic DNA and **Figure 1** shows one of the alignments with genomic sequence file NC_000019.9.

```
Query  241   QYQFLEDAVRNQRKLLASLVKRLGDKHATLQKSTKEVRSS--------------------  280
             +YQFLEDAVRNQRKLLASLVKRLGDKHATLQKSTKEVRSS
Sbjct  22939 RYQFLEDAVRNQRKLLASLVKRLGDKHATLQKSTKEVRSS*VWVLGLWGWPRAARPYLA*  23118


Query  281   ------IRQVSDVQKRVQVDVKMAIL  300
                   IRQVSDVQKRVQVDVKMAIL
Sbjct  23119 PAVSPRIRQVSDVQKRVQVDVKMAIL  23196
```

Figure 1 Adjacent exons separated by a small intron.

Why are there two areas of identity between the query and the genomic DNA and a large gap? This is one contiguous genomic DNA sequence, the Subject (Sbjct) going from 22939 to 23196. It encodes query amino acids 242–280 AND 281–300. There was insufficient distance between these two exons on the genomic DNA sequence for BLAST to separate them into two separate alignments: the gap penalties did not exceed the benefit of including a distant 20 out of 20 match. Instead, you see two regions of identity separated by (translated) genomic DNA. The "*" symbols in the Sbjct line represent encountered stop codons in the intron.

Figure 6.25 Evenly spaced alignment patterns. Human leucine zipper 4 protein NP_057467 is aligned with an unknown sea anemone protein XP_001640018 by BLASTP. Note the patterns for serines, histidine–glycine pairs, and threonine–glutamine pairs created by small internal repeats in both sequences.

```
Score = 84.3 bits (207),  Expect = 2e-14
 Identities = 63/261 (24%), Positives = 93/261 (35%), Gaps = 15/261 (5%)

Query  107  NGQPLIEQEKCSDNYEAQAEKNQGQSEGNQHQSEGNPDKSEESQGQPEENHHSERSRNHL  166
             +     Q  S+    +      S       S        S     G    H    +++ +
Sbjct  139  SNT----QHDMSNTQHDMSNTQHDMSNTQHGMSNTQHGMSNTQHGMSNTQHGMSNTQHGM  194

Query  167  ERSL-SQSDRSQGQLKRHHPQYERSHGQYKRSHGQSERSHGHSERSHGHSERSHGHSERS  225
             +    S+   G    H   + HG     HG S     HG S     H    S    HG S
Sbjct  195  SNTQHGMSNTQHGMSNTQHDMSKTQHGMSNTQHGMSNTQHGMSNTQHDMSNTQHGMSNTQ  254

Query  226  HGHSKR----SRSQGDLVDTQSDLIATQRDLIATQKDLIATQRDLIATQRDLIVTQRDLV  281
             H S         S +Q D+   TQ D+   TQ  +   TQ D+ +TQ  +   TQ D+ +TQ  +
Sbjct  255  HDMSSTQHGMSNTQHDMSSTQHDMSNTQHGMSNTQHDMSSTQHGMSNTQHDMSITQHGMC  314

Query  282  ATERDLIN----QSGRSHGQS  298
              T+ D+ N     S   HG S
Sbjct  315  NTQHDMSNTQHGMSNTQHGMS  335
```

specify membership to a family or function. But remember, not everything has been discovered.

These key amino acids should not be confused by similarity due to composition—for example, one proline-rich sequence aligning with another proline-rich sequence, with no related function. But having domains rich in one or more amino acids is a property, too. It may be required to have a domain that is rich in hydrophobic amino acids, for example.

Look at the alignment between a human and sea anemone protein in **Figure 6.25**. The percentage identity is low (24%) but the distance between certain amino acids has been maintained for over one billion years of evolution. This similarity deserves a closer look! Maybe these two proteins are folded the same way and have the same or related function.

6.7 SUMMARY

This chapter marks the end of formal coverage of the BLAST programs in this book. Chapter 3 covered BLASTN, Chapter 4 explored BLASTP, and both BLASTX and TBLASTN were covered in Chapter 5. Chapter 6 advanced your knowledge of using BLAST through workflows, specialized topics, parameter adjustment, and frequently encountered problems. For the remainder of this book, BLAST can and will be used as a handy tool to call upon as you address sequence analysis problems with other bioinformatics applications.

EXERCISES

Exercise 1: Simple sequences

This first set of exercises explores very simple sequences: two-nucleotide combinations such as CTCTCT.

1. What are all the possible two-nucleotide combinations?

2. Considering that BLASTN searches both strands of DNA, what is the minimum number of searches required to look for all these possible two-nucleotide combinations in the human genome?

3. Perform BLASTN searches of all the possible two-nucleotide combinations: search the human RefSeq mRNA sequences with queries of length ~30 nucleotides. Was it necessary to adjust any parameters?

4. Looking at the coordinates of the hits, how does BLASTN handle the alignments between the query and the hits?

5. For each search, look at one mRNA sequence record of the hit and, by eye, find the two-nucleotide element found with BLASTN. This may be easier if you change the "Display Settings" from GenBank to FASTA.

Exercise 2: Reciprocal BLAST

Major urinary proteins (MUPs) are found in the urine of mammals and have been recently shown to play a role in the fear response between species. Mice can detect and are afraid of the smell of isolated rat and cat MUPs. They even fear the smell of these animal proteins produced in genetically engineered bacteria. The role of MUPs in animal interactions is complex, as they have also been shown to induce aggression and other behaviors within the same species. MUPs are detected by receptors in a specialized nasal structure called the vomeronasal organ. It will be interesting to discover how intraspecies and interspecies detection of MUPs shapes the behavior of animals.

The number of MUP genes varies between mammals, ranging from a single copy (for example, dog) to over 20 copies (mouse). Although several primates appear to have functional MUP genes, humans only have a *pseudogene*, defective because of a single base-pair change. MUPs are also part of a larger family, lipocalins, which extends well beyond mammals.

You are a scientist who has been studying the role of mouse MUP26 in social behavior and wonder if the MUP genes of the opossum (*Monodelphis domestica*) would be interesting. The opossum is a marsupial, a distinct branch of mammals, and sequence analysis of its genes would represent a departure from many of the other MUP studies. You wonder if opossums may have a complex social communication with other animals, mediated by MUPs. As a start, you are interested in finding the ortholog to mouse MUP26 in the newly sequenced opossum genome.

1. Retrieve the mouse MUP26 RefSeq protein sequence and conduct a BLASTP search against the *Monodelphis domestica* non-redundant proteins.

2. When the results are returned, survey the top hits and pick one or more proteins that may be an opossum MUP26 ortholog.

3. Perform a reciprocal BLASTP search against *Mus musculus* RefSeq proteins.

4. Based on the opossum sequence that has the top hit to mouse MUP26 in a reciprocal BLAST, identify the opossum ortholog to the mouse MUP26.

Exercise 3: Exon identification with TBLASTN

In this exercise, you will use TBLASTN to identify the exons of a human gene with a human protein. You will be asked to navigate between databases to retrieve information.

1. Retrieve the RefSeq record for NP_005753.

2. Examine this sequence by eye. Considering that you will be using TBLASTN to identify human exons with this protein sequence, predict what parameter you may need to change in order to find all the exons.

3. From the RefSeq protein record, navigate to the CCDS database and identify the amino acids encoded by each exon. Construct a table, much like Figure 6.10, where each exon's amino acids are on separate lines. How many exons encode protein sequences?

4. From the RefSeq protein record, navigate to the RefSeq mRNA record and identify how many exons are spliced together to form the mRNA.

5. From the annotation found with the RefSeq mRNA record, how long are the 5P and 3P UTRs?

ⓘ Playing possum

In reaction to extreme fear the American opossum will collapse and reach a coma-like state for up to four hours. This is the origin of the phrase, "playing possum," or feigning death. During that time, the animal's heart and respiratory rates decrease, there is excessive salivation and the release of urine and feces, and the animal emits a terrible odor. This death-like state often works as a defense, as predators soon lose interest and leave.

6. Using the default settings (reset the form if necessary) of the NCBI TBLASTN form, perform a pairwise TBLASTN of NP_005753 and genomic sequence AC016630.

7. Record and compare your predictions to the CCDS sequences. Is any sequence missing?

8. Change parameters to try to find any missing sequence. What parameter(s) did you change?

9. Were you able to find all the exons based on the protein sequence?

Exercise 4: Identification of orthologous exons with TBLASTN

By a number of measures, the platypus is a very unusual mammal. It appears to have features, both physical and genetic, from birds, reptiles, and fish. It is so bizarre that when it was first discovered by scientists, it was thought of as an elaborate hoax. It has webbed feet like a duck, and its face is dominated by a specialized structure, reminiscent of a duck's bill. The genus name of the platypus, *Ornithorhynchus*, should remind you of ornithology, the study of birds, and *rhynchus* is Latin for beak. The platypus "beak" is an organ that is specialized for hunting. While diving in water, the platypus closes its eyes, ears, and nostrils and relies entirely on the electrosensory receptors on its bill to locate the minute electrical signals generated by the muscular contractions of prey.

Other features of the platypus are reminiscent of other animals. Remarkably, the platypus lays eggs rather than giving birth to live young, and the eggs are laid, like birds, through the single opening for both the digestive tract and reproduction. Once hatched, the young platypus is fed milk, not through nipples (the platypus has none) but through skin pores. The male platypus has a venom gland and barb-like structures on its hind legs to deliver the toxin. This venom protein suggests a case of convergent evolution with reptilian venoms. Instead of maintaining its body temperature at 37°C like most other mammals, the platypus is a constant 32°C. The platypus has 52 chromosomes, 10 of which are sex chromosomes. Sequence analysis has shown that it has genes previously found only in birds, amphibians, and fish.

But what about mammalian features that have been lost or radically changed: do we see changes at the gene level? This exercise explores a platypus gene that may no longer be under evolutionary selection.

There are at least five mammalian genes responsible for the creation of tooth enamel. In Section 6.4, the coding regions of a mouse enamel protein, amelo-tin, were identified using the human protein. Now you will look for another gene responsible for tooth formation in the platypus genome. At birth the platypus has several teeth but soon loses these and is toothless for the rest of its life. Could a gene responsible for a fundamental tooth component lose function because it is no longer needed? You will address this and other questions in your analysis.

1. Using the human enamelin protein and TBLASTN, search the reference platypus genome using the default settings (except change the maximum number of target sequences to 10). When completed, identify the possible exons for the platypus enamelin protein and paste the appropriate align-ments into a word processor document (in order from N- to C-terminus). This will be referred to as your "gene model." This search may take a while so continue working on the next two questions below. Note that the term "gene model" usually refers to a predicted cDNA, not a collection of alignments or other evidence.

2. Look up the CCDS database entry for your query and identify the exons and encoded protein (by exon) to be expected should your query be similar to the platypus gene and protein. Construct a table, much like Figure 6.10, where the

amino acids encoded by each exon are on separate lines and place this at the top of your gene model.

3. Read about the enamelin protein function by searching PubMed. There are other genes related to enamelin and these may also be found with this first TBLASTN. Remember: always work within the same genomic fragment (for example, NW_12345678), make sure the alignments you pick are all on the same strand, and check the genomic coordinates are sequential.

4. Vary the TBLASTN parameters and approaches, as demonstrated in this and other chapters, to search for coding exons that do not appear with the default TBLASTN parameters. When doing this step, be sure to perform pairwise TBLASTN alignments between your query and the appropriate platypus genomic fragment identified in step 1, above, to speed up your analysis and be a good neighbor of shared computer resources. Should you find additional exons, record what parameter changes led to your discoveries.

5. Add to and edit your gene model as you identify candidate exons. If necessary, examine and manually translate DNA sequence outside of alignments to add amino acids not aligned by BLAST.

6. From the information in your gene model, splice together what you think is the platypus enamelin protein. Perform another pairwise TBLASTN alignment between your model and the platypus genome to check your work.

7. Perform a BLASTP search with your prediction and identify the predicted platypus enamelin already in RefSeq. How do these compare?

8. Perform another pairwise BLASTP alignment between your model and the original query protein. How do they compare? Would any edits of your model improve the alignment, remembering that the platypus protein sequence is different from your query? Is there any sequence missing? From the standpoint of protein sequence, does it appear that the platypus gene might be full length and functional?

9. If necessary, examine the platypus genomic sequence by eye by retrieving the DNA sequence of your subject. Is there any unsequenced genomic DNA in the vicinity of where missing exons may be?

FURTHER READING

Apte S (2009) A disintegrin-like and metalloprotease (reprolysin-type) with thrombospondin type 1 motif (ADAMTS) superfamily: functions and mechanisms. *J. Biol. Chem.* 284, 31493–31497. A nicely illustrated mini-review article on this large and biochemically important family.

Catón J & Tucker AS (2009) Current knowledge of tooth development: patterning and mineralization of the murine dentition. *J. Anat.* 214, 502–515. A wonderful article on the genes and tissues of tooth development.

Goodier JL & Kazazian HH (2008) Retrotransposons revisited: the restraint and rehabilitation of parasites. *Cell* 135, 23–35. A nice review article on this class of repetitive elements, frequently (if not accidentally) encountered in sequence analysis.

Logan DW, Marton TF & Stowers L (2008) Species specificity in major urinary proteins by parallel evolution. *PLoS One* 3, e3280 (DOI:10.1371/journal.pone.0003280). A good discussion of this large family, with nice illustrations of gene organization and evolutionary relationships.

Mikkelsen TS, Wakefield MJ, Aken B et al. (2007) Genome of the marsupial *Monodelphis domestica* reveals innovation in non-coding sequences. *Nature* 447, 167–177. A thorough and interesting genome paper with comparisons to other model genomes.

Shih C & Weinberg RA (1982) Isolation of a transforming sequence from a human bladder carcinoma cell line. *Cell* 29, 161–169. This paper describes the isolation of a human oncogene with the help of repetitive elements.

Terakita A (2005) The opsins. *Genome Biol.* 6, 213–221. A good review article covering the opsin family members, along with their structure and function.

Warren WC, Hillier LW, Marshall Graves JA et al. (2008) Genome analysis of the platypus reveals unique signatures of evolution. *Nature* 453, 175–184. This is a wonderful article that draws direct comparisons between the unusual features of the platypus and the genomic discoveries by the authors.

Yu Y-K & Altschul SF (2005) The construction of amino acid substitution matrices for the comparison of proteins with non-standard compositions. *Bioinformatics* 21, 902–911. This article discusses the mathematical details of modifying substitution matrices to accommodate differences in protein composition.

CHAPTER 7

Bioinformatics Tools for the Laboratory

Key concepts

- Mapping restriction enzymes with the NEBcutter tool
- Converting DNA sequence with Reverse Complement
- Translation of DNA sequences with the ExPASy Translate tool
- Identifying open reading frames with the NCBI ORF Finder
- Using Primer3 software for designing PCR primers
- Measuring DNA composition with DNA Stats
- Measuring protein composition with the Composition/Molecular Weight Calculation Form
- The Sequence Retrieval System (SRS), a database-querying tool
- Graphically visualizing sequence similarity with DotPlot

7.1 INTRODUCTION

As you work your way through this book, and your own analyses, you will encounter times when you need a small utility that does one thing and does it well. This chapter will focus on sequence analysis utilities that come in handy for work in the average academic, biotechnology, or pharmaceutical laboratory. You've already discovered how a word-processing program can save you time and minimize errors through the "Find and Replace" function. These additional programs described in this chapter are more sophisticated and specific for sequence analysis. They will help you design and implement experiments as well as analyze and make sense of the data collected. They will often make the difference between knowing answers and just guessing. If you are ever doing anything manual, repetitive, and prone to error, you should ask yourself "is there a simple tool which can do this for me?"

There are literally hundreds of bioinformatics utilities on the Internet and it would be impossible to present them all here. This chapter will survey some practical as well as academic applications. The interfaces can be radically different in approach. Some utilities will be simple to understand and use while others will require more work to get at answers. Help documents will range from polished publications to no help at all. Some sites are maintained by commercial entities while others are kindly offered by academic laboratories. You should get used to the variety and be prepared to seek replacements. The utilities presented here are good but you may find others more suited to your needs. There are also commercial desktop applications, which may provide most of what you need in one place.

It is important to remember that there is unsatisfactory software out there; Web pages that don't work, pages that generate ugly or confusing results, and sites that use outdated approaches or generate incorrect data. As you explore the Internet, remember to apply scientific methods of evaluation. You may have to experiment with sequences that you understand (controls) to verify that the Website you just discovered is working properly.

7.2 RESTRICTION MAPPING AND GENETIC ENGINEERING

Open up a vendor catalog that caters to laboratories that clone, manipulate, and sequence genes and you see pages of maps of plasmids and viruses, the vectors of genetic engineering. Also included are lists of restriction enzymes that cleave these DNAs. Some of the first sequence analysis software was developed to identify restriction enzyme sites, generate maps, and assist in the planning of pioneering recombinant DNA experiments. Before describing a good example of software for this purpose, we will extensively review topics of "wet" laboratory work: restriction enzymes and the DNA sequences they cleave.

Restriction enzymes

One of the critical advances that led to the explosive growth of genetic engineering was the discovery and utilization of **restriction enzymes**. Restriction enzymes are bacterial proteins that cleave DNA into pieces. Rather than doing this randomly, restriction enzymes recognize a very specific short sequence, for example GAATTC, and cleave both strands of DNA at this site. These restriction sites are palindromes: the sequence, read 5P to 3P, is the same on both strands of DNA, as seen in **Figure 7.1**.

When describing the recognition sequence and cleavage site of a restriction enzyme, the shorthand is to only show one strand and to place a caret (^) to mark the spot where cleavage takes place (see **Table 7.1**). The EcoRI cut is asymmetric, G^AATTC, leaving a short single-stranded DNA sequence protruding. If a 5P end protrudes after cleavage, then this is described as a "5P overhang." If a 3P end protrudes, then this is described as a "3P overhang." When the cleavage takes place in the center of the recognition site, for example CCC^GGG, this leaves a "blunt" end with no overhang. Table 7.1 shows a sampling of restriction enzymes with a six-nucleotide recognition site.

The recognition site of commercially available restriction enzymes varies from 4 (for example, AluI, AG^CT) to 13 nucleotides (for example, SfiI, GGCCNNNN^NGGCC). There is another commercially available enzyme, CspCI, which recognizes a relatively short sequence but then cleaves far upstream of this site (**Figure 7.2**). As Figure 7.2 demonstrates, some enzymes do not require

Figure 7.1 The EcoRI restriction enzyme site.
(A) The EcoRI site, GAATTC, with the flanking sequence depicted as Ns. Note that the site is palindromic: GAATTC is on the top strand while the complementary strand, reading 5P to 3P, is identical: GAATTC. (B) The enzyme cleaves after the G on both strands. Once the backbone of the DNA is broken in two places, the short stretch of complementary bases is not enough to keep the strands together. (C) Separated, the two 5P ends have unpaired bases or "sticky" ends of "AATT."

(A)
```
5P-NNNNNGAATTCNNNNNN-3P
3P-NNNNNCTTAAGNNNNNN-5P
```

(B)
```
5P-NNNNNG    AATTCNNNNNN-3P
3P-NNNNNCTTAA    GNNNNNN-5P
```

(C)
```
5P-NNNNNG-3P              5P-AATTCNNNNNN-3P
3P-NNNNNCTTAA-5P              3P-GNNNNNN-5P
```

Table 7.1 A sampling of restriction enzymes, their recognition sites, and species of origin

Enzyme	Recognition site	Bacteria species
BamHI	G^GATCC	*Bacillus amyloliquefaciens*
EcoRI	G^AATTC	*Escherichia coli*
HindIII	A^AGCTT	*Haemophilus influenzae*
KpnI	GGTAC^C	*Klebsiella pneumoniae*
PstI	CTGCA^G	*Providencia stuartii*
SacI	GAGCT^C	*Streptomyces achromogenes*
SalI	G^TCGAC	*Streptomyces albus*
SmaI	CCC^GGG	*Serratia marcescens*
SphI	GCATG^C	*Streptomyces phaeochromogenes*
XbaI	T^CTAGA	*Xanthomonas badrii*

Restriction enzymes are pronounced with different rules. For some, every letter is said. For others, syllables are used. Here are the common ways to pronounce the list: "bam-H-1," "echo-are-1," "hin-dee-3," "kay-p-en-1," "p-s-t-1," "sack-1," "sal-1," "smah-1," "s-p-h-1," and "x-b-a-1."

NN^NNNNNNNNNNNN**CAANNNNNGTGG**NNNNNNNNNNNNNNN

Figure 7.2 The recognition site for restriction enzyme CspCI. The enzyme recognizes the internal specific nucleotides, CAANNNNNGTGG, but then reaches upstream from these bases to cleave the DNA backbone (depicted with ^).

a specific sequence for the entire recognition site, but only require that some nucleotides be present. The other nucleotides can be any base (depicted as N).

The 3P and 5P overhangs are also known as "sticky ends" and, under the right circumstances, allow joining with other DNA molecules that have the same sticky ends. For example, DNA fragments having protruding AATT sticky ends can be joined to other fragments having the same protruding and complementary bases. Genetic engineers quickly recognized this power of restriction enzymes and designed many experiments and procedures around the use of these enzymes. A common approach was to prepare inserts with restriction enzyme cleavage and clone them into corresponding "sticky ends" of a cloning vector. As an introduction to these steps, consider a scenario where you are cloning genomic DNA into a vector using just two restriction enzymes. In **Figure 7.3**, a piece of genomic DNA has been trimmed with SphI and EcoRI.

Using the DNA in Figure 7.3 as an example insert, a vector is cut with SphI and EcoRI, generating sticky ends to accept the corresponding ends of the insert (**Figure 7.4**).

(A)
Sphl: GCATG^C
EcoRI: G^AATTC

(B)   Sphl   EcoRI
       cnnnnnnnnnnnnnnnnnnnnnnnnnnnnnnnnnnnnnnnnnnnnng
       gtacgnnnnnnnnnnnnnnnnnnnnnnnnnnnnnnnnnnnnnnnnnncttaa

Figure 7.3 An insert with SphI and EcoRI sticky ends. (A) The recognition sequences for SphI and EcoRI restriction enzymes. (B) DNA was digested with a combination of SphI and EcoRI restriction enzymes. After digestion, the complementary bases of these restriction sites are exposed. The nucleotides here are depicted in lowercase in order to later follow their insertion into an uppercase plasmid sequence. Nucleotides recognized by the enzymes are underlined.

Alu repeats

In Section 6.5, the short interspersed repetitive sequences called Alu elements were described. Where did this name come from? If you digest human genomic DNA with the restriction enzyme AluI and size-fractionate the fragments on an agarose gel, you observe a general gradient of sizes from very large to very small. But embedded in this relatively smooth gradient of sizes is a prominent band of DNA fragments which all have approximately the same small size. This band caught the eye of scientists and upon analysis they discovered these thousands of repetitive elements that shared a conserved AluI restriction site. These were then named "Alu elements."

Figure 7.4 Cloning of an SphI–EcoRI fragment. (A) A vector sequence with SphI and EcoRI sites underlined. Once cleaved (B), removal of the intervening sequence leaves (C). (D) The extending nucleotides of the insert, depicted in Figure 7.3, are complementary to the extending nucleotides of the vector. The insert (lowercase) is accepted into the vector (uppercase) and a ligase is used to restore the backbone of the DNA. (E) The two restriction enzyme sites (underlined), SphI and EcoRI, are restored and this vector plus insert could be cut again by these two enzymes.

(A)
```
        SphI                                                              EcoRI
AAGCTTGCATGCCTGCAGGTCGACTCTAGAGGATCCCCGGGTACCGAGCTCGAATTCTTAT
TTCGAACGTACGGACGTCCAGCTGAGATCTCCTAGGGGCCCATGGCTCGAGCTTAAGAATA
```

(B)
```
        SphI                                                              EcoRI
AAGCTTGCATG|CCTGCAGGTCGACTCTAGAGGATCCCCGGGTACCGAGCTCG|AATTCTTAT
TTCGAAC|GTACGGACGTCCAGCTGAGATCTCCTAGGGGCCCATGGCTCGAGCTTAA|GAATA
```

(C)
```
        SphI                                                              EcoRI
AAGCTTGCATG                                                        AATTCTTAT
TTCGAAC                                                                GAATA
```

(D)
```
        SphI                                                              EcoRI
AAGCTTGCATG cnnnnnnnnnnnnnnnnnnnnnnnnnnnnnnnnnnnnnnnnnnng AATTCTTAT
TTCGAAC gtacgnnnnnnnnnnnnnnnnnnnnnnnnnnnnnnnnnnnnnnnnncttaa GAATA
```

(E)
```
        SphI                                                              EcoRI
AAGCTTGCATGcnnnnnnnnnnnnnnnnnnnnnnnnnnnnnnnnnnnnnnnnnnngAATTCTTAT
TTCGAACgtacgnnnnnnnnnnnnnnnnnnnnnnnnnnnnnnnnnnnnnnnnncttaaGAATA
```

Restriction enzyme mapping: the polylinker site

Using some of the earliest bioinformatics software, scientists simulated the digestion of DNA using many different restriction enzymes to find which would do the job. This procedure is called restriction enzyme mapping. For a simple demonstration of software that generates a restriction map, we'll use a short DNA sequence known as a **polylinker**.

There are often many different steps to genetic engineering, with foreign inserts being manipulated a number of times to achieve the experimental goals. One of the greatest advancements in the field of genetic engineering was the construction of artificial vectors which carried features that allowed easy manipulation in the laboratory. To accommodate the many different possible insertions and approaches, a vector polylinker cloning site was invented. A polylinker is a marvel of engineering; in one example, within the span of 57 nucleotides there are 10 different commonly used restriction enzyme sites. This generates 53 different cloning site possibilities. To see these restriction sites, we'll use a restriction mapping tool called the NEBcutter.

NEBcutter

There are choices of restriction mapping tools on the Internet, but the NEBcutter tool at the New England Biolabs Website is easy to use and generates a map with a clean look (tools.neb.com/NEBcutter2/index.php). For an example of a polylinker, a sequence can be found in GenBank file HQ418395, coordinates 1504–1560.

This sequence is shown in **Figure 7.5A** in double-stranded form. In these 57 nucleotides are the sites for 10 restriction enzymes listed in Figure 7.5B, all of which have a six-nucleotide enzyme recognition site.

Go to the GenBank file and copy the polylinker sequence as clean text (no numbers or spaces). Navigate to the NEBcutter Website, paste in the sequence, and hit "Submit." When the screen refreshes you see a very complicated map with many more restriction enzyme sites than the 10 listed in Figure 7.5B. Many of these additional sites are not practical to use for digesting the polylinker site because they would cut the plasmid elsewhere too frequently. For example, AluI recognizes a four-base sequence for cutting, AG^CT. AluI cuts the 9921 nucleotide vector,

(A)

```
AAGCTTGCATGCCTGCAGGTCGACTCTAGAGGATCCCCGGGTACCGAGCTCGAATTC
TTCGAACGTACGGACGTCCAGCTGAGATCTCCTAGGGGCCCATGGCTCGAGCTTAAG
```

(B)

Enzyme	Recognition site
BamHI	G^GATCC
EcoRI	G^AATTC
HindIII	A^AGCTT
KpnI	GGTAC^C
PstI	CTGCA^G
SacI	GAGCT^C
SalI	G^TCGAC
SmaI	CCC^GGG
SphI	GCATG^C
XbaI	T^CTAGA

Figure 7.5 A polylinker cloning site. (A) Both strands of the 57 nucleotide sequence that contains 10 different restriction enzyme sites. (B) A list of the 10 restriction enzymes. (C) A screenshot, generated by the New England Biolabs NEBcutter, showing the location of all 10 sites.

(C)

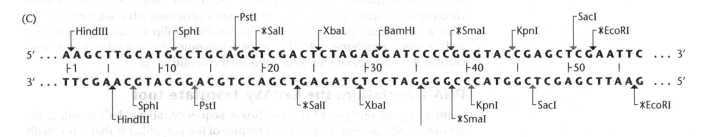

mentioned as a source of the polylinker sequence above, 41 times (see **Box 7.1**). To illustrate just the 10 sites listed in Figure 7.5B, click on "Custom Digest" (after the first "Submit") in the NEBcutter form and use the check boxes to select the 10 enzymes. Then hit the "Digest" button to view the map seen in Figure 7.5C.

The map generated by the NEBcutter tool shows the sequence, now both strands, and has arrows and labels to indicate the locations in the DNA backbones where the enzymes cleave. Notice how the enzyme recognition sites overlap.

Box 7.1 How often do restriction enzymes cut DNA?

The frequency of restriction enzyme sites in a DNA sequence is dependent on the sequence composition, but for the most part it is a function of the number of nucleotides in the enzyme recognition site. If DNA sequence were completely random, a given four-nucleotide sequence will be found once every 256 nucleotides. This is calculated by multiplying the probabilities of finding each base in each position:

$$\tfrac{1}{4} \times \tfrac{1}{4} \times \tfrac{1}{4} \times \tfrac{1}{4} = \tfrac{1}{256}$$

In the example shown in Figure 7.5, 41 AluI sites were found in a 9921 nucleotide vector. By calculation, 39 sites were expected, quite close to this estimation. The enzyme recognition sites in Figure 7.5B are six nucleotides long. These sites are expected to come up once per 4096 nucleotides. The vector is 9921 nucleotides long so we would estimate approximately two sites from an enzyme that recognizes six bases. But remember, this is a cloning vector and some restriction sites may have been purposefully destroyed so the polylinker site would contain the only EcoRI site, for example.

Input:	5P–ACTG–3P
Reverse-complement:	5P–CAGT–3P
Reverse:	3P–GTCA–5P
Complement:	3P–TGAC–5P
Desired outcome:	5P–ACTG–3P
	3P–TGAC–5P

Figure 7.6 Output from the "Reverse Complement" utility on the Sequence Manipulation Suite Website. 5P and 3P labels have been added to more easily follow the result of the form's actions. Of the three options, "complement" was needed to generate the desired outcome.

Generating reverse strand sequences: Reverse Complement

In addition to assisting in laboratory experiments, simple utilities can be used to generate illustrations of sequence manipulations. This can be very important in documenting the planning, execution, and result of genetic engineering procedures. For example, a simple utility called Reverse Complement was used to generate Figure 7.5A.

When obtaining DNA sequences from most Websites, only one strand is provided. To show the second strand of the polylinker in Figure 7.5A, a utility was needed to generate the complement. In this case, the "Sequence Manipulation Suite," listing many utilities, was used, www.bioinformatics.org/sms2/index.html. A Web form, "Reverse Complement," takes DNA sequence as input and gives the choice of reverse-complement, reverse, and complement. To illustrate these choices, consider a simple sequence, ACTG, and the desired outcome, seen in **Figure 7.6.**

Reverse-complement refers to the complementary strand of the input, displayed in a 5P to 3P manner. Reverse literally reverses the order of the nucleotides, showing the sequence 3P to 5P. Complement generates what we need for the illustration. It generates the complement but doesn't flip the orientation as seen with reverse-complement. In the desired outcome, the second strand is now correct: the T is opposite the A, G opposite the C, and so on.

DNA translation: the ExPASy Translate tool

Continuing our analysis of the polylinker sequence, above, let's simulate the translation of this short sequence. A feature of the polylinker is that it is usually placed within a gene, in a region that tolerates a short insertion, and allows the read-through translation in that gene's reading frame. The insertion of almost any DNA sequence into the polylinker disrupts the translation of the "reporter" gene and inactivation of this gene is utilized in screening clones.

To translate the polylinker, go to the Translate tool on the ExPASy Website, web.expasy.org/translate. Simply paste in the single strand of the polylinker and select "Includes nucleotide sequence" from the drop-down menu. Click on "Translate Sequence" and the output shown in **Figure 7.7** is seen.

Figure 7.7 ExPASy Translate tool. The six reading frames, translated by the ExPASy Translate tool.

5′3′ Frame 1
```
aagcttgcatgcctgcaggtcgactctagaggatccccgggtaccgagctcgaattc
 K  L  A  C  L  Q  V  D  S  R  G  S  P  G  T  E  L  E  F
```

5′3′ Frame 2
```
aagcttgcatgcctgcaggtcgactctagaggatccccgggtaccgagctcgaattc
  S  L  H  A  C  R  S  T  L  E  D  P  R  V  P  S  S  N
```

5′3′ Frame 3
```
aagcttgcatgcctgcaggtcgactctagaggatccccgggtaccgagctcgaattc
   A  C  M  P  A  G  R  L  -  R  I  P  G  Y  R  A  R  I
```

3′5′ Frame 1
```
gaattcgagctcggtacccggggatcctctagagtcgacctgcaggcatgcaagctt
 E  F  E  L  G  T  R  G  S  S  R  V  D  L  Q  A  C  K  L
```

3′5′ Frame 2
```
gaattcgagctcggtacccggggatcctctagagtcgacctgcaggcatgcaagctt
  N  S  S  S  V  P  G  D  P  L  E  S  T  C  R  H  A  S
```

3′5′ Frame 3
```
gaattcgagctcggtacccggggatcctctagagtcgacctgcaggcatgcaagctt
   I  R  A  R  Y  P  G  I  L  -  S  R  P  A  G  M  Q  A
```

All six reading frames, three forward and three in reverse, are presented with one-letter abbreviations for the amino acids. Dashes represent termination codons. Note that four of the six reading frames are open from end to end, allowing read-through translation at the insertion site in these frames.

7.3 FINDING OPEN READING FRAMES

Although the above Translate tool can be used to identify open reading frames, there are other tools that give you graphical output to help you choose reading frames from larger sequences. This can be a frequently encountered problem; does an EST or a section of genomic DNA encode a protein? Perhaps the identification of large reading frames will help.

If you started with an unknown DNA sequence and knew that a protein-coding region was present, what properties in the DNA sequence could lead you to identifying this coding region? Similarity to known genes or proteins is one property, and this has been covered extensively in Chapters 3 (BLASTN) and 5 (BLASTX). An additional approach is to look for **open reading frames** (ORFs). In order to encode a protein, the DNA sequence must be free of translation termination codons for an extended length of one reading frame. This "open" reading frame would then end with a termination codon marking the 3P boundary of the coding region. If you examine the other forward reading frames in the same region, you almost always see frequent termination codons that could remove these frames from consideration as coding sequence. If DNA sequence were completely random, you would encounter termination codons three times out of every 64 codons (see Figure 4.2) or about once every 64 nucleotides.

Of course, very small real coding regions could be hidden among random stretches of DNA sequence that happen to have very few stop codons. Although many mammalian proteins are hundreds or thousands of amino acids long, there are many real proteins that are less than 100 amino acids in length. Also, remember that the first nucleotide of all three termination codons is T (TAG, TGA, TAA), so a DNA sequence with few Ts could have many long stretches of sequence without any terminators but not necessarily code for a protein. Other data must be gathered to predict if a short reading frame is real: for example, does the encoded protein sequence show any similarity to anything in the database?

In small genomes, space is often limiting so you see very little noncoding sequence. Unlike mammalian genomes where you may have thousands of nucleotides between genes, bacterial or viral genomes can have genes separated by 100 nucleotides or less. Overlapping genes are not uncommon. Evolutionary constraints on these overlapping coding regions must be very strong since a single nucleotide change could alter two protein sequences at once.

The NCBI ORF Finder

To demonstrate the capabilities of the NCBI ORF Finder, let's look at some genes that were studied in earlier chapters. Go to www.ncbi.nlm.nih.gov/gorf/gorf.html, and enter NM_181744, which is the accession number for a human opsin-5 transcript (first encountered in Section 6.2). Notice that you may also enter a sequence in FASTA format. Selecting the standard genetic code (the default in the drop-down menu), click on the "OrfFind" button.

When the window refreshes, this tool shows you a graphic of ORFs in all six frames as colored boxes. By using a drop-down menu, you can raise or lower the minimum length of the ORFs shown (you have to hit the "Redraw" button to refresh the view). This filtering can reduce the number of small ORFs from your view since there are many that are not biologically real. But if you don't find what you are looking for with a large window, you can lower this parameter and examine smaller coding regions. By clicking on the graphic of the open reading frame, the predicted translation will then be displayed below it. In addition, the table listing

Figure 7.8 The NCBI ORF Finder tool. A transcript of human opsin-5, NM_181744, was analyzed and shown are ORF predictions, both in graphic and table form. The sequence goes 5P to 3P, left to right, respectively, and each reading frame gets a separate row in the graphic. The largest ORF of this transcript was clicked on, changing its color, marking the table listing with the same color, and showing the translation of that ORF below (only a portion of the translation is shown in this figure).

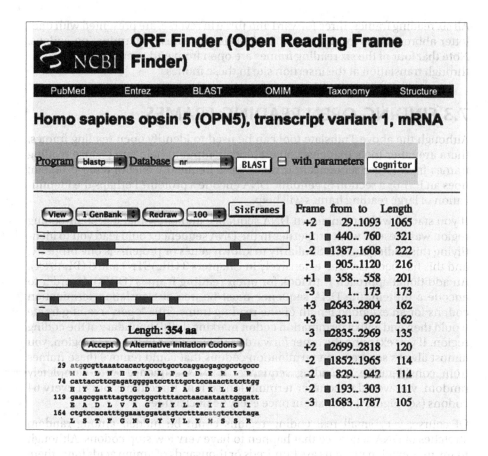

the ORFs refreshes and indicates which ORF you have selected (**Figure 7.8**). A nice feature is that you can use a built-in BLASTP function to test your ORF for sequence similarity to anything already in a small handful of databases.

Looking at the results carefully, a 1065 nucleotide ORF is the largest and was found at the extreme 5P end. The total length of the transcript is 3496 nucleotides, the ORF representing only 30% of the total length. This is typical for vertebrate genes, with a relatively short 5P UTR (in this case, 28 nucleotides) and an extensive 3P UTR.

Now, let's compare three other transcripts: NM_014244, NM_198904, and NM_001492. As seen in **Figure 7.9A** and Figure 7.9B, the largest ORF is at the 5P end of the transcript. Checking the annotation of these two genes, *ADAMTS2* and GABA A receptor gamma 2, confirms that these large ORFs are the correct choices. However, Figure 7.9C shows a transcript requiring contemplation. This transcript, NM_001492, is a clear deviation from the pattern seen so far and is from a very unusual gene. This gene, *GDF1*, encodes two different proteins from a single transcript and is referred to as "bicistronic" to reflect this. The largest ORF at the 5P end encodes the LASS1 protein, and the largest ORF at the 3P end encodes the GDF1 protein. There are two RefSeq records: NM_001492 and NM_021267 for *GDF1* and *LASS1*, respectively. In the annotation of each, there is a "misc feature" which points out that the other ORF and coding region exist. Even without the complication of the second large ORF, there are four other ORFs each over 500 nucleotides. Using the convenient BLASTP function within the ORF Finder shows that there is no record of these other ORFs producing a protein.

As mentioned before in Chapter 5, because of the tendency for reverse transcriptase to fall off mRNA prematurely, there is a 3P bias in many cDNA libraries. The result is that the 5P UTR of many cDNAs may be very short and not the correct length. Only with careful and often different cDNA synthesis techniques are the full-length 5P UTRs identified. Regardless, the ORF will tend to be at the 5P end of the cDNA, with a 5P UTR of several hundred nucleotides or less. Of course,

(A)

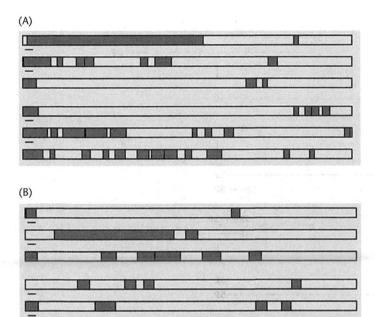

(B)

(C)

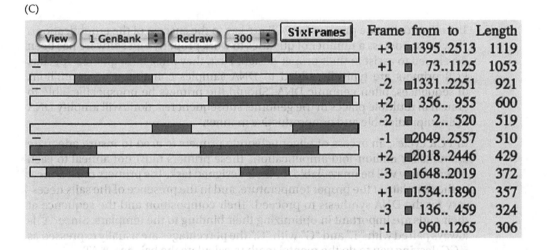

Figure 7.9 ORF Finder graphics of three human transcripts. (A) *ADAMTS2*, NM_014244, a 6772 nucleotide transcript with a 3636 nucleotide coding region shown as the largest ORF in the first reading frame. (B) GABA A receptor gamma 2, NM_198904, a 3957 nucleotide transcript with a 1428 nucleotide coding region shown as the largest ORF in the second reading frame. (C) A bicistronic transcript, NM_001492, encoding both *LASS1* (coordinates 73–1125) and *GDF1* (coordinates 1395–2513). The total length of the transcript is 2558 nucleotides.

there are exceptions. Some 3P UTRs can be incredibly long, easily numbering in the thousands of nucleotides. Historically, some cDNAs were annotated as encoding novel proteins but they were later found to be all 3P UTR sequence! Only later, when full-length cDNAs were identified, was the true identity of these DNAs determined, and the "novel" coding regions turned out to be short open reading frames in the 3P UTR.

7.4 PCR AND PRIMER DESIGN TOOLS

The **polymerase chain reaction** (**PCR**) is one of the major tools of molecular biology and this genomic era. Double-stranded DNA is denatured to single strands through the elevation of incubation temperature, and then cooled to allow oligonucleotide **primers** to anneal to the now single-stranded templates. DNA polymerase uses these primers to synthesize the complementary strands for the templates, and then the double-stranded DNA is denatured again to start a new cycle. Only 20 rounds of PCR are needed to increase the amount of starting material by one million-fold, making rare sequences accessible and creating new approaches to cloning and detection. One round of PCR is illustrated in **Figure 7.10**.

Figure 7.10 The polymerase chain reaction. In this cartoon, one round of amplification is illustrated. (A) One double-stranded DNA template. (B) This is heated to melt it to single strands. (C) At cooler temperatures, oligonucleotide primers then anneal to the single-stranded DNA in a sequence-specific manner. (D) The 3P ends of these oligonucleotides act as primers for DNA synthesis by a thermally stable DNA polymerase, generating double-stranded copies of the original starting material.

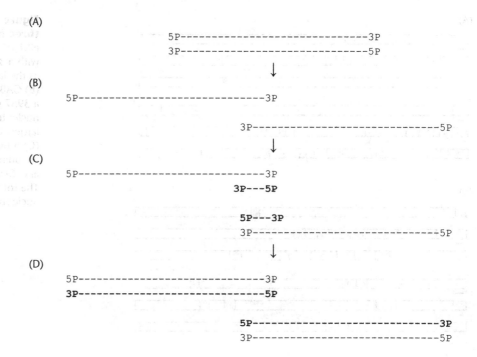

The oligonucleotide primers for PCR are key to the success of the amplifications. They must possess a number of qualities to work properly and software has been developed to assist in their proper design. First, they must be sequence-specific. PCR primers are typically added to DNA samples containing many millions of sequences, often genomic DNA. Should the primers be nonspecific, able to anneal in multiple places in the genome, subsequent reactions will amplify DNA in an unpredictable and nonproductive manner.

In PCR cycles, an excess of oligonucleotide primers is used to insure adequate amounts for million-fold amplification. These primers must not anneal to each other or they will be unavailable for their designed task. The primers must anneal to the template at the proper temperature, and in the presence of the salts necessary for the DNA synthesis to proceed. Their composition and the sequence at their ends are important in optimizing their binding to the template. Since "A" is always paired with "T," and "G" with "C," the percentages are usually expressed as %GC, leaving you to do the mental math to calculate the balance, %AT.

Primer3

Primer3 (frodo.wi.mit.edu/primer3) is a well-respected and widely used Web form for primer design. Two pages of parameters are available for primer refinement should the default settings not provide results. After entering your DNA sequence (the target or "source" sequence for priming), you are given the opportunity to mask common repeat units from a drop-down menu (Mispriming Library) by choosing the species. With literally millions of Alu and L1 repeats in many genomes, you MUST avoid these sequences when performing PCR cycles on genomic DNA or you may end up amplifying sequences from all over the target genome.

To demonstrate the capabilities of the Primer3 Website, paste the DNA sequence for the human beta globin gene, NG_000007, into the source sequence text field. If you paste in the FASTA format (with the annotation line), the form will include the annotation in the results page. For this example, the goal is to design two primers that flank the second exon, nucleotide coordinates 273–495 in this sequence file. To input this information into Primer3, go to the "Targets" field and enter 273,223 representing the starting nucleotide and the length of the exon, respectively. Another way to designate the exon is to surround it with square brackets within the field where you pasted in your sequence. Finally, you also have the option to identify specific regions to avoid ("Excluded Regions"). You may want to exclude sequences of adjacent exons, for example. **Figure 7.11B** shows the results from this example. Note that it is a simple, easy-to-understand text graphic that can easily be pasted into an electronic record of your laboratory work.

Five primer pairs (this is the default amount) are listed in Figures 7.11A and 7.11D, and one pair is displayed on the sequence in Figure 7.11B. Coordinates, melting temperature, and the primer sequences are displayed along with statistics in Figure 7.11E, including how many possibilities were considered (in this case, in the thousands!). Notice that the "Right" (downstream) primer design encountered many more problem sequences than the "Left" primer; downstream primers were rejected because of "bad GC%" and the melting temperature was a problem for many candidates.

> **STS: What's that?**
> Looking at the annotation of the human beta globin gene region (NG_000007) there are many, many locations annotated as "STS." These are Sequence Tagged Sites, which are short (100–500 nucleotides) sequences that are unique in the genome and were identified and designated for the purposes of generating large genome maps. If you have enough STSs, you can put together a map with accurate distances. Once they are placed on the map, they can be used (in pairs or more) to identify distances, insertions or deletions, or chromosomal rearrangements.

(A)

```
PRIMER PICKING RESULTS FOR gi|28380636|ref|NG_000007.3| Homo sapiens
beta globin region (HBB@); and hemoglobin, beta (HBB); and hemoglobin,
delta (HBD); and hemoglobin, epsilon 1 (HBE1); and hemoglobin, gamma A
(HBG1); and hemoglobin, gamma G (HBG2), RefSeqGene on chromosome 11

No mispriming library specified
Using 1-based sequence positions
OLIGO           start  len     tm     gc%    any    3' seq
LEFT PRIMER       225   20   60.19  50.00   5.00   3.00 CCTAAGCTGATTCGGCCATA
RIGHT PRIMER      522   20   59.57  45.00   8.00   2.00 CAATGCATGCAGAAAGAAGC
SEQUENCE SIZE: 2870
INCLUDED REGION SIZE: 2870

PRODUCT SIZE: 298, PAIR ANY COMPL: 4.00, PAIR 3' COMPL: 2.00
TARGETS (start, len)*: 273,223
```

(B)

```
  1 GGATCCTCACATGAGTTCAGTATATAATTGTAACAGAATAAAAAATCAATTATGTATTCA

 61 AGTTGCTAGTGTCTTAAGAGGTTCACATTTTTATCTAACTGATTATCACAAAAATACTTC

121 GAGTTACTTTTCATTATAATTCCTGACTACACATGAAGAGACTGACACGTAGGTGCCTTA

181 CTTAGGTAGGTTAAGTAATTTATCCAAAACCACACAATGTAGAACCTAAGCTGATTCGGC
                                              >>>>>>>>>>>>>>>>>

241 CATAGAAACACAATATGTGGTATAAATGAGACAGAGGGATTTCTCTCCTTCCTATGCTGT
    >>>>                 ************************************

301 CAGATGAATACTGAGATAGAATATTTAGTTCATCTATCACACATTAAACGGGACTTTACA
    ***********************************************************

361 TTTCTGTCTGTTGAAGATTTGGGTGTGGGGATAACTCAAGGTATCATATCCAAGGGATGG
    ***********************************************************

421 ATGAAGGCAGGTGACTCTAACAGAAAGGGAAAGGATGTTGGCAAGGCTATGTTCATGAAA
    ***********************************************************

481 GTATATGTAAAATCCACATTAAGCTTCTTTCTGCATGCATTGGCAATGTTTATGAATAAT
    ***************          <<<<<<<<<<<<<<<<<<<<

541 GTGTATGTAAAAGTGTGCTGTATATTCAAAAGTGTTTCATGTGCCTAGGGGTGTCAAATA

601 CTTTGAGTTTGTAAGTATATACTTCTCTGTAATGTGTCTGAATATCTCTATTTACTTGAT

661 TCTCAATAAGTAGGTATCATAGTGAACATCTGACAAATGTTTGAGGAACAATTTAGTGTT
```

(remaining sequence deleted from this figure)

Figure 7.11 Primer3 results for NG_000007. In this figure, the results of Primer3 have been divided and pasted into five panels. (A) The "best" primer, or at least one of many that are acceptable. (B) The input sequence, decorated to indicate the primer locations upstream and downstream of the target sequence. (continued overleaf)

Figure 7.11 Primer3 results for NG_000007 (continued). (C) The legend for this part of the figure. (D) Four other primers along with important numbers and the sequences. (E) The statistics of the primer analysis.

(C)
```
KEYS (in order of precedence):
****** target
>>>>>> left primer
<<<<<< right primer
```

(D)
```
ADDITIONAL OLIGOS

                  start  len      tm    gc%    any     3'   rep seq

1 LEFT PRIMER       229   21   59.76  57.14   5.00   2.00  11.00 GGCACTGACTCTCTCTGCCTA
  RIGHT PRIMER      514   20   59.17  55.00   3.00   0.00  10.00 AAGCGTCCCATAGACTCACC
  PRODUCT SIZE: 286, PAIR ANY COMPL: 5.00, PAIR 3' COMPL: 0.00

2 LEFT PRIMER       229   21   59.76  57.14   5.00   2.00  11.00 GGCACTGACTCTCTCTGCCTA
  RIGHT PRIMER      523   21   59.95  42.86   2.00   2.00  11.00 GAAAACATCAAGCGTCCCATA
  PRODUCT SIZE: 295, PAIR ANY COMPL: 4.00, PAIR 3' COMPL: 2.00

3 LEFT PRIMER       216   21   60.64  47.62   3.00   1.00  12.00 TGGGTTTCTGATAGGCACTGA
  RIGHT PRIMER      514   20   59.17  55.00   3.00   0.00  10.00 AAGCGTCCCATAGACTCACC
  PRODUCT SIZE: 299, PAIR ANY COMPL: 4.00, PAIR 3' COMPL: 2.00

4 LEFT PRIMER       229   21   59.76  57.14   5.00   2.00  11.00 GGCACTGACTCTCTCTGCCTA
  RIGHT PRIMER      516   21   60.27  52.38   3.00   0.00  10.00 TCAAGCGTCCCATAGACTCAC
  PRODUCT SIZE: 288, PAIR ANY COMPL: 5.00, PAIR 3' COMPL: 1.00
```

(E)
```
Statistics
ADDITIONAL OLIGOS

                  start  len      tm    gc%    any     3' seq

1 LEFT PRIMER       229   20   59.81  45.00   5.00   0.00 AGCTGATTCGGCCATAGAAA
  RIGHT PRIMER      522   20   59.57  45.00   8.00   2.00 CAATGCATGCAGAAAGAAGC
  PRODUCT SIZE: 294, PAIR ANY COMPL: 4.00, PAIR 3' COMPL: 1.00

2 LEFT PRIMER       228   20   59.81  45.00   5.00   0.00 AAGCTGATTCGGCCATAGAA
  RIGHT PRIMER      522   20   59.57  45.00   8.00   2.00 CAATGCATGCAGAAAGAAGC
  PRODUCT SIZE: 295, PAIR ANY COMPL: 4.00, PAIR 3' COMPL: 2.00

3 LEFT PRIMER       230   20   59.67  50.00   5.00   0.00 GCTGATTCGGCCATAGAAAC
  RIGHT PRIMER      522   20   59.57  45.00   8.00   2.00 CAATGCATGCAGAAAGAAGC
  PRODUCT SIZE: 293, PAIR ANY COMPL: 4.00, PAIR 3' COMPL: 0.00

4 LEFT PRIMER       225   20   60.19  50.00   5.00   3.00 CCTAAGCTGATTCGGCCATA
  RIGHT PRIMER      523   20   59.42  45.00   8.00   0.00 CCAATGCATGCAGAAAGAAG
  PRODUCT SIZE: 299, PAIR ANY COMPL: 4.00, PAIR 3' COMPL: 1.00

Statistics
          con   too    in    in          no    tm    tm  high  high        high
          sid  many   tar  excl   bad    GC   too   too   any    3'  poly   end
          ered   Ns   get   reg   GC% clamp   low  high compl compl     X  stab    ok
Left      2413    0     0     0   246     0  1667   100     0     0     5     9   386
Right    22074    0     0     0  1596     0 13230  3046     0     0    96   136  3970
Pair Stats:
considered 1539386, unacceptable product size 1539380, ok 6
primer3 release 1.1.4
```

The Primer3 software is available for download to other computers and so implementations of this software are found elsewhere on the Web.

Primer-BLAST

The NCBI version of the above Primer3 program is called "Primer-BLAST," and can be found within the Resource List on the NCBI home Web page. The same sequence used in the Primer3 example, above, is entered in the NCBI form for easy comparison and the results appear in **Figure 7.12**. Entering the target sequence (the second exon) is different: you enter the "Range" for each primer in the boxes to the right of the sequence entry field. They are 1 through 273, and 495 through 1606, to flank the second exon (nucleotides 273–495).

A nice feature at the NCBI version of Primer3 is the ability to design primers based on the location of exon junctions. The NCBI Web form will take an accession number as input, so the annotation from that GenBank file is utilized to design primers which will span or avoid exon junctions in mRNA. Designing a primer that spans an exon is particularly useful when studying alternative splicing in genes. Primers can be designed which, based on the spanning of exons, can only generate a PCR product if a certain splice form is present in an mRNA or cDNA population. Is there no product after the PCR is run? Then that splice form is not found in that tissue, for example.

In addition to making primer predictions using the Primer3 software, it performs (and, therefore, you must wait longer for the results) a BLASTN search using your source DNA to insure the specificity of the primers. Be sure to enter the species in the Organism field so this feature works properly. As seen in the graphic appearing in the results window (Figure 7.12), the NCBI BLAST search found five acceptable primer pairs. Clicking on "Advanced parameters" will reveal additional options for the design.

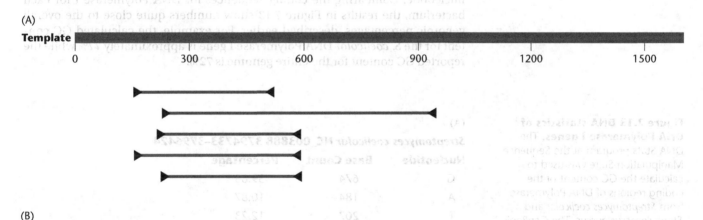

(A)

(B)

Primer pair 1

	Sequence (5'->3')	Template strand	Length	Start	Stop	Tm	GC%	Self complementarity	Self 3' complementarity
Forward primer	AGAGACTGACACGTAGGTGCCTT	Plus	23	157	179	57.21	52.17	6.00	2.00
Reverse primer	TGCCAATGCATGCAGAAAGAAGC	Minus	23	525	503	57.14	47.83	8.00	2.00
Product length	369								

Figure 7.12 NCBI Primer-BLAST output for the human beta globin gene, with exon two as the target. (A) The NCBI implementation of Primer3 has a simple graphic showing the primer locations indicated as triangles pointing in the direction of primer extension. The target sequence lies in the region between primers, indicated as a gray bar. (B) Important information is also displayed for each primer: the primer sequences, product length, location, melting temperature (Tm), and %GC.

7.5 MEASURING DNA AND PROTEIN COMPOSITION

DNA Stats

The base-pair composition of genomes varies significantly, with much debate about what causes this variability. Theories include codon usage, optimal growth temperature, and genome size, with comparisons done at the exon, gene, and genome level. Some discoveries in this area are not well understood. For example, the well-known bands on human chromosomes have a low GC content, but the connection to genome structure is not clear. Base-pair composition sometimes has a practical meaning. For example, as mentioned in Section 7.4, GC content is also a prominent parameter for the proper design of PCR primers.

If you were interested in determining the GC content of a piece of DNA, there are simple utilities on the Web to perform these calculations. One program is DNA Stats, which is found in the Sequence Manipulation Suite, visited earlier with Reverse Complement, www.bioinformatics.org/sms2/index.html.

As a demonstration, let's examine the GC content of a gene, DNA Polymerase I, conserved between two species of bacteria. According to the BioNumbers Website, bionumbers.hms.harvard.edu/default.aspx, the bacterium *Streptomyces coelicolor* has a genomic GC content of 72% while for another bacterium, *Staphylococcus aureus*, it is 33%. This difference is so large you can easily see this by visual examination of DNA sequences. The input is DNA sequence in FASTA format and this is pasted into the single input window (the accession numbers and genomic coordinates for this example appear in **Figure 7.13**). Be sure to pay attention to what appears after the ">" sign for your sequence as these words or description will conveniently appear in the results. After hitting submit, DNA Stats quickly returns a table of percentages for each nucleotide. Comparing the coding sequences for DNA Polymerase I for each bacterium, the results in Figure 7.13 show numbers quite close to the overall genomic percentages, described earlier. For example, the calculated GC content for the *S. coelicolor* DNA Polymerase I gene is approximately 77% while the reported GC content for the entire genome is 72%.

Figure 7.13 DNA statistics of DNA Polymerase I genes. The DNA Stats program at the Sequence Manipulation Suite was used to calculate the GC content of the coding regions of DNA Polymerase I from *Streptomyces coelicolor* and *Staphylococcus aureus*. The GenBank accession numbers and genomic coordinates appear next to the species. The output from DNA Stats was modified for this figure.

(A)

Streptomyces coelicolor NC_003888 3794733–3796424

Nucleotide	Base Count	Percentage
G	674	39.83
A	184	10.87
T	207	12.23
C	627	37.06

Total GC content: 39.83 + 37.06 = 76.89

(B)

Staphylococcus aureus NC_002952 1835366–1832736

Nucleotide	Base Count	Percentage
G	500	19.00
A	976	37.10
T	807	30.67
C	348	13.23

Total GC content: 19 + 13.23 = 32.23

Composition/Molecular Weight Calculation Form

A related type of analysis is to determine the amino acid composition of protein sequences. Let's say you are annotating a protein sequence and you suspect that a particular region may have different properties than the rest of the protein. Could this region be rich in a particular amino acid? The Protein Information Resource (PIR) at Georgetown University has a utility called "Composition/ Molecular Weight Calculation Form" (pir.georgetown.edu/pirwww/search/ comp_mw.shtml) which gives you the percentage of each amino acid for a given protein sequence.

To demonstrate this form, **Figure 7.14** shows the amino acid profile of a region of zinc finger MIZ domain-containing protein 1, UniProtKB sequence Q6P1E1, which is a 223 amino acid domain annotated as "proline rich." What proline composition is needed for the label "proline rich"? The PIR utility indicates that this region is 26% proline (Figure 7.14A). This is more than double the value for the next most abundant amino acids, glycine or serine. Compare this to the rest of the protein (Figure 7.14B), avoiding another 134 amino acid domain that is also annotated as proline rich: the remainder of this protein is only 8.7% proline. Being rich in one amino acid comes at the expense of showing less diversity in other amino acids. Comparing Figures 7.14A and 7.14B, many of the other amino acids have a higher percent in panel B (but note that the vertical scale is different between panels).

To take the composition comparison to a much higher level, the "Release Notes" of the Swiss-Prot database (www.expasy.ch/sprot/relnotes/relstat.html) have a chart of the amino acid frequencies for the entire database. **Figure 7.15** shows that if you look at over 530,000 protein sequences, proline is found at 4.70% of all amino acids.

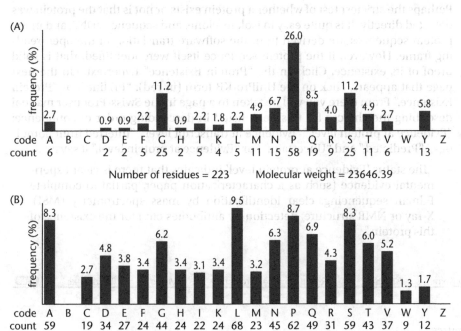

Figure 7.14 Amino acid composition of zinc finger MIZ domain-containing protein 1 (Q6P1E1) from the Protein Information Resource (PIR). (A) The Composition/Molecular Weight Calculation Form at the PIR was used to calculate the percent proline composition in a 223 amino acid domain of Q6P1E1 annotated as "proline rich." (B) The percentages of proline differ by twofold compared to the rest of the same protein.

Ala (A) 8.26	Gln (Q) 3.93	Leu (L) 9.66	Ser (S) 6.55
Arg (R) 5.53	Glu (E) 6.75	Lys (K) 5.85	Thr (T) 5.34
Asn (N) 4.06	Gly (G) 7.08	Met (M) 2.42	Trp (W) 1.08
Asp (D) 5.46	His (H) 2.27	Phe (F) 3.86	Tyr (Y) 2.92
Cys (C) 1.36	Ile (I) 5.97	Pro (P) 4.70	Val (V) 6.87

Figure 7.15 Amino acid composition as the percent of the complete Swiss-Prot database. This table is adapted from the Release Notes of Swiss-Prot database 2012_01.

7.6 ASKING VERY SPECIFIC QUESTIONS: THE SEQUENCE RETRIEVAL SYSTEM (SRS)

Suppose you want to ask a very specific question such as, "How many human proteins are less than 100 amino acids long and involved in ion transport?" The European Bioinformatics Institute (EBI) provides a wonderful service called the Sequence Retrieval System (SRS) (srs.ebi.ac. uk), copyrighted by Lion bioscience AG, where you can ask very specific questions such as this. SRS is a relational database, allowing queries to be made upon multiple fields at once. Of course, SRS gives you the ability to retrieve sequences found by your queries for later analysis. Although most of this book centers on the analysis of one or several sequences at a time, retrieval tools such as SRS let you start thinking big and very specifically.

At the SRS Website, search parameters are entered via several tabbed pages as shown in **Figure 7.16A**. Starting with the "Library Page," you click the checkboxes for the databases you wish to search, shown in Figure 7.16B. In this example, the UniProtKB/Swiss-Prot database is chosen to insure a high-quality curated list of proteins. Next, click on the "Query Form" tab. Although many searches can be launched from here, for this example the "Extended Query" form link under "Tips" (as seen in Figure 7.16C) within the left sidebar is clicked.

When the page refreshes, you will see about two pages of "fields" you can specify. It is here you can make a query very specific by entering values in multiple fields and, therefore, require that the proteins found contain all the features you specify. For this example, let's start by entering "Homo sapiens" under the field name "Species." Under "Sequence Length," enter "100" in the "<=" field. These entries are seen in **Figure 7.17A**.

Perhaps the strictest test of whether a protein exists or not is that the protein was detected directly. It is quite easy to isolate, clone, and sequence cDNA and many protein sequences are derived from the software translation of the open reading frame. However, if the protein sequence itself were identified, that is solid proof of its existence. Click on the "Protein Existence" hypertext. On the next page that appears, click on the UniProtKB term (in red): "PE line" for "Protein Existence." From there you will be taken to a page in the Swiss-Prot user manual describing the choices for this field, listed in descending order of confidence: "Evidence at protein level," "Evidence at transcript level," "Inferred from homology," "Predicted," and "Uncertain." For "Evidence at protein level," it says:

> "The status 'Evidence at protein level' indicates that there is clear experimental evidence (such as a characterization paper, partial to complete Edman sequencing, clear identification by mass spectrometry (MSI), X-ray or NMR structure, detection by antibodies etc.) for the existence of this protein."

Figure 7.16 Navigation within the SRS Website. (A) Primary navigation within the SRS Website (srs.ebi.ac.uk) is through tabbed pages. (B) On the Library Page, choose the database you wish to search. (C) Navigate to the Extended Query Form from the Query Form tab and a link on the left sidebar.

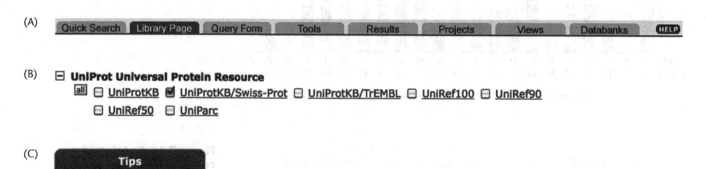

(A)   Quick Search   Library Page   Query Form   Tools   Results   Projects   Views   Databanks   HELP

(B)   ⊟ **UniProt Universal Protein Resource**
   [all] ⊟ UniProtKB   ☑ UniProtKB/Swiss-Prot   ⊟ UniProtKB/TrEMBL   ⊟ UniRef100   ⊟ UniRef90
      ⊟ UniRef50   ⊟ UniParc

(C)   **Tips**

To do more advanced queries, use the **Extended Query** Form.

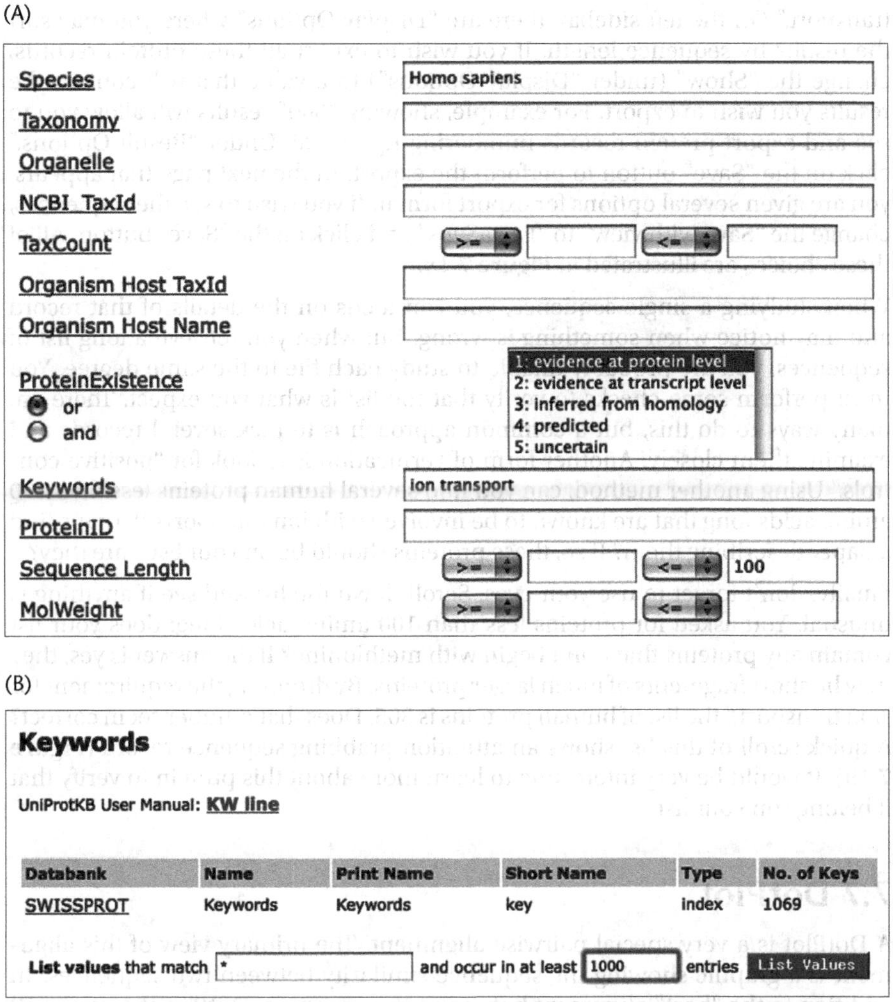

(A)

Species	Homo sapiens
Taxonomy	
Organelle	
NCBI_TaxId	
TaxCount	>= / <=
Organism Host TaxId	
Organism Host Name	

ProteinExistence
- or
- and

```
1: evidence at protein level
2: evidence at transcript level
3: inferred from homology
4: predicted
5: uncertain
```

Keywords	ion transport
ProteinID	
Sequence Length	>= / <= 100
MolWeight	>= / <=

(B)

Keywords

UniProtKB User Manual: **KW line**

Databank	Name	Print Name	Short Name	Type	No. of Keys
SWISSPROT	Keywords	Keywords	key	index	1069

List values that match [*] and occur in at least [1000] entries [List Values]

Figure 7.17 Field entry in SRS. (A) In the Extended Query Form, you have a choice of many search terms. Populate one or more of these fields to query the chosen database for these criteria. (B) The Keywords field may be populated with one or more terms used by the database. To see the choices, search the Keyword list by name or number of entries as shown here.

But it would also be a mistake to not use or trust protein sequences derived from sequenced cDNAs or derived from homology. Nevertheless, in this example, in the extended form, click on the "evidence at protein level."

Each field name is hypertext and you can click here for a list of acceptable terms for each field or an explanation of what a field name means. One of the most powerful fields to use is called "Keywords." Since we chose the Swiss-Prot database to search, we can take advantage of the very extensive annotation provided by the curators of this database. The functions and structures for each protein are listed in the database files. Click on "Keywords" and, in the next page that appears, click on the red "KW line" hypertext to take you to the Swiss-Prot user manual and an explanation of what "Keyword" means.

To see the list of keywords, go back to the query form and, again, click on "Keywords." Instead of clicking on the red hypertext on the next page, click on the "List Values" button and see the multiple pages of keywords that are available for queries. The default list view shows keywords even if the word was used only once. For a shorter list, and the more common keywords, enter 1000 in the entries field before clicking on the "List Values" button, as seen in Figure 7.17B. The keyword "ion transport," and the targeted proteins of our query, appears in over 15,000 Swiss-Prot records (all species). Go back to the form and enter "ion transport" in the Keywords field.

Submit the search and you are presented with a page listing the proteins. They all satisfy the criteria that you required in the query form: they are from *Homo sapiens*, the evidence for the protein existence is at protein level, the length is less than or equal to 100 amino acids, and all the files have the keyword "ion

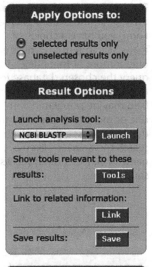

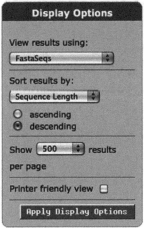

Figure 7.18 Customizing the view or export of the SRS results. By making choices from items on the left sidebar, the results may be sorted, changed in format, displayed on one page or many, and exported.

transport." On the left sidebar, there are "Display Options" where you may sort the results by sequence length. If you wish to export all these protein records, change the "Show" (under "Display Options") to a value that will contain the results you wish to export. For example, showing "500" results will allow you to see and export protein records numbering up to 500. Under "Result Options," click on the "Save" button to perform the export. In the next page that appears, you are given several options for export format. If you wish to see the sequences, change the "Save with view" to "FastaSeqs" and click on the "Save" button. All of these choices are illustrated in **Figure 7.18**.

When studying a single sequence, you can focus on the details of that record and may notice when something is wrong. But when you retrieve a long list of sequences, you are probably unable to study each file to the same degree. You must perform some checks to verify that the list is what you expect. There are many ways to do this, but a common approach is to pick several records and examine them closely. Another form of verification is to look for "positive controls." Using another method, can you find several human proteins less than 100 amino acids long that are known to be involved with ion transport? Perhaps find a paper describing them? If so, these proteins should be on your list—are they?

Finally, don't forget to use your eyes. Scroll down the list and see if anything is unusual. You asked for proteins less than 100 amino acids long: does your list contain any proteins that don't begin with methionine? If the answer is yes, they may be short fragments of much larger proteins. By dropping the requirement for "ion transport," the list of human proteins is 305. Does that number seem correct? A quick scroll of this list shows an attention-grabbing sequence record (**Figure 7.19**). It would be very interesting to learn more about this protein to verify that it belongs on your list.

7.7 DotPlot

A DotPlot is a very special pairwise alignment. The primary view of this alignment is a graphic showing the sequence similarity between two sequences. In addition to the "best" alignment between two sequences, DotPlot shows you all the secondary similarities between a vertical and horizontal line. These similarities are represented as dots on the graph, with good alignments shown as an arrangement of dots in one or more diagonal lines.

Consider the simple alignments in **Figure 7.20** (in real practice, you would never consider such a short sequence). Figure 7.20A shows a DotPlot-like alignment between two identical sequences "ATG" which are arrayed on the vertical and horizontal axes. Wherever the intersection of a row and a column value is identical, a dot appears in the graph. A perfect alignment between two sequences is a diagonal. Figure 7.20B shows the consequence when one sequence (on the horizontal plane) has a small insertion of two Cs which are not found in the other sequence: there is a short diagonal but the dot from the intersection of G and G (G×G) is pushed to the right. Figure 7.20C shows several situations. Like Figure 7.20A, there is a diagonal of two dots showing the similarity between the two sequences, but there is an interruption (two Cs) in the horizontal sequence. DotPlot picks up another alignment, AT aligned with AT, in the corner. Finally, there are two other points of identity (C×C and G×G) that DotPlot detects and shows you as single dots, not part of any diagonal. If you didn't know the

```
>sp|Q156A1|ATX8_HUMAN Ataxin-8;
MQQQQQQQQQQQQQQQQQQQQQQQQQQQQQQQQQQQQQQQQQQQQQQQQQQQQQQQQQQQQQQQQQQQQQQQQQQQQ
QQQQQQQQQQ
```

Figure 7.19 An unusual sequence found with SRS. This sequence appears in the results list of the second SRS search.

sequences of these two DNA records, what can you conclude from this graphic? The 5P end of the horizontal sequence aligns with both the 5P and 3P ends of the vertical sequence. Some similarity exists past the AT alignments, but there is no trend and the dots appear to be random. With these simple examples, you now have the foundation for understanding DotPlots.

Real DotPlots do not compare base by base as shown in the simple examples in Figure 7.20. With only four nucleotides possible, the alignment of two non-identical sequences would result in many meaningless dots spread over your plot. Instead, DotPlots use "windows" or lengths of sequences. For example, the default window size in the DotPlot we'll be using is nine nucleotides. That means that the first nine nucleotides of sequence one are compared to the first nine nucleotides of sequence two. If the nucleotides in the window of nine are identical, then a dot is drawn at the intersection of these two windows. If they are not identical, then no dot is drawn.

For the next comparison, the window must shift. Rather than jump ahead nine nucleotides in each sequence, the next window of nine is started by advancing a single nucleotide in one sequence while keeping the other sequence fixed (**Figure 7.21**). In the absence of a shifting window, nearly identical sequences would fail to generate a diagonal of dots if they were misaligned by one base. Notice that the two sequences in Figure 7.21 are almost identical, but it takes two shifts of the window to align the twelve nucleotides perfectly (Figure 7.21C).

DotPlot of alternative transcripts

Next, let's perform a DotPlot comparison of two transcripts from the same gene, where alternative splicing results in one transcript having an extra exon (**Figure 7.22A**). Here, one transcript of a human potassium channel gene (NM_032115) uses all six exons, drawn as adjacent rectangles. Another transcript (NM_001135107) skips exon five and this is shown as a missing rectangle. Using the DotPlot Web form at Colorado State University (www.vivo.colostate.edu/molkit/dnadot/index.html), place the sequence of NM_032115 in the window labeled "DNA Number 1" (horizontal axis) and the sequence of NM_001135107 in the window labeled "DNA Number 2" (vertical axis). Note that the input is just sequence, so do not include the annotation line from a FASTA format file. Click on the "Make Plot" button and you get the output seen in Figure 7.22B.

It is easier to understand what is going on in this DotPlot by looking again at the exons in Figure 7.22A. These rectangular exons are in perfect alignment until exon five, and then exon six of NM_001135107 is pushed to the right. The same thing happens in the DotPlot in Figure 7.22B; the diagonal line is perfect until a region is pushed to the right a certain distance but then resumes the solid diagonal line again. The displacement location and distance can be obtained by clicking on the breaks in the diagonal lines, as instructed in text at the bottom of the plot. Also note that there are some dots off the diagonal in the DotPlot. These are short stretches, perhaps no more than two or three adjacent dots, representing 18–27 (two or three windows of nine) nucleotides in a row that are identical.

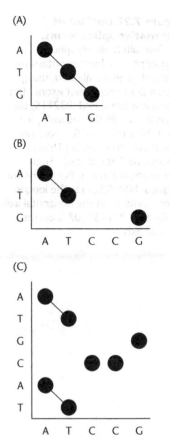

Figure 7.20 Principles of a DotPlot. (A) When nucleotides are identical, a dot is placed in the graph. (B) When the horizontal sequence has two nucleotides that are not present on the vertical sequence, the last dot is pushed to the right. (C) Two sequences compared when similarity is weak. Certain trends are seen (diagonals), representing short regions of identity, but other times there are only single bases in common (unconnected dots) generating a more random pattern.

(A)

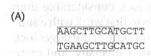

(B)

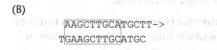

(C)

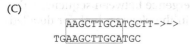

Figure 7.21 The shifting window comparison of a DotPlot. The panels simulate the movement of one sequence while keeping the other fixed. Then windows of nine nucleotides are scored: a perfect match results in a dot. If the match is not nine out of nine, then no dot is drawn.

Figure 7.22 DotPlot of alternative splice forms.
(A) Two alternatively spliced transcripts of a human potassium channel gene are aligned, using rectangles to represent exons. The first splice form, NM_032115, uses exon five while the second splice form, NM_001135107, does not. Not drawn to scale. (B) Using the DotPlot tool at Colorado State University, the two splice forms are aligned. NM_032115, the longer splice form, is on the horizontal axis, while NM_001135107 is on the vertical axis.

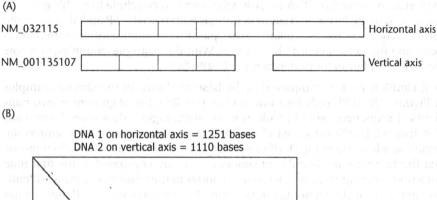

(A)

NM_032115 — Horizontal axis

NM_001135107 — Vertical axis

(B)

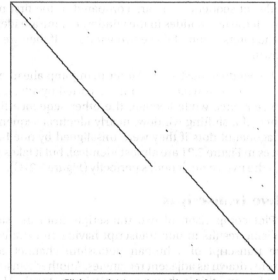

DNA 1 on horizontal axis = 1251 bases
DNA 2 on vertical axis = 1110 bases

Click on plot to get positional data

DotPlot is not meant to be used alone. Although conclusions can be drawn from the graphics, details of the alignment and the junction of the exons can be better seen with some other pairwise alignment tool. Used in conjunction with BLASTN, it should be easier to interpret both results (from DotPlot and BLASTN) to come up with the drawing in Figure 7.22A.

DotPlots of orthologous genes

Let's now consider a more difficult problem. Throughout Chapter 3, you saw how members of the globin family are quite similar, and it was easy to find them using BLAST. We also saw that between species, globin genes could be easily detected. Comparing the human and rabbit beta globin transcripts, there is approximately 87% identity. But we should expect the noncoding parts of the genes to have less identity. Orthologous introns of different sizes or the presence of repeats such as LINEs and SINEs may complicate the analysis further. A DotPlot is an ideal place to start for this type of analysis. Here we will compare the rabbit and human beta globin genes. Note that when comparing much larger loci, considerable time may be needed for each DotPlot. A suggested approach is to first work with a subset of both sequences, for example just aligning a common gene of the larger loci, to save time and computing resources at Colorado State University, and more quickly arrive at the right parameters for the larger comparison.

In this example, a 1620 nucleotide rabbit sequence (GenBank accession number V00882) will be plotted against a 1993 nucleotide human sequence (accession number EF450778). There is a size difference between these two genes, not attributable to alternative splicing. Since there is divergence between sequences, the parameters will be varied several times to attain the cleanest look for detailed interpretation.

Paste the rabbit sequence into the "DNA Number 1" box and the human sequence in "DNA Number 2." Leave the Window Size at the default (nine) and click on the "Make Plot" button. Within moments, the window refreshes and you get the image in **Figure 7.23A**. Even between distantly related human and rabbit, there are many nine-nucleotide sequences with perfect identity as seen by the hundreds of dots. Although some of the identities are random and scattered about the plot, you can see several diagonal lines going from the upper left to lower right. These represent long stretches of sequence that are similar between species and where there are many nine out of nine nucleotides being identical.

Now, let's raise the stringency of the comparison. Just by increasing the window size from 9 to 13 (Figure 7.23B), you see a drop in the random matches (noise) off

Figure 7.23 DotPlots comparison of rabbit and human beta globin loci. Genomic regions of the rabbit (V00882) and human (EF450778) loci are on the horizontal and vertical axes, respectively. Each panel is labeled with the parameters of the alignment.

(A) Window Size 9

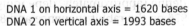

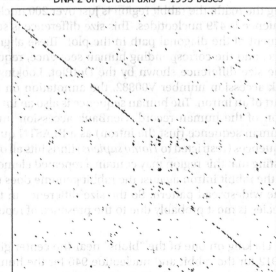

Click on plot to get positional data

(B) Window Size 13

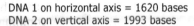

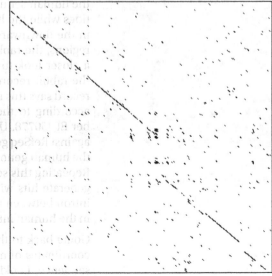

Click on plot to get positional data

(C) Window Size 9, 1mismatch

DNA 1 on horizontal axis = 1620 bases
DNA 2 on vertical axis = 1993 bases

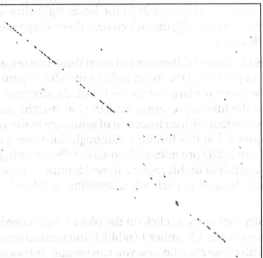

Click on plot to get positional data

(D) Window Size 13, 3 mismatches

DNA 1 on horizontal axis = 1620 bases
DNA 2 on vertical axis = 1993 bases

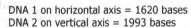

Click on plot to get positional data

the diagonal. Note that only window sizes of odd numbers are accepted by the tool. You are now asking for 13 out of 13 nucleotides to be an exact match, and there are many fewer of them that happen by chance between non-orthologous regions.

A window size of 13 (Figure 7.23B) is cleaner than a window size of 9, but the diagonal lines are weaker. Looking at the largest diagonal (lower third of plot), there are now wide gaps between dots. As a start, experiment with the Mismatch Limit. Let's go back to the window size of nine, but allow one mismatch within the nine-nucleotide window. Enter these numbers into the appropriate fields in the DotPlot tool and hit the Make Plot button. Figure 7.23C shows that noise of the plot goes up again, but the diagonal lines are now much stronger and sometimes longer. Finally, re-plot the comparison with a window size of 13 and three mismatches (Figure 7.23D). There is a slight increase in the noise over the last plot, but comparing Figure 7.23C to Figure 7.23D, the largest diagonal is more continuous with the larger window and mismatch of three. This last window–mismatch combination generates a plot that strikes a balance between signal (diagonals) and noise. Notice that the small set of diagonals in the lower right shows more detail including a small change in the alignment between these diagonals, suggesting small insertions or deletions.

Near the center of the plot is the largest difference between these two sequences. Moving your eyes from the upper left to the lower right of the plot, Figure 7.23D clearly shows that identity between the two sequences is lost, the diagonal is displaced downward, and then the identity resumes until the lower-right corner of the plot is reached. This is consistent with an insertion of sequence in the vertical axis (the longer human sequence) at this location. Although this is seen earlier with the default settings, Figure 7.23D provides additional details including some identities between human and rabbit in this region. These identities can occur in multiple diagonals but closely located to each other, resulting in "blobs" of dots rather than clean lines.

The Colorado State University tool lets you click on the plot and get coordinates for both the horizontal sequence, DNA Number 1 (rabbit), and vertical sequence, DNA Number 2 (human). With those coordinates, you can identify subsections of the full-length sequences and compare these visually or with other tools. For a set of coordinates, let's use the two places where the diagonal across the plot breaks (place one) and then resumes (place two). Clicking on these two places on the plot window, you get two pairs of coordinates (in a yellow bar, just below the window and mismatch boxes). These define (approximately, depending on exactly where you click) coordinates 761–976 on the rabbit sequence and 832–1310 for the human sequence. Doing the math, the rabbit region is just over 200 nucleotides while the human sequence is 479 nucleotides. This size difference is seen in the downward displacement of the diagonal path in the plot. These aligned regions, the rabbit sequence, and the corresponding human sequence, require a closer look to explain the size difference shown by the DotPlot. Looking at the rabbit record, GenBank accession number V00882, the annotation on the record says this region is part of an intron. The human sequence is also an intron according to the annotation of the human record, GenBank accession number EF450778. Using the human sequence (just the intron) as a BLASTN query against RefSeq genomic sequences (restricted to *Homo sapiens*) finds hits all over the human genome, suggesting that this region may contain a repeated element. Repeating this search with the rabbit intron against the rabbit genome does not generate hits with the same widespread pattern. So the size difference in this intron between the two species is most probably due to the presence of repeats in the human intron.

Going back to the DotPlot, clicking on one of the "blobs" near the center gives coordinates of nucleotide 912 for the rabbit and nucleotide 946 for the human sequence. Looking at the two GenBank records at these locations, you see the sequences in **Figure 7.24**.

Rabbit	TTATTTTCTTTTCATTTTCTGTAACTTTTT
Human:	TTTTAGTTTCTTTTATTTGCTGTTCATAAC

Figure 7.24 Comparison of rabbit and human sequences. These two regions are able to align with themselves in multiple ways, sliding a window of 13 nucleotides and allowing three mismatches.

These sequences are rich in Ts. Remembering the sliding windows used in DotPlot and the stringency of 13 identities and three mismatches, this region would score as dots in a number of alignments on multiple diagonals (a blob) due to the alignment of these stretches of Ts.

7.8 SUMMARY

The central theme of this chapter was the use of specialized tools to accomplish individual tasks encountered in sequence analysis. These tools can range from quite simple, for example Reverse Complement and DNA Stats, to quite powerful, such as Primer3 and the SRS database-searching tool. Through the description and demonstration of these utilities, your box of analysis tools has grown quite a bit and perhaps makes you brave enough to collect others on your own. These tools can complement the others described in this text, allowing you to proceed in laboratory experiments, understand more about your sequences of interest, or begin new projects.

EXERCISES

Spider silk: a workflow of analysis

In the following series of exercises, you will be applying a number of utilities to one gene and its protein.

Long admired for its beauty and texture, silk produced by the domestic silkworm, *Bombyx mori*, has played a major part in trade, exploration, and cultural practices for centuries. This biopolymer is harvested from the insects' cocoons and woven into textiles. This includes clothes, upholstery, and other forms of cloth that are dyed to enhance the reflective nature of the material.

Arachnids, notably spiders, produce several forms of silk, each with a different purpose. These vary in stickiness, strength, and elasticity, accommodating a lifestyle of web building, capture and confinement of prey, reproduction, and locomotion. Produced by special glands within the spider, silks are composed of proteins that change from a liquid to a solid fine thread, moments after leaving the spider's body.

With many thousands of spider species in the world, and only relatively few studied, this biopolymer has great interest and potential for scientific, medical, and commercial interests. Agnarsson, Kuntner, and Blackledge recently conducted a directed search for the strongest spider silk, choosing that of Darwin's bark spider as a leading candidate. This spider builds a massive web, up to 2.8 m^2 (10,000 square inches) in size. This web, large enough to cover a small automobile, is suspended in air on individual threads up to 25 meters long. Although extremely lightweight, this spider's silk threads have tremendous strength and elasticity, and (by weight) have been measured to be 50 times stronger than steel.

A combination of factors makes spider silk very strong. First, the primary protein structure reveals many repeated domains. These domains fold and can stack on top of each other and form intra- and intermolecular hydrogen bonds. Although these chemical bonds are quite weak individually, in large quantities they produce a material that has great strength and elasticity. These properties are quite important for withstanding wind or the entanglement of prey.

Unfortunately, the silk protein genes for Darwin's bark spider have not been determined yet. In fact, the cloning and sequencing of the silk protein genes of any spider is very difficult due to the highly repetitive nature of both the DNA and

> **Opportunity knocks**
>
> Are you a biologist or a computer programmer? Most institutions have people skilled in these fields and there may be ample opportunity to work together on developing your own bioinformatics utility. Many biologists pick up a programming language such as Perl and apply it to the problems they face. Computer science departments are often looking for projects for their students or staff. This is a great opportunity for collaboration, where everyone wins. If your utility does something you can't find elsewhere, consider sharing it with the world. Make sure it is "bulletproof": try to break it, or stress it with a large problem. If it is a solid performer, please share it. Every year, the journal *Nucleic Acids Research* dedicates an entire issue to bioinformatics software, large and small. Browsing this journal and others may inspire ideas and encourage you to publish your work.
>
> Don't forget a help document! When it comes to the Internet, people often have a short attention span, especially for software with no instructions. Do you want to know the most helpful button you could create for your utility? A button that demonstrates what your Web page can do! This button automatically fills in an interesting sequence so you can just push the "submit" button. This "picture" is worth a thousand words.

protein sequences. In this exercise, you will be analyzing a silk gene and protein of *Latrodectus hesperus*, the western black widow spider. The major ampullate spidroin 1 protein was cloned by Ayoub and co-workers and is strikingly different from the genes studied thus far in the book.

1. Retrieve ABR68856, the sequence record of the major ampullate spidroin 1 protein, cloned by Ayoub and co-workers. Copy the FASTA format of this sequence and paste it into a word processor document. Using the "Find and Replace" function in your word processor, remove all the carriage returns and spaces.

2. Examine this sequence by eye. Do you see a pattern that ends with a string of alanines? Most of these repeats end with at least five As so you can use this sequence to break the entire sequence record into individual lines for each repeat. Place a carriage return after each stretch of alanines. It will look something like this

   ```
   VGGAGQGGYGRGGAGQGGAAAAAAAAAA
   GSGQGGYGGQGAGQGGAGAAAAAAAA
   GGAGQGGQGGYGGGGYGQGGAGQGGAGAAAAAAAA
   GGAGQGGYGRGGAGQGGAAAAAAAAA
   GAGQGGYGGQGAGQGGSGAAAAAAAA
   Etc.
   ```

3. Examine these repeats by eye: what are the patterns that you see?
   For an extra and challenging project, behind the protein repeats, the DNA sequence is highly repeated, too. It is very complex to analyze in this fashion, but put the DNA sequence for this silk gene into the Colorado State University DotPlot tool. Using a big window, visualize the repeated domains using a DotPlot.

4. Using the PIR Composition/Molecular Weight Calculation Form, determine the composition of the major ampullate spidroin 1 protein, ABR68856. Be sure to paste the FASTA format into the Web form but do not use the annotation line. Compare the result to the amino acid composition of the entire Swiss-Prot database (Figure 7.15).

5. Using the NCBI ORF Finder Web form, determine the open reading frames for the DNA record, GenBank accession number EF595246, for this protein. By clicking on the ORF Finder graphic, and examining the encoded protein sequence by eye, determine which ORF is the correct one.

6. Having two very large open reading frames is quite unusual. Perhaps this is due to the composition of the DNA sequence. Obtain just the coding region from the DNA record, GenBank accession number EF595246. Using the DNA Stats program from the Sequence Manipulation Suite, determine the base composition of the correct open reading frame, and based on the results propose a reason for the presence of two large reading frames.

7. Since this gene is so unusual, examine the codon usage by using the Codon Usage tool from the Sequence Manipulation Suite. Considering the flexibility of the genetic code, allowing multiple codons to be used for individual amino acids, are there any codons used considerably more than others?

8. You are now interested in cloning the coding region of this silk protein gene. Using the New England Biolabs NEBcutter Web page, enter the DNA sequence for GenBank accession number EF595246. Identify two restriction enzyme sites that would allow you to clone the silk protein open reading frame into the polylinker seen in Figure 7.5. Be sure to identify the sites in the polylinker that would be used in the cloning. The goal is to include the least amount of flanking (noncoding) sequence and you must use commercially available restriction enzymes.

9. Using PCR, you now want to amplify a region of the spider silk protein gene. Place the coding region only of EF595246 into Primer3 and predict five pairs of primers.

10. Using the pair of primers which would generate the largest amplification product, translate the amplified DNA sequence in the ExPASy Translate tool.

11. Place the sequence obtained into the SmaI cloning site of the polylinker in Figure 7.5. Would this sequence be in frame with the frame 1 translation of the polylinker?

FURTHER READING

Agnarsson I, Kuntner M & Blackledge TA (2010) Bioprospecting finds the toughest biological material: extraordinary silk from a giant riverine orb spider. *PLoS One* 5, e11234 (DOI:10.1371/journal.pone.0011234). This is the scientific reference describing the very strong spider silk, to supplement the exercise for this chapter.

Ayoub NA, Garb JE, Tinghitella RM et al. (2007) Blueprint for a high-performance biomaterial: full-length spider dragline silk genes. *PLoS One* 2, e514 (DOI:10.1371/journal.pone.0000514). This reference describes the gene and protein product studied in this chapter's exercises. This is the first published full-length spider silk gene.

Gasteiger E, Gattiker A, Hoogland C et al. (2003) ExPASy: the proteomics server for in-depth protein knowledge and analysis. *Nucleic Acids Res.* 31, 3784–3788. A brief overview of this fantastic proteomics resource. This publication is part of an annual issue of *Nucleic Acids Research* which is dedicated to bioinformatics programs and utilities. Other papers from this same issue appear below.

International Rice Genome Sequencing Project (2005) The map-based sequence of the rice genome. *Nature* 436, 793–800. An easily digestible (pun intended) description of the Rice Genome Project. This work was from the effort of over 200 authors and describes the methods, analysis, and findings.

Milo R, Jorgensen P, Moran U et al. (2010) BioNumbers—the database of key numbers in molecular and cell biology. *Nucleic Acids Res.* 38 (Database issue): D750–D753. A brief description of the Website that may have the number you are looking for.

Stothard P (2000) The Sequence Manipulation Suite: JavaScript programs for analyzing and formatting protein and DNA sequences. *Biotechniques* 28, 1102–1104. This publication gives a brief description of many of the programs generously shared by the Stothard group.

Strachan T & Read AP (1999) Human Molecular Genetics 2, Chapter 6. Garland Science. Much has been written about PCR, conceived of over 25 years ago. This textbook chapter provides an excellent overview of the reaction (and variations, thereof), advantages, disadvantages, applications, and references. Available on the NCBI Bookshelf, www.ncbi.nlm.nih.gov/books.

Vincze T, Posfai J & Roberts RJ (2003) NEBcutter: a program to cleave DNA with restriction enzymes. *Nucleic Acids Res.* 31, 3688–3691. A short paper describing this easy-to-use Website.

Wu CH, Yeh LS, Huang H et al. (2003) The Protein Information Resource. *Nucleic Acids Res.* 31, 345–347. The PIR is home of the Composition/Molecular Weight tool used in this chapter.

Zdobnov EM, Lopez R, Apweiler R & Etzold T (2002) The EBI SRS server-new features. *Bioinformatics* 18, 1149–1150. A short paper describing the SRS, a tool for making very specific database queries.

ACAAGGGACTAGAGAAACCAAAA
AGAAACCAAAACGAAAGGTGCAGAA
AACGAAAGGTGCAGAAGGGGAAACAGATGCAGA
GAAGGGGAAACAGATGCAGAAAGCATC
AGAAAGCATC
ACAAGGGACTAGAGAAACCAAAACGAAAGGTGCAGAAGGGGAAACAGATGCAGAAAGCATC
CTAGAGAAACCAAAA
AGAAACCAAAACGAAAGGTGCAGAA
AACGAAAGGTGCAGAAGGGG
GAAGGGG

CHAPTER 8

Protein Analysis

Key concepts

- Detecting patterns and motifs associated with function
- Visual representations of diversity and structure
- Visualizing and manipulating protein structures
- Detecting whole domains and/or protein families
- Predicting post-translational processing and protein topology

8.1 INTRODUCTION

Whole books have been written on protein analysis but what is covered in this single chapter are some of the major problems and opportunities for studying proteins. The amino acid sequence of a protein can provide clues to its function. The recognition of sequence patterns in protein-coding genes can allow predictions, many of them quite accurate, as to family membership and the role in the cell. Once translated, proteins undergo modification, which adds a further layer of expression and regulation to these sequences. Some common post-translational modifications such as phosphorylation are reversible, allowing "switches" to be turned on and off, repeatedly. These changes, in turn, influence structure, function, and translocation, making the world of proteins a very dynamic place. Amino acids, both adjacent and at distant sites, work together to create the three-dimensional form of the protein, bind other proteins or molecules, and create active sites where enzymatic reactions take place. In this chapter, several protein structures will be studied, manipulated, and optimized to better understand the relationships between sequence, structure, and function.

8.2 FINDING FUNCTIONAL PATTERNS

Nature has been very effective at creating a huge variety of protein sequences to perform the many thousands of functions in living cells. Through trial and error, and natural selection, certain sequences have become the solution to life's problems at the cellular level. Nature is very efficient at reusing sequences when a function is needed. Genes carry the evidence that once something has been invented, it is used again and again, or modified to accommodate a new function. Although some variety is tolerated, a functional unit retains conserved amino acids and that signature can be recognized, either by eye or using software, and used to identify a probable function and membership to protein families.

If you have a collection of proteins that all have the same function, a comparison of the sequences of these proteins can lead to the definition of a signature or **motif** that is necessary, and perhaps sufficient, to produce that function. Closely related orthologs and paralogs are problematic in these studies, since they have so much sequence in common. However, more distantly related proteins that share a common function can more easily point to the essence of that function in the form of a sequence pattern or profile.

A repeating pattern within a zinc finger

To illustrate the discovery of patterns, we'll start with a protein named "zinc finger protein 74" (UniProtKB Q16587). Look carefully at the sequence in **Figure 8.1**. Do you see any repeating patterns? Would you be surprised to learn that there are 12 tandemly repeated sequences present in this protein? Whatever is repeating, the default width of 60 characters for the UniProtKB file is not lining up the amino acids in an obvious pattern.

Figure 8.2 has the same protein sequence with the returns removed and the margins changed to narrow the text field. Now the repeating pattern is becoming obvious because many amino acids are lined up vertically.

If you look at this sequence long enough, a clear pattern is revealed. There is a repeating sequence that ends with a pair of histidines separated by three amino acids. Searching for all of these pairs and placing a return after the second histidine in each repeat creates lines that are 28 amino acids long (**Figure 8.3**).

These are not just simple repeating patterns. In this case, something obvious about the sequence led us to a functional unit, known as a "zinc finger," that is widespread in nature. The repetition in zinc finger protein 74 and the degree of conservation between repeats made it easier to identify the functional unit. The zinc finger has a specific function: it binds nucleic acids and is found in some

Figure 8.1 Zinc finger protein 74, with 12 repeating domains. The default width of 60 amino acids on each line does not align the repeats, reducing our ability to detect them by eye.

```
>sp|Q16587|ZNF74_HUMAN Zinc finger protein 74 OS=Homo
sapiens GN=ZNF74 PE=2 SV=3
MEIPAPEPEKTALSSQDPALSLKENLEDISGWGLPEARSKESVSFKDVAVDFTQEEWGQL
DSPQRALYRDVMLENYQNLLALGPPLHKPDVISHLERGEEPWSMQREVPRGPCPEWELKA
VPSQQQGICKEEPAQEPIMERPLGGAQAWGRQAGALQRSQAAPWAPAPAMVWDVPVEEFP
LRCPLFAQQRVPEGGPLLDTRKNVQATEGRTKAPARLCAGENASTPSEPEKFPQVRRQRG
AGAGEGEFVCGECGKAFRQSSSLTLHRRWHSREKAYKCDECGKAFTWSTNLLEHRRIHTG
EKPFFCGECGKAFSCHSSLNVHQRIHTGERPYKCSACEKAFSCSSLLSMHLRVHTGEKPY
RCGECGKAFNQRTHLTRHHRIHTGEKPYQCGSCGKAFTCHSSLTVHEKIHSGDKPFKCSD
CEKAFNSRSRLTLHQRTHTGEKPFKCADCGKGFSCHAYLLVHRRIHSGEKPFKCNECGKA
FSSHAYLIVHRRIHTGEKPFDCSQCWKAFSCHSSLIVHQRIHTGEKPYKCSECGRAFSQN
HCLIKHQKIHSGEKSFKCEKCGEMFNWSSHLTEHQRLHSEGKPLAIQFNKHLLSTYYVPG
SLLGAGDAGLRDVDPIDALDVAKLLCVVPPRAGRNFSLGSKPRN
```

Figure 8.2 Zinc finger protein 74 with some of the repeating domains aligned. A change of line width has allowed some of the repeated domains to align.

```
>sp|Q16587|ZNF74_HUMAN Zinc finger protein 74 OS=Homo
sapiens GN=ZNF74 PE=2 SV=3
MEIPAPEPEKTALSSQDPALSLKENLEDISGWGLPEARSKESVSFKDVAVDFTQEE
WGQLDSPQRALYRDVMLENYQNLLALGPPLHKPDVISHLERGEEPWSMQREVPRGP
CPEWELKAVPSQQQGICKEEPAQEPIMERPLGGAQAWGRQAGALQRSQAAPWAPAP
AMVWDVPVEEFPLRCPLFAQQRVPEGGPLLDTRKNVQATEGRTKAPARLCAGENAS
TPSEPEKFPQVRRQRGAGAGEGEFVCGECGKAFRQSSSLTLHRRWHSREKAYKCDE
CGKAFTWSTNLLEHRRIHTGEKPFFCGECGKAFSCHSSLNVHQRIHTGERPYKCSA
CEKAFSCSSLLSMHLRVHTGEKPYRCGECGKAFNQRTHLTRHHRIHTGEKPYQCGS
CGKAFTCHSSLTVHEKIHSGDKPFKCSDCEKAFNSRSRLTLHQRTHTGEKPFKCAD
CGKGFSCHAYLLVHRRIHSGEKPFKCNECGKAFSSHAYLIVHRRIHTGEKPFDCSQ
CWKAFSCHSSLIVHQRIHTGEKPYKCSECGRAFSQNHCLIKHQKIHSGEKSFKCEK
CGEMFNWSSHLTEHQRLHSEGKPLAIQFNKHLLSTYYVPGSLLGAGDAGLRDVDPI
DALDVAKLLCVVPPRAGRNFSLGSKPRN
```

```
>sp|Q16587|ZNF74_HUMAN Zinc finger protein 74 OS=Homo sapiens GN=ZNF74 PE=2 SV=3
MEIPAPEPEKTALSSQDPALSLKENLEDISGWGLPEARSKESVSFKDVAVDFTQEE
WGQLDSPQRALYRDVMLENYQNLLALGPPLHKPDVISHLERGEEPWSMQREVPRGP
CPEWELKAVPSQQQGICKEEPAQEPIMERPLGGAQAWGRQAGALQRSQAAPWAPAP
AMVWDVPVEEFPLRCPLFAQQRVPEGGPLLDTRKNVQATEGRTKAPARLCAGENAS
TPSEPEKFPQVRRQRGAG
AGEGEFVCGECGKAFRQSSSLTLHRRWH
SREKAYKCDECGKAFTWSTNLLEHRRIH
TGEKPFFCGECGKAFSCHSSLNVHQRIH
TGERPYKCSACEKAFSCSSLLSMHLRVH
TGEKPYRCGECGKAFNQRTHLTRHHRIH
TGEKPYQCGSCGKAFTCHSSLTVHEKIH
SGDKPFKCSDCEKAFNSRSRLTLHQRTH
TGEKPFKCADCGKGFSCHAYLLVHRRIH
SGEKPFKCNECGKAFSSHAYLIVHRRIH
TGEKPFDCSQCWKAFSCHSSLIVHQRIH
TGEKPYKCSECGRAFSQNHCLIKHQKIH
SGEKSFKCEKCGEMFNWSSHLTEHQRLH
SEGKPLAIQFNKHLLSTYYVPGSLLGAGDAGLRDVDPIDALDVAKLLCVVPPRAGRNFSL
GSKPRN
```

Figure 8.3 Zinc finger protein 74, with the 12 repeats revealed by introducing returns after a histidine pair.

important classes of proteins. For example, the zinc fingers in transcription factors bind DNA elements upstream of and within genes, thus unwinding chromatin and exposing sequences to RNA polymerase for transcription to commence.

Although the repeats are quite similar, the multiple sequence alignment in Figure 8.3 shows that there is not an absolute conservation of sequence. The histidine pair is always present, as is a pair of cysteines toward the N-terminus of these repeats. But many of the other amino acids are variable. If our multiple sequence alignment were to be expanded to include zinc fingers from other proteins, and from other organisms, we would discover the variation tolerated for the motif to still bind nucleic acids. But, does this variation represent allowed substitutions or could the function of each zinc finger be slightly different? The answer is yes for both questions.

Visual inspection of zinc finger protein 74 with easily recognizable repeats was easy enough, and reflects the expansion of one repeat to many within a single gene. But searching for patterns manually is not always that easy. Luckily we have software and databases to help us with more difficult patterns. For years, dedicated scientists have compared proteins with similar functions and have defined patterns for those functions. Laboratory experiments have confirmed that if you change a single amino acid within a pattern, function is lost or severely disabled. Many of biology's patterns and functions have been cataloged and made available in the PROSITE database, www.expasy.ch/prosite, which is at the ExPASy Website, home of UniProtKB and the manually curated SwissProt databases. The PROSITE database, and its associated search tools, contains over one thousand patterns and protein family signatures. Like Swiss-Prot, PROSITE is a very high quality and indispensable tool for analyzing proteins. It can be used to analyze your sequence for the many signatures that have already been defined. Matches to these signatures are critical for designing laboratory experiments and understanding the functions of unknown proteins.

If you take the sequence for zinc finger protein 74 (Q16587) and paste it into the PROSITE form, you would learn that this sequence contains 12 zinc finger C2H2-type domains defined by both a *profile* and a *pattern* (more on the profile, later). The pattern (PS00028, click on this hypertext found on the result page) is defined as follows:

```
C-x(2,4)-C-x(3)-[LIVMFYWC]-x(8)-H-x(3,5)-H
```

The way to interpret this code is as follows:

C-x(2,4)-C	"C, followed by another C, 2 to 4 amino acids away"
x(3)	"3 of any amino acid"
[LIVMFYWC]	"A single amino acid, but it must be L, I, V, M, F, Y, W, or C"
x(8)	"8 of any amino acid"
H	"A single amino acid, and it is H"
x(3,5)	"3 to 5 of any amino acid"
H	"A single amino acid, and it is H"

For additional help in interpreting the code, there is extensive documentation at the PROSITE Website, prosite.expasy.org/prosite_doc.html.

By eye, we found a repeating pattern of 28 amino acids (Figure 8.3). The sequences were not identical, but according to the PROSITE definition for the C2H2-type domains, only four key amino acids, two histidines and two cysteines, are absolutely conserved. In zinc finger protein 74, we see two amino acids between the cysteines, and three amino acids between the histidines. Yet, the PROSITE pattern indicates that two to four amino acids can be seen between cysteines, and three to five amino acids can be seen between histidines. Structural analysis has shown that zinc fingers start two amino acids upstream of the first cysteine. So correcting the spacing of the repeats in Figure 8.3, the pattern is still 28 amino acids long, and the important cysteines and histidines are bold in **Figure 8.4**.

Each PROSITE pattern has an identifier and the zinc finger pattern is PS00028. Click on this hypertext, and PROSITE has information about this signature, including a figure of how the cysteines and histidines team up to bind a zinc atom (**Figure 8.5**).

Also present on this information page, under the "Technical Section," is another PS00028 hyperlink which takes you to additional information, including how reliable this motif is in identifying proteins correctly and assigning function. For this C2H2 zinc finger motif, here is a summary of some of the information provided by PROSITE:

- There are over 13,000 occurrences of this pattern in over 2100 different sequences in UniProtKB. In fact, one sequence has over 35 zinc fingers.

Figure 8.4 Zinc fingers of the zinc finger protein 74.

```
>sp|Q16587|ZNF74_HUMAN Zinc finger protein 74 OS=Homo sapiens GN=ZNF74 PE=2 SV=3
MEIPAPEPEKTALSSQDPALSLKENLEDISGWGLPEARSKESVSFKDVAVDFTQEEWGQLDSPQRALYRDVMLENYQNL
LALGPPLHKPDVISHLERGEEPWSMQREVPRGPCPEWELKAVPSQQQGICKEEPAQEPIMERPLGGAQAWGRQAGALQR
SQAAPWAPAPAMVWDVPVEEFPLRCPLFAQQRVPEGGPLLDTRKNVQATEGRTKAPARLCAGENASTPSEPEKFPQVRR
QRGAGAG
EGEFVCGECGKAFRQSSSLTLHRRWHSR
EKAYKCDECGKAFTWSTNLLEHRRIHTG
EKPFFCGECGKAFSCHSSLNVHQRIHTG
ERPYKCSACEKAFSCSSLLSMHLRVHTG
EKPYRCGECGKAFNQRTHLTRHHRIHTG
EKPYQCGSCGKAFTCHSSLTVHEKIHSG
DKPFKCSDCEKAFNSRSRLTLHQRTHTG
EKPFKCADCGKGFSCHAYLLVHRRIHSG
EKPFKCNECGKAFSSHAYLIVHRRIHTG
EKPFDCSQCWKAFSCHSSLIVHQRIHTG
EKPYKCSECGRAFSQNHCLIKHQKIHSG
EKSFKCEKCGEMFNWSSHLTEHQRLHSE
GKPLAIQFNKHLLSTYYVPGSLLGAGDAGLRDVDPIDALDVAKLLCVVPPRAGRNFSLGSKPRN
```

- There are almost 300 instances, in over 200 proteins, where this pattern was detected in proteins not known to function as zinc fingers. These are "false positives".

- There are 80 times when the pattern was not found in known zinc fingers. These are "false negatives."

Two values are provided directly by PROSITE:

- Precision (true hits/(true hits + false positives)): 97.70%

- Recall (true hits/(true hits + false negatives)): 99.34%

These numbers reveal that a small fraction of the known zinc finger proteins are missed, and there are some false positives. The exceptions can be interesting. In the false positives of any pattern, other factors must be contributing to the function, such as the location of these patterns, the surrounding amino acids, and the location of the protein. For example, should proteins containing zinc fingers get secreted outside the cell, there would be much less opportunity for these patterns to function like other zinc fingers and bind nucleic acids. With false negatives, it would appear that nature has found another way to bind nucleic acids without the pattern being recognized by us.

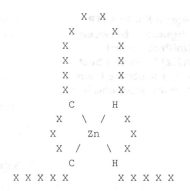

Figure 8.5 The PROSITE figure of PS50157, the C2H2-type zinc finger. Cysteines and histidines are shown binding to a zinc atom. "X" represents any amino acid. Modified from the original PROSITE pattern.

8.3 ANNOTATING AN UNKNOWN SEQUENCE

Now let's use the PROSITE database to help us identify an unknown protein. The sequence is UniProt record A3IZJ7, a "putative uncharacterized protein" from *Cyanothece* sp. CCY0110, a cyanobacterial strain found off the coast of Zanzibar, Africa. Using this 237 amino acid sequence in a BLASTP search against the non-redundant protein database, the best hit is a protease from *Streptococcus uberis*. The alignment from this search is shown in **Figure 8.6**. Although promising, the hit has a low percentage identity (32%), a high Expect value (0.006), and the annotation on this *Streptococcus* sequence is limited.

Compare this to analysis using PROSITE. Entering the UniProt accession number and clicking "Scan" gives a hit to one pattern, "PS00142 ZINC_PROTEASE Neutral zinc metallopeptidases, zinc-binding region signature," and one profile, "PS50215 ADAM_MEPRO ADAM type metalloprotease domain profile." These hits are consistent with the BLASTP result (in which a protease was the best hit), but provide much more information.

> **ⓘ Elvis lives**
> In 1991, Kaper and Mobley wrote a tongue-in-cheek letter which was published in *Science* magazine about the discovery of a sequence motif found in numerous records of the protein database. The motif is glutamic acid-leucine-valine-isoleucine-serine, or ELVIS, and they claimed that this was proof that the deceased singer with the same name was immortalized in protein sequences. They searched the National Biomedical Research Foundation's Protein Identification Resource protein database that had about 25,000 sequence records at the time. For controls, they shuffled the order of the five amino acids, from "ELVIS" to "LIVES" and said that LIVES was not present in the database. In addition, another control of similar size, HAYDN, was also absent, thus supporting their claim.
> Alas, databases have grown since 1991 and their findings have not stood the test of time. The current size of the UniProt database is over 523,000 records and a search of the Swiss-Prot division today using ELVIS as a query does indeed find over 100 hits. But the same is now true for the motifs LIVES and HAYDN so their controls no longer support their theory. However, research by the author of this book shows that a slightly longer motif, HENDRIX (X symbolizing any amino acid, of course), only finds one hit in today's database: the "O-antigen ligase" from *Escherichia coli*. Is this significant? And will it stand the test of time?

```
>ref|YP_002562420.1| protease [Streptococcus uberis 0140J]
Length=243

 Score = 45.1 bits (105),  Expect = 0.006, Method: Compositional matrix adjust.
 Identities = 33/105 (32%), Positives = 52/105 (50%), Gaps = 13/105 (12%)

Query  134  DSSVDYLGVASFSSRKVTNMAVVR-IDQICNASLRADLSINQLSKLLANTIAHEIGHTLG  192
            D+S+D G+A  S+++  + V+R  D  NA   D     S+ + +T HE+GH++G
Sbjct  150  DASLDAAGIAKVQSQEL--LKVIRHADVYLNAYYLLDNQYGYNSERIVHTAEHELGHSIG  207

Query  193  LDHSDLDTDVMQDGVDHRVHCLMPPSFHGEQITLMNHAISKYKDK  237
            LDH D  VMQ        SF+G Q T M   + Y+++
Sbjct  208  LDHKDDKESVMQS---------SGSFYGIQETDMEAVRALYQNE  242
```

A zinc protease pattern

Click on the PROSITE entry number PS00142 found in the above result page. The zinc protease pattern found above is as follows:

```
[GSTALIVN]-{PCHR}-{KND}-H-E-[LIVMFYW]-{DEHRKP}-H-{EKPC}-
[LIVMFYWGSPQ]
```

Like the zinc finger protein, the histidines in this pattern bind a zinc atom. In addition to several strict requirements, that is the two histidines and the glutamic acid, there are three positions (square brackets) where a limited set of amino acids is acceptable. There are additional symbols in this PROSITE pattern, not seen in the zinc finger pattern: the curly brackets. These symbolize the exclusion of amino acids from being found in a position. For example, {PCHR} indicates "any amino acid may be found here except proline, cysteine, histidine, or arginine." There are no Xs (indicating "any amino acid") in any position in this pattern, so the requirements are strict.

This pattern is complex but is only 10 amino acids long. If BLASTP were used in this type of analysis, the BLOSUM62 matrix can find many substitutions for each position, and can be quite successful at finding other ADAM-type metalloproteases. In fact, further down the BLASTP list of weak hits with this query is a metalloprotease with a zinc-binding site (E value 0.16). But the PROSITE pattern implements some strict requirements, for example a histidine must be at certain positions, and only ADAM metalloprotease-specific substitutions can be found using this pattern. The BLASTP search may find these proteases, but could also find other sequences that are missing key amino acids for metalloprotease function. However, these could be sequences that have recently lost the ADAM metalloprotease signature, a topic for another analysis project.

It is interesting that both the zinc finger protein studied above and this protease rely on a zinc atom for structure and function. Except for the two important histidines separated by three amino acids, there is little in common between these two patterns. For function, the zinc finger binds nucleic acids while the protease digests other proteins.

The ADAM_MEPRO profile

The ADAM_MEPRO profile hit (PS50215, found on the original results page) in the above cyanobacteria protein is much more complicated. Unlike a PROSITE pattern, which has a fairly constrained sequence to match proteins, a PROSITE profile is usually built from multiple sequence alignments of larger functional domains. There is often considerable variability seen in each position and, rather than represent it as a very complex pattern, a matrix is built. An example matrix from the PROSITE User Manual (available on the Website under the "Documents" link) appears in **Figure 8.7**.

```
            F   K   L   L   S   H   C   L   L   V
            F   K   A   F   G   Q   T   M   F   Q
            Y   P   I   V   G   Q   E   L   L   G
            F   P   V   V   K   E   A   I   L   K
            F   K   V   L   A   A   V   I   A   D
            L   E   F   I   S   E   C   I   I   Q
            F   K   L   L   G   N   V   L   V   C
```

A	-18	-10	-1	-8	8	-3	3	-10	-2	-8
C	-22	-33	-18	-18	-22	-26	22	-24	-19	-7
D	-35	0	-32	-33	-7	6	-17	-34	-31	0
E	-27	15	-25	-26	-9	23	-9	-24	-23	-1
F	60	-30	12	14	-26	-29	-15	4	12	-29
G	-30	-20	-28	-32	28	-14	-23	-33	-27	-5
H	-13	-12	-25	-25	-16	14	-22	-22	-23	-10
I	3	-27	21	25	-29	-23	-8	33	19	-23
K	-26	25	-25	-27	-6	4	-15	-27	-26	0
L	14	-28	19	27	-27	-20	-9	33	26	-21
M	3	-15	10	14	-17	-10	-9	25	12	-11
N	-22	-6	-24	-27	1	8	-15	-24	-24	-4
P	-30	24	-26	-28	-14	-10	-22	-24	-26	-18
Q	-32	5	-25	-26	-9	24	-16	-17	-23	7
R	-18	9	-22	-22	-10	0	-18	-23	-22	-4
S	-22	-8	-16	-21	11	2	-1	-24	-19	-4
T	-10	-10	-6	-7	-5	-8	2	-10	-7	-11
V	0	-25	22	25	-19	-26	6	19	16	-16
W	9	-25	-18	-19	-25	-27	-34	-20	-17	-28
Y	34	-18	-1	1	-23	-12	-19	0	0	-18

Figure 8.7 An example PROSITE profile, taken from the User Manual. The multiple sequence alignment appears across the top while the 20 amino acids start rows down the left side. The scoring for each amino acid in each position appears at the intersection between the columns and rows.

In this profile, seven rows from a larger multiple sequence alignment are shown on top. For example, the first sequence is "FKLLSHCLLV." Looking at the sequence rows, you can see the levels of variability. The first position in these sequences is phenylalanine, tyrosine, or leucine (other amino acids may appear in this position in the remainder of the protein sequence alignment, not shown). Below this top section of the matrix are rows for each amino acid (labeled on the left) and scoring at the intersection of each column and row. Phenylalanine, tyrosine, or leucine in the first position are given scores of 60, 34, and 14, respectively. Based on the physicochemical properties of the present amino acids, others are given scores based on compatibility (positive score), incompatibility, or infrequency seen in other substitutions (negative score). This scoring is used to judge the quality of matches to the profile and only hits above a threshold are shown in the results.

With a PROSITE profile, you will not see a cartoon as we saw with the zinc finger (Figure 8.5). Instead, there is a link to a "Sequence Logo." The Logo view, first created by Crooks and colleagues in 2004, is a clever way of representing the diversity at a position. The consensus sequence of a Logo goes from left to right (**Figure 8.8** and color plates). At each position, one-letter codes for the amino acids are stacked vertically, with the height of each proportional to the frequency of those amino acids found in that position. At a glance, conserved and nonconserved positions can be picked out of the sequence. Figure 8.8 is a section of the 197 amino acid ADAM type metalloprotease profile, PS50215. Position ten is usually an aspartic acid (D), while position one is most often an arginine (R) or lysine (K), two physicochemically related amino acids. As you can see from this figure, the amino acids that are found in low frequencies are not always legible. Nevertheless, a Logo is an excellent way of presenting a lot of information in a simple display and you will see this used in other analysis tools.

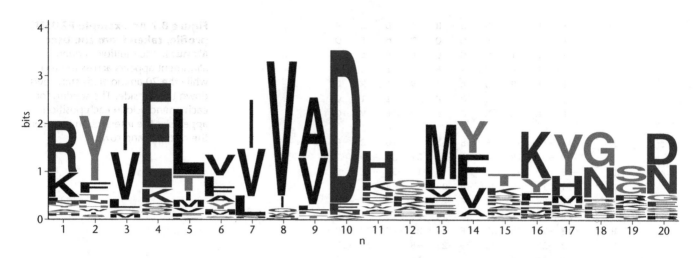

Figure 8.8 Logo for a section of the ADAM type metalloprotease profile, PS50215. The sequence is read from left to right. The variability in a position is represented by stacked one-letter amino acid codes, the height of each being proportional to the frequency found in that position. See color plates for a color version of this figure.

8.4 LOOKING AT THREE-DIMENSIONAL PROTEIN STRUCTURES

Zif268 is another zinc finger protein and has been used as a model system in numerous studies. Also known as "Early growth response protein 1," Zif268 is a mouse transcription factor that binds to CGCCCCGC and other closely related DNA sequences. When the entire 533 amino acid sequence (UniProtKB P08046) is used as a BLASTP query, there is good conservation across species: 89% identity between the mouse and human sequences. However, when using just the three zinc finger domains of Zif268 as a BLASTP query, there is 100% identity between the 89 amino acid mouse sequence and sequences from species as distant as the African clawed frog, *Xenopus laevis*. The three-dimensional structures of the three zinc finger domains of Zif268 have been determined by **X-ray crystallography** and will nicely illustrate the structural features that orchestrate the binding of the zinc atom as well as the overall relationship between the protein sequence and sequence of the DNA to which it binds.

To see the protein structure, navigate directly to the home page of the Research Collaboratory for Structural Bioinformatics (RCSB), www.rcsb.org, the group that maintains the Protein Data Bank (PDB). The PDB is "the single worldwide repository of information about the 3D structures of large biological molecules, including proteins and nucleic acids." Once there, enter the PDB identifier "1aay" (no quotes) into the search field at the top of the page, and click on the return key.

Typically, PDB identifiers are a mix of numbers and letters seen in uppercase. However, it may be confusing at times to interpret the text in uppercase (is that the letter "O" or a zero; is that a number "1" or the letter "L"?). To reduce the confusion, in this book the PDB identifier is shown with lowercase letters.

The first page you see is the "Summary" which includes basic but important information about the structure. For 1aay, there is a publication abstract available, as well as a description of what is contained in the structure. In this case, there are three molecules present: the zinc finger protein plus two short complementary DNA sequences. The tabs across the top of the window provide access to other categories of information. Be sure to explore the other tabs, especially "Literature" for a very extensive list of related papers and structures.

Clicking on the "Sequence" tab will show you detailed information about the protein and DNA sequences. This page has a cartoon illustrating the protein sequence and a summary of the substructures found within the Zif268 sequence (**Figure 8.9** and color plates). The boundaries of the three zinc fingers are labeled across the top of the figure (d1aaya1, d1aaya2, and d1aaya3). The eye-catching parts of this cartoon are the substructures (rows labeled "dssp"). Shown here are the positions of **beta sheets** (arrows), **alpha helices** (spirals), hydrogen-bonded turns (loops), and regions with no defined structure (straight lines). Looking at

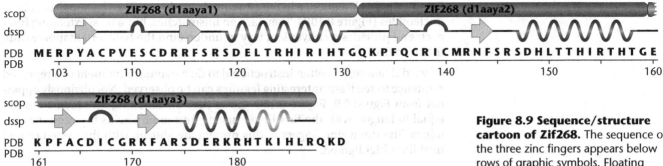

the sequence underneath the symbols, you observe that the histidine pairs are within helical regions while the cysteine pairs are between two beta sheets, separated by a loop.

Across the bottom of the cartoon are amino acid coordinates that deserve some discussion. First, the amino acid numbering in the PDB file (Figure 8.9) does not match the amino acid coordinates (333–421) of the three zinc fingers within the Zif268 of the UniProtKB file, P08046 (see **Figure 8.10A**). Proteins to be crystallized are often not purified directly from their host organism, but instead are genetically engineered and produced in large quantity, for example as a bacterial clone. Looking closely at the figure of the PDB file (Figure 8.9), the first coordinate of the PDB file is amino acid 101, and it is a methionine instead of the normal histidine (Figure 8.10A). So it appears that the three mouse zinc fingers are embedded in some other sequence, probably for the purposes of protein production and purification. This is very common practice and only direct correspondence with the authors or careful reading of their papers will give you the complete story. Nevertheless, it is important to understand the details of the sequence to help interpret the structure.

Second, the structure (Figure 8.9) is derived from just a portion of the complete mouse protein (the region is highlighted in Figure 8.10A). It is also common practice to not crystallize an entire protein. Growing protein crystals is difficult, especially for long sequences, and isolated domains of interest are often crystallized instead.

(A)

```
>sp|P08046|EGR1_MOUSE Early growth response protein 1 OS=Mus musculus GN=Egr1
PE=1 SV=2
MAAAKAEMQLMSPLQISDPFGSFPHSPTMDNYPKLEEMMLLSNGAPQFLGAAGTPEGSGG
NSSSSTSSGGGGGGGSNSGSSAFNPQGEPSEQPYEHLTTESFSDIALNNNEKAMVETSYPS
QTTRLPPITYTGRFSLEPAPNSGNTLWPEPLFSLVSGLVSMTNPPTSSSSAPSPAASSSS
SASQSPPLSCAVPSNDSSPIYSAAPTFPTPNTDIFPEPQSQAFPGSAGTALQYPPPAYPA
TKGGFQVPMIPDYLFPQQQGDLSLGTPDQKPFQGLENRTQQPSLTPLSTIKAFATQSGSQ
DLKALNTTYQSQLIKPSRMRKYPNRPSKTPPHERPYACPVESCDRRFSRSDELTRHIRIH
TGQKPFQCRICMRNFSRSDHLTTHIRTHTGEKPFACDICGRKFARSDERKRHTKIHLRQK
DKKADKSVVASPAASSLSSYPSPVATSYPSPATTSFPSPVPTSYSSPGSSTYPSPAHSGF
PSPSVATTFASVPPAFPTQVSSFPSAGVSSSFSTSTGLSDMTATFSPRTIEIC
```

(B)

```
Polymer 1:    5P-AGCGTGGGCGT-3P
Polymer 2:    3P-CGCACCCGCAT-5P
```

(C)

```
ME
RPYACPVESCDRRFSRSDELTRHIRIHTG
 QKPFQCRICMRNFSRSDHLTTHIRTHTG
  EKPFACDICGRKFARSDERKRHTKIHLR
QKD
```

Figure 8.9 Sequence/structure cartoon of Zif268. The sequence of the three zinc fingers appears below rows of graphic symbols. Floating your mouse over the graphics will reveal their meaning: the spirals are alpha-helical regions, arrows are beta sheets, loops are hydrogen-bonded turns, and triangles are isolated beta-bridges (not shown). See color plates for a color version of this figure.

Figure 8.10 Protein and nucleic acid sequences in the PDB file 1aay. The structure comprises three molecules or "chains," the sequences of which are depicted in these panels. (A) The section of Zif268 that was crystallized is highlighted, showing that the structure of only a section of the full-length protein was determined. (B) Both strands of the DNA sequence are in the structure; these are called "polymers" in the PDB file. Note that Polymer 2 was "flipped" so it can be shown base-pairing with Polymer 1. A single base protrudes at both ends of the double-stranded DNA. (C) This region can be subdivided into three zinc fingers (aligned on their right).

Finally, the DNA sequence in the structure is made up of two base-pairing oligonucleotides (Figure 8.10B). There are no mismatches, but a base extending from each end provides a convenient way to understand the helix orientation within the structure.

As we did before, it is often instructional to do a manual alignment of a repeated sequence to see if any interesting features can be observed. Not obviously apparent from Figure 8.9, Figure 8.10C shows that the zinc finger domains are not equal in length, with the first finger having four amino acids between the cysteines. The three zinc fingers of this protein are shown, with the cysteines and histidines highlighted.

Jmol: a protein structure viewer

Now that you have an understanding of the sequences that were crystallized and the structures that were revealed, it is time to look at and manipulate a high-quality crystal structure image. The default viewer for the PDB is a powerful application, Jmol. We will only touch on a fraction of its abilities here, and you are encouraged to seek more information on the Web such as "An introduction to Jmol Scripting" by Silva and Marcey (www.callutheran.edu/BioDev/omm/scripting/molmast.htm), and the extensive command descriptions at "Jmol interactive scripting documentation" (chemapps.stolaf.edu/jmol/docs).

To launch Jmol, go to the right panel on the PDB "Summary" tab for 1aay and click on the "View in Jmol" button. During the launch, if you are asked if you trust a Java applet's certificate, answer yes. Within seconds, you are then shown a window containing the structure of the Zif268 zinc finger–DNA complex. The DNA strands, complete with bases pointing toward each other, are purple helices. The protein structure follows the icons you already saw in the earlier cartoon: arrows, helices, and regions (now white against the black background) where the structure is not fixed or certain. Look closely, and you can see the gray zinc atoms (spheres).

Each zinc finger has one helical region and two beta sheets so you can easily pick out the three fingers. Float your mouse over almost any part of the structure and (with a pause), a small window pops up and identifies the molecule. For example, floating your mouse over one end (a white line) will find [ARG]103:A.CA #449. This is arginine 103, chain A, alpha carbon number 449. The "A" chain is the protein sequence. Floating your mouse over the other end of the protein will find [ARG]187:A.CA #1172. These two steps identified the N and C termini, respectively, with the same numbering as seen in Figure 8.9.

But the real treat arrives by clicking anywhere, holding the mouse button down, and then moving the mouse. The structure moves! Through this movement, you can fully appreciate the three-dimensional aspect of the structure. You can now see how the protein sequence wraps around the DNA helix, following the major groove. The protein sequence is no longer a linear display of letters, but a polymer that folds back onto itself, bringing nonadjacent regions into close proximity.

There are some additional methods of manipulating the structure:

- Pressing the shift key while moving the mouse down or up will zoom in or out, respectively.

- Pressing the shift key while moving the mouse left or right will cause the structure to rotate in the X-Y plane.

- Press the shift key and double-click the mouse on the structure, holding the mouse button down on the second click. Now, dragging the mouse to the right or left will move the structure to the right or left in the window.

- Pressing the shift key and double-clicking the mouse on the background will center the structure on the screen and reset the position seen when you first launched Jmol.

Until now, you have been observing the structure in one style. Below the black screen are drop-down menus where you can explore different views. The present "Cartoon" style may give you the impression that there are gaps between atoms, but other views (CPK style and Surface choices) will better show how tightly packed the atoms are. On the left sidebar of the Web page, there is a help system with a glossary for brief explanations of the menu choices. Visit www.jmol.org and you will see the extensive documentation available for this tool. After exploring these options, bring the structure back to the default by hitting the "Reset Display" button.

Exploring and understanding a structure

With the basic skills outlined above, you can now explore the details of the structure (best seen with Jmol but visible in **Figure 8.11** and the color plates). Here is a list of four features, easily explored with the Zif268 structure.

1. Orientation of the DNA. Each end of the double helix has a protruding base. According to the annotation within the PDB file and Figure 8.10B, one protruding 5P end is an A (adenine) and the other a T (thymine). As any biochemistry book will show, adenine has two rings, a hexagon and pentagon, while the thymine has a single hexagon. Looking closely at the DNA strands of Zif268 (zoom in if necessary), you can find an unpaired double ring on one end of the helix and an unpaired single ring at the other. With this information, and the bases shown in Figure 8.10B, you can understand the order of the bases on the strands.

2. Orientation of the protein. Floating the mouse over the (white) ends of the protein will reveal the first amino acid (arginine 103) and the last (arginine 187). A more obvious method, seen at a glance in the "Cartoon" view, is to use the symbols for the helices and beta sheets. They are arrow shaped, always pointing toward the C-terminus. You may have to rotate the structure to see the orientation clearly.

3. Observe. Only the helical regions of the protein are interacting with the DNA helix while the beta sheets are to the outside of the protein structure. The three zinc ions (gray spheres) are located between the helical region of each zinc finger and the loop that is between the two beta sheets.

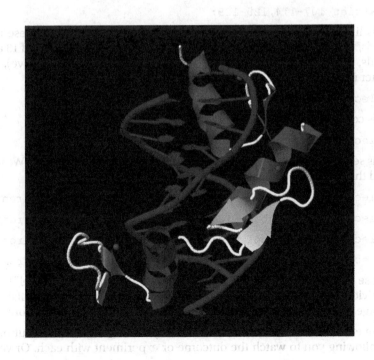

Figure 8.11 Zif268 PDB structure, 1aay. Each zinc finger consists of a helical region, interacting directly with the central DNA helix. External to these helices are two beta sheets that are parallel, joined by a loop. The zinc atoms (spheres) lie between the helix and loop. See color plates for a color version of this figure, although the colors described and shown there may differ from what is seen on your screen.

4. Nucleotide–amino acid relationship. The paper that reported this structure (by Elrod-Erickson et al.) describes how key amino acids of the zinc fingers (that is, the arginines and aspartic acids) make direct hydrogen bonds to the DNA bases. Viewing the structure in "Backbone" style and zooming in shows the amino acids that are closest to the nucleic acids. Float your mouse over the amino acids to identify them.

Jmol scripting

Until now, you have only manipulated the image using your mouse. But there are literally pages of commands that you can execute within Jmol to refine and customize your view of the structure. These commands are entered in the one-line text window below the image. Exploring "An introduction to Jmol Scripting" and "Jmol interactive scripting documentation" mentioned above will either impress you or leave you overwhelmed. Here is a brief demonstration of a **script** used to make a good picture. It will focus on the relationship between the zinc atom and the surrounding protein helix and loop. Together, they form a rigid structure with a conformation that allows the helix to sit in and bind the major groove of the DNA sequence. The Jmol script below will generate a view which is focused on that exact region of the zinc finger.

First, we need to know the region of study. As you recall from Figure 8.5, a pair of cysteines and a pair of histidines are responsible for binding zinc. These amino acids reside in the loop and helical region, respectively, presented in Figure 8.9. Focusing on the first zinc finger of Figure 8.9, the amino acids involved are numbers 107, 112, 125, and 129. Floating your mouse over the protein structure will identify where this finger resides in the structure. While you are there, float your mouse over the zinc near these amino acids and find the atom number. It is number 1183.

Now, in the text entry field labeled "enter advanced Jmol commands..." (below the structure), we will enter a number of commands to limit the view to just this region, and to customize the view. Note that each command ends with a semicolon and this is critical.

1. `select 125,129,112,107; wireframe;`

    This changes these four amino acids from the default "Cartoon" to "Wireframe." This will show the projecting side chains of the amino acids.

2. `restrict 107-114,125-129;`

    This limits the view and the rest of the manipulations to just these amino acids, making everything else disappear. Notice that this is a total of 13 amino acids. Only four will be changed to the wireframe style (step 1, above), so the other nine will appear in the cartoon style.

3. `select all; color amino;`

    This colors the amino acids (for example, cysteines are yellow).

4. `select atomno=1183; spacefill;`

    This selects the zinc atom and enlarges it to the proper atomic size. We identified the zinc atom by floating the mouse over it.

5. `select 125.CA; label His; font label 12; set labelfront;`

    `select 129.CA; label His; font label 12; set labelfront;`

    `select 107.CA; label Cys; font label 12; set labelfront;`

    `select 112.CA; label Cys; font label 12; set labelfront;`

    These commands put labels (any word past the command "label") on the individual amino acids and keep the labels in front while manipulating the image. The ".CA" labels the carbon atom. The "12" refers to size 12 font.

You can enter the above commands separately, hitting the "Submit" button after each, allowing you to watch the outcome or experiment with each. Or you can

(A)

```
select 125,129,112,107; wireframe; restrict 107-114,125-129; select
all; color amino; select atomno=1183; spacefill; select 125.CA; label
His; font label 12; set labelfront; select 129.CA; label His; font
label 12; set labelfront; select 107.CA; label Cys; font label 12; set
labelfront; select 112.CA; label Cys; font label 12; set labelfront;
```

(B)

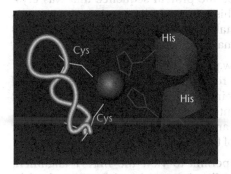

Figure 8.12 A Jmol script for 1aay. (A) The script used. (B) The result of the script. The rest of the structure is invisible due to the "restrict" command. The two histidines and two cysteines were chosen to display in Wireframe view, while the other amino acids are still in Cartoon. These four amino acids were also labeled. The zinc atom was changed to Spacefill view to better represent its size within the structure. See color plates for a color version of this figure.

enter all of these as one command (see **Figure 8.12A**, a single line without carriage returns). It is recommended that you create/troubleshoot/learn with single commands, keeping them in a text document, then later copying and pasting them into the window. Remember, the "Reset Display" button will get you out of trouble should something unexpected happen.

Upon entering this script, an image appears (Figure 8.12B and color plates). Now the relationship between the four amino acids (the cysteines and the histidines) and the zinc atom is clear, especially in the "live" Jmol where you can rotate the structure. The side chains of the amino acids are projecting toward the zinc, as if reaching out and holding the zinc in place. Actually, the interaction is mutual; the presence of the zinc creates the rigid structure between the metal and the protein. By mixing the cartoon, spacefill, and wireframe styles, we see the details necessary to discover the relationships between substructures. Isolating this region from the other amino acids allows a view from any angle without obscuring the details.

8.5 ProPhylER

As sequencing and discovery have become easier, and as we more broadly compare related proteins, it is easier to detect what nature has allowed and, by extrapolation, forbidden. This history of a protein's sequence and structural evolution is recorded in what we see today. This is the view of Binkley and co-authors, developers of ProPhylER, a wonderful data reduction and visualization tool that shows us at a glance the freedom and constraints of evolution upon protein sequences. Through the use of color and icons, ProPhylER simplifies the presentation of complex information. Behind the views that ProPhylER provides is the power of comparative analysis. High-quality and curated multiple sequence alignments of homologs are analyzed and the variation of each position in a protein is captured and shown to us in two ways: the ProPhylER "Interface" and "CrystalPainter."

To get started with ProPhylER, we'll first go to the RCSB Protein Data Bank (PDB) Website, www.rcsb.org/pdb, for some additional information. Enter "1ru0" as the PDB identifier in the search window at the top of the screen and this will bring up the record of a protein called DCoH2. Exploring the annotation of the PDB record, we see that this is a mouse protein and this structure was determined to test if DCoH2 can substitute for a close protein family member tied to diabetes. The resolution of the structure is 1.60 Å, which is excellent (the smaller the

The music of proteins

Many people have recognized a similarity between protein sequences and music. The 20 amino acids, the underlying DNA codons, the structures, the biochemical properties, and the sometimes repetitive or domain nature of protein sequences can be reinterpreted as notes, chords, movements, and the voices of music. Some very creative people have taken protein sequences and turned them into something you can listen to. Examples can be found all over the Internet but there is a nice collection called the "Genetic Music Sourcepage," www.whozoo.org/mac/Music/Sources.htm. The "Music Samples Page" presents an extensive list of proteins that you can hear as beautiful music. This page includes the protein sequences along with their musical interpretation, including beta globin, rhodopsin, and spidroin, proteins we have studied in this book. After clicking on that link, be sure to navigate to the "Protein Primer: A Musical Introduction to Protein Structure" for a wonderful explanation of their methods which are quite sophisticated. Especially enjoyable is a body of work called "Life Music," by composers Clark and Dunn (and Mother Nature), whozoo.org/mac/Music/CD.htm.

number, the finer the detail). Four molecules of DCoH2 protein form a tetramer. It is expected that the faces of the proteins that come in contact with each other will have evolutionary constraints to insure proper interactions, and with ProPhylER we should be able to see the amino acids under the most pressure to be conserved for these interactions. Although normally a tetramer inside cells, this crystal structure is of a dimer that will show us the interacting amino acids between two DCoH2 proteins.

Clicking on the "Sequence" tab, you see the protein sequence that was crystallized (**Figure 8.13**). As seen earlier, this cartoon indicates the substructures that were determined by the X-ray crystallography. You should keep this PDB window open to help interpret the information displayed in ProPhylER.

Now go to the ProPhylER Website, www.prophyler.org, and click on the side-bar "search" hypertext. When the page refreshes, you are given the option of searching with a query sequence, a protein database (PDB) identifier, a UniProt identifier, or other accession numbers. Enter 1ru0 as the PDB identifier and click the "Search" button. The page refreshes and you are given the option to look at the "Interface" or "CrystalPainter" view of the protein.

On this same page, there is also a hyperlink to view the BLAST alignment between the sequence in 1ru0 (the crystallized protein) and Q9CZL5, the full-length UniProt sequence. This is a critical view of information (**Figure 8.14**). Notice that the sequence of 1ru0 is missing 34 amino acids of the N-terminus. As we saw earlier with the structure of Zif268, only a portion of the DCoH2 protein was crystallized and you may get confused about the coordinates. Rather than do the math all the time, you may find it easier to refer to the BLASTP alignment. But note that the coordinates of the query sequence do not exactly match the coordinates of the sequence in the cartoon (Figure 8.13) but are an approximate reference. For example, the fourth, fifth, and sixth amino acids in Figure 8.13 (DAQ) correspond to positions six, seven, and eight in the BLASTP alignment.

The Interface view

Click the ProPhylER "Launch" button for the Interface view and a separate window will appear. Your Internet browser may request permission to launch the window script, which you should allow.

The Interface view shows us a representation of the full-length DCoH2 sequence, with analysis displayed in two main panels (**Figure 8.15** and color plates). At the top of the figure is a plot of the conservation. The sequence goes from left to right, and the degree of conservation deduced from the multiple sequence alignments is shown as a line rising and falling with conserved and nonconserved scores, respectively. On the top of this plot is a row labeled "ECRs"; these are the conserved regions on a gray scale, the darkest representing the most conserved, and the peaks of conservation are numbered (1 and 2 appear in Figure 8.15). In addition to conservation, other measurements such as hydropathy and polarity can be plotted by clicking on the relevant item in the list on the left.

Figure 8.13 Protein Data Bank (PDB) cartoon representation of the DCoH2 sequence and structure, 1ru0.

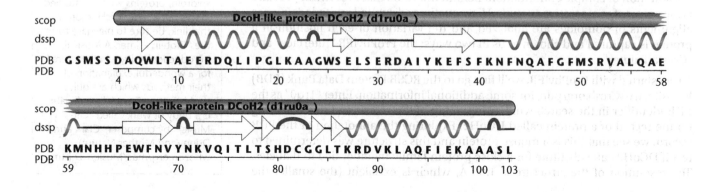

```
Query:      2 SMSSDAQWLTAEERDQLIPGLKAAGWSELSERDAIYKEFSFKNFNQAFGFMSRVALQAEK 61
              +MSSDAQWLTAEERDQLIPGLKAAGWSELSERDAIYKEFSFKNFNQAFGFMSRVALQAEK
Sbjct:     33 AMSSDAQWLTAEERDQLIPGLKAAGWSELSERDAIYKEFSFKNFNQAFGFMSRVALQAEK 92

Query:     62 MNHHPEWFNVYNKVQITLTSHDCGGLTKRDVKLAQFIEKAAASL 105
              MNHHPEWFNVYNKVQITLTSHDCGGLTKRDVKLAQFIEKAAASL
Sbjct:     93 MNHHPEWFNVYNKVQITLTSHDCGGLTKRDVKLAQFIEKAAASL 136
```

Figure 8.14 BLASTP alignment between the crystallized protein sequence from 1ru0 (Query) and the full-length sequence (Subject (Sbjct): UniProtKB Q9CZL5) that appears in the ProPhylER displays. A helpful translation: Query = CrystalPainter coordinates; Subject = Interface coordinates. The underlined sequence is the most conserved location.

Below the plot is the sequence panel of the full-length protein, Q9CZL5. The sequence coordinates are on top, and individual amino acids are now shown. Below that is a "Logo" summary of the major amino acids found in each position, derived from a multiple sequence alignment. The colored window within the top panel, described above, controls the sequence displayed within this bottom panel. Click on the colored window, hold the mouse button down, then move the mouse. By sliding this window left or right, the sequence displayed in the sequence panel will change.

ProPhylER is rich with features and only some can be described here. Click on the "help" or "documentation" hypertext on the ProPhylER home page for additional instruction. The publication describing ProPhylER (see Further Reading) also provides a thorough explanation.

If you slide the colored window and center over amino acid coordinate 100, the most conserved amino acid positions are shown. In Logo view of this region, most of the letters are full height, with no additional stacking present. Clicking on the histidine at position 95 will cause a vertical red line to appear on the top

Figure 8.15 Interface view of mouse DCoH2 in ProPhylER. See color plates for a color version of this figure.

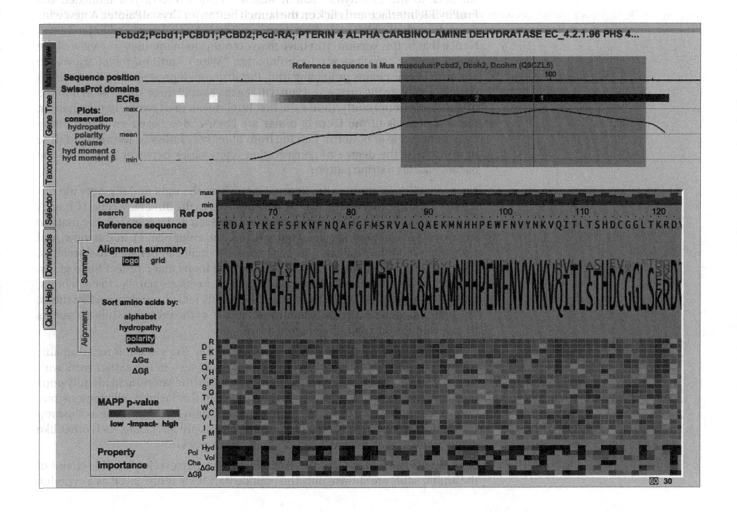

panel, as shown in Figure 8.15. This red line is very close to the "1" on the ECRs bar at the top of the upper panel, also indicating that this location has the highest conservation score.

Below the Logo alignment is a colorful section where all 20 amino acids, labeled on the left, are given a MAPP (multivariate analysis of protein polymorphism) p-value and colored according to this score. MAPP is a measure of the biochemical variability seen in a position and this translates to an "impact" factor: what would happen if this amino acid were substituted here? For regions that are very conserved, such as this one, the predominant color will be a shade of red or yellow, indicating that substitutions of the corresponding amino acids from these colored rows would make a large (negative) impact on the structure and, presumably, function of the protein. Sliding the colored window to coordinates 30–60 would reveal a section that is largely blue: many amino acid substitutions are tolerated here. You should keep this Interface window open as well for later reference.

The CrystalPainter view

So now we have established that one region of DCoH2 is highly conserved, peaking around amino acid 100. This is valuable information. If you were trying to establish the function of this protein, you would have to incorporate the importance nature has placed on this region. Where are the most important regions in the folded protein structure? Here is where a powerful aspect of ProPhylER comes into play. If a crystal structure of a protein is in their database, the conservation scores are "painted" upon the crystal structure, hence the name of the other view, CrystalPainter.

Go back to the ProPhylER Search Results window where you launched the ProPhylER Interface and click on the launch button for CrystalPainter. A new window is launched which includes a version of Jmol, the tool used earlier for Zif268. Notice that in this version, you have many buttons to manipulate the view of the structures. The default view is a combination, "Wire + Cartoon," which shows the amino acid side chains projecting from the structure represented by the Cartoon view. Manipulate the image and you will see the overall three-dimensional shape of the dimer. Importantly, unlike the first Jmol viewer used earlier in this chapter, the amino acids of the DCoH2 dimer are color-coded based on evolutionary constraint (scale at bottom), ranging from blue (very little variation) to red (free to evolve). As the degree of conservation often varies between neighbors, the overall effect is a stripe pattern.

Next, let's click on some of the Button Controls on the right and discover the different views and effects. The two DCoH2 proteins in the dimer are called Chain A and Chain B. Click on the "Cartoon" button for each and it is easy to see that the structure is of two proteins (see **Figure 8.16A** and color plates). In this view, and with some rotation, you can now easily see the small gap between the proteins. Also visible are the helices, beta sheets, and the loops that connect the regions. Rotate the molecules and you see how the beta sheets are parallel to each other, and the helices are so perfect that you can literally look through them like tubes. The helical regions are on the outer rounded face of the proteins while the sheets form an almost flat plane (see Figure 8.16B).

Notice the symmetry of the dimer in Figure 8.16A. The same internal domains of each protein are interacting with each other. As seen in the earlier work with Jmol, by moving the mouse arrow over any position, the amino acid identity pops up (there may be a delay). Exploring the dimer in this fashion will demonstrate that the coordinates of the small loop on top of Chain A are the same as the loop on the bottom of Chain B, for example. The chains are not facing each other like a mirror reflection. Instead, one is rotated 180 degrees.

Now explore the Button Controls. Backbone is the sparsest view: no side chains of the amino acids are shown; only the peptide backbone is displayed as a cylinder.

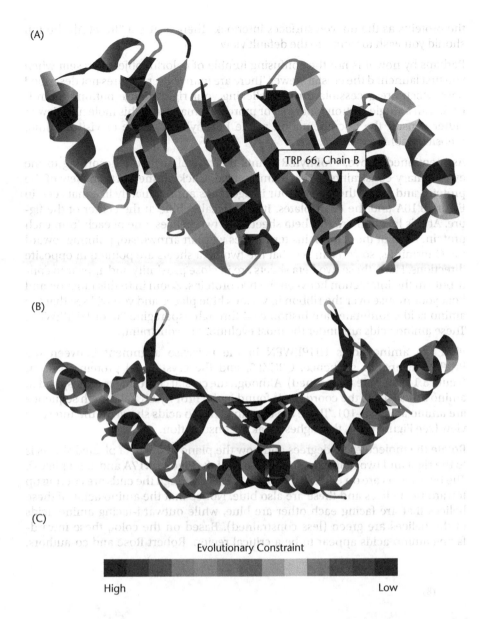

(A)

TRP 66, Chain B

(B)

(C)

Evolutionary Constraint

High   Low

**Figure 8.16 "Cartoon" views of
the DCoH2 dimer in ProPhylER.**
(A) In this view, there is a clear
divide between the two proteins
in this dimer. Floating your mouse
over a position will provide the
identification of that amino acid as
shown for tryptophan 66 on Chain B.
(B) Another view of the structure,
showing how the beta sheets form
an almost flat region while the helices
are on the outside of the protein.
(C) The color-coding of the positions
is based on the conservation of
sequence revealed by a multiple
sequence alignment. See color plates
for a color version of this figure.

But this view is excellent for following the path of the molecules. Wireframe shows
the backbone plus the side chains. A yellow double ring of tryptophan at coordinate seven is on the outside of the proteins and easily found because it sticks out
above the other side chains. Spacefill shows a more realistic size of each atom
and you see that there is little empty space visible in the proteins. Alanine 101 is
noticeably red; according to the Evolutionary Constraint scale on the bottom of
the CrystalPainter, this amino acid coordinate is very free to change in evolution.
A reminder: if you want to compare the coordinates of the Interface view with the
CrystalPainter view, you should refer to the BLASTP alignment between the two
sequences (Figure 8.14).

It should be noted that the ProPhylER Interface and CrystalPainter views use the
color scale in a different manner. In the CrystalPainter view, the color red indicates very little evolutionary constraint (see Figure 8.16C and color plates) while
in the Interface view, red indicates that the impact (MAPP p-value) would be high
should a substitution take place here (Figure 8.15 and color plates).

"Wire + Cartoon" is the default view and is a combination of the Wireframe and
Cartoon views. For added detail, you can zoom in by holding down the shift key
and moving the mouse vertically. Finally, the Surface view shows the solvent-
accessible surfaces of a protein. In this view, it is hard to see a visible gap between

the proteins as the uneven surfaces interlock. There is also a "Reset All" button should you wish to return to the default view.

Perhaps by now it is not the confusing jumble of colorful objects present when you first launched the crystal viewer. There are many other features not described here which are accessible through clicking your right mouse button (control-click with a single-button mouse). For more information on this molecule viewer, either consult the documentation at the ProPhylER Website or visit the Jmol Website, jmol.sourceforge.net.

As mentioned above, ProPhylER paints the crystal structure according to the evolutionary constraints on each amino acid. Click on the Cartoon view of the proteins and rotate them until your image looks approximately like that seen in Figure 8.16A and the color plates. It is noticeably blue at the center of the figure. At this location are two beta sheets and two helices, one of each from each protein. Each of these domains terminates with an arrowhead, pointing toward the C-terminus, so you can see that the two beta sheets are pointed in opposite directions. Here, these two beta sheets are in close proximity and appear to contribute to the interaction between the two proteins. Zoom in to this junction and float your mouse over the ribbon in various blue places and you will see that the amino acid coordinates are histidine 64 through asparagine 70, or HHPEWFN. These amino acids are under the most evolutionary constraint.

Find the amino acids HHPEWFN in the pairwise alignment between the ProPhylER anchor sequence, Q9CZL5, and the crystallized protein, 1ru0, in Figure 8.14 (they are underlined). Although the crystal coordinates correspond to amino acids 64–70, the coordinates found in the ProPhylER full-length sequence are amino acids 95–101. These are the same amino acids shown in the Interface view (see Figure 8.15), the highest peak in conservation.

Rotate the molecules 90 degrees and now the plane of the beta pleated sheets is to the right and two helical regions are to the left (**Figure 8.17A** and color plates). The beta sheets are on edge and difficult to see, although the ends are curving up toward the helices and these are also blue. Notice that the amino acids of these helices that are facing each other are blue, while outward-facing amino acids of the helices are green (less constrained). Based on the color, these inward-facing amino acids appear to be a critical region. Robert Rose and co-authors,

Figure 8.17 Close-up of the DCoH2 dimer structure. (A) Side view of conserved regions of both proteins. From this angle, you are looking down the helices that are in close proximity. The beta sheets are on edge, but their ends curve toward the helices. Although numbering in the opposite direction, this region appears like a mirror image, with one chain on top and the other on the bottom. (B) Isolated amino acids, showing side chains; Chain A is on top, Chain B at the bottom. The side chains at the far right and far left belong to histidine 62 of the two chains. All other groups are pointing toward the other chain. See color plates for a color version of this figure.

(A)

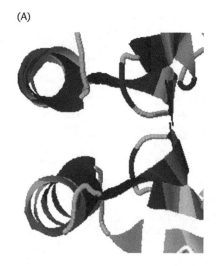

(B)

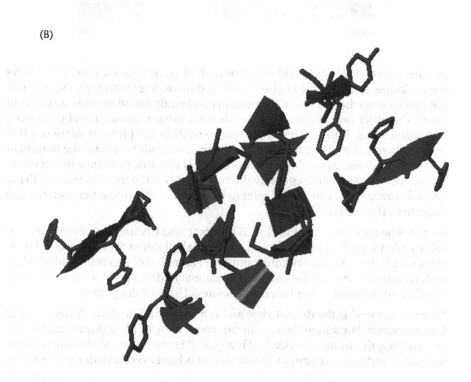

who solved the crystal structure, identify the helices as the primary points of interaction between the monomers.

Could the most conserved amino acids in this region be interacting closely? Probably they are but it is hard to see the side chains when all of the amino acids of both proteins are visible. Using a Jmol script, we can alter the view so only certain amino acids are visible. Resetting the Jmol window returns the view to "Wire + Cartoon," with the side chains visible. Right-clicking on the window (or control-click with a single-button mouse) will bring up a menu and the choice "Console" will make a small window appear, allowing the entering of Jmol script commands. Entering the following script, followed by the return key, will isolate these listed amino acids:

```
restrict 43, 47, 50, 62, 63, 64, 66, 68, 70;
```

Now, the side chains from the listed amino acids are visible (Figure 8.17B). Also visible are the small, flat pieces of the cartoon: helices and beta sheets. From this angle, and better inspected "live" so you can rotate and zoom, many blue side chains point toward the opposite protein. An exception is histidine 62, projecting outward on the far right and left of Figure 8.17B. These groups point to another helix within the protein, perhaps providing stability in that interaction. Although the helices of the two proteins appear further apart than the sheets (Figure 8.17A), the side chains of these and other amino acids reach toward the opposite protein and are highly conserved. Notice that the coordinates of helical amino acids (43, 47, and 50) are not adjacent to each other numerically. Since they are on spiraling alpha helices, these amino acids point in the same direction (toward the other protein) when they are on the same side of the helix.

To get the full story on these structures, take a look at the article by Rose and co-workers. People who are fully trained and experienced in crystallography can make the best interpretation of these complicated structures.

8.6 THE IMPACT OF SEQUENCE ON STRUCTURE

Now that you are familiar with seeing and manipulating protein structures with Jmol, we will now compare orthologous structures. Distant orthologs from different species perform the same function but differ by sequence. The sequence difference may change the activity of the protein (for example, the optimal conditions for the conversion of substrate) but the function is unchanged. Since structure is so tightly linked to function we would expect the structures to be conserved despite differences in amino acid sequence.

To compare orthologous structures, lysozyme will be examined. Lysozyme was the first enzyme to have its crystal structure determined and it has been studied for many years. In fact, a text search of the RCSB structure database identifies over 1100 lysozyme structures from a large number of organisms. Lysozyme catalyzes the hydrolysis of polysaccharide chains and acts as a natural antibiotic, causing the destruction of bacterial cell walls. It is abundant in egg white, tears, mucus, and saliva.

Lysozyme sequences from the Australian black swan (*Cygnus atratus*) and Atlantic cod fish (*Gadus morhua*) were chosen to demonstrate what could be seen with distantly related proteins. A BLASTP alignment (**Figure 8.18**) between PDB 1gbs and GenBank ACB20803 shows 57% identity and 73% similarity between these sequences. The alignment covers almost the entire length of both sequences and only one gap of four amino acids was introduced for these two proteins which are separated by approximately 400 million years of evolution.

Figure 8.19 (see also color plates) shows the swan (left) and cod (right) lysozyme structures. After launching the Jmol viewers for these proteins on the RCSB Website, they were manipulated to be approximately the same size and orientation. The similarities in the structures, such as the long vertical helical region at the center and the beta sheets in the upper left, were used to position them. Their

Figure 8.18 BLASTP alignment between the black swan and Atlantic cod lysozyme protein sequences. The query is the swan sequence (RCSB 1gbs) and the cod sequence (GenBank ACB20803) is the subject.

```
Score =  216 bits (549), Expect = 2e-61
Identities = 106/185 (57%), Positives = 136/185 (73%), Gaps = 4/185 (2%)

Query: 5    YGNVNRIDTTGASCKTAKPEGLSYCGVPASKTIAERDLKAMDRYKTIIKKVGEKLCVEPA 64
            YG++ +++T+GAS KT++ + L Y GV AS T+A+ D  M++YK+ I  V +K  V+PA
Sbjct: 3    YGDITQVETSGASSKTSRQDKLEYDGVRASHTMAQTDAGRMEKYKSFINNVAKKHVVDPA 62

Query: 65   VIAGIISRESHAGKVLKN----GWGDRGNGFGLMQVDKRSHKPQGTWNGEVHITQGTTIL 120
            VIA IISRES AG V+ N    GWGD NGFGLMQVDKR H+P+G WN E HI Q T IL
Sbjct: 63   VIAAIISRESRAGNVIFNTTPPGWGDNYNGFGLMQVDKRYHEPRGAWNSEEHIDQATGIL 122

Query: 121  TDFIKRIQKKFPSWTKDQQLKGGISAYNAGAGNVRSYARMDIGTTHDDYANDVVARAQYY 180
             +FI+ IQKKFPSW+ +QQLKG I+AYN G G V SY  +D  TT  DY+NDVVARAQ+Y
Sbjct: 123  VNFIQLIQKKFPSWSTEQQLKGAIAAYNTGDGRVESYESVDSRTTGKDYSNDVVARAQWY 182

Query: 181  KQHGY 185
            K++G+
Sbjct: 183  KKNGF 187
```

Figure 8.19 The swan and cod fish lysozyme structures. Both structures were generated by the RCSB Jmol viewer and are from (A) the swan (PDB identifier 1gbs) and (B) the cod fish (PDB identifier 3gxk). See color plates for a color version of this figure.

similarities are striking. Despite the differences in sequence, helices, turns, and beta sheets are folded into very similar structures. There are subtle differences, such as the length of a helix, or the angles of the domains, but these two structures are highly conserved.

The beta sheets (yellow; see color plates) are in a region where there are multiple turns and no predicted structure (white). Floating the mouse over these white loops reveals the protein coordinates, amino acids 77–87 in both structures. Looking back at the pairwise alignment (Figure 8.18), this region includes the single gap. This suggests that this region may be more free to evolve than others as indels are allowed, the region appears to be less structured, and it is found on the outside of the protein.

This approach to comparing two structures by the manual turning of side-by-side structures (Figure 8.19) is only suitable for gross examination. The RCSB Website

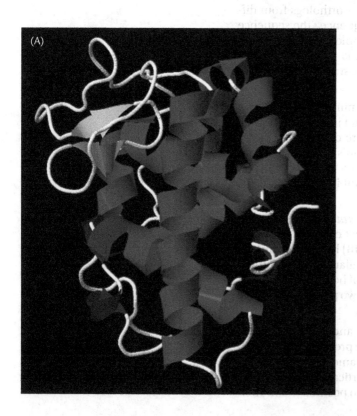

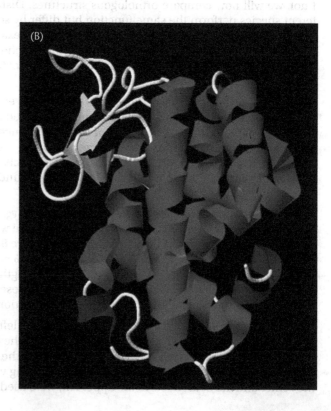

has a set of tools for more sophisticated comparisons. On the left sidebar on any page within the Website, click on the "Compare Structures" link under the Tools section. You are taken to a new page where you are prompted to enter two structure identifiers. After entering 1gbs and 3gxk, you then choose a comparison method. In addition to three pairwise sequence comparison methods, there are seven structural comparisons to choose from, with some redundancy. The method jFATCAT-rigid aligns the structures with a DotPlot-like approach to optimize the alignments, allowing regions to twist, much like the natural flexion of a protein. Clicking on the "Compare" button launches the comparison.

Once the window refreshes, the top panel shows that, although there is a calculated 56.8% sequence identity, there is a 98% structural similarity. Below this panel is a window (**Figure 8.20** and color plates). The two structures are now superimposed upon each other and can be manipulated like any Jmol structure. In this figure, the swan structure is orange and dark gray while the cod structure is cyan and light gray. The structural alignment has been manipulated to be similar to the two figures seen in Figure 8.19. Notice how well aligned the two structures are in this figure. Regions quite similar such as the center helix are almost perfect superimpositions, with the cyan and orange colors blending together. Other regions show strong similarity with structures appearing as parallel twists and turns. The gray regions are structures that align poorly. One such region is the light gray loop in the upper left. The color identifies this loop as a cod structure and floating the computer mouse over this region identifies it as the TTPP sequence in the gap of the pairwise alignment (see Figure 8.18). That is, this region, having no swan counterpart, appears as a light gray loop.

Despite only 57% identity at the sequence level, these structures are almost structurally identical and this comparison tool allows easy recognition of regions of similarity and difference. The structures are so similar because, as mentioned in Section 8.5, amino acids possess properties which favor their location within certain structures. Numerous substitutions are tolerated because they either contribute to the structure or do not disrupt it. For example, amino acids like alanine and leucine are favored in alpha helices while proline and glycine are

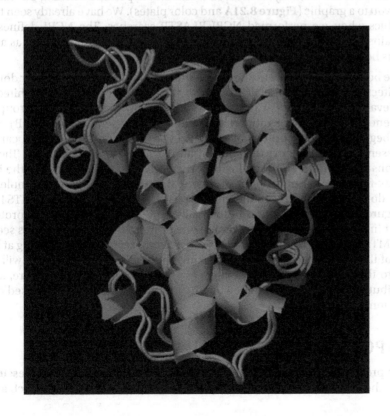

Figure 8.20 Structural alignment of the swan and cod fish lysozymes using the jFATCAT-rigid method. The swan and cod lysozyme structures are displayed in contrasting colors (see color plates for a color version of this figure). Gray regions (for example the loop in the upper left) signify where the two structures are too divergent to align.

considered "helix breakers." These amino acid propensities were recognized in the 1970s by pioneers such as Chou and Fasman. Using crystal structures, algorithms were developed which could predict structural elements from protein sequence. Modern algorithms trained with the many crystal structures now available can achieve approximately 70% accuracy for predicting beta sheets and alpha helices.

8.7 BUILDING BLOCKS: A MULTIPLE DOMAIN PROTEIN

Using other tools, we'll now examine a protein that has multiple but different domains and nicely illustrates the combinations and substructures within proteins. These **domains** are found in other proteins as well, and in different combinations. If these domains are thought of as solutions to problems (such as, how do I fold a protein into a rigid structure? How do I bind nucleic acids?), then it is clear from this protein and many thousands of others that nature reuses domains again and again to solve problems. These building blocks often make up an entire protein, or just a tiny region, but these parts are merged together, swapped, truncated, or rearranged to form new proteins using pieces of the old standards.

ADAMTS4, also known as "a disintegrin and metalloproteinase with thrombospondin motifs 4," is a proteinase that cleaves cartilage proteins. Although it is most probably part of the turnover process for the constant rebuilding of cartilage, it may also be responsible for the breakdown of cartilage that leads to osteoarthritis. Examining the annotation of the UniProtKB record, O75173, reveals the presence of multiple domains including a peptidase, disintegrin, and TSP type-1 (thrombospondin type 1) domain. Clicking on the hypertext for these domains (under "General annotation") within the O75173 record takes you to a listing of other proteins; hundreds to thousands of proteins have a peptidase domain, disintegrin domain, or a TSP domain.

The NCBI will show you a similar analysis of the domains for this protein. The RefSeq equivalent for human ADAMTS4 is NP_005090. On the right sidebar of this record is an "Identify Conserved Domains" hyperlink. Clicking on this will take you to a graphic (**Figure 8.21A** and color plates). We have already seen these graphics when we performed NCBI BLASTP searches. The NCBI defines the domains, as seen in this figure, and also provides analysis details such as alignments between multiple proteins (not shown in Figure 8.21).

At the bottom of Figure 8.21A there is a button called "Search for similar domain architectures" which is the NCBI link for the Conserved Domain Architecture Retrieval Tool (CDART). Clicking on this button brings you to groups of sequences having one or more of these domains present (Figure 8.21B). Each line begins with the number of sequences in that group and has a cartoon representing the basic organization of all the proteins in that group. The use of icons for the different domains is particularly effective at showing the huge variety found in nature in both numbers and arrangement. For example, the TSP1 domain (thrombospondin type 1 repeats) is found once in ADAMTS4 (top of Figure 8.21B), and is found individually and in multiples in other proteins. Other lines on other pages are missing one or more of the five domains seen in ADAMTS4. There are dozens of pages of variations for sequences having at least one of these ADAMTS4 domains! Clicking on the plus sign on the left will take you to the collection of those sequences sharing a similar architecture, again distributed over many pages showing the multiple times nature has used combination of these domains.

8.8 POST-TRANSLATIONAL MODIFICATION

Once proteins are synthesized, many undergo **post-translational processing** to reach the mature and functional state. This can take on many forms, such as the

(A)

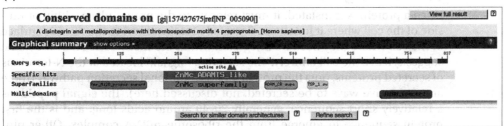

(B)

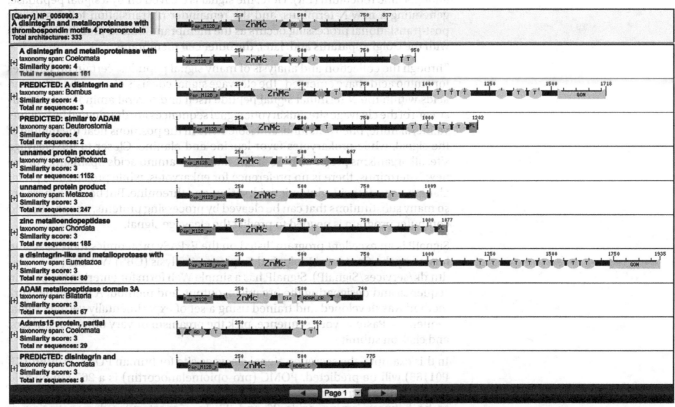

creation of disulfide bonds, or the digestion by proteases to generate chains or smaller peptides of the original translation product. Other common modifications include the addition of chemical groups such as sugars, phosphates, and methyl groups, to name just a few. These and other modifications change the conformation of the proteins, mark them for transport or destruction, or put them on the pathway to cause one or more biochemical events in a cell. These additional layers of regulation are critical to function, and their presence gives hints to those functions.

Many of these modifications can be predicted by bioinformatics software tools which are listed on the ExPASy Resources Web page, www.expasy.ch/resources. Some of these tools are looking for the presence of an absolute amino acid sequence, while others are looking for biochemical properties driven by regional amino acid composition. It is important to realize that these are predictions and factors such as the location of these sequences within the larger protein, regional folding, and other post-translational modification events may prevent predicted events from taking place.

Below are two examples of post-translational modifications and software to predict their presence.

Figure 8.21 The NCBI Conserved Domain Architecture Retrieval Tool. (A) The conserved domains found in NP_005090, ADAMTS4. Clicking on these icons will take you to a page dedicated to that single domain, listing other proteins containing this domain and presenting a multiple sequence alignment of these many sequences. (B) Clicking on the "Search for similar domain architectures" button shown in (A) takes you to a presentation of related proteins, sharing at least one domain with your query. See color plates for a color version of this figure.

Secretion signals

Once a protein is translated, it must relocate to the proper place within or outside of the cell where it functions. Secreted proteins travel to the outside surface of the cell where they become integrated within the membrane or, as the name suggests, leave the cell entirely. The secretion "signal," located at the N-terminus of secreted proteins, puts the protein on this traditional secretion pathway (there are alternative ways to be secreted, not discussed here). The **signal peptide** is a relatively short sequence, generally 20–40 amino acids long, and is the first protein sequence to emerge from the ribosome–mRNA complex. Other proteins recognize this sequence and insert the peptide into the membrane of the endoplasmic reticulum (ER). Here, the signal is cleaved off by a signal peptidase, generating a new N-terminus, and the remainder of translation follows. Other post-translational processing occurs as the membrane-bound protein then joins with the Golgi apparatus and, later, the outer cell membrane.

Through the collection and analysis of many signal peptides, certain preferences for amino acids have emerged. The cleavage signal consists of a number of amino acids within the N-terminal signal peptide itself and several amino acids downstream of the cleavage site. Eukaryotic signal sequences are dominated by leucine, valine, alanine, phenylalanine, and isoleucine in the positions near the middle of the signal, while prokaryotes favor leucine and alanine. Closer to the cleavage site, all organisms prefer alanine at one and three amino acids upstream. For the new N-terminus, there is no preference for eukaryotes, while prokaryotes prefer alanine, aspartic acid, glutamic acid, serine, and threonine. But because there are so many substitutions that can be cleaved by processing proteins, a sophisticated computer program is needed to predict the cleavage signal.

SignalP is an excellent program, listed on the ExPASy proteomics tools page and hosted by the Center for Biological Sequence Analysis (CBS) Website (www.cbs.dtu.dk/services/SignalP). SignalP has a simple Web form for entering the protein sequence and designating the organism, output, and method. A neural network method was developed and trained using a set of experimentally verified signal sequences. Paste in your sequence, pick the organism or vary other parameters, and click on submit.

In this example, the secretion signal cleavage site for human POMC (UniProtKB P01189) will be predicted. POMC (pro-opiomelanocortin) is a 267 amino acid protein with a secretion signal cleavage site that was experimentally determined to be between amino acids 26 and 27. In a spectacular demonstration of additional post-translational processing, POMC undergoes extensive cleavage by other proteases to create at least nine biologically active peptides: NPP (N-terminal peptide of POMC; 76 amino acids), melanotropin alpha, beta, and gamma (13, 18, and 11 amino acids, respectively), corticotropin (39 amino acids), lipotropin beta and gamma (89 and 56 amino acids, respectively), beta-endorphin (31 amino acids), and met-enkephalin (5 amino acids). These various peptides participate in pathways throughout the body for the regulation of appetite, pigmentation, blood pressure, and other forms of signaling.

Figure 8.22 (and color plates) shows the graphic and table from the analysis. It shows a peak, significantly higher than other predictions, indicating that the cleavage site lies between glycine 26 and tryptophan 27 of POMC. In addition, scores from various calculations which generated the graphic are also displayed: S is the probability that an amino acid is part of the signal peptide, C is the probability that an amino acid is the N-terminus of the processed protein, and Y is a measure of the correlation between the S and C scores. A peak of the C score should coincide (Y) with the plunge in the S score.

SignalP–4.0 prediction (euk networks): sp_P01189_COLI_HUMAN

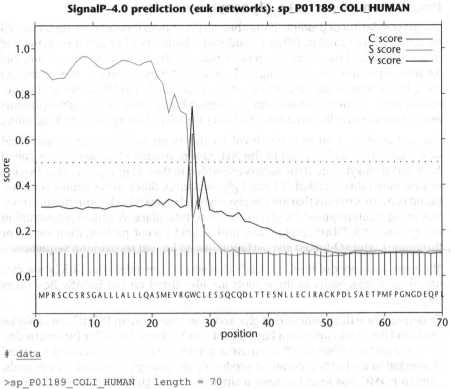

```
# data
>sp_P01189_COLI_HUMAN    length = 70
# Measure    Position    Value    Cutoff    signal peptide?
  max. C        27        0.625
  max. Y        27        0.745
  max. S         9        0.965
  mean S       1-26       0.888
     D         1-26       0.823     0.450     YES
Name=sp_P01189_COLI_HUMAN SP='YES' Cleavage site between pos. 26 and 27:
VRG-WC D=0.823 D-cutoff=0.450 Networks=SignalP-noTM
```

Figure 8.22 SignalP prediction of secretion signal peptide cleavage sites. The SignalP-NN prediction, graphically showing a sudden drop in the probability for being part of the signal peptide (S score), and increases in the probability of being the new N-terminus after cleavage (C score) and the correlation between the two (Y score). See color plates for a color version of this figure.

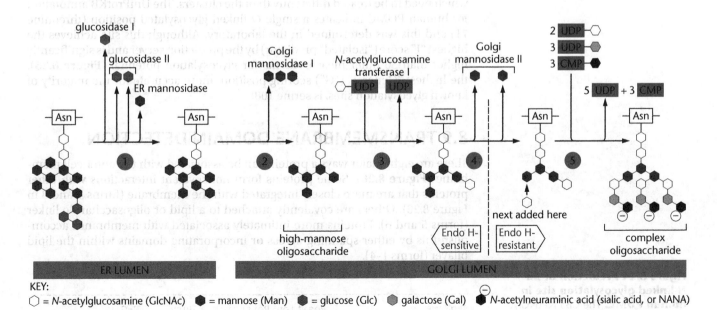

KEY:

◯ = *N*-acetylglucosamine (GlcNAc) ⬢ = mannose (Man) ⬢ = glucose (Glc) ⬢ galactose (Gal) ⬢ *N*-acetylneuraminic acid (sialic acid, or NANA)

Figure 8.23 Complex oligosaccharide processing following the glycosylation of proteins. While in the ER, *N*-linked glycosylation takes place, and the sugars are further processed during the transport through the compartments of the Golgi apparatus. From, Alberts B, Johnson A, Lewis J et al. (2007) Molecular Biology of the Cell, 5th ed. Garland Science.

Prediction of protein glycosylation sites

The **glycosylation** of proteins is another very common form of post-translational modification of proteins. Often considered a hallmark of secreted proteins, the initial addition of the sugar group takes place in the ER and may be responsible for the proper transport, folding, and stability of many proteins. The sugars are complex structures that undergo a series of enzymatic modifications in the successive compartments of the Golgi (see **Figure 8.23**). Once mature, glycoproteins on the surfaces of cells play a variety of roles in cell–cell recognition and signaling.

The initial glycosylation event involves the modification of an amino acid side chain. The modification of the NH$_2$ group, usually of asparagine, is called *N*-linked glycosylation. If the sugars are added to the OH group, usually of serine or threonine, this is called *O*-linked glycosylation. Since these amino acids are found in many proteins that are not glycosylated, there is both sequence and context recognition required for glycosylation to take place. *N*-linked glycosylation recognizes NXS/T (asparagine, any amino acid except proline, then serine or threonine), while *O*-linked glycosylation has no known recognition sequence.

There are a number of Internet prediction servers hosted by the CBS (www.cbs. dtu.dk/services). Many of these tools are also listed on the ExPASy Resources page.

To demonstrate the prediction of glycosylation sites, human POMC can again be used, and the results appear in **Figure 8.24**. The CBS tool NetNGlyc (www.cbs.dtu. dk/services/NetNGlyc) is 86% accurate at predicting laboratory-verified *N*-linked glycosylation and 61% accurate at predicting the non-glycosylated amino acids. Human POMC was found to have a single *N*-linked glycosylation site based on laboratory work and this was precisely predicted by the NetNGlyc server.

O-linked glycosylation of proteins is much more challenging to predict. As indicated earlier, there is no consensus recognition site, yet the CBS NetOGlyc server (www.cbs.dtu.dk/services/NetOGlyc) achieves 76% accuracy at predicting laboratory-verified *O*-linked modified amino acids and 93% accuracy in predicting nonmodified amino acids. The scoring methods are based on amino acid composition as well as surface accessibility, and these methods were built on a foundation of known *O*-linked glycosylation sites to design and test the software. *O*-linked glycosylation often occurs in clusters but there are also isolated sites which need to be treated differently than the clusters. The UniProtKB annotation for human POMC indicates a single *O*-linked glycosylated position (threonine 71) and this was determined in the laboratory. Although this site achieves the highest "I" score ("isolated" predictor) by the prediction server and is significantly higher than 30 other measurements for glycosylation potential (**Figure 8.25**), the highest "General" ("G") scoring position, more accurate for the majority of known glycosylation sites, is serine 108.

8.9 TRANSMEMBRANE DOMAIN DETECTION

There are eight major ways a protein can be associated with the outer cell membrane (**Figure 8.26**). Some proteins form noncovalent interactions with other proteins that are more closely integrated with the membrane (forms 7 and 8 in Figure 8.26). Others are covalently attached to a lipid or oligosaccharide linker (forms 5 and 6). Proteins more intimately associated with membranes accomplish this by either spanning across or incorporating domains within the lipid bilayer (forms 1–4).

Figure 8.24 Prediction of an *N*-linked glycosylation site in human POMC. The CBS NetNGlyc server output indicates that the asparagine at position 91 in the protein is predicted to be *N*-linked glycosylated.

```
-----------------------------------------------------------------------
SeqName                 Position Potential  Jury      N-Glyc
                                            Agreement result
-----------------------------------------------------------------------
sp_P01189_COLI_HUMAN    91 NSSS  0.7487     (9/9)      ++
-----------------------------------------------------------------------
```

Name	S/T	Pos	G-score	I-score	Y/N	Comment
sp_P01189_C	S	4	0.273	0.019	.	-
sp_P01189_C	S	7	0.241	0.049	.	-
sp_P01189_C	S	9	0.218	0.045	.	-
sp_P01189_C	S	21	0.111	0.040	.	-
sp_P01189_C	S	31	0.149	0.052	.	-
sp_P01189_C	S	32	0.152	0.023	.	-
sp_P01189_C	T	38	0.222	0.072	.	-
sp_P01189_C	T	39	0.217	0.174	.	-
sp_P01189_C	S	41	0.187	0.039	.	-
sp_P01189_C	S	55	0.265	0.082	.	-
sp_P01189_C	T	58	0.390	0.058	.	-
sp_P01189_C	T	71	0.365	0.361	.	-
sp_P01189_C	S	92	0.240	0.030	.	-
sp_P01189_C	S	93	0.278	0.027	.	-
sp_P01189_C	S	94	0.287	0.037	.	-
sp_P01189_C	S	95	0.298	0.037	.	-
sp_P01189_C	S	97	0.305	0.060	.	-
sp_P01189_C	S	98	0.298	0.031	.	-
sp_P01189_C	S	108	0.504	0.128	S	-
sp_P01189_C	S	125	0.493	0.071	.	-
sp_P01189_C	S	138	0.305	0.050	.	-
sp_P01189_C	S	140	0.261	0.043	.	-
sp_P01189_C	S	168	0.261	0.021	.	-
sp_P01189_C	T	181	0.403	0.052	.	-
sp_P01189_C	S	208	0.258	0.034	.	-
sp_P01189_C	S	230	0.200	0.060	.	-
sp_P01189_C	T	242	0.361	0.069	.	-
sp_P01189_C	S	243	0.273	0.071	.	-
sp_P01189_C	S	246	0.202	0.022	.	-
sp_P01189_C	T	248	0.211	0.175	.	-
sp_P01189_C	T	252	0.236	0.071	.	-

Figure 8.25 Prediction of an *O*-linked glycosylation site in human POMC. The CBS NetOGlyc server predicts that position108 (serine) is glycosylated. The glycosylation site discovered in the laboratory is position 71, the highest-scoring "isolated" prediction.

The protein domains that span a membrane have amino acid properties that can be predicted by bioinformatics software. Alpha-helical **transmembrane** domains (forms 1 and 2) must stretch the depth of the membrane's lipid bilayer, making them approximately 20–30 amino acids long, while a beta barrel (form 3)

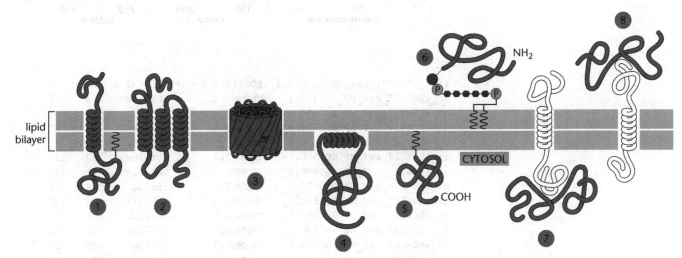

Figure 8.26 Membrane proteins. Pictured are the eight major ways a protein can be associated with a membrane. Forms 1–4 traverse or are embedded in the membrane, forms 5 and 6 are covalently linked to membrane molecules, while forms 7 and 8 are only associated with membrane proteins. From, Alberts B, Johnson A, Lewis J et al. (2007) Molecular Biology of the Cell, 5th ed. Garland Science.

only requires approximately 10 amino acids in this extended beta-sheet conformation. Since they are embedded in a hydrophobic environment, the amino acids must be hydrophobic.

To examine a protein that spans the membrane, we'll analyze human rhodopsin, NP_000530, which was discussed in Chapter 6. Within the span of 348 amino acids, rhodopsin crosses the membrane seven times, making it part of a large and important class of proteins called seven transmembrane receptors. They are also called G protein-coupled receptors because a cytosolic domain of the receptor interacts with members of another family, G proteins. When the exterior domains of the receptor bind to a ligand (for example, a hormone) the receptor–G protein interaction inside the cell is perturbed. In turn, this starts a signaling cascade within the cell, often involving protein modifications, protein–protein interactions, and changes in gene expression.

For the prediction of transmembrane domains, we'll use another tool listed on the CBS Web services page called TMHMM. After entering the human rhodopsin protein sequence into the Web form and clicking on "Submit," the results page presents a graphic plus a table with the predicted domains. **Figure 8.27A** (see also color plates) shows a cartoon with the amino acid coordinates increasing from left to right. Across the top, red blocks indicate predicted transmembrane domains with connecting lines showing the intervening sequences, predicted to be either inside (lower line) or outside (upper line) of the cell. The figure is dominated by red columns of probability scores, indicating the location of the

Figure 8.27 Transmembrane domain prediction using TMHMM. (A) Within the span of 348 amino acids, human rhodopsin is predicted to pass through the cell outer membrane seven times. The result is four sections on the exterior of the cell and four in the interior. See color plates for a color version of this figure. (B) A table documents the predictions and provides coordinates for the various domains.

(A)

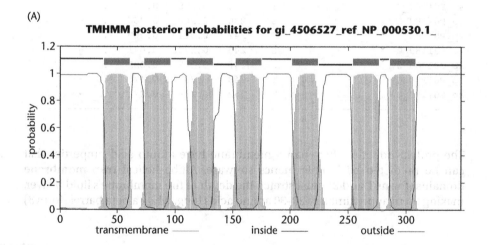

(B)

```
# gi_4506527_ref_NP_000530.1_    POSSIBLE N-term signal sequence
gi_4506527_ref_NP_000530.1_      TMHMM2.0      outside       1     38
gi_4506527_ref_NP_000530.1_      TMHMM2.0      TMhelix      39     61
gi_4506527_ref_NP_000530.1_      TMHMM2.0      inside       62     73
gi_4506527_ref_NP_000530.1_      TMHMM2.0      TMhelix      74     96
gi_4506527_ref_NP_000530.1_      TMHMM2.0      outside      97    110
gi_4506527_ref_NP_000530.1_      TMHMM2.0      TMhelix     111    133
gi_4506527_ref_NP_000530.1_      TMHMM2.0      inside      134    152
gi_4506527_ref_NP_000530.1_      TMHMM2.0      TMhelix     153    175
gi_4506527_ref_NP_000530.1_      TMHMM2.0      outside     176    201
gi_4506527_ref_NP_000530.1_      TMHMM2.0      TMhelix     202    224
gi_4506527_ref_NP_000530.1_      TMHMM2.0      inside      225    253
gi_4506527_ref_NP_000530.1_      TMHMM2.0      TMhelix     254    276
gi_4506527_ref_NP_000530.1_      TMHMM2.0      outside     277    285
gi_4506527_ref_NP_000530.1_      TMHMM2.0      TMhelix     286    308
gi_4506527_ref_NP_000530.1_      TMHMM2.0      inside      309    348
```

transmembrane domains. There are seven predicted transmembrane domains, with four cytosolic domains and an equal number of domains on the outside of the cell. The table (Figure 8.27B) also documents these predictions and provides coordinates for the various domains.

What do transmembrane domains look like? **Figure 8.28** shows the predicted sequences of human rhodopsin. As can be seen by this figure, the sequences are equal in length but do not "align" with amino acids in common. However, most of these amino acids have nonpolar side chains. Embedded in the hydrophobic lipid environment of the membrane, these amino acids form hydrogen bonds with each other to form alpha helices.

8.10 SUMMARY

In this chapter, we explored methods of analyzing proteins. The functional domains of proteins were introduced and identified using tools at the PROSITE Website. We got our first look at protein structures, first in cartoon form at the Research Collaboratory for Structural Bioinformatics Website and then in a three-dimensional representation that could be manipulated using Jmol. The Web tool ProPhylER allowed us to see levels of conservation at each amino acid position and then "painted" these upon crystal structures of proteins. A comparison tool at the RCSB Website was used to structurally compare lysozyme orthologs. The presence of multiple domains in proteins and related family members was displayed in the NCBI CDART database. Finally, we explored post-translational modification, such as secretion signals and glycosylation, along with protein topology using prediction applications found at the Center for Biological Sequence Analysis Website.

EXERCISES

Aquaporin-5

Aquaporins are a very large and ubiquitous family of proteins that form channels for the flow of water across outer cellular membranes. Controlling this substantial flow of water is an active process that is selective, excluding ions and only allowing the passage of water molecules in order to maintain osmotic gradients. There are 12 aquaporin family members in humans and they are responsible for both general "housekeeping" maintenance of cellular hydration and function-specific channeling of water that is more obvious in our daily lives. In the kidneys, aquaporins concentrate urine, reducing the daily filtrate of 180 liters per day to an incredible 1 liter. They are also responsible for the production of sweat, tears, and saliva. Aquaporins were discovered in 1990 by Peter Agre who would later receive a Nobel Prize in Chemistry for his work on these important proteins which play a central role in the normal physiology of cells.

Several human diseases are associated with mutations in aquaporin proteins or their dysregulation. Aquaporins lack a secretion signal and rely on other cellular events to move them from the intracellular membranes to the outer cell membrane. For example, the hormone vasopressin triggers the movement of aquaporin-2 to the plasma membranes of renal collecting ducts. However, defects in the hormone, its receptor, or aquaporin-2 can all lead to diabetes insipidus, which is characterized by excessive thirst and very dilute urine. Defects in the proper trafficking of aquaporin-5 to the membrane result in Sjögren's syndrome, a disease causing both dry eyes and mouth, making it difficult to speak or swallow. Disruption of normal aquaporin-5 function may also cause hyperhidrosis, or excessive sweating. Defects in aquaporin-0 are linked to congenital cataracts, although the protein may be contributing to the structure of the lens in addition to water regulation.

So what does a protein that channels the flow of water in and out of cells look like? In the exercises below, you will be applying multiple applications to understand

1- MLAAYMFLLIVLGFPINFLTLYV
2- YILLNLAVADLFMVLGGFTSTLY
3- NLEGFFATLGGEIALWSLVVLAI
4- AIMGVAFTWVMALACAAPPLAGW
5- SFVIYMFVVHFTIPMIIIFFCYG
6- VIIMVIAFLICWVPYASVAFYIF
7- IFMTIPAFFAKSAAIYNPVIYIM

Figure 8.28 The seven predicted transmembrane domains of human rhodopsin. The CBS prediction server TMHMM predicts seven transmembrane domains for human rhodopsin, NP_000530. As can be seen by the sequence of these seven domains, they are all equal length (23 amino acids) and few of the amino acids align.

features concerning the sequence and properties of aquaporin-5. This process can be typical in analyzing a protein: look for physical features which give hints to the function. Importantly, the crystal structure of aquaporin-5 has been determined and you will be examining this protein visually.

1. Find the human aquaporin-5 protein sequence, NP_001642, and copy the FASTA format to the clipboard.

2. Using the CBS prediction server "NetNGlyc," determine where glycosylation sites might be located on the protein. Keep this result window open.

3. Enter the PDB identifier "3d9s" into ProPhylER and bring up both the Interface and CrystalPainter views of human aquaporin-5. Rotate and change the view of the protein, and what do you see? What are the obvious characteristics of the protein and the pore-forming structure, and can you speculate on the selective passage of just water through the pore?

4. What is that non-protein molecule or ligand in the center of the structure? Where would you find this information?

Jmol renders the CrystalPainter view in ProPhylER and you can manipulate the structure using Jmol scripts. To enter the scripting commands, you have to bring up the Jmol console in ProPhylER. You do this by moving your cursor over the Jmol window in CrystalPainter and either control-mouse click (single-button mouse) or right click (two-button mouse) to see a menu.

5. Select "Console" to bring up a small window where you can enter Jmol scripting commands.

6. Using the Button Controls, change the CrystalPainter view to Cartoon. Use the Jmol console to label the alpha carbon of the predicted glycosylated amino acid (determined in question 2, above) with the word "Glycosylated." Be sure to verify the amino acid coordinate with the BLASTP alignment provided by ProPhylER.

7. Using the CBS prediction server "TMHMM," predict the transmembrane domains of human aquaporin-5. Using these results, and by identifying the location of N-terminal or C-terminal amino acids in the ProPhylER CrystalPainter view, orient the structure to have exterior-facing amino acids at the top of the window, and cytoplasmic amino acids at the bottom.

8. Use the ScanPROSITE tool to identify any patterns or profiles in human aquaporin-5. Describe what you find. Keep this results window open.

9. Using the Interface view in ProPhylER, find this region and describe the level of conservation in this region.

10. When the structure was determined, the location of water molecules within the pore was also defined. These steps will allow you to see the path of water through the pore.

- "Reset All" in the Jmol window of ProPhylER.

- Change the style to backbone for all chains.

- Hide the ligand and show the solvent. The solvent, in this case, is water. You can now see where the water molecules interact with and travel through the pore.

- Use Jmol scripting commands in the console to change the amino acids of the identified pattern (question 8, above) into spacefill. Be sure to verify the amino acid coordinates with the BLASTP alignment provided by ProPhylER. These amino acids are key to regulating water molecules flowing through the protein.

- Hide three of the four chains to focus on one protein and its water molecules.

- Manipulate the image in order to show a nice view of the water traveling through the pore and by these key amino acids.

- Your view of the water is obscured by several helices. Identify their coordinates by floating your mouse over them.

- Use the Jmol script command "hide" to remove these helices from view so you can clearly see the water molecules within the pore. For example, "hide 1-10;" causes these amino acids to disappear.

11. Use the CBS server for "NetPhos" to predict the phosphorylation sites on human aquaporin-5. You will see that a number of predicted sites appear in the results. Identify the serine with the highest score. Keep this result window open. Note: this was not covered in this chapter. However, this form is very similar to the other CBS prediction servers.

12. Horsefield et al. speculate that the phosphorylation of the amino acid identified in question 11, above, might disrupt the hydrogen bonding between arginine 153, phenylalanine 226, and proline 227 and this change in conformation might trigger the movement of aquaporin-5 to the membrane. Generate a new view in ProPhylER where you restrict the view to these four amino acids and view them. Label the alpha carbons of these four amino acids with the three-letter code for each amino acid and generate a nice picture of the relationships between these amino acids. You may have to change the background color of the window using the ProPhylER Button Controls.

FURTHER READING

Alberts B, Johnson A, Lewis J et al. (2007) Molecular Biology of the Cell, 5th ed. Garland Science, New York.

Berman HM, Westbrook J, Feng Z et al. (2000) The Protein Data Bank. *Nucleic Acids Res.* 28, 235–242. This is the primary citation for the Protein Data Bank (PDB).

Binkley J, Karra K, Kirby A et al. (2010) ProPhylER: a curated online resource for protein function and structure based on evolutionary constraint analyses. *Genome Res.* 20, 142–154. ProPhylER compares related proteins and visually shows you regions of evolutionary constraint or freedom.

Blom N, Gammeltoft S & Brunak S (1999) Sequence and structure-based prediction of eukaryotic protein phosphorylation sites. *J. Mol. Biol.* 294, 1351–1362. The citation for a CBS-based prediction server for phosphorylation sites.

Callewaert L & Michiels CW (2010) Lysozymes in the animal kingdom. *J. Biosci.* 35, 127–160. An extensive review article, with numerous figures and broad coverage of this gene family.

Crooks GE, Hon G, Chandonia JM & Brenner SE (2004) WebLogo: a sequence logo generator. *Genome Res.* 14, 1188–1190. WebLogo represents sequence diversity at a position by stacking the single-letter codes of either amino acids or nucleotides. You will find WebLogo in use in a number of applications in addition to ProPhylER, described in this chapter.

Elrod-Erickson M, Rould MA, Nekludova L & Pabo CO (1996) Zif268 protein–DNA complex refined at 1.6 Å: a model system for understanding zinc finger–DNA interactions. *Structure* 4, 1171–1180. This paper has wonderful figures showing the details of the nucleotide–amino acid interactions of this transcription factor binding the consensus binding site. For structure PDB identifier 1aay.

Geer LY, Domrachev M, Lipman DJ & Bryant SH (2002) CDART: protein homology by domain architecture. *Genome Res.* 12, 1619–1623. A nice article describing the NCBI Conserved Domain Architecture Retrieval Tool.

Gupta R & Brunak S (2002) Prediction of glycosylation across the human proteome and the correlation to protein function. *Pac. Symp. Biocomput.* 7, 310–322. The original citation for NetNGlyc.

Horsefield R, Nordén K, Fellert M et al. (2008) High-resolution x-ray structure of human aquaporin 5. *Proc. Natl Acad. Sci. USA* 105, 13327–13332. This paper describes the crystal structure of human aquaporin-5, the subject of the exercises at the end of this chapter. For structure PDB identifier 3d9s.

Julenius K, Mølgaard A, Gupta R & Brunak S (2005) Prediction, conservation analysis, and structural characterization of mammalian mucin-type O-glycosylation sites. *Glycobiology* 15, 153–164. The citation for NetOGlyc.

Kaper JB & Mobley HLT (1991) Immortal sequence. *Science* 253, 951–952. A letter to the Editor describing how the motif ELVIS is found in the protein database.

Nielsen H, Engelbrecht J, Brunak S & von Heijne G (1997) Identification of prokaryotic and eukaryotic signal peptides and prediction of their cleavage sites. *Protein Eng.* 10, 1–6. This article describes the development, testing, and use of the neural network SignalP algorithm, described in this chapter.

Petersen TN, Brunak S, von Heijne G & Nielsen H (2011) SignalP 4.0: discriminating signal peptides from transmembrane regions. *Nat. Methods* 8, 785–786. This is the paper describing the secretion signal prediction software used in this chapter.

Porter S, Clark IM, Kevorkian L & Edwards DR (2005) The ADAMTS metalloproteinases. *Biochem. J.* 386, 15–27. A tremendous review article describing this protein family. Several nice illustrations and a rich description of individual family members, their involvement in diseases, and functions, and a number of references.

Raffin-Sanson ML, de Keyzer Y & Bertagna X (2003) Proopiomelanocortin, a polypeptide precursor with multiple functions: from physiology to pathological conditions. *Eur. J. Endocrinol.* 149, 79–90. A wonderful review article on POMC, richly illustrated and describing, in detail, this interesting protein that is processed into many peptides with multiple and very different functions.

Rose RB, Pullen KE, Bayle JH et al. (2004) Biochemical and structural basis for partially redundant enzymatic and transcriptional functions of DCoH and DCoH2. *Biochemistry* 43, 7345–7355. This is the scientific paper describing a crystal structure described in this chapter. For structure PDB identifier 1ru0.

Sigrist CJA, Cerutti L, de Castro E et al. (2010) PROSITE, a protein domain database for functional characterization and annotation. *Nucleic Acids Res.* 38, D161–D166. A recent update on the PROSITE database.

Sonnhammer ELL, von Heijne G & Krogh A (1998) A hidden Markov model for predicting transmembrane helices in protein sequences. *Proc. Int. Conf. Intell. Syst. Mol. Biol.* 6, 175–182. The citation for TMHMM, the tool described in this chapter which predicts transmembrane domains.

Tortorella M, Pratta M, Liu RQ et al. (2000) The thrombospondin motif of aggrecanase-1 (ADAMTS-4) is critical for aggrecan substrate recognition and cleavage. *J. Biol. Chem.* 275, 25791–25797. This is a very good article describing the domains and function of ADAMTS4.

Ye Y & Godzik A (2003) Flexible structure alignment by chaining aligned fragment pairs allowing twists. *Bioinformatics* 19 (Suppl. 2), ii246–255. This is the reference for jFATCAT, the program that aligns protein structures.

Internet resources

Jmol: an open-source Java viewer for chemical structures in three dimensions, www.jmol.org.

This is the official Web page for Jmol, the structure viewer discussed in this chapter. See also their wiki page, wiki.jmol.org.

CHAPTER 9

Explorations of Short Nucleotide Sequences

- Examining and predicting transcription factor binding sites
- The translation start site (Kozak) sequence
- Exon splicing donor and acceptor sites
- Polyadenylation signals
- Alternative splicing

9.1 INTRODUCTION

In earlier chapters we studied DNA sequences that were hundreds or thousands of bases in length. In this chapter, we'll be studying sequences that are as small as four nucleotides long. The fine details of DNA sequence are important in defining key elements of gene structure. This chapter will describe examples where large collections of short sequences can create a consensus sequence, and small samples that reveal how variable nature can be. Some of the most exciting discoveries can be those where a sequence appears to "break the rules" and does not possess the usual signals for gene structure and expression.

The small sequences that this chapter will focus on tend to be regulatory in nature, providing the master switches for turning genes on or off and modulating functions such as transcription, exon splicing, or translation. Variability in sequence creates additional layers of gene regulation, so recognizing this variability is a necessary part of the effort to understand gene expression. Importantly, the study of these short sequences helps train your eye to notice the small things when performing sequence analysis.

When you discover a gene or cDNA, you have many problems to solve: "Where does the coding region begin?" "Where does it end?" "Is the 5P end of this cDNA missing?" "Is this an exon?" "Is this an intron?" Knowing that cDNAs have issues with sequencing accuracy, using and understanding fine details are essential to problem solving. Sequencing accuracy is an issue with cDNAs, and much of what is found in the public databases is from high-throughput efforts having very little human supervision. Thus, you may encounter sequence or assembly problems. You may also be in the exciting position of being the first person on the planet to take a close look at a sequence and you want to make the most of it.

The title of this chapter is deliberate. Rather than provide a thorough survey of what defines these small sequences, the goal is to explore the variation seen between sequences. Just a small sample will be examined, often focusing on sequences that have already been studied in previous chapters. These genes were initially chosen because of their size, domain structure, or the interesting biochemical properties of the encoded protein. Because we already know something about these genes, the smaller details will help complete the picture. We'll start this chapter at the extreme 5P end of the eukaryotic gene, where transcription initiates.

9.2 TRANSCRIPTION FACTOR BINDING SITES

The structure of chromatin is fundamentally built upon the strong association between DNA and protein. Histone proteins associate to form nucleosomes and DNA wraps itself tightly around these complexes. Higher-order structures called heterochromatin form, and become tightly wound and generally inaccessible to transcription. In models of gene regulation, profound changes in chromatin occur causing the unwinding of these superstructures, reaching a state referred to as euchromatin. The many players of transcription now have access to DNA and the process can commence and proceed.

Before, during, and following transcription, molecules in the vicinity of the promoter region of a gene undergo constant chemical modifications. Histones lose or gain methyl or acetyl groups, and there is swapping of histone proteins within the nucleosomes. Nucleosomes shift positions, changing their distribution patterns on the DNA backbone. Early events of gene activation involve the initial binding of only a few **transcription factors** to the promoter (initiation), but this is quickly followed by the addition of many others (continuation). As RNA polymerase II moves along the DNA, proteins modify the chromatin in advance of the polymerase. In the wake of the moving polymerase, histones reassociate with the DNA, again with the help of other proteins.

All of these events are preceded by some kind of biological trigger. To illustrate this, we will study the binding sites of the chicken estrogen receptor. Estrogen is a steroid hormone produced by reproductive tissues of all vertebrates. Estrogen diffuses into cells where it is bound by the estrogen receptor, a transcription factor protein containing two zinc finger domains. This estrogen receptor complex enters the nucleus where it binds specific DNA sequences in the promoter regions of estrogen-responsive genes. Work by Pierre Chambon, Ronald Evans, and Elwood Jensen on the actions of estrogen and other steroid hormones led to their receiving the 2004 Lasker Basic Medical Research Award.

Transfac

To study the DNA sequences bound by the estrogen receptor complex, we will use the database and tools available at www.gene-regulation.com. This site is the home of the Transfac database, an industry standard for a curated collection of **transcription factor binding sites** that were detected experimentally in the laboratory. A variety of methods can detect the binding of protein to DNA and map the exact location: the bound DNA will have different migration rates in a gel, or be protected from modification or degradation.

The promoter we'll be studying is that of the chicken ovalbumin gene. Ovalbumin is a storage protein in the white of the egg, and this and other components of the egg provide nourishment to the developing chick prior to hatching. This ovalbumin gene promoter was the object of many early studies of steroid regulation of gene expression and is a well-characterized model for transcription factor binding sites.

The regulatory DNA elements of interest for this promoter region are within several hundred nucleotides of the beginning of transcription, but other transcription regulatory elements are located many thousands of nucleotides away from the transcription start site of genes. It is remarkable that sequences as short as five to ten nucleotides, in the proper context, can influence gene expression at such a linear distance. Genomic secondary structures most probably bring these elements physically closer to the transcription start site, aided by proteins that help bridge the gap. Alternatively, large sections of chromatin must be activated to "turn on" these genes, and these distant elements are only a small part of this large picture that is not totally understood.

From the Transfac database home page, create an account (it is free). Then go to the left sidebar, part of which can be seen in **Figure 9.1A**, and click on the Databases link which brings you to a list of resources available from this Website. The Transfac database listing, shown in Figure 9.1B, provides links to their extensive documentation. Clicking on >Search< brings you to a simple form where you can choose which Transfac table to search (Figure 9.1C). We will be exploring records from the Factor, Gene, Matrix, and Site tables and these will be discussed below. A Cell record will provide annotation on the cell/tissue of origin for the protein factors that bind DNA sites. The transcription factors described in the Transfac database are divided into 50 different classes and these can be searched using the Class button.

(A)

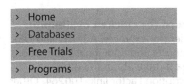

(B)

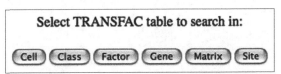

Figure 9.1 Transfac navigation.
(A) The Databases link on the left sidebar of the Transfac Website.
(B) Selecting Databases takes you to a new page where a >Search< link is seen. (C) Selecting Search, the page refreshes, and the database entry-point choices are presented. (D) Clicking on the "Gene" button takes you to this form.

(C)

(D)

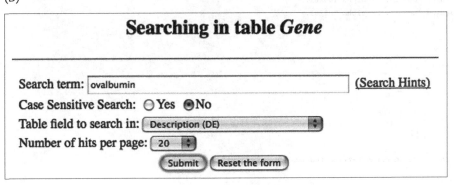

To reach the record of the chicken ovalbumin gene, click on the "Gene" button (Figure 9.1C) and you are brought to a simple form (Figure 9.1D) where you can enter the query (ovalbumin) and select "Description (DE)" from the drop-down menu. Click on "Submit" and you are brought to the ovalbumin gene record (**Figure 9.2**).

As can be seen in Figure 9.2, a Transfac Gene record has a number of rows, each with a two-letter code at the far left. Clicking on these hypertext row labels (for example, DE) will generate a help page with explanations of all row labels used in that record. The record begins with some basics: an accession number (AC G000071), an identifier (ID CHICK$OA) providing more information about this record (from a chicken), a brief description (DE ovalbumin), and the organism (OS chick, Gallus gallus). In between these lines and others are spacers, "XX," providing welcome visual breaks for easier viewing of the record.

The next major section contains binding site (BS) positions of the transcription factors, shown as negative numbers emphasizing their locations as upstream of the start of transcription. Each binding site sequence has a length, so it is

Figure 9.2 A Transfac Gene record for chicken ovalbumin. Each line begins with a field identifier (for example, AC, DE, and OS). To see the estrogen receptor binding site and additional information, click on the Site link, R03567. To learn more about the estrogen receptor protein, click on the transcription Factor link, T00264. Other transcription factor binding sites identified in this ovalbumin Gene record include NF-1, HNF-3alpha, and COUP-TF1.

```
AC   G000071
XX
ID   CHICK$OA
XX
DT   14.06.1995 (created); dbo.
DT   25.04.1996 (updated); dbo.
CO   Copyright (C), Biobase GmbH.
XX
SD   Ov
XX
DE   ovalbumin
XX
OS   chick, Gallus gallus
OC   eukaryota; animalia; metazoa; chordata; vertebrata; aves; neornithes; neognathae; galliformes; phasianidae
XX
BC   6.1.3.1
XX
BS    -1124    -1095   R01192; CHICK$OA_01; Binding factors: NF-1 (-like proteins) T00601.
BS     -900     -732   R01618; CHICK$OA_05.
BS     -898     -874   R08890; CHICK$OA_16; Binding factors: HNF-3alpha T00371, HNF-3beta T03256, HNF-3gamma T01050.
BS     -784     -765   R03200; CHICK$OA_11.
BS     -655     -617   R03567; CHICK$OA_13; Binding factors: ER-alpha T00264.
BS     -518     -483   R03568; CHICK$OA_14; Binding factors: ER-alpha T00264.
BS     -345     -320   R01853; CHICK$OA_06.
BS     -310     -274   R03569; CHICK$OA_15; Binding factors: ER-alpha T00264.
BS     -290     -275   R01854; CHICK$OA_07.
BS     -268     -240   R01855; CHICK$OA_08.
BS     -241     -218   R01856; CHICK$OA_09.
BS     -165     -129   R03566; CHICK$OA_12; Binding factors: ER-alpha T00264.
BS     -160     -140   R01857; CHICK$OA_10.
BS      -90      -66   R01193; CHICK$OA_02;Binding factors COUP-TF1 T00148,COUP-TF1 T00149,COUP-TF2 T00045,PPAR-alpha T00991,RXR-alpha T01345.
BS      -83      -71   R01194; CHICK$OA_03; Binding factors: ER-alpha T00261.
BS      -52      -41   R01195; CHICK$OA_04; Binding factors: ER-alpha T00261.
XX
DR   EMBL: J00895; GGOVAL01.
DR   EMBL: M29020; GDOVAL5A.
DR   EMBL: V00437; GGOV03.
DR   TRRD: 00049.
DR   TRANSPATH: G000071.
XX
RN   [1]; RE0014431.
RX   PUBMED: 9832435.
RA   Dean D. M., Berger R. R., Sanders M. M.
RT   A winged-helix family member is involved in a steroid hormone-triggered regulatory circuit
RL   Endocrinology 139:4967-4975 (1998).
```

expressed as a pair of negative coordinates. For example, the first estrogen receptor binding site is located between 655 and 617 nucleotides upstream of the transcription start site (the 5P end of the mRNA). This binding site has an accession number, R03567, and identifies the molecule that was shown to bind to this site, the estrogen receptor alpha protein, ER-alpha T00264. "ER" is a common abbreviation for estrogen receptor and should not be confused with the identical abbreviation for the cellular structure, the endoplasmic reticulum.

The accession number R03567 is a hyperlink to the Site (transcription factor binding site) record which contains more information, including the sequence which is shown in **Figure 9.3**. As described in the Transfac documentation (available from the link seen in Figure 9.1B), the sequence is derived from published research. A larger sequence (that is, the 38 nucleotides seen in Figure 9.3) may represent the resolution of the ER binding, but the authors identified TGACCT as a key element and these bases appear in uppercase.

Identifying other binding sites for the estrogen receptor

As can be seen in Figure 9.2, protein factors binding the transcription factor binding sites are labeled with accession numbers that begin with T. Clicking on T00264 takes you to the record for chicken estrogen receptor protein. In addition to providing sequence and domain details about this protein, this Factor record contains a listing of all the other gene promoter regions where chicken estrogen receptor was shown to bind. The list from this record appears in **Figure 9.4**.

In addition to the four sites in the ovalbumin gene (notice the "Ov" symbol in the middle of the BS lines in Figure 9.4) seen earlier in the Gene record (Figure 9.2), there are two other genes, symbolized by apoVLDL II and VIT-II. If you navigate to their Gene accession numbers, G000048 and G000081, respectively, you can find Swiss-Prot accession numbers. The following functions of these three proteins in chicken can be found on the ExPASy Website.

1. Apo very low density lipoprotein. "Protein component of the very low density lipoprotein (VLDL) of egg-laying females. Potent lipoprotein lipase inhibitor,

```
SQ    gactatgaactcacatccaaaggagctTGACCTgatac
```

Figure 9.3 An estrogen receptor alpha binding site in the chicken ovalbumin promoter region. This sequence was obtained from the Transfac Site record, R03567, and represents the sequence found between −655 and −617.

```
BS    R03564 AS$ER_03; Quality: 6.
BS    R03565 AS$TR_03; Quality: 6.
BS    R00153 CHICK$APOVLDL_01; Quality: 4; apoVLDL II, G000048; chick, Gallus gallus.
BS    R00154 CHICK$APOVLDL_02; Quality: 4; apoVLDL II, G000048; chick, Gallus gallus.
BS    R03566 CHICK$OA_12; Quality: 6; Ov, G000071; chick, Gallus gallus.
BS    R03567 CHICK$OA_13; Quality: 6; Ov, G000071; chick, Gallus gallus.
BS    R03568 CHICK$OA_14; Quality: 6; Ov, G000071; chick, Gallus gallus.
BS    R03569 CHICK$OA_15; Quality: 6; Ov, G000071; chick, Gallus gallus.
BS    R01570 CHICK$VIT2_03; Quality: 4; VIT-II, G000081; chick, Gallus gallus.
BS    R01571 CHICK$VIT2_04; Quality: 4; VIT-II, G000081; chick, Gallus gallus.
BS    R01573 CHICK$VIT2_06; Quality: 2; VIT-II, G000081; chick, Gallus gallus.
BS    R01577 CHICK$VIT2_10; Quality: 4; VIT-II, G000081; chick, Gallus gallus.
BS    R01579 CHICK$VIT2_12; Quality: 4; VIT-II, G000081; chick, Gallus gallus.
BS    R02150 ER$CONS; Quality: 6.
```

Figure 9.4 Other estrogen receptor binding sites in chicken. This listing of binding sites was obtained by clicking on the transcription Factor link, T00264, as seen in the "BS" lines in Figure 9.2.

preventing the loss of triglycerides from VLDL on their way from the liver to the growing oocytes."

2. Ovalbumin. "Storage protein of egg white."

3. Vitellogenin II. "Precursor of the major egg-yolk proteins that are sources of nutrients during early development of oviparous organisms."

Since no other genes are listed, it is easy to conclude that in the chicken, the estrogen receptor's sole purpose is to regulate these three proteins involved in egg production. However, this would be incorrect. Remember that the database records are based on publications of factor binding. More likely, there are no published findings outside of these three genes. The estrogen receptor could be binding hundreds of chicken genes but if these have not been studied and published, they will be absent from the Transfac database.

Let's return to the ovalbumin Gene record as shown in Figure 9.2. The Factor T00261 is labeled "ER-alpha" yet this number is different from the ER-alpha number seen earlier, T00264. There is a simple explanation: T00264 is the chicken estrogen receptor while T00261 is the human estrogen receptor. Remember, the Transfac database is built from laboratory evidence of binding and function experiments. In an assay using the chicken promoter region and human estrogen receptor protein, someone showed that binding could take place in these regions. This does not necessarily mean that the chicken receptor cannot bind here, but perhaps this wasn't tested (this could be discussed in the publication associated with this assay). Click on the binding Site hyperlink R01194 and you will navigate to the details about this site and the published evidence demonstrating the binding to this sequence.

There are additional transcription factor binding sites listed in Figure 9.2. These include sites for the factors NF-1, HNF-3alpha, HNF-3beta, HNF-3gamma, COUP-TF1, COUP-TF2, PPAR-alpha, and RXR-alpha. The presence of other sites will sometimes reflect co-regulation of expression. For example, COUP-TF1 also binds the vitellogenin II gene, suggesting cooperation between these factors and two genes involved in egg production. Note that when multiple factors bind the same site they are listed on the same line.

There are some additional transcription factor binding sites (R01618, R03200, R01853, R01854, R01855, R01856, and R01857) which have no binding factor name listed. That is, experiments showed that something is binding these regions, but the binding factor was not identified.

Let's return to the human estrogen receptor Factor record, T00261, the link for which can be seen in Figure 9.2. Clicking here will take you to a page listing all the sequences that were shown to be bound by the human estrogen receptor (**Figure 9.5**). This list is much longer than the list for chicken estrogen receptor, reflecting a greater history of experiments with the human receptor protein, rather than the chicken having fewer genes responsive to estrogen.

The list of sites bound by the human protein includes the chicken elements, as expected, along with two *Xenopus* vitellogenin genes. The list includes a number of genes associated with cancer: BRCA1, cathepsin D, c-fos, EBAG9, and p53. BRCA1 is a tumor suppressor and mutations in this gene are responsible for approximately 80% of the inherited forms of breast and ovarian cancers, two tissues responsive to estrogen. As scientists identify genes modulated by estrogen, they can build a biochemical pathway including these and other proteins, drawing a picture of the cascading effects of estrogen on gene regulation and cellular functions.

Predicting transcription factor binding sites

Since the sequences of transcription factor binding sites are known, could they be predicted within a query sequence, much like finding restriction enzyme sites? There are many challenges to overcome; perhaps the greatest is that many sites will be predicted, and identifying the real and functional sites is difficult.

```
BS   R02715 AS$ER_02; Quality: 2.
BS   R03564 AS$ER_03; Quality: 6.
BS   R03869 AS$ER_04; Quality: 2.
BS   R13349 AS$ER_05; Quality: 3.
BS   R14257 AS$ER_06; Quality: 2.
BS   R14198 AS$ERalpha1_01; Quality: 1.
BS   R03565 AS$TR_03; Quality: 6.
BS   R01194 CHICK$OA_03; Quality: 1; Ov, G000071; chick, Gallus gallus.
BS   R01195 CHICK$OA_04; Quality: 1; Ov, G000071; chick, Gallus gallus.
BS   R01581 CHICK$VIT2_14; Quality: 2; VIT-II, G000081; chick, Gallus gallus.
BS   R02150 ER$CONS; Quality: 1.v
BS   R00144 HS$APOE_03; Quality: 1; apoE, G000207; human, Homo sapiens.
BS   R14229 HS$BRCA1_01; Quality: 6; BRCA1, G002124; human, Homo sapiens.
BS   R14230 HS$BRCA1_02; Quality: 6; BRCA1, G002124; human, Homo sapiens.
BS   R13265 HS$CATHD_06; Quality: 2; CATH-D, G001123; human, Homo sapiens.
BS   R14291 HS$CATHD_07; Quality: 2; CATH-D, G001123; human, Homo sapiens.
BS   R14225 HS$CC3_03; Quality: 1; C3, G000230; human, Homo sapiens.
BS   R14226 HS$CC3_04; Quality: 1; C3, G000230; human, Homo sapiens.
BS   R14227 HS$CC3_05; Quality: 1; C3, G000230; human, Homo sapiens.
BS   R16116 HS$CFOS_35; Quality: 1; c-fos, G000218; human, Homo sapiens.
BS   R14231 HS$COX7A2L_01; Quality: 3; COX7A2L, G008909; human, Homo sapiens.
BS   R14232 HS$EBAG9_01; Quality: 3; EBAG9, G008910; human, Homo sapiens.
BS   R03958 HS$GHA_11; Quality: 6; CGA, G000271; human, Homo sapiens.
BS   R14217 HS$HSP27_02; Quality: 3; HSP27, G008805; human, Homo sapiens.
BS   R14203 HS$NQO1_02; Quality: 1; NQO1, G006920; human, Homo sapiens.
BS   R16684 HS$P53_15; Quality: 3; p53, G001075; human, Homo sapiens.
BS   R14211 HS$PGR_01; Quality: 1; PGR, G003936; human, Homo sapiens.
BS   R14194 HS$PS2_04; Quality: 4; PS2, G000373; human, Homo sapiens.
BS   R14207 HS$PTMA_01; Quality: 1; human, Homo sapiens.
BS   R14209 HS$PTMA_02; Quality: 1; human, Homo sapiens.
BS   R14199 HS$SLC9A3R1_01; Quality: 1; SLC9A3R1, G003777; human, Homo sapiens.
BS   R14200 HS$SLC9A3R1_02; Quality: 1; SLC9A3R1, G003777; human, Homo sapiens.
BS   R14235 HS$TF_14; Quality: 1; Tf, G000405; human, Homo sapiens.
BS   R14286 HS$TGFA_04; Quality: 1; TGFalpha, G002920; human, Homo sapiens.
BS   R14228 MOUSE$CFOS_16; Quality: 2; c-fos, G000486; mouse, Mus musculus.
BS   R14292 MOUSE$CHAT_01; Quality: 2; Chat, G008976; mouse, Mus musculus
BS   R03366 RABBIT$UG_15; Quality: 2; Ug, G000679; rabbit, Oryctolagus cuniculus.
BS   R14214 RAT$BCK_02; Quality: 3; BCK, G001914; rat, Rattus norvegicus.
BS   R14197 RAT$JUN_01; Quality: 2; Jun, G008873; rat, Rattus norvegicus.
BS   R03480 RAT$LHB_01; Quality: 1; LHbeta, G000764; rat, Rattus norvegicus.
BS   R14237 RAT$OXT_01; Quality: 2; Oxt, G008925; rat, Rattus norvegicus.
BS   R10107 RAT$VEGF_01; Quality: 1; VEGF, G002589; rat, Rattus norvegicus.
BS   R10108 RAT$VEGF_02; Quality: 1; VEGF, G002589; rat, Rattus norvegicus.
BS   R01586 XENLA$VITA2_01; Quality: 1; VIT-A2, G000864; clawed frog, Xenopus laevis.
BS   R02665 XENLA$VITB1_05; Quality: 2; VIT-B1, G000865; clawed frog, Xenopus laevis.
```

Figure 9.5 List of human estrogen receptor binding sites. These were obtained from the human estrogen receptor Factor record, T00261.

Two programs at the Transfac Website will be used to demonstrate transcription factor binding site prediction. Let's use the chicken ovalbumin promoter region as a query since we know the number and location of the known estrogen receptor binding sites.

An experiment with MATCH

First, let's look at the sequences of the binding sites that are listed in the estrogen receptor Factor record, T00264. These are the sites that the chicken receptor is known to bind (**Figure 9.6**). As can be seen in the figure, there is diversity in sequence for the estrogen receptor binding sites. The core sequence (that is, TGACCT) can be easily aligned but the flanking sequence differs in every position. They are also different lengths; data are either missing from the short sequences or the longer sequences may contain information that is less important. Another

Figure 9.6 Chicken estrogen receptor binding sites. The above sequences were taken from the individual Transfac Site records, listed in the chicken Factor Table, T00264 (see Figure 9.4). They were manually aligned based on the core sequence of TGACCT. The uppercase nucleotides were identified as critical to factor binding. The estrogen receptor binding site consensus was derived from sites obtained from human, rat, mouse, cow, frog, and chicken.

```
R00153                            GGGGCTCAGTGACCC
R00154                            AGGTCAGACTGACCT
R03566            gcagcagagccttagcTGACCTtttcttgggacaag
R03567    gactatgaactcacatccaaaggagctTGACCTgatac
R03568              attcgacattcatctgTGACCTgagcaaaatgattta
R03569              aagaggcttTGACCTgtgagctcacctggacttcata
R01571                            GGTCAGCGTGACC
R01573                              GCGTGACCGGAGCTGAAAGAACAC
R01577                            GGTCAACATAACC
R01579                            GCCAGCCGTGACC
R02150  (ER consensus)           AGGTCANNNTGACCT
```

possibility is that the different sites require different amounts of sequence flanking the core in order to be functional. The core sequence, in the context of the surrounding bases, is in the right location and is adjacent to the sites of other transcription factors, all of which come into play to act as an estrogen receptor binding site. Therefore, we should not simply discard the information that surrounds these core elements.

In previous chapters, we encountered similar problems of matching variable sequences which were solved with matrices. For example, the BLOSUM62 matrix is used for scoring similarity with BLASTP, BLASTX, and TBLASTN. Prosite profiles are matrices representing many possibilities for sequences found in motifs. The creators of Transfac have built matrices to act as queries when searching promoter regions for transcription factor binding sites, and have built a tool for searching with these matrices called MATCH. **Figure 9.7** shows the matrix for estrogen receptor binding sites. This matrix was built from sequences of human, rat, mouse, and chicken binding sites.

Before using MATCH, we need to retrieve the chicken ovalbumin promoter sequence used by Transfac in their Gene record. The chicken ovalbumin Gene page (Figure 9.2) has several links to sequence files in rows labeled "DR" for "Database References." Looking at the Site records, four of the estrogen receptor binding sites identified in the Figure 9.2 list indicate that M29020 is the DNA sequence that contains these binding sites. The other two binding sites were identified by human estrogen receptor binding. A good test of the matrices built for the MATCH program would be for MATCH to find these four binding sites within M29020.

Figure 9.7 The matrix of estrogen receptor binding sites. Transfac matrix M00191 was built from human, rat, mouse, and chicken binding sites. The sequence is read top to bottom, 5P to 3P, with the nucleotides appearing in the last column. The numbering in the first column corresponds to matrix nucleotide 01, 02, and so on. The middle columns are labeled A, C, G, and T and the count of each base in each position is recorded as numbers in those columns. For example, the third base (03) was A 12 times, while other nucleotides were found four times or less. Based on this ratio, an A appears in the last column for this row. For a full description of how the matrix was built, see the documentation within Transfac.

PO	A	C	G	T	
01	6	4	4	6	N
02	4	9	3	4	N
03	12	4	3	1	A
04	6	1	11	2	R
05	3	2	11	4	G
06	3	3	4	10	N
07	3	10	3	4	C
08	11	2	4	3	A
09	4	9	3	4	N
10	3	6	3	8	N
11	2	3	9	6	N
12	0	1	0	19	T
13	1	0	19	0	G
14	20	0	0	0	A
15	0	18	2	0	C
16	1	18	1	0	C
17	2	7	2	9	Y
18	7	1	8	4	N
19	8	5	3	4	N

```
BS      -655     -617    R03567; CHICK$OA_13; Binding factors: ER-alpha T00264.
BS      -518     -483    R03568; CHICK$OA_14; Binding factors: ER-alpha T00264.
BS      -310     -274    R03569; CHICK$OA_15; Binding factors: ER-alpha T00264.
BS      -165     -129    R03566; CHICK$OA_12; Binding factors: ER-alpha T00264.
```

Figure 9.8 The four chicken estrogen receptor binding sites within the ovalbumin promoter. These lines were taken from the chicken ovalbumin Gene record, which appears in Figure 9.2.

A look at the annotation for this EMBL record indicates that this sequence is over 5300 nucleotides long and described as "Chicken ovalbumin gene, 5′flank." How much of the 5P region is included? Does it end at the transcription start site? Is it in the 5P to 3P orientation? Was it correctly annotated? We cannot be sure about the answers to any of these questions. The four sites listed in **Figure 9.8** are within 655 nucleotides upstream of the transcription start site. Should we just use the last 655 nucleotides of M29020 and assume that this is the correct sequence? No.

To make sure the comparison between known sites and those found with MATCH is fair, only the identical section of M29020 should be used instead of the entire 5300 nucleotides. To identify the location of this sequence within the larger sequence record, the sequences of R03567 and R03566 were used as BLASTN queries against M29020, acting as upstream and downstream markers, respectively. The location of the R03567 sequence will provide the 5P boundary of the sequence and the location of R03566 will provide the 3P boundary along with the orientation of the sequence. R03567 aligns with M29020 starting at nucleotide 2885 and R03566 aligns with nucleotides 3375 to 3410. Therefore, a DNA fragment of M29020 from 2885 to 3410 (526 nucleotides) should be used in MATCH.

Using the hyperlink within the Transfac Site record (for example, R03567), navigate to the EMBL record for M29020. The DNA record that appears is in FASTA format and this is not a convenient way to copy the sequence we have identified, above, because of the absence of coordinates. However, at the top of the page are five links, one of which is called "Text Entry." Clicking here changes the sequence to EMBL format and you can now exactly copy nucleotides 2885–3410 to the clipboard, including the numbers and spaces.

Returning to the Transfac home page, click on the "Programs" link on the left sidebar (Figure 9.1A). The page refreshes, listing a variety of programs available to use at this Website along with a brief description. From this list, choose MATCH.

On the left side of the form that appears, paste the sequence into the query window. This Transfac query form conveniently accepts what you copied to the clipboard, which is a mix of sequence, numbers, spaces, and carriage returns. Click the radio button labeled "OR take a new sequence and enter a name for it," and enter a name in the text field adjacent to this label, for example "chicken ovalbumin promoter region." On the right side of the form, choose "vertebrates" from the short list to make your search more specific. "Use high quality matrices only" should be checked. Finally, click on the radio button "to minimize the sum of both error rates." This setting should minimize both the false positive and false negative hits in the search results, raising the stringency of the search. In the lower left of the window, click on "Submit the form."

Your results are returned in a few moments (**Figure 9.9**). There are 29 transcription factor binding sites predicted for this 526 nucleotide sequence (this summary appears at the bottom of the results). There is a row for each hit. The columns contain matrix identifiers, positions of the prediction (including a label indicating the positive or negative strand), two scores of the match to the matrix, the sequence found, and the name of the factor. The "core" represents the most conserved region within a site sequence, while the "matrix" match looks at the similarity over the length of the site. For example, the rows with a predicted estrogen receptor sequence show a core (uppercase) of TGACC. A match of

matrix identifier	position (strand)	core match	matrix match	sequence (always the (+)-strand is shown)	factor name
V$OCT1_Q6	11 (+)	0.824	0.827	tcacatCCAAAggag	Oct-1
V$BARBIE_01	15 (+)	0.979	0.937	atccAAAGGagcttg	Barbie Box
V$ER_Q6	17 (+)	1.000	0.949	ccaaaggagctTGACCtga	ER
V$CDPCR1_01	100 (+)	0.793	0.834	cACTGAtggc	CDP CR1
V$COMP1_01	120 (−)	1.000	0.763	aggtatgcaatggCAATCcattcg	COMP1
V$CHOP_01	122 (+)	1.000	0.929	gtaTGCAAtggca	CHOP-C/EBPalpha
V$OCT1_Q6	125 (+)	0.770	0.807	tgcaatGGCAAtcca	Oct-1
V$CDPCR1_01	132 (−)	0.910	0.917	gcaaTCCATt	CDP CR1
V$ER_Q6	144 (+)	1.000	0.936	acattcatctgTGACCtga	ER
V$GATA_C	168 (−)	1.000	0.968	atgatTTATCt	GATA-X
V$CAAT_01	184 (−)	0.953	0.943	tgAATGGttgct	CCAAT box
V$OCT1_Q6	202 (+)	0.893	0.831	cctcatGAAAggca	Oct-1
V$CREL_01	213 (+)	1.000	0.860	ggcaaTTTCC	c-Rel
V$OCT1_Q6	214 (−)	0.888	0.804	gcaaTTTCCacactc	Oct-1
V$OCT1_02	228 (+)	1.000	0.968	cacaaTATGCaacaa	Oct-1
V$FOXJ2_01	239 (+)	1.000	0.946	acaaagacAAACAgagaa	FOXJ2
V$HFH3_01	242 (−)	1.000	0.950	aagacAAACAgag	HFH-3
V$SOX9_B1	251 (+)	1.000	0.942	agagaACAATtaat	SOX-9
V$SOX5_01	253 (+)	1.000	0.991	agaACAATta	Sox-5
V$NFY_Q6	336 (−)	1.000	0.973	gtgATTGGata	NF-Y
V$CAAT_01	337 (−)	1.000	0.974	tgATTGGataag	CCAAT box
V$GATA_C	342 (+)	1.000	0.971	gGATAAgaggc	GATA-X
V$ER_Q6	344 (+)	1.000	0.950	ataagaggcttTGACCtgt	ER
V$HNF4_01	345 (−)	1.000	0.836	taagaggCTTTGacctgtg	HNF-4
V$COUP_01	355 (+)	1.000	0.879	TGACCtgtgagctc	COUP-TF / HNF-4
V$CMYB_01	434 (+)	1.000	0.944	tcatgagactGTTGGttt	c-Myb
V$FOXD3_01	440 (+)	0.936	0.911	gaCTGTTggttt	FOXD3
V$OCT1_06	463 (+)	0.964	0.978	cacaaTGAAAtgcc	Oct-1
V$ER_Q6	496 (+)	1.000	0.929	agagccttagcTGACCttt	ER

Figure 9.9 The MATCH search results of the chicken ovalbumin promoter region.

1.000 is 100% identical and this is a perfect match to the core sequence seen in Figure 9.7. Our query sequence should show that the first estrogen receptor site starts at position one, but instead nucleotide 17 is listed. Comparing the R03567 sequence (see Figure 9.6) to that shown in the MATCH results table (Figure 9.9), it appears that MATCH truncated the site to include only some of the sequence flanking the core.

A total of four estrogen receptor (ER) binding sites were predicted with MATCH. Based on the distance between real sites, and the location of and distance between the predicted sites, MATCH has identified all four experimentally derived sites and no other sites. In this single test where we knew the location of real sites, MATCH was impressive, but remember that these are predictions. These predictions can certainly guide laboratory experiments, but it would be risky to assume that all of the predictions are 100% correct.

An experiment with PATCH

PATCH is another prediction program available at the Transfac Website and you can find it in the listing of programs where you found MATCH. Whereas MATCH searches query sequences for matches to a matrix, PATCH looks for sequences that are identical to the core sequences of known sites, collectively referred to as PATterns. MATCH could be considered very specific because the sequence match goes beyond the core binding site, and could perhaps exclude binding sites of interest. PATCH relaxes the stringency by using smaller sequence matches and will find all sequences that are identical to any known core sequence. The trade-off is that a smaller pattern could mean more false positives.

Barbie Box

Look at the results in Figure 9.9, under "factor name." Barbie??? No, not the doll. This is short for "Barbiturate-inducible element."

```
Identifier          Position    Mismatches    Score    Binding Factor         Seq Pattern
XENLA$VITA2_02      28 (+)          0         100.00   ER-alpha, ER-beta1,    TGACCT
                                                       NHP-1, T3R-alpha1,
                                                       T3R-beta1

CHICK$OA_12         28 (+)          0         100.00   ER-alpha               TGACCT
CHICK$OA_13         28 (+)          0         100.00   ER-alpha               TGACCT
CHICK$OA_14         28 (+)          0         100.00   ER-alpha               TGACC
CHICK$OA_15         28 (+)          0         100.00   ER-alpha               TGACCT
CHICK$OA_03         28 (-)          0         100.00   ER-alpha               GGTCA
CHICK$OA_04         28 (-)          0         100.00   c-Fos, c-Jun,          GGTCA
                                                       ER-alpha

XENLA$VITA2_02      28 (-)          0         100.00   ER-alpha, ER-beta1,    AGGTCA
                                                       NHP-1, T3R-alpha1,
```

Using MATCH, 29 sites were predicted for the 526 nucleotide chicken ovalbumin promoter region. PATCH finds 579 sites using the same promoter sequence, although the number of differences is not as large as it seems. Consider the 5P-most laboratory-verified site at PATCH coordinate 28 (**Figure 9.10**). Since the sites are reduced to short core elements, this site is identified by six different chicken estrogen receptor sites and two *Xenopus* sites. The other laboratory-identified sequences are also identified by multiple site patterns. But even if we consider that every site found by MATCH is found by eight PATCH patterns (Figure 9.10), there are still several hundred pattern matches to account for to bring the number of hits to 579.

PATCH successfully found the four known sites in the chicken promoter, also found by MATCH. In addition, three novel sites were identified (see **Table 9.1** for a comparison of MATCH and PATCH hits). These sites were not found by any chicken patterns. The first and second novel sites, at coordinates 161 and 266, respectively, were identified by a rat pattern, RAT$VEGF_02 (see **Figure 9.11**). The third site, at coordinate 489, was found by the same rat pattern in addition to a second rat pattern and a human pattern. These patterns are from the VEGF and cathepsin D genes, both of which were shown to be estrogen responsive so these new chicken sites could be real. It has already been established that the chicken ovalbumin promoter has multiple estrogen receptor binding sites so it may not be surprising that additional sites can be found. But experimental evidence for these sites is absent from this database. Perhaps these are divergent sites in the

Figure 9.10 Multiple patterns identifying a single transcription factor binding site. The position of all these PATCH hits is identical: the site at nucleotide 28. What differs is the transcription factor that is able to bind this position, and these are listed in the first column.

Table 9.1 Comparison of experimentally determined results with the MATCH and PATCH predictions

Gene coordinates*	Distance from 5P end of M29020**	MATCH	PATCH
−655 to −617	17	Found	Found
−518 to −483	144	Found	Found
	161		New
	266		New
−310 to −274	344	Found	Found
	489		New
−165 to −129	496	Found	Found

*Locations of the estrogen receptor binding sites within the chicken ovalbumin promoter region (listed in Figure 9.8). The Gene records express these values as negative numbers, upstream of the transcription start site.

**Gene coordinates were converted to distances from the 5P end of this region to be compatible with MATCH and PATCH coordinates.

Identifier	Position	Mismatches	Score	Binding Factor	Seq Pattern
RAT$VEGF_02	161 (+)	0	100.00	ER-alpha, ER-beta	GAGCA
RAT$VEGF_02	266 (-)	0	100.00	ER-alpha, ER-beta	GAGCA
HS$CATHD_01	489 (+)	0	100.00	ER-alpha, Sp1	GGGCA
RAT$VEGF_01	489 (+)	0	100.00	ER-alpha, ER-beta	GGGCA
RAT$VEGF_02	489 (-)	0	100.00	ER-alpha, ER-beta	TGCCC

Figure 9.11 Novel estrogen receptor binding sites found with PATCH. These five lines were taken from the much longer PATCH results.

chicken, no longer functioning or only weak binders of the estrogen receptor and not identified in the laboratory experiments.

What can we make of the other, non-estrogen receptor complex, predicted sites from MATCH and PATCH? In the absence of laboratory experiments, a case must be built based on other findings before serious consideration of these sites as "real." Behind each transcription factor is a signaling pathway, and these pathways should be considered along with publications about the genes downstream of these promoters. What is known about the transcription factors? What is known about the target promoters (and the attached genes)? Is there any biology that links them together? Considering that promoter elements are small and embedded in sequences (promoter regions) that are not as conserved as coding regions, are these predicted sites found in orthologous promoters? Are they found adjacent to other transcription factor binding sites that are known to create composite elements of function? Could nature be selecting for the presence of these elements? Taken together, this type of evidence would help build a case that a prediction may have biological merit.

As can be gathered from the above section, placing importance on very short sequences at variable (and sometimes long) distances from the beginning of translation can be a very complicated topic of analysis. We will now focus on another small sequence which is constrained by location: the exact place where translation begins.

9.3 TRANSLATION INITIATION: THE KOZAK SEQUENCE

In eukaryotes, protein translation begins with the binding of the mRNA 5P end by the small ribosomal subunit along with additional protein co-factors. This complex then moves 3P-ward along the mRNA, searching for the translation initiation codon, AUG. (Many sections of this chapter are about mRNA, rather than DNA, so U will be used instead of T.)

There may be several translation initiation codons to choose from. Specific recognition of the correct one is facilitated by sequence surrounding this first codon and this region is often referred to as the **Kozak sequence**, named after Marilyn Kozak who first described it. After studying hundreds of sequences, GCCRCCAUGG emerged as the consensus for this region. The R is shorthand for purine, either an A or a G. For a listing of other, widely used symbols for groups of nucleotides, see the table in Box 3.1. For reference, each position in the consensus is assigned a number that relates it to the position of the A in AUG. For example, the first base in the Kozak sequence, G, is six nucleotides upstream of the A and is referred to as –6. The G after the AUG is referred to as +4.

It is important to remember that GCCRCCAUGG is a consensus sequence and that individual sequences may differ significantly. Work by Kozak and others demonstrates that single base changes of these nucleotides can affect the initiation of translation by as much as twentyfold. Not all genes have evolved to possess high rates of translation. This reflects the requirement that the levels of translation be customized for some level between the highest and the lowest possible. After millions of years of evolution, and significant opportunity to optimize, the sequences surrounding the AUG have reached a state that works for the cell, tissue, organ, developmental state, and organism.

To appreciate the possible levels of variation, the Kozak sequences of 15 mRNAs studied in earlier chapters were collected and appear in **Table 9.2**. This collection is not a balanced survey as almost half of the transcripts encode hemoglobins, and most of these are human. The hemoglobin cluster provides an opportunity to see the variation or similarity within sequences that are clearly related. The remaining transcripts on this list provide a sampling of what is seen in very different families.

At least in this very small sample, the base in the +4 position is usually a G, matching the consensus. The –3 position is always a purine, either A or G, and in the consensus it is most often an A. Although the sequences resemble the consensus, almost any base can be found in the remaining five positions. Notice the sequence for the GABA receptor; it is very A-rich in comparison to the other sequences.

Kozak sites are recognized by the translation apparatus as a "start here" instruction. What are the competing sequences? Looking at the 5P UTR of the human GABA A receptor, there are four other AUGs upstream of this translation start site

Table 9.2 Examples of Kozak sequences

RefSeq identifier	Transcript definition	Kozak*
NM_000798	*Homo sapiens* dopamine receptor D5 (*DRD5*), mRNA	CCCGAAAUGC
NM_000518	*Homo sapiens* hemoglobin, beta (*HBB*), mRNA	GACACCAUGG
NM_008220	*Mus musculus* hemoglobin, beta adult major chain (*Hbb-b1*), mRNA	GACAUCAUGG
NM_033234	*Rattus norvegicus* hemoglobin, beta (*Hbb*), mRNA	GACACCAUGG
NM_000519	*Homo sapiens* hemoglobin, delta (*HBD*), mRNA	GACACCAUGG
NM_005330	*Homo sapiens* hemoglobin, epsilon 1 (*HBE1*), mRNA	GGCAUCAUGG
NM_000184	*Homo sapiens* hemoglobin, gamma G (*HBG2*), mRNA	GACGCCAUGG
NM_000559	*Homo sapiens* hemoglobin, gamma A (*HBG1*), mRNA	GACGCCAUGG
NM_000207	*Homo sapiens* insulin (*INS*), transcript variant 1, mRNA	UCUGCCAUGG
NM_123204	*Arabidopsis thaliana* ribulose bisphosphate carboxylase small chain1B / RuBisCO small subunit 1B (*RBCS-1B*) (ATS1B) (AT5G38430) mRNA, complete cds	GAAGUAAUGG
NM_000537	*Homo sapiens* renin (*REN*), mRNA.	GGAAGCAUGG
NM_181744	*Homo sapiens* opsin 5 (*OPN5*), transcript variant 1, mRNA	AACAGAAUGG
NM_014244	*Homo sapiens* ADAM metallopeptidase with thrombospondin type 1 motif, 2 (*ADAMTS2*), transcript variant 1, mRNA	GCUGCCAUGG
NM_198904	*Homo sapiens* gamma-aminobutyric acid (GABA) A receptor, gamma 2 (*GABRG2*), transcript variant 1, mRNA	AAAGCGAUGA
NM_021267	*Homo sapiens* LAG1 homolog, ceramide synthase 1 (*LASS1*), transcript variant 1, mRNA	GGCGGUAUGG

*The "AUG" start codon within the Kozak sequence is underlined. The Kozak consensus sequence is GCCRCCAUGG. The Ts of the RefSeq record sequences were changed to U for this table.

```
AGGGGCAUGA
CTCCGUAUGA
GCCUCGAUGA
GCUUUGAUGG
AAAGCGAUGA  CDS
GCCRCCAUGG  Kozak consensus
```

Figure 9.12 Alignment of AUG sequences in the 5P UTR of the human GABA A receptor. The 359 nucleotide 5P UTR of NM_198904 has four other AUGs; shown are the surrounding seven bases aligned with the translation start AUG (coding sequence, CDS) and the consensus Kozak sequence.

(**Figure 9.12**). Generally, translation initiates at the first AUG encountered by the ribosome unless the surrounding sequence is not favorable. However, some level of translation can initiate outside of the "normal" AUG and this provides additional layers of regulation. Short, upstream ORFs (uORFs) can be translated and it is thought that the resulting protein is of no consequence, but the act of translating these uORFs diminishes translation initiation at the correct AUG. However, should the upstream AUG be in frame with the rest of the coding region, an alternative N-terminus could arise. Nature is constantly experimenting so it should not be surprising that these and other features of the UTR are utilized to regulate gene expression or create proteins with new features. **Box 9.1** provides some ideas for collecting Kozak sequences, both automatically and manually.

9.4 VIEWING WHOLE GENES

Before examining other small sequences within genes, let's explore additional ways to view basic gene structures—in this case, exons and introns.

The Ensembl genome browser, www.ensembl.org, contains an enormous amount of information that can be viewed in the context of genes and their

---

Box 9.1 How do you collect these Kozak sequences?

Your approach depends on the size of your study. To see the full extent of variability and to build a strong consensus (or prove there is none), the more sequences the better. Hundreds or thousands of sequences would provide a solid body of analysis. Perhaps a database of these sequences could be identified. Or, if you have programming skills (or know someone who does), you could write a simple program to identify the coordinates of the Kozak sequence, for example, and automatically collect these sequences. If your study is small, there is no database, or you have no programming skills, collecting them manually is your only choice. A small pilot study may generate clues that something interesting is going on, providing impetus for a larger, more labor-intensive study. Although more prone to error, care can be taken and a streamlined approach be developed to minimize the work, distractions, and opportunities for problems. For example, here is an efficient approach for manually gathering the Kozak sequences:

- Using the accession number, open the nucleotide record at the NCBI Website.

- Identify the coordinate of the beginning of the CDS using the annotation (the "A" of "ATG").

- Subtract six from the coordinate identified, above. Remember this number or write it down.

- Click on the FASTA format hypertext to change the format. Although this is not absolutely necessary, this reduces the subsequent scrolling down the sometimes extensive annotation to find the sequence.

- Click on the "Change Region Shown" button on the right sidebar.

- Enter the remembered number, above, in the "begin" text box and hit the return key.

- When the sequence refreshes, the first 10 nucleotides are the Kozak sequence. The "ATG" starting at nucleotide 7 is a good landmark: you need one more base beyond this triplet. This can be copied to the clipboard and pasted in your document.

Opportunities for error in this procedure and what you can do to reduce this risk (not a comprehensive list)

- Getting the accession number wrong. If you are collecting different sequences from the same DNA record, consider making the accession number hypertext in your word processor or spreadsheet document so you can click on it to navigate to the same Web page each time.

- Failing to read the annotation carefully. Take the time to do it right.

- Doing the math wrong or not remembering the number. Go "low-tech": pencil and paper. If this is a full-blown research project, you should practice good scientific procedures and record everything in your laboratory notebook.

- Copying the incorrect sequence. Take the time to do it right.

- Forgetting to copy! You end up pasting the sequence that you pasted in the last time. Stay focused and think about what you are doing.

- Finally, look at the sequences. Is it what you expected? Is there something obviously incorrect, like no "ATG"? Are sequential sequences identical (did you paste the same sequence twice)?

products. In fact, Chapter 12 is dedicated to **genome browsers**. A primary entry point from the Ensembl home page is a drop-down search menu. Rather than searching all species, narrow the search to "human" and enter "beta globin" (no quotes). A results summary is returned and from the list, click on the hits under "gene." One of those is beta hemoglobin, symbolized with "*HBB*." Click on that link and you are brought to a page with all the transcripts, translation products, and a summarized view of the gene structure in the genomic environment. On the left sidebar is a link called "sequence." Clicking there will return a view of the gene sequence, similar to that seen in **Figure 9.13**. Note that two noncoding 5P exons and their surrounding introns are not included in this figure to focus on the coding exons. In addition, deletion of this upstream sequence creates a better match to a later figure.

In this figure, the exons are shaded and stand out above the introns or noncoding sequences. From this view, you can appreciate the sizes of the exons and the introns, see the distances between these features, and easily see the sequences that occur at the junctions between introns and exons. But this is a small gene. Some genes could go on for pages, with most of the view being intronic sequence.

Figure 9.13 Human beta globin exons viewed from the Ensembl genome browser. The exons are shaded. Some 5P sequence was removed to make this view more similar to the sequence represented in Figure 9.14. ATG and TAA are underlined in exons 1 and 3, respectively.

```
---Sequence deleted---                       GTTTGAAGTCC
AACTCCTAAGCCAGTGCCAGAAGAGCCAAGGACAGGTACGGCTGTCATCACTTAGACCTC
ACCCTGTGGAGCCACACCCTAGGGTTGGCCAATCTACTCCCAGGAGCAGGAGGGGCAGGA
GCCAGGGCTGGGCATAAAAGTCAGGGCAGAGCCATCTATTGCTTACATTTGCTTCTGACA
CAACTGTGTTCACTAGCAACCTCAAACAGACACCATGGTGCATCTGACTCCTGAGGAGAA
GTCTGCCGTTACTGCCCTGTGGGGCAAGGTGAACGTGGATGAAGTTGGTGGTGAGGCCCT
GGGCAGGTTGGTATCAAGGTTACAAGACAGGTTTAAGGAGACCAATAGAAACTGGGCATG
TGGAGACAGAGAAGACTCTTGGGTTTCTGATAGGCACTGACTCTCTCTGCCTATTGGTCT
ATTTTCCCACCCTTAGGCTGCTGGTGGTCTACCCTTGGACCCAGAGGTTCTTTGAGTCCT
TTGGGGATCTGTCCACTCCTGATGCTGTTATGGGCAACCCTAAGGTGAAGGCTCATGGCA
AGAAAGTGCTCGGTGCCTTTAGTGATGGCCTGGCTCACCTGGACAACCTCAAGGGCACCT
TTGCCACACTGAGTGAGCTGCACTGTGACAAGCTGCACGTGGATCCTGAGAACTTCAGGG
TGAGTCTATGGGACGCTTGATGTTTTCTTTCCCCTTCTTTTCTATGGTTAAGTTCATGTC
ATAGGAAGGGGATAAGTAACAGGGTACAGTTTAGAATGGGAAACAGACGAATGATTGCAT
CAGTGTGGAAGTCTCAGGATCGTTTTAGTTTCTTTTATTTGCTGTTCATAACAATTGTTT
TCTTTTGTTTAATTCTTGCTTTCTTTTTTTTTCTTCTCCGCAATTTTTACTATTATACTT
AATGCCTTAACATTGTGTATAACAAAAGGAAATATCTCTGAGATACATTAAGTAACTTAA
AAAAAAACTTTACACAGTCTGCCTAGTACATTACTATTTGGAATATATGTGTGCTTATTT
GCATATTCATAATCTCCCTACTTTATTTTCTTTTATTTTTAATTGATACATAATCATTAT
ACATATTTATGGGTTAAAGTGTAATGTTTTAATATGTGTACACATATTGACCAAATCAGG
GTAATTTTGCATTTGTAATTTTAAAAAATGCTTTCTTCTTTTAATATACTTTTTTGTTTA
TCTTATTTCTAATACTTTCCCTAATCTCTTTCTTTCAGGGCAATAATGATACAATGTATC
ATGCCTCTTTGCACCATTCTAAAGAATAACAGTGATAATTTCTGGGTTAAGGCAATAGCA
ATATCTCTGCATATAAATATTTCTGCATATAAATTGTAACTGATGTAAGAGGTTTCATAT
TGCTAATAGCAGCTACAATCCAGCTACCATTCTGCTTTTATTTTATGGTTGGGATAAGGC
TGGATTATTCTGAGTCCAAGCTAGGCCCTTTTGCTAATCATGTTCATACCTCTTATCTTC
CTCCCACAGCTCCTGGGCAACGTGCTGGTCTGTGTGCTGGCCCATCACTTTGGCAAAGAA
TTCACCCCACCAGTGCAGGCTGCCTATCAGAAAGTGGTGGCTGGTGTGGCTAATGCCCTG
GCCCACAAGTATCACTAAGCTCGCTTTCTTGCTGTCCAATTTCTATTAAAGGTTCCTTTG
TTCCCTAAGTCCAACTACTAAACTGGGGGATATTATGAAGGGCCTTGAGCATCTGGATTC
TGCCTAATAAAAAACATTTATTTTCATTGCAATGATGTATTTAAATTATTTCTGAATATT
TTACTAAAAAGGGAATGTGGGAGGTCAGTGCATTTAAAACATAAAGAAATGAAGAGCTAG
TTCAAACCTTGGGAAAATACACTATATCTTAAACTCCATGAAAGAAGGTGAGGCTGCAAA
CAGCTAATGCACATTGGCAACAGCCCCTGATGCATATGCCTTATTCATCCCTCAGAAAAG
GATTCAAGTAGAGGCTTGATTTGGAGGTTAAAGTTTTGCTATGCTGTATTTTACATTACT
TATTGTTTTAGCTGTCCTCATGAATGTCTTTTCACTACCCATTTGCTTATCCTGCATCTC
TCAGCCTTGACTCCACTCAGTTCTCTTGCTTAGAGATACCACCTTTCCCCTGAAGTGTTC
CTTCCATGTTTTACGGCGAGATGGTTTCTCCTCGCCTGGCCACTCAGCCTTAGTTGTCTC
TGTTGTCTTATAGAGGTCTACTTGAAGAAGGAAAAACAGGGGTCATGGTTTGACTGTCCT
GTGAGCCCTTCTTCCCTGCCTCCCCCACTCACAGTGACCCGGAATCTGCAGTGCTAGTCT
CCCGGAACTATCACTCTTTCACAGTCTGCTTT
```

If you were studying a large gene, with huge introns and many exons, you may find it cumbersome to navigate and view, particularly if the details you seek are only at the intron–exon junctions. Notice that other than shading, the exons are not labeled.

Another view at the Ensembl Website solves this problem nicely. From this same page where the view in Figure 9.13 is seen, look at the table at the top of the page that includes transcripts and translation products and click on the transcript hypertext for ENST00000335295, the primary transcript from the human beta globin gene. The page will refresh and you will see a page dominated by a box-like drawing of the exons. Go to the left sidebar and click on the hyperlink for "exons" and you will navigate to a page that includes an image similar to that seen in **Figure 9.14**. In this view, the intron bases have been reduced to short sequences just flanking the exons.

Finally, another view of the exons and an even smaller amount of the flanking introns is shown by Splign, a tool hosted by the NCBI Website (www.ncbi. nlm.nih.gov/sutils/splign/splign.cgi). Unlike the static views found in Ensembl, Splign performs an alignment between cDNA and genomic sequence that you provide and will search for the exon splice junctions, directed by the alignment. It will even generate alignments with orthologous sequences, for example if you have a cDNA from one organism and the genomic DNA from a related species. Splign provides a map of the exons on the genomic sequence, and clicking on the map changes the view in the window. The view is the pairwise alignment between cDNA and genomic DNA, along with the translation of the exon (**Figure 9.15**). In addition to the graphic output that we see from this Web page, Splign coordinates are used by the NCBI to generate the exon/intron annotations that appear in RefSeq records. For example, you may see something like this in a record's annotation:

`"/inference="alignment:Splign:1.39.8"."`

Figure 9.14 Ensembl transcript view of human beta globin exons. This is an abbreviated view of the gene, with some intron sequence deleted and replaced by dots. When examining genes with long introns, this view allows you to focus on regions immediately flanking exons. The exons are numbered and the other sequences identified (for example, 5P upstream, intron 1–2). Note that the columns containing genomic coordinates and exon lengths were removed to create this figure.

	Sequence
5′ upstream sequence	.........ggtttgaagtccaactcctaagccagtgccagaagagccaaggacaggta
Exon 1	CGGCTGTCATCACTTAGACCTCACCCTGTGGAGCCACACCCTAGGGTTGGCCAATCTACT CCCAGGAGCAGGGAGGGCAGGAGCCAGGGCTGGGCATAAAAGTCAGGGCAGAGCCATCTA TTGCTTACATTTGCTTCTGACACAACTGTGTTCACTAGCAACCTCAAACAGACACC**ATG**G TGCATCTGACTCCTGAGGAGAAGTCTGCCGTTACTGCCCTGTGGGGCAAGGTGAACGTGG ATGAAGTTGGTGGTGAGGCCCTGGGCAG
Intron 1–2	gttggtatcaaggttacaagacagg.........tattggtctattttcccacccttag
Exon 2	GCTGCTGGTGGTCTACCCTTGGACCCAGAGGTTCTTTGAGTCCTTTGGGGATCTGTCCAC TCCTGATGCTGTTATGGGCAACCCTAAGGTGAAGGCTCATGGCAAGAAAGTGCTCGGTGC CTTTAGTGATGGCCTGGCTCACCTGGACAACCTCAAGGGCACCTTTGCCACACTGAGTGA GCTGCACTGTGACAAGCTGCACGTGGATCCTGAGAACTTCAGG
Intron 2–3	gtgagtctatgggacgcttgatgtt.........catacctcttatcttcctcccacag
Exon 3	CTCCTGGGCAACGTGCTGGTCTGTGTGCTGGCCCATCACTTTGGCAAAGAATTCACCCCA CCAGTGCAGGCTGCCTATCAGAAAGTGGTGGCTGGTGTGGCTAATGCCCTGGCCCACAAG TATCAC**TAA**GCTCGCTTTCTTGCTGTCCAATTTCTATTAAAGGTTCCTTTGTTCCCTAAG TCCAACTACTAAACTGGGGGATATTATGAAGGGCCTTGAGCATCTGGATTCTGCCTAATA AAAACATTTATTTTCATTGCAA
3′ downstream sequence	tgatgtatttaaattatttctgaatattttactaaaaagggaatgtgggga.........

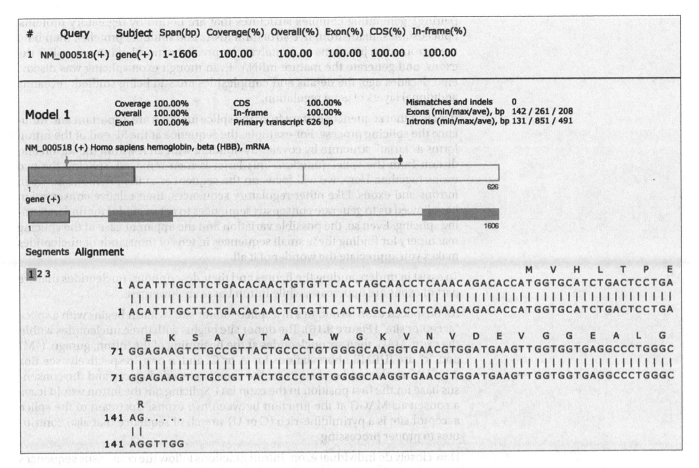

Query   Subject   Span(bp) Coverage(%) Overall(%) Exon(%) CDS(%) In-frame(%)

1 NM_000518(+) gene(+) 1-1606   100.00   100.00   100.00   100.00   100.00

Model 1 Coverage 100.00% CDS 100.00% Mismatches and indels 0
Overall 100.00% In-frame 100.00% Exons (min/max/ave), bp 142 / 261 / 208
Exon 100.00% Primary transcript 626 bp Introns (min/max/ave), bp 131 / 851 / 491

NM_000518 (+) Homo sapiens hemoglobin, beta (HBB), mRNA

gene (+)

Segments Alignment

M V H L T P E
1 ACATTTGCTTCTGACACAACTGTGTTCACTAGCAACCTCAAACAGACACCATGGTGCATCTGACTCCTGA
||||||||||||||||||||||||||||||||||||||||||||||||||||||||||||||||||||||
1 ACATTTGCTTCTGACACAACTGTGTTCACTAGCAACCTCAAACAGACACCATGGTGCATCTGACTCCTGA

E K S A V T A L W G K V N V D E V G G E A L G
71 GGAGAAGTCTGCCGTTACTGCCCTGTGGGGCAAGGTGAACGTGGATGAAGTTGGTGGTGAGGCCCTGGGC
||||||||||||||||||||||||||||||||||||||||||||||||||||||||||||||||||||||
71 GGAGAAGTCTGCCGTTACTGCCCTGTGGGGCAAGGTGAACGTGGATGAAGTTGGTGGTGAGGCCCTGGGC

R
141 AG......
||
141 AGGTTGG

To look at the human beta globin alignment, navigate to the Splign Web page, and launch the form for "online" analysis. The input is either a GenBank accession number or FASTA sequence; we'll use both. First, bring up the Web page for the cDNA sequence for human beta hemoglobin, NM_000518. From the right sidebar of the nucleotide record, navigate to the "Gene" Web page for this gene. From there, obtain the FASTA format for the genomic sequence using the link in the "Genomic regions, transcripts, and products" section. The NCBI defines the gene as beginning with the first base of the cDNA. Ensembl will sometimes go upstream of this location, including rare exons not seen in the NCBI RefSeq records. Even if you included more upstream genomic sequence, Splign will trim it away to just include the sequence that aligns with the provided cDNA.

Enter the cDNA accession number into the Splign online form. For genomic sequence, paste in the FASTA format you obtained from the Gene record (note that this form will sometimes have trouble if the annotation line is more complicated than a ">" sign and a simple name). Click on the "Align" button and the page will generate what can be seen in Figure 9.15. Splign generates a nice-looking graphic, and the added translation product does not make the information too busy for easy interpretation. It is unfortunate that Splign shows only five bases past the end of the exon; this is one base short of showing the entire consensus splice site, which is discussed below. This last base is easily seen in the Ensembl genome browser figures.

Figure 9.15 The NCBI Splign tool. The RefSeq cDNA NM_000518 for human beta globin is aligned to the genomic sequence from the Gene record. In this view, only the first exon is shown because the first graphical exon (above the word "Segments") was clicked. Note that this box is a different shade, indicating which exon sequence you are viewing. The translation of the exon is shown above the alignment.

9.5 EXON SPLICING

Messenger RNA processing to the mature form is a complex series of events involving multiple players. RNA polymerase transcribes a gene, generating a precursor RNA sometimes reaching tens of thousands of bases long, made up of only four nucleotides: A, U, G, and C. The precursor RNA folds onto itself (base pairing), generating complex structures that are bound by regulatory proteins. Spliceosomes, made from five structural RNAs and multiple proteins, then bind the intron–exon junctions and catalyze the precise removal of introns, splice the exons, and generate the mature mRNA. Even though exon splicing was discovered decades ago, the details and complexities are still being studied, revealing additional layers of gene regulation.

It is clear that sequences distant from the splice junction are important and influence the splicing process. For example, the sequence at the 5P end of the intron forms a "lariat" structure by covalently bonding with an A within the intron and distant from the splice junction. This lariat is then removed, splicing the two exons together. Here, we will focus on the sequences at the junctions between introns and exons. Like other regulatory sequences, their relative conservation has allowed us to generate consensus sequences to act as guides for understanding splicing. Even so, the possible variation and the apparent ease of the splicing machinery for finding these small sequences in tens of thousands of nucleotides makes you appreciate the wonder of it all.

To assist in understanding the figures and their descriptions, nucleotides that are part of exons are uppercase, while intron bases are lowercase.

An unspliced exon terminates in a splice "donor site" while it begins with a splice "acceptor site" (**Figure 9.16**). The donor site begins with three nucleotides within the exon, MAG, then six nucleotides at the beginning of the intron, guragu. ("M" and "r" are IUBMB abbreviations for 'A or C' and 'a or g,' respectively; see Box 3.1). At the other end of the intron, the exon is preceded by cag and the consensus base for the first position in the exon is G. Splicing out the intron would leave a consensus MAGG at the junction between two exons. Upstream of the splice acceptor site is a pyrimidine-rich (C or U) stretch of sequence that also contributes to proper processing.

How closely do individual exon–intron junctions follow the consensus sequences for donor and acceptor sites shown in Figure 9.16? Again, we'll turn to our set of transcripts used earlier to explore Kozak sequences. **Table 9.3** lists the sequences at the splice sites, with a row for each donor–acceptor pair per transcript. The exception is the dopamine receptor that has only a single exon. Scanning down the list, it appears that the donor and acceptor sites are fairly close to the consensus sequences. Differences from the consensus are seen but, overall, the list has several positions that are almost invariant. For example, the donor site is almost always MAG|gur and the last two bases of the intron (within the acceptor site) are usually ag.

The table also includes the number of consensus sites within the transcribed portion of the gene. The frequency of acceptor sites is much higher than donor sites, most probably reflecting the lengths of the consensus sequences (four versus nine nucleotides, respectively). Based on this small sample, it appears that consensus donor sites are rare, but there are a number of potential acceptor sites. Context must be important, such as having acceptor sites downstream of polypyrimidine stretches. Or maybe these nonstandard splice forms are short-lived, because these potential splice products are usually not detected. Of course, Table 9.3 also demonstrates that sequences divergent from the consensus can act as splice junctions, too. Thus, additional potential sites are not represented in this table.

How were these consensus sequences found within the gene? The Sequence Manipulation Suite (www.bioinformatics.org/sms2/index.html), also covered in Chapter 7, has a program called "DNA Pattern Find." The FASTA format of the gene sequences (exons plus introns) was pasted into the form and the pattern was entered following the instructions on the Web page.

Figure 9.16 Consensus exon splice sites. Introns are flanked by splice donor sites and splice acceptor sites. The convention in this book section is to represent intron sequence in lowercase and exon sequence in uppercase.

```
             Splice donor site              Splice acceptor site
--EXON-------MAG|guragu------intron-------cag|G--------EXON--
```

Table 9.3 Exon splice junctions for a small gene set

RefSeq identifier	Transcript definition	Splice donor site	Splice acceptor site	Consensus donor sites in gene	Consensus acceptor sites in gene
NM_000798	*Homo sapiens* dopamine receptor D5 (*DRD5*), mRNA	single-exon gene	single-exon gene	0	11
NM_000518	*Homo sapiens* hemoglobin, beta (*HBB*), mRNA	**CAG**guuggu	uag**G**	0	8
		AGGgugagu	cag**C**		
NM_008220	*Mus musculus* hemoglobin, beta adult major chain (*Hbb-b1*), mRNA	**CAG**guuggu	uag**G**	0	7
		AGGgugagu	cag**C**		
NM_033234	*Rattus norvegicus* hemoglobin, beta (*Hbb*), mRNA	**CAG**guuggu	uag**G**	0	6
		AGGgugagu	cag**C**		
NM_000519	*Homo sapiens* hemoglobin, delta (*HBD*), mRNA	**CAG**guuggu	cag**A**	0	7
		AGGgugagu	cag**C**		
NM_005330	*Homo sapiens* hemoglobin, epsilon 1 (*HBE1*), mRNA	**CAG**guaagc	uag**A**	1	11
		AAGgugagu	cag**C**		
NM_000184	*Homo sapiens* hemoglobin, gamma G (*HBG2*), mRNA	**AAG**guaggc	cag**G**	1	13
		AAGgugagu	cag**C**		
NM_000559	*Homo sapiens* hemoglobin, gamma A (*HBG1*), mRNA	**AAG**guaggc	cag**G**	1	13
		AAGgugagu	cag**C**		
NM_000207	*Homo sapiens* insulin (*INS*), transcript variant 1, mRNA	**AGG**gugagc	cag**U**	0	23
NM_123204	*Arabidopsis thaliana* ribulose bisphosphate carboxylase small chain1B / RuBisCO small subunit 1B (*RBCS-1B*) (ATS1B) (AT5G38430) mRNA, complete cds	**AAG**guaaug	uag**G**	0	0
		GAGguaaua	uag**C**		
NM_000537	*Homo sapiens* renin (*REN*), mRNA	**ACG**guaauu	cag**G**	0	95
		GACgugagu	cag**A**		
		GUGgugaga	cag**U**		
		ACCguaagu	cag**G**		
		CAGgugggg	cag**A**		
		CGAguaagg	uag**G**		
		GGGgucaga	cag**G**		
		GAUguaaga	cag**U**		
		CAGgugagg	cag**G**		
NM_181744	*Homo sapiens* opsin 5 (*OPN5*), transcript variant 1, mRNA	**UUG**guaagc	uag**G**	3	198
		CAGguaaca	uag**U**		
		AUGguaagu	cag**G**		
		AAGguaagu	cag**G**		
		CAGguaaaa	uag**G**		

The splice junctions were collected from a number of RefSeq gene records described elsewhere in this book, and represented here with a donor–acceptor pair per line. Nucleotides of exons are bold and uppercase while intron sequences are lowercase. The consensus sequences for the donor and acceptor sites are MAGguragu and cagG, respectively. Also listed is the number of sequences that match the consensus sites within the transcribed portion of the genes. For example, the consensus cagG is found 95 times in the human renin gene (introns plus exons).

Renin: a striking example of a small exon

Human renin, especially exon 6, can serve as an interesting example of gene structure. **Figure 9.17** shows an Ensembl browser view of this exon along with flanking introns and exons. Remarkably, exon 6 is only nine nucleotides long and is flanked by 2029 nucleotides of intron upstream and 564 nucleotides downstream. Also remarkable is that in an ocean of 2602 nucleotides, the splicing machinery recognizes this tiny sequence as being an exon. Of the nine nucleotides, four are part of the two splice-recognition signals. Not considering the

Figure 9.17 Human renin, exons 5, 6, and 7. The exons are shaded. Exon 6 is only nine nucleotides long and four of the nine nucleotides contribute to splicing signals.

```
CCTCTGGTCCTTCCTCCCACAGGTGGGTGGAATCACGGTGACACAGATGTTTGGAGAGGT
CACGGAGATGCCCGCCTTACCCTTCATGCTGGCCGAGTTTGATGGGGTTGTGGGCATGGG
CTTCATTGAACAGGCCATTGGCAGGGTCACCCCTATCTTCGACAACATCATCTCCCAAGG
GGTGCTAAAAGAGGACGTCTTCTCTTTCTACTACAACAGgtGGGGACTGGGACTCCAAGG
GCTGAGGTGGGGGGACAGGAGGGGAGAAGAGATGGGGAGTGGAAGGAGAGTCTGGGCCAG
AATTGTAAAGTGTTTGTAATTAGGTGACAGCCAATCAATATCTAGAGCTGTACTAGCCAA
TATGGAAGGCACTATTGAAATTTAAATTAATTAAATACAGTTAAGCATCAATTAAGCATT
CAACTGGTGGCTCTTAGTTGTACTAGCCACACGTCAAATGCCTGGCAGCCACGGTGGCTA
GTAACTACAGTCTTATGACAGTGCAGATAGAATATTCCCAGCATGACAGGACATTCTA
ATAGACAGCGCCACTCTGGAGCAAGAGGAGATGCAAGGTGGGGGCGATGGTAAATAAGGG
ATTACTGTGACCTGTAGCCCTGCCTGTTAGGGCCATGGCTCCTCCCACACAGAGACAGCC
AACTTCAGTCATCCATTAGATCCTTCATTCGTTTGTTTGCTCACTCATCAGTTCAGTAAA
TGCTATGTGCCAAGCACTGTGGTAGGCTCTGGGGGTGCAGCAGTGAACACAGTGAACAAG
GCAGAATCTGTACTCCCCTACCCACATAGAGCTTACAGGCTAACAGGGAAGACAAGACAT
ATTCCAACATAAAGAGTGTCACAGGCAGGCAGCAAGTGTGGTGCTGAAAACCATGGATGC
TTTTCAATTCTAGGCTGAGCTTATATGCAGCTCAGCCAGCCTTGGGGAAGCTCTTGAGCA
GGGTTGGGCTCTACTCCAAACTGCTGGGCTTAGAAAGATGGCATGAGTTGGAGACAAGAG
AGCTGGAGGCAAAAGGGGCTGGGTGCAGTGGCTCATGCCTGTAATCCCAGCACTTTGGGA
GGCCAAGGCGAGAGGATCGCATGAGCCCAGGAGTTAAGGCTTCAGTGAGCAGTGATTGTG
CCACTGCACTCCAGCTAAGGCAACAGAGTGAGATCCAGTCTCAAAAAAAAAAAAAAAAAA
AAAAAGTCACAAGGGTAAGAACATGAGGCCAGTGGCAAAAAGAATAGAGGAGAGGATCA
GAGTTCAGAGAAATCTCACAGTAAAATGGAGAGGAGTCTCCGGTTTGGTGATAGAAAGTG
AGGCCTTGAGAAAAGGCCAATTGGCGGCTCTGCATTCAGGGGTGGTCTTTAGAAGAACTG
TTTTAGAGGAAGTGGGGGCAAGGCCAGATGGCAAGAAGTTAAGAGGTGGACGACGTGGGT
GTCAGGAAGTGGAGGTCATGAGATGTAGGCTGCCCTGGGACATTCAACAGGGAAGGGAAT
GGGGGGTGGCGTGGGGGGTGAGATCCAGAAGCAGAAGAGGAAGGGTGGGTGTTTTTAAAT
GCTAGAGGATGCTCGAGTGATGCCTGTAGGTGGAGGAAGAAGCCAATGGAAAGAAAGAGA
TTAAAAATGTGGAAAGAAGAGGAGCTAAATGGGGGCACTGGAGTTTGGAGGCCTTGAAAG
AGATGAGGTTCCAGCAGACAGGAAGAAGCCAGGTTTTGCAGAGGAGAGGGCTGGCCTCTT
CTTTTATCTTGGGATGGGAAGGAGGGAACATCCAGAGAGATACTGAAGTGTTGAGAGACA
GGCAGGAGGGAATTTGTGCTAGCATATACACATACATTCCGAATTTATAAAAACACAAGT
AGTTTGCAGTTGCACAAAATAACATATGCACACCTACACACCCATGCACACATGTGCATG
TGTGAATTCTAGTATGAATTCTGGAAAAACACATCACACACACAGGCATGCCCTGGAGAC
TAGGCCTACAGTAGTCCCTGAGCCAAGTGCAGTGAGGAGGAAAGGAAGGTGAGGGGAATC
AGCTCCAGACGGGGCACCAGGAGCCTGGCTCCAGTCCCCCACTTGTTCACTCATGGACTG
GGTAACTTCAGGCAAGTGACTTCGCCTCTTGGTGACTCCATTGCCTGAAGGGCAAAGAGA
GTACATAACACCCACCCTGCCAAACAGCAGGGCTGATGAGGCTGGCATGAAATGAAGCTT
CCTTTCTGCTGTCTCTCTTTCTCTGCAGAGATTCCGAGTAAGGAGACAAAACCCCCACAT
GGCTGTGACCTTCCAGTACTCCCCGAGCACCTGACCTAGAATTACACACGCCACCGGCCC
AAAACTCACATCAGCAAGCCCAGCCTCCGCTAGATGCCGAAGTTCTCTGTCTCTCCTTCC
TGCTCTCTCCATGCCACCTGCCCACCCCATACCCAATAGCCTCCCCAGGGTCCCCTCCCA
TGCACCTGCTCAATCAGCAGCAACCCAAGAGTGAGGGGTGTCCATTTGTGTCTTGTTCAC
ATCCACTCACTGTCCTTGTACCTGCTCCTTTTCTGTGACCTCTCTGGGGATGCTTTTTGG
GGGAACAGCTGGACTACCCTGGAACAACCTCTGGTTGGTCTTGGGGAGGGGAAGAAAGGC
AGAGAAGCAGTATGTTCTGCATGCTTCCCAACGACAGCTCCGAGCCTGGCTGTCTGTCCC
ACATTCCTCTGCTCTAGAGCCCTCTGTCCTCCCCTGCACCCTTGTGCAACCTTCCCCAAT
TGCCTGAGTTGCTGGGTCCTGGAGGTTATGGGTTTCCAAGAGCTTCTGATCTTTCCTTTA
GGAATTCCCAATCGCTGGGAGGACAGATTGTGCTGGAGGCAGCGACCCCCAGCATTACG
AAGGGAATTTCCACTATATCAACCTCATCAAGACTGGTGTCTGGCAGATTCAAATGAAGG
GGTCAGAAATCCTCAACCCTCCCCGGGCTCCAAAAAAATGCTGCCGTCACTGGGGGTTGGGG
```

reading frame, an exon as small as nine nucleotides can code for a minimum of three amino acids or be part of four codons:

AGAUUCCGA

Can be divided into codons three ways:

AGA|UUC|CGA

A|GAU|UCC|GA

AG|AUU|CCG|A

As demonstrated in Table 9.3, splicing signals are small and variable. Consider the consensus acceptor signal, cagG: there are a total of 20 cagG sequences in this region of the gene. Yet the splicing machinery recognizes this tiny exon and these precious nine nucleotides are spliced into the 1447 nucleotide mRNA (ENST00000272190). Perhaps it is no accident that the length is nine, so if it is skipped, the rest of the renin transcript stays in frame. Indeed, according to the Ensembl genome browser, there are two transcripts for human renin, and one does not include this tiny exon. It is also absent in the mouse.

Another striking splice: human *ISG15* ubiquitin-like modifier

The NCBI hosts a database called the Consensus CDS Protein Set (CCDS). Links to this database appear in the annotation of RefSeq records where the coding sequence (CDS) is defined and it is listed in the NCBI A–Z site map. On a CCDS Web page for a gene, the nucleotides of the coding region and the corresponding translation are featured, with the exons and their translations alternating colors. For example, in **Figure 9.18A**, the human beta globin transcript and protein are

(A)
Human beta globin, nucleotide sequence (444 nucleotides):

**ATGGTGCATCTGACTCCTGAGGAGAAGTCTGCCGTTACTGCCCTGTGGGGCAAGGTGAACGTGGATGAAG
TTGGTGGTGAGGCCCTGGGCAG**GCTGCTGGTGGTCTACCCTTGGACCCAGAGGTTCTTTGAGTCCTTTGG
GGATCTGTCCACTCCTGATGCTGTTATGGGCAACCCTAAGGTGAAGGCTCATGGCAAGAAAGTGCTCGGT
GCCTTTAGTGATGGCCTGGCTCACCTGGACAACCTCAAGGGCACCTTTGCCACACTGAGTGAGCTGCACT
GTGACAAGCTGCACGTGGATCCTGAGAACTTCAGG**CTCCTGGGCAACGTGCTGGTCTGTGTGCTGGCCCA
TCACTTTGGCAAAGAATTCACCCCACCAGTGCAGGCTGCCTATCAGAAAGTGGTGGCTGGTGTGGCTAAT
GCCCTGGCCCACAAGTATCACTAA**

Translation (147 amino acids):

MVHLTPEEKSAVTALWGKVNVDEVGGEALGRLLVVYPWTQRFFESFGDLSTPDAVMGNPKVKAHGKKVLG
AFSDGLAHLDNLKGTFATLSELHCDKLHVDPENFR**LLGNVLVCVLAHHFGKEFTPPVQAAYQKVVAGVAN
ALAHKYH**

(B)
Human *ISG15* ubiquitin-like modifier, nucleotide sequence (498 nucleotides):

ATGGGCTGGGACCTGACGGTGAAGATGCTGGCGGGCAACGAATTCCAGGTGTCCCTGAGCAGCTCCATGT
CGGTGTCAGAGCTGAAGGCGCAGATCACCCAGAAGATCGGCGTGCACGCCTTCCAGCAGCGTCTGGCTGT
CCACCCGAGCGGTGTGGCGCTGCAGGACAGGGTCCCCCTTGCCAGCCAGGGCCTGGGCCCCGGCAGCACG
GTCCTGCTGGTGGTGGACAAATGCGACGAACCTCTGAGCATCCTGGTGAGGAATAACAAGGGCCGCAGCA
GCACCTACGAGGTACGGCTGACGCAGACCGTGGCCCACCTGAAGCAGCAAGTGAGCGGGCTGGAGGGTGT
GCAGGACGACCTGTTCTGGCTGACCTTCGAGGGGAAGCCCCTGGAGGACCAGCTCCCGCTGGGGGAGTAC
GGCCTCAAGCCCCTGAGCACCGTGTTCATGAATCTGCGCCTGCGGGGGAGGCGGCACAGAGCCTGGCGGGC
GGAGCTAA

Translation (165 amino acids):

MGWDLTVKMLAGNEFQVSLSSSMSVSELKAQITQKIGVHAFQQRLAVHPSGVALQDRVPLASQGLGPGST
VLLVVDKCDEPLSILVRNNKGRSSTYEVRLTQTVAHLKQQVSGLEGVQDDLFWLTFEGKPLEDQLPLGEY
GLKPLSTVFMNLRLRGGGTEPGGRS

(C)

Splice Donor
...<u>ACAGCCATG</u>ggctgg...
Kozak

Figure 9.18 The CCDS display of exons and their corresponding translations. The NCBI CCDS database shows the exon distribution of the translation product for a gene. On the Web page, the alternating exons are color coded, here they are bold and non-bold. (A) The CCDS file (CCDS7753) of the three-exon human beta globin gene. (B) The two-exon CCDS file (CCDS6) for the *ISG15* ubiquitin-like modifier gene (NM_005101). Note that the initiation codon is the only codon found in the first exon. (C) The overlap between the Kozak sequence (underlined) and the splice donor site (overlined) for the *ISG15* ubiquitin-like modifier gene.

color coded (shown in bold and non-bold here) to show the codons of the three exons and their corresponding translations.

If you examine the CCDS record for the human *ISG15* ubiquitin-like modifier gene, you see that this gene has two coding exons and the first coding exon encodes only the first amino acid (Figure 9.18B). In this figure, only the ATG and the first methionine are bold in the nucleotide and translation fields, respectively. Exon 1 is 110 nucleotides long, and the last three bases are the ATG. In this case, the Kozak sequence (consensus GCCRCCAUGG) and splice donor site (consensus MAG|guragu) overlap, having four nucleotides in common over this 15 nucleotide sequence (Figure 9.18C). The evolutionary constraints upon this region must be extreme, because single nucleotide changes could disrupt translation initiation, splicing, or both. How common is this gene structure? In humans alone, there are over 300 other genes where the initiation codon is the only codon found in the first coding exon.

Alternative splicing

Prior to the sequencing of the entire human genome, estimates of the number of protein-coding genes varied considerably. Both commercial and noncommercial laboratories were studying ESTs and, by clustering these into transcriptional units, generated estimates that were 100,000 genes or greater. Proteomic studies also saw tremendous diversity at the protein level, consistent with the gene number being at the high end of the estimates.

When the human genome was published in 2001 and the numbers were more definitively determined, it was shown that we had approximately 26,000 protein-coding genes. Since that time, the number has been adjusted downward and it is thought that we have approximately 21,000 protein-coding genes. Imagine the shock when our gene count was compared with that of the fruit fly (13,800). With all of our complexity of structure, behavior, and intellect, how is this possible that we have just 7000 more genes than a fly?

It would be a great exaggeration to say that we now know the answer to this question. But what is clear is that the majority of our protein-coding genes are **alternatively spliced** to generate two of more different proteins per gene. Realizing that we have many more thousands of proteins than genes encoding them helps explain our rank above the fly. However, there is alternative splicing in *Drosophila*, too! In fact, one of the most spectacular examples of alternative splicing is the *Drosophila Dscam* gene that may generate many thousands of alternatively spliced mRNAs. For an eye-catching display of splices, visit the NCBI Gene database page for this gene.

Alternative splicing can take many forms. For some genes, alternative promoters are used to generate different 5P UTR sequences, with no apparent effect on the translation product. Other genes can have alternative 5P ends that do alter the N-terminal protein sequence. When "internal" exons are changed, the translation product can have changes that range from minor to major. Finally, genes can be found that have alternative 3P ends that may change the C-terminus of the protein.

Different translation products from the same gene can generate new structures and functions, although the latter is not always apparent. These different proteins can vary in tissue distribution, temporal expression, or stability. Sometimes the functional difference is obvious, such as one form being secreted while another is not. Some of the earliest examples of this kind were discovered in the examination of immunoglobulin proteins where alternative splicing created either a membrane protein or one that is released into the blood.

What causes a gene transcript to be alternatively spliced? Differences in the transcription start site will, at least, generate transcripts with different 5P sequences that could contain different splicing signals. If the 3P UTR of a transcript is extended or reduced, that will also change the number of splicing signals. But internal alternative splices are more challenging to understand. The precursor RNA contains all the introns and potential exons so all the signals are present simultaneously. What will cause some exons to be utilized while others are skipped?

What can vary by tissue, developmental timing, or stability are the regulatory proteins that are involved in the splicing process. Proteins regulate the recognition and binding of spliceosome complexes to the intron–exon junctions in the unspliced precursors. Perhaps different combinations of regulatory proteins can lead to the simultaneous expression of multiple transcript forms in one tissue. But should a cell type only express one set of regulatory proteins, then only the transcript specific for that cell type will be present.

Human plectin: alternative splicing at the 5P end

To explore another set of small sequences, let's examine the Kozak and splicing sites of the human plectin gene. Plectin is a large gene, covering over 61,000 nucleotides (contrast this to beta globin at 1600 nucleotides long) and containing over 30 exons. The translation product, over 4500 amino acids long, is a structural protein playing a central role in the cytoskeleton. Plectin proteins form networks with actin and intermediate filaments to mediate functions such as the change and maintenance of cell shape, movement, and the transport of molecules and vesicles within the cell. Mutations in the plectin gene are linked to disorders where the skin blisters in response to minor injury, and to a form of muscular dystrophy.

At least eight different human plectin transcripts have been identified and these are expressed in a wide range of cell types and tissues. Each transcript has a different 5P exon that is alternatively spliced onto the second coding exon. The remaining 31 exons are identical in these eight transcripts. These small 5P exons are scattered within the first 37,000 nucleotides of the gene (**Figure 9.19**), and all of the other exons are found in the remaining 24,000 nucleotides.

There are numerous Websites that specialize in alternatively spliced genes, but here fast DB (www.fast-db.com) will be demonstrated. A free account must be created in order to use this application. **Figure 9.20** shows a fast DB result image.

The alternative splicing of plectin affords the opportunity to examine eight different Kozak and splice donor sites within the same gene. These will demonstrate the variation of sequences that is possibly linked to their expression and specific roles in the cell. **Figure 9.21** shows three small sequences for each transcript: the 10-nucleotide Kozak sequence, the splice donor site from the alternatively spliced 5P exon, and the splice acceptor site for exon two (common to all).

As described earlier, the Kozak consensus is GCCRCCAUGG. None of the eight Kozak sequences of plectin (Figure 9.21) matches the consensus sequence precisely. They range from a nine out of ten match (gccggcAUGG) to a six out of ten (uccccgAUGA). Presumably, these differences have an influence on the rate of recognition by the translational machinery.

The consensus for the alternative exon donor splice site sequence is MAG|guragu. As seen with the Kozak sequences, none of the donor sites (Figure 9.21) matches

NCBI RefSeq	Coordinates
NM_000445	1–43
NM_000445	1371–1569
NM_201378	3217–3343
NM_201379	22826–23020
NM_201380	25870–26562
NM_201381	32009–32098
NM_201382	32804–32967
NM_201383	34222–34354
NM_201384	37156–37396

Figure 9.19 Nucleotide coordinates of the alternatively spliced first coding exon of plectin. This figure shows the locations of the eight differentially spliced exons. The exon at the extreme 5P end of the gene (nucleotides 1–43) is untranslated.

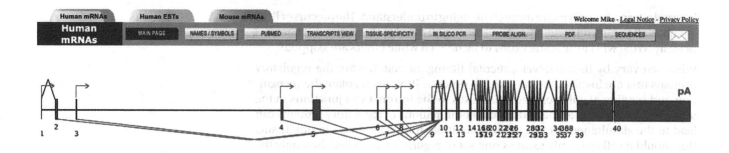

Figure 9.20 fast DB graphic of human plectin alternative splicing. Gene exons are portrayed as boxes on the horizontal line, going 5P to 3P, left to right. Arrows indicate transcription start sites. Alternative splicing events are drawn as thin lines below the gene, while invariant splicing is drawn as thin lines above the gene. Exon numbers appear below the exons.

the consensus perfectly and presumably they would differ in their recognition by the splicing machinery.

The alternative N-termini found on these isoforms direct these proteins to different locations within the cell. These relatively small signals within these huge proteins direct the proteins to the nucleus, mitochondria, hemi-desmosomes, and perinuclear regions, in addition to generating proteins with a diffuse distribution. Rather than create eight separate nearly identical genes to handle the required localization, nature generates eight different versions by varying the 5P end of the transcripts through alternative splicing of the primary gene transcript. However, the evolutionary pressure on these first exons is high. The insertion or deletion of any bases that are not within a multiple of three would shift the reading frame and the ensuing splice to exon two would result in a sequence not resembling plectin.

Figure 9.22 shows the translation products of the first and second exons for the human isoforms. The sequences in this figure were obtained from the NCBI CCDS records for each isoform. Note that the contributions of the alternative exons vary in length but each ends with an aspartic acid (D). This amino acid is encoded by the codon GAU, which is created with contributions of nucleotides from the first (G) and second exon (AU). This U is outside of the exon 2 acceptor site (shown in gray in Figure 9.21).

At the time this book was written, the NCBI plectin Gene record indicated that the Human Genome Organisation (HUGO) Gene Nomenclature Committee (HGNC) gene symbol was *PLEC*. At the other Websites, entering "PLEC" to find the alternative splicing record for this gene gave no hits. The NCBI Gene record indicated that plectin was also known as "PLEC1" and this symbol worked at the other Websites. As more family members are found and as annotation is updated, gene names and symbols change and it is challenging for Websites to keep up with this moving target. Be prepared to use "synonyms" when you can't find a gene using gene names or symbols. The synonyms of proteins were explored in Section 2.4.

Consensus splice junctions, translated

Let's consider the consequences of the consensus splice sites on translation and the last amino acid of the exon. During splicing, the consensus donor site, MAG|guragu, is joined with the consensus acceptor site, cag|G (the contribution

Figure 9.21 Kozak and donor splice sites of plectin 5P exons. The isoform as well as the transcript accession numbers label each row. Kozak, splice donor, and splice acceptor sites for alternative exons 1 and 2 are shown. Junctions between introns and exons are marked with a vertical bar, "|," both within the sequences and in the map across the top. The last exon base (U) is gray to indicate that it is not part of the consensus range but can participate in the translation across the junction.

```
                          Kozak|<-------Exon---------|----------Intron------|----Exon-->
   NP_000436   NM_000445   gaggccAUGU..        ..CAG|guaacu..         ..cag|AU
   NP_958780   NM_201378   gccggcAUGG..        ..AAG|guacgg..         ..cag|AU
   NP_958781   NM_201379   gcagacAUGG..        ..AAG|gugcgg..         ..cag|AU
   NP_958782   NM_201380   gcagccAUGG..        ..CAG|gucagc..         ..cag|AU
   NP_958783   NM_201381   uccccgAUGA..        ..CCG|guaggu..         ..cag|AU
   NP_958784   NM_201382   gcggcuAUGG..        ..AAG|guaagc..         ..cag|AU
   NP_958785   NM_201383   gccggcAUGU..        ..UAG|gugggu..         ..cag|AU
   NP_958786   NM_201384   ggcagcAUGU..        ..AAG|guagca..         ..cag|AU
```

```
From CCDS
NP_000436   MSGEDAEVRAVSEDVSNGSSGSPSPGDTLPWNLGKTQRSRRSGGGAGSNGSVLDPAERAVIRIADERDRVQKKTFTKWVNKHLIK
NP_958780                                          MAGPLPDEQDFIQAYEEVREKYKDERDRVQKKTFTKWVNKHLIK
NP_958781                                           MDPSRAIQNEISSLKDERDRVQKKTFTKWVNKHLIK
NP_958783                                                  MKIVPDERDRVQKKTFTKWVNKHLIK
NP_958784                                MEPSGSLFPSLVVVGHVVTLAAVWHWRRGRRWAQDEQDERDRVQKKTFTKWVNKHLIK
NP_958785                               MSGAGGAFASPREVLLERPCWLDGGCEPARRGYLYQQLCCVDERDRVQKKTFTKWVNKHLIK
NP_958786                              MSQHQLRVPQPEGLGRKRTSSEDNLYLAVLRASEGKKDERDRVQKKTFTKWVNKHLIK
```

by the next exon of the mature mRNA is underlined). After the intron sequence is spliced out, this leaves CAGG or AAGG at the place where the two exons are joined. These two sequences can generate four possible codons in the three forward frames that approach or span the junction:

CAGGN can be read as

CAG

AGG

GGN

and

AAGGN can be read as

AAG

AGG

GGN

These four codons encode the following amino acids:

CAG: glutamine

AAG: lysine

AGG: arginine

GGN: glycine

What amino acids do we see at the end of each (coding) exon? In order to answer this, the human CCDS protein sequences were downloaded from the NCBI Website and the first and second exons of over 23,000 human proteins were extracted. After eliminating single-exon genes (they have no second exon and can be easily identified), the last amino acids encoded by the first exons were captured and tallied (**Figure 9.23**). Glycine is the most abundant, probably reflecting the four possible codons that would encode it (GGN) in a splice junction. The other amino acids near the top of the list include those predicted above: lysine, glutamine, and arginine. The exception is glutamic acid, which is almost as abundant as arginine. Glutamic acid is encoded by GAA and GAG, and these codons are noticeably absent from the consensus possibilities. But so are the other codons that encode the remaining 15 amino acids on the list; all amino acids are represented. The presence of all the amino acids suggests that other nucleotides in these critical positions will satisfy the criteria of being potential splicing signals. Indeed, two of the splice sites seen in Table 9.3, the single *Arabidopsis* donor site and the fifth site for renin, result in a glutamic acid at the junction of these exons. Aspartic acid, seen at the end of each exon 1 of the eight plectin isoforms, is further down the list.

9.6 POLYADENYLATION SIGNALS

Take a moment and imagine the transcriptional landscape. Double-stranded DNA, wrapped around nucleosomes and bound with other regulatory molecules, is being transcribed. RNA polymerase II is transcribing the DNA, moving along at over 1000 nucleotides per minute. All the while, the growing mRNA is undergoing

Figure 9.22 N-termini of human plectin isoforms. Shown are the first two exons from seven of the eight isoforms. The missing isoform, NCBI accession number NP_958782, is too long to show in this figure. Protein sequence from the invariant second exon is underlined.

Amino acid	# Last amino acid, first exon
Glycine	4009
Lysine	2477
Glutamine	2404
Arginine	2303
Glutamic acid	2224
Alanine	1369
Aspartic acid	1274
Valine	1123
Serine	1070
Leucine	773
Methionine	691
Threonine	428
Tryptophan	415
Proline	368
Asparagine	236
Cysteine	189
Isoleucine	187
Tyrosine	131
Histidine	95
Phenylalanine	94

Figure 9.23 Count of amino acids at the end of the first human exon. The protein sequences encoded by human exons were downloaded from the NCBI CCDS database. After excluding single-exon genes, the last amino acids of all first-exon sequences were captured and tallied.

processing by the spliceosome. Otherwise, the RNA polymerase would have to drag around a very long precursor RNA, particularly for genes that are tens of thousands of bases long. The RNA grows, then loses an intron, then grows again, and then another intron gets spliced out.

Simultaneous to transcription, the growing RNA is being read by another protein complex, the cleavage and polyadenylation specificity factor (CPSF). This complex is looking for a very small signal, usually AAUAAA, which triggers the cleavage of the mRNA 10–35 nucleotides downstream from the signal. Once cleaved, additional proteins, one of which is the poly(A) polymerase, bind the end of the mRNA. It is this enzyme that adds approximately 200 A nucleotides to the 3P end of the newly cleaved mRNA. This poly(A) tail is added without a complementary template; that is, there is no stretch of T nucleotides for the purpose of generating a complementary poly(A). It should not be forgotten that the RNA polymerase II enzyme is still transcribing downstream after the cleavage and polyadenylation event, but this downstream RNA product is rapidly degraded and the polymerase eventually stops transcribing.

A transcribed gene is a very busy place; for those not keeping track, there are at least five complexes involved with the movement of RNA polymerase. Described here are proteins that (a) modify and unwind chromatin in advance of the RNA polymerase; (b) modify chromatin after RNA polymerase has passed; (c) splice out introns; (d) scan the mRNA for poly(A) signals; and (e) polyadenylate mRNA.

Almost all RNA polymerase II transcripts have poly(A) tails. An mRNA without a poly(A) tail is rapidly degraded before leaving the nucleus. Poly(A) tails are not always present in database cDNA records. Laboratories differ in their policies, with some routinely trimming tails off while others leave them on when submitting cDNA sequences to databases. Cloning may also limit the length of the tail so, unless an experiment is specifically designed to preserve them, nothing definitive can be concluded by studying the length of poly(A) tails in cDNA records.

What you can do is look upstream from the end of cDNA records and try to find the poly(A) signal. It is supposed to be 10–35 nucleotides upstream from the poly(A) tail. But is the entire 3P UTR found in the cDNA record? Remember, the focus of many laboratories is the protein-coding region and they may not have tried to make sure they had the entire 3P UTR. The signal is usually AAUAAA but there are exceptions; AUUAAA is also seen.

Again, we'll turn to our gene set to get a sense of how common AAUAAA could be in mRNAs. Considering its importance, nature may select against the presence of this six-nucleotide signal should it arise in the wrong place. If DNA sequence were random, we would expect the frequency to be $\frac{1}{4} \times \frac{1}{4} \times \frac{1}{4} \times \frac{1}{4} \times \frac{1}{4} \times \frac{1}{4}$, or once every 4096 nucleotides. As can be seen in **Table 9.4**, the AAUAAA is found once per gene in half of our small gene set. In other genes it is found multiple times; five times in the almost 4000 nucleotide GABA A receptor, and four times in the 583 nucleotide gamma G hemoglobin. In these and other records, most of the additional polyadenylation signals are found in the 3P UTR of the mRNAs. It would appear that there are potentially multiple places for polyadenylation, perhaps for a functional reason. For many transcripts, there is growing evidence that UTR length varies by tissue and developmental stage. Perhaps this variation is tightly tied to the forces regulating which polyadenylation signal is utilized.

Be aware that the translation of poly(A) tails (by software) will inappropriately be performed and even enter the public databases. To find this yourself, see the example in **Box 9.2**.

9.7 SUMMARY

In this chapter, we focused on the smallest details of sequence analysis by exploring very short DNA or RNA sequences. Transcription factors bind DNA elements upstream of the start of transcription, and applications within the Transfac

Table 9.4 Polyadenylation signals

RefSeq identifier	Transcript definition	cDNA length	Poly(A) signal	Sequence location	Distance from end of mRNA**
NM_000798	*Homo sapiens* dopamine receptor D5 (*DRD5*), mRNA	2398	AAUAAA	2016, 2356*	20
NM_000518	*Homo sapiens* hemoglobin, beta (*HBB*), mRNA	626	AAUAAA	602	24
NM_008220	*Mus musculus* hemoglobin, beta adult major chain (*Hbb-b1*), mRNA	626	AAUAAA	604	≥22
NM_033234	*Rattus norvegicus* hemoglobin, beta (*Hbb*), mRNA	620	AAUAAA	596	≥24
NM_000519	*Homo sapiens* hemoglobin, delta (*HBD*), mRNA	774	AAUAAA	745	16, 20, ≥29
NM_005330	*Homo sapiens* hemoglobin, epsilon 1 (*HBE1*), mRNA	816	AAUAAA	170, 791*	25
NM_000184	*Homo sapiens* hemoglobin, gamma G (*HBG2*), mRNA	583	AAUAAA	31, 554, 557*, 561	≥26
NM_000559	*Homo sapiens* hemoglobin, gamma A (*HBG1*), mRNA	584	AAUAAA	31, 555, 558*	≥26
NM_000207	*Homo sapiens* insulin (*INS*), transcript variant 1, mRNA	469	AAUAAA	446	23
NM_123204	*Arabidopsis thaliana* ribulose bisphosphate carboxylase small chain1B / RuBisCO small subunit 1B (*RBCS-1B*) (ATS1B) (AT5G38430) mRNA, complete cds	808	???	???	???
NM_000537	*Homo sapiens* renin (*REN*), mRNA	1493	AAUAAA	1442	24
NM_181744	*Homo sapiens* opsin 5 (*OPN5*), transcript variant 1, mRNA	3496	AAUAAA	1971, 2304, 2396, 3461*	22
NM_014244	*Homo sapiens* ADAM metallopeptidase with thrombospondin type 1 motif, 2 (*ADAMTS2*), transcript variant 1, mRNA	6772	AUUAAA	6719	35
NM_198904	*Homo sapiens* gamma-aminobutyric acid (GABA) A receptor, gamma 2(*GABRG2*), transcript variant 1, mRNA	3957	AAUAAA	593, 2067, 2385, 2824, 3920*	20, 24, 37
NM_021267	*Homo sapiens* LAG1 homolog, ceramide synthase 1 (*LASS1*), transcript variant 1, mRNA	2558	AAUAAA	2543	≥15

The sequence and location of the polyadenylation signal, if present, were taken from the RefSeq annotation. The gene sequences (exons and introns) for these mRNAs were then searched for the presence of AAUAAA and the coordinates were recorded. If more than one location was found, the poly(A) signal coordinate found in the annotation was marked with an asterisk (*). The *Arabidopsis* RuBisCO record did not have any polyadenylation signal coordinate in the annotation.

**The distance between the annotated signal and the poly(A) tail. If no poly(A) tail was present in the record, it could only be estimated that the location of the tail was greater than or equal to (≥) the distance between the signal sequence and the end of the cDNA.

Box 9.2 An easy experiment

Open the NCBI BLASTP form and enter a string of 20 lysines (K) as the query. Make sure simple sequence filtering is off and run a BLASTP against the nr human proteins. You will find many hits and most, if not all, are at the C-terminus of proteins based on the alignment coordinates. Find the DNA record of these proteins and you will discover that they all have a poly(A) tail. It is the same tail that was translated into KKKKKKKKKKKKKKKKKKKK... (one of the codons for lysine is AAA). What happened here is that these sequences were automatically translated and, without human supervision, nobody was around to say "Wait a minute, there was no termination codon in the cDNA so our software just kept on translating, right through the poly(A) tail."

database Website allowed us to find or predict these elements. The start of translation is regulated by the 10 nucleotide Kozak sequence, and a number of these were compared to show the diversity seen at this position in mRNA. Genome browsers were just touched upon to graphically show gene structures and introduce the signals for RNA splicing. These signals were further explored by examining the splice sites of a number of genes. Finally, polyadenylation signals at the ends of mRNAs were introduced and examined in detail.

EXERCISES

Inhibitor of Kappa light polypeptide gene enhancer in B-cells (*IKBKAP*)

Familial dysautonomia (FD) is an autosomal recessive disorder primarily affecting individuals descended from Ashkenazi Jews. Those with the disease suffer severe problems with their sensory and autonomic nervous systems, but no decrease in intelligence. Through effects of the disorder on the sensory nerves, FD patients have reduced sensitivity to pain, temperature, and taste. The impaired autonomic nervous system leads to problems with their smooth muscle function (heart and internal organs) as well as problems in controlling gland function. As a consequence of the disorder, affected individuals develop heart, respiratory, blood pressure, and digestive problems. Symptoms gradually worsen and many patients die by their early thirties.

In 2001, Anderson and colleagues published a paper where they described the cause of FD as a defect in the *IKBKAP* (Inhibitor of Kappa light polypeptide gene enhancer in B-cells, Kinase complex-Associated Protein) gene. According to the RefSeq annotation, this protein regulates three kinases in response to inflammatory cues. The *IKBKAP* gene is almost 67,000 nucleotides long and has 37 exons, but a single base mutation at the donor splice site of the 20th intron causes the skipping of an exon during RNA processing. Rather than the normal CAAgtaagtg, the mutated splice donor sequence is CAAgtaagcg. This simple t-to-c change causes the downstream acceptor site of exon 21 to be favored as the next exon to be joined to the maturing message. The resulting deletion of exon 20 from the mature message causes a frameshift in the translation, and a truncated and nonfunctional protein is generated.

In this set of exercises, you will be studying the small sequences of IKBKAP.

1. Find the human *IKBKAP* RefSeq mRNA sequence and identify the Kozak sequence. Do the same for the mouse, rat, *Xenopus*, and chicken orthologs. Align these Kozak sequences manually and compare them to each other and to the consensus Kozak sequence.

2. Using the Ensembl genome browser, identify the first six nucleotides of the first 25 introns of human *IKBKAP* and align them. These represent the intron bases of the exon splice donor sites. To the alignment, add the intron nucleotides of the mutated splice donor (see above) underneath the normal intron nucleotides for this splice site. In addition, add the consensus splice donor sequence. Be sure to label the rows of the alignment. Observe the variation of the sequences and compare the affected splice site to the consensus.

3. Using the RefSeq record of the human *IKBKAP* transcript, go to the CCDS record and examine the protein sequence. When amino acid codons span the junction between exons, the amino acids they encode are colored red in these database records. Tally the red amino acids from the CCDS protein sequence. How do these numbers compare with those expected by translating the consensus nucleotides at these junctions (Figure 9.23)?

4. In this part of the exercise, you will collect the 500 nucleotides immediately upstream of the transcription start site of the human *IKBKAP* and we'll refer to this as the promoter region. Here, this is defined as the 500 nucleotides upstream of the 5P end of the NCBI Gene record. This can be accomplished

by first going to the FASTA format sequence for the NCBI Gene record. Once there, use the "Change the region shown" box of this Web page to bring the 500 nucleotide promoter region sequence into view.

5. Go to the gene-regulation.com Website and navigate to the MATCH form. There, paste in the human promoter region. Give your sequence a name, choose "vertebrates" from the list of matrices, and "minimize the sum of both error rates" by checking the appropriate radio button. Submit the sequence for analysis.

6. Using another browser window, repeat steps 4 and 5 and perform the MATCH analysis on the mouse *IKBKAP* promoter region.

7. By comparing the two results, generate a list of transcription factor binding sites held in common between the two promoters. Since they are conserved between distant species, they may be important to the transcriptional regulation of the gene.

8. Click on the hypertext Factor name from the MATCH results and explore the biology of the transcription factors that bind the sites identified in step 7. Where are these transcription factors active? Some will be found in all tissues or their expression will be unknown.

9. Based on disease symptoms of familial dysautonomia, which transcription factor binding sites identified in this exercise look more interesting to explore?

FURTHER READING

Anderson SL, Coli R, Daly IW et al. (2001) Familial dysautonomia is caused by mutations of the IKAP gene. *Am. J. Hum. Genet.* 68, 753–758. The original paper on the cause of familial dysautonomia. Nicely illustrated and describes the discovery process.

Chen M & Manley JL (2009) Mechanisms of alternative splicing regulation: insights from molecular and genomics approaches. *Nat. Rev. Mol. Cell Biol.* 10, 741–754. A wonderfully illustrated review article on alternative splicing, and the proteins involved in this process.

de la Grange P, Dutertre M, Correa M & Auboeuf D (2007) A new advance in alternative splicing databases: from catalogue to detailed analysis of regulation of expression and function of human alternative splicing variants. *BMC Bioinformatics* 8, 180. This paper describes fast DB, an alternative splicing database.

Flicek P, Amode MR, Barrell D et al. (2011) Ensembl 2011. *Nucleic Acids Res.* 39, D800–D806. A nice overview of this enormous and valuable bioinformatics Website.

Hannenhalli S (2008) Eukaryotic transcription factor binding sites—modeling and integrative search methods. *Bioinformatics* 24, 1325–1331. A great review article, discussing transcription factor binding sites along with their prediction.

Heldring N, Pike A, Andersson S et al. (2007) Estrogen receptors: how do they signal and what are their targets. *Physiol. Rev.* 87, 905–931. The title says it all in this extensive review article, filled with references, which is widely cited.

Kadonaga JT (2004) Regulation of RNA polymerase II transcription by sequence-specific DNA binding factors. *Cell* 116, 247–257. This is a fantastic review on the topic, describing the many thousands of proteins that regulate gene transcription, and filled with many references.

Kapustin Y, Souvorov A, Tatusova T & Lipman D (2008) Splign: algorithms for computing spliced alignments with identification of paralogs. *Biol. Direct* 3, 20. A paper, rich in data, describing this powerful application for aligning cDNA to genomic DNA, with emphasis in predicting splicing junctions.

Kozak M (1987) An analysis of 5′-noncoding sequences from 699 vertebrate messenger RNAs. *Nucleic Acids Res.* 15, 8125–8148. This classic paper describes Dr Kozak's work which defined the consensus bearing her name.

Li B, Carey M & Workman JL (2007) The role of chromatin during transcription. *Cell* 128, 707–719. A wonderful review article, illustrating the molecular events during the activation and expression of genes. Packed with references.

Loke JC, Stahlberg EA, Strenski DG et al. (2005) Compilation of mRNA polyadenylation signals in *Arabidopsis* revealed a new signal element and potential secondary structures. *Plant Physiol.* 138, 1457–1468. The polyadenylation signaling for plant genes is complex and this topic is reviewed in this article.

Matys V, Kel-Margoulis OV, Fricke E et al. (2006) TRANSFAC and its module TRANSCompel: transcriptional gene regulation in eukaryotes. *Nucleic Acids Res.* 34, D108–D110. From the *Nucleic Acids Research* database issue, a nice introduction to this bioinformatics standard.

Proudfoot NJ, Furger A & Dye MJ (2002) Integrating mRNA processing with transcription. *Cell* 108, 501–512. Wonderfully illustrated review article, covering mRNA processing from the 5P cap to the polyadenylation site.

Pruitt KD, Harrow J, Harte RA et al. (2009) The consensus coding sequence (CCDS) project: Identifying a common protein-coding gene set for the human and mouse genomes. *Genome Res.* 19, 1316–1323. This is an often-overlooked resource at the NCBI Website.

Rezniczek GA, Abrahamsberg C, Fuchs P et al. (2003) Plectin 5´-transcript diversity: short alternative sequences determine stability of gene products, initiation of translation and subcellular localization of isoforms. *Hum. Mol. Genet.* 12, 3181–3194. This excellent paper describes both the gene structure of plectin as well as the biology behind the spliceforms.

Schmucker D, Clemens JC, Shu H et al. (2000) *Drosophila* Dscam is an axon guidance receptor exhibiting extraordinary molecular diversity. *Cell* 101, 671–684. Through alternative splicing, this member of the immunoglobulin superfamily may generate many thousands of isoforms.

Zhang T, Haws P & Wu Q (2004) Multiple variable first exons: a mechanism for cell- and tissue-specific gene regulation. *Genome Res.* 14, 79–89. This paper describes the identification of multiple genes, including plectin, which have variable first exons and provides a broad view of splicing and gene expression.

ACAAGGACTAGAGAAACCAAAA
AGAAACCAAAACGAAAGGTGCAGAA
AACGAAAGGTGCAGAAGGGGAAACAGATGCAGA
GAAGGGGAAACAGATGCAGAAAGCAT
AGAAAGCAT
ACAAGGACTAGAGAAACCAAAACGAAAGGTGCAGAAGGGGAAACAGATGCAGAAAGCAT
CAAGGACTAGAGAAACCAAAA
AGAAACCAAAACGAAAGGTGCAGAA
AACGAAAGGTGCAGAAGGG
GAAGGG

CHAPTER 10

MicroRNAs and Pathway Analysis

Key concepts

- Function of miRNAs
- Processing of miRNAs
- Genomic organization of miRNA genes
- Biological networks and their regulation by miRNAs
- The prediction of miRNA target sites
- The links between miRNAs and human disease

10.1 INTRODUCTION

Most genes studied in earlier chapters encoded proteins. In this chapter, we turn to a group of genes that are transcribed but not translated. MicroRNAs (**miRNA**, pronounced either microRNA or em-eye-RNA) are approximately 22-nucleotide sequences that repress the translation of protein-coding mRNAs. MicroRNA genes are far from being minor curiosities. In fact, the study of miRNAs and their effect on biology and human disease has been one of the most active areas of research in recent years, generating over 10,000 publications since the discovery of miRNAs in 1993. Testament to their importance, miRNAs are unchanged in sequence for at least 450 million years of evolution.

10.2 miRNA FUNCTION

The **Central Dogma** of biology describes the flow of information within a cell. DNA is transcribed to mRNA which, in turn, is translated into protein. Decorating this information tree are pathways of regulation and function as outlined by **Figure 10.1**. DNA cannot be transcribed and RNA cannot be translated without protein. Proteins are modified by other proteins (or themselves) and their interactions and functions have been the center of attention for decades. In addition to providing a template for translation, priming RNA starts DNA synthesis, tRNAs shuttle amino acids to the ribosome, and RNAs have important structural and functional roles in complex structures such as ribosomes and spliceosomes.

An important new role for RNA was discovered when miRNAs were shown to provide an additional layer of gene regulation. The transcripts of miRNA genes are trimmed into a mature form approximately 22 nucleotides in length, and then, in a complex with the Argonaute protein, bind to mRNAs. MicroRNAs bind

Figure 10.1 Enhanced Central Dogma. Every biology student learns that genetic information flows (straight arrows) through a cell from DNA to mRNA to protein. Other forms of information and action, depicted as curved arrows, are not following this route.

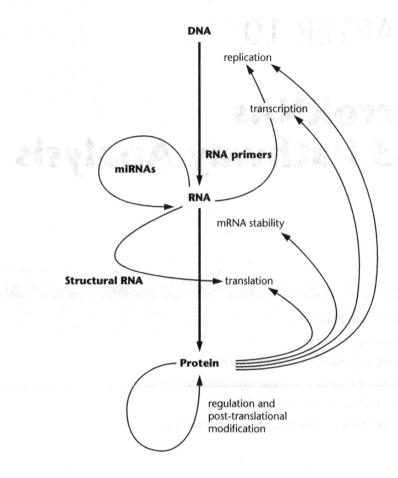

target sequences of similar length in the 3P UTR of messenger RNA and repress translation. This binding is primarily through 100% base pairing between a "seed" region, seven nucleotides in length at the 5P end of the miRNA, and target UTR (**Figure 10.2**). In animals, additional binding downstream of the seed region is imperfect and may be improved via secondary structure formation. However, the exact role of this 3P sequence is not clear.

The binding to specific target sites in the 3P UTR of mRNA triggers cleavage of the mRNA and concomitant mRNA instability, or a partial inhibition of translation. This new decoration to the Central Dogma and regulation of gene expression is not a minor contributor to how cells work. What has become clear, through the thousands of publications on miRNA, is that in every higher organism, hundreds of miRNAs are repressing the translation of thousands of mRNAs. Rather than turning translation off completely, one or more miRNAs are dampening the translation of individual mRNAs. Collectively, these miRNAs are orchestrating changes in cellular and organism developmental events and function.

Figure 10.2 Predicted base pairing between miRNAs and target sites. Shown here is the predicted base pairing between two miRNAs, miR-25 and miR-106b, and their respective target sites in *CDKN1C* and *CDKN1A*. Beyond the shaded seed regions, base pairing appears to be limited.

miR-25 and *CDKN1C*

```
3P UTR 5P   ...AAAUUUUGAAAACUGUGCAAUG...  3P
               |    || |   |||||||||
miRNA  3P      AGUCUGGCUCUGUUCACGUUAC     5P
```

hsa-miR-106b and *CDKN1A*

```
3P UTR 5P   ...AAGUAAACAGAUGGCACUUUG...  3P
               | |  || |   |||||||
miRNA  3P      UAGACGUGACAGUCGUGAAAU     5P
```

10.3 miRNA NOMENCLATURE

At first glance, the naming conventions of miRNAs can be quite confusing. Here are examples.

hsa-miR-121
hsa-let-7b
mmu-miR-1-1
mmu-miR-1-2
rno-miR-10a
rno-miR-10b

The first three letters are a code for the organism: "hsa" stands for *Homo sapiens*, "mmu" for *Mus musculus*, and "rno" for *Rattus norvegicus*. When speaking, say each letter of the species abbreviation. There are almost 170 organisms represented in miRBase so the three-letter abbreviations will quickly get obscure if you wander beyond your species of interest. The next three-letter field is usually miR (pronounced "meer"), but "let" (pronounced as "let" and not ell-ee-tee) is preserved for historical reasons; these were the first miRNAs discovered. The "r" in mir should be lowercase if you are referring to the gene but uppercase (R) if you are referring to the mature sequence. The number that follows represents the gene designation. Should two or more distinct genes be processed to an identical mature sequence, they are numbered, for example mmu-miR-1-1 and mmu-miR-1-2. However, if the mature sequences are only quite similar, they are designated with letters such as rno-miR-10a and rno-miR-10b.

10.4 miRNA FAMILIES AND CONSERVATION

MicroRNA genes belong to families with members possessing an identical seven-nucleotide signature ("seed," mentioned above) from positions 2–8 in the mature miRNA that is essential for miRNA binding. **Figure 10.3A** shows a multiple sequence alignment of mature members of the human miR-25 family.

If you expand the comparison to include the precursor sequences, the alignment quality does not improve. In fact, Figure 10.3B demonstrates that the family signature (seed sequence) may reside in different locations within the precursors. Outside of the family signature, similarity searches for family members would be challenging.

Across species, however, there is nearly perfect sequence conservation for a given mature miRNA. In **Figure 10.4A**, miR-25 is 100% identical in animals ranging from human to fish, a range spanning 450 million years of evolution. When aligning the full-length genes (Figure 10.4B), conservation is extremely high. Below, we'll learn that the entire sequence is crucial to forming a base-pairing structure prior to processing into the mature form. Clearly, miRNA sequences are critical to the survival of higher organisms and allow little room for sequences to drift apart.

miRNA: a perfect weapon

Virally infected cells will often undergo apoptosis, or programmed cell death, in response to the infected state. However, herpes simplex virus-1 (HSV-1) inhibits apoptosis, thereby insuring survival to emerge at a later date. This is accomplished by expression of the viral latency-associated transcript (LAT). Studies have shown that the HSV-1 LAT gene encodes a miRNA (miR-LAT) which represses the translation of two cellular transcripts responsible for apoptosis: TGF beta-1 and *SMAD3*. Because it is not a protein, miR-LAT evades normal antibody and cellular defenses that a cell usually mounts in response to an infection. After establishment in human cells, HSV-1 infection enters a state of latency when the virus can lie dormant and undetected by the immune system for years before emerging as a cold sore.

Figure 10.3 The hsa-miR-25 family. (A) Mature miRNAs from this family. (B) Unprocessed members of the same family. Where possible, the shaded seed sequences are aligned. The sequence orientation is 5P to 3P, left to right. Notice that the seed sequence of hsa-mir-32 is from the 5P end of the strand. Beyond the precursor sequence region, similarity is poor.

(A)

```
hsa-miR-25      5P-CAUUGCACUUGUCUCGGUCUGA-3P
hsa-miR-32      5P-UAUUGCACAUUACUAAGUUGCA-3P
hsa-miR-92a-1   5P-UAUUGCACUUGUCCCGGCCUGU-3P
hsa-miR-92b     5P-UAUUGCACUCGUCCCGGCCUCC-3P
hsa-miR-363     5P-AAUUGCACGGUAUCCAUCUGUA-3P
hsa-miR-367     5P-AAUUGCACUUUAGCAAUGGUGA-3P
```

(B)

```
hsa-mir-25       GGCCAGUGUUGAGAGGCGGAGACUUGGGCAAUUGCUGGACGCUGCCCUGGGCAUUGCACUUGUCUCGGUCUGACAGUGCCGGCC
hsa-mir-32     GGAGAUAUUGCACAUUACUAAGUUGCAUGUUGUCACGGCCUCAAUGCAAUUUAGUGUGUGUGAUAUUUUC
hsa-mir-92a-1      CUUUCUACACAGGUUGGGAUCGGUUGCAAUGCUGUGUUUCUGUAUGGUAUUGCACUUGUCCCGGCCUGUUGAGUUUGG
hsa-mir-92b     CGGGCCCCGGGCGGGCGGGAGGGACGGGACGCGGUGCAGUGUUGUUUUUUCCCCGCCAAUAUUGCACUCGUCCCGGCCUCCGGCCCCCCCGGCCC
hsa-mir-363     UGUUGUCGGGUGGAUCACGAUGCAAUUUUGAUGAGUAUCAUAGGAGAAAAUUGCACGGUAUCCAUCUGUAAACC
hsa-mir-367     CCAUUACUGUUGCUAAUAUGCAACUCUGUUGAAUAUAAAAUUGGAAUUGCACUUUAGCAAUGGUGAUGG
```

(A)

```
hsa-miR-25   cauugcacuugucucggucuga
mmu-miR-25   cauugcacuugucucggucuga
rno-miR-25   cauugcacuugucucggucuga
dre-miR-25   cauugcacuugucucggucuga
ggo-miR-25   cauugcacuugucucggucuga
ppa-miR-25   cauugcacuugucucggucuga
ppy-miR-25   cauugcacuugucucggucuga
mml-miR-25   cauugcacuugucucggucuga
lla-miR-25   cauugcacuugucucggucuga
mne-miR-25   cauugcacuugucucggucuga
fru-miR-25   cauugcacuugucucggucuga
tni-miR-25   cauugcacuugucucggucuga
xtr-miR-25   cauugcacuugucucggucuga
bta-miR-25   cauugcacuugucucggucuga
mdo-miR-25   cauugcacuugucucggucuga
cfa-miR-25   cauugcacuugucucggucuga
eca-miR-25   cauugcacuugucucggucuga
```

(B)

```
hsa-mir-25   ggccaguguugagaggcggagacuugggcaauugcuggacgcugcccugggcauugcacuugucucggucugacagugccggcc
mmu-mir-25   ggccaguguugagaggcggagacuugggcaauugcuggacgcugcccugggcauugcacuugucucggucugacagugccggcc
ggo-mir-25   ggccaguguugagaggcggagacuugggcaauugcuggacgcugcccugggcauugcacuugucucggucugacagugccggcc
ppa-mir-25   ggccaguguugagaggcggagacuugggcaauugcuggacgcugcccugggcauugcacuugucucggucugacagugccggcc
ppy-mir-25   ggccaguguugagaggcggagacuugggcaauugcuggacgcugcccugggcauugcacuugucucggucugacagugccggcc
mml-mir-25   ggccaguguugagaggcggagacuugggcaauugcuggacgcugcccugggcauugcacuugucucggucugacagugccggcc
lla-mir-25   ggccaguguugagaggcggagacuugggcaauugcuggacgcugcccugggcauugcacuugucucggucugacagugccggcc
mne-mir-25   ggccaguguugagaggcggagacuugggcaauugcuggacgcugcccugggcauugcacuugucucggucugacagugccggcc
eca-mir-25   ggccaguguugagaggcggagacuugggcaauugcuggacgcugcccugggcauugcacuucgucucggucugacagugccggcc
bta-mir-25   ggccaguguugagaggcggagacuugggcaauugcuggacgcugcccgggcauugcacuugucucggucugacagugccggcc
rno-mir-25   ggccaguguugagaggcggagacacgggcaauugcuggacgcugcccugggcauugcacuugucucggucugacagugccggcc
mdo-mir-25   ggccaguguugagaggcggagacuugggcaauugcugaacucgcccugggcauugcacuugucucggucugacagugcuggc
```

Figure 10.4 Multiple sequence alignment of mir-25 orthologs.
(A) Multiple sequence alignment of mature miR-25 orthologs. Although there have been millions of years since they shared a common ancestor, the sequence is perfectly preserved between human and species as distant as fish. (B) Alignment of full-length orthologous mir-25 family members. There are very few changes in the sequence, demonstrating that conservation goes beyond the mature miRNA, even though the rest of the sequence is discarded. The miRNAs are from human (hsa), mouse (mmu), rat (rno), zebrafish (dre), gorilla (ggo), bonobo (ppa), orangutan (ppy), rhesus macaque (mml), woolly monkey (lla), pigtail macaque (mne), pufferfish (fru), green pufferfish (tni), western clawed frog (xtr), cow (bta), opossum (mdo), dog (cfa), and horse (eca).

10.5 STRUCTURE AND PROCESSING OF miRNAs

MicroRNA genes are transcribed by RNA polymerase II, the same **multimeric** enzyme that transcribes protein-coding genes. Indeed, miRNA genes found between the exons of protein-coding genes are transcribed as intronic RNA that happens to have a gene within. Once transcribed, the single-stranded miRNA assumes a secondary structure, commonly described as a hairpin or stem–loop. **Figure 10.5** shows the structure of hsa-mir-32, which is located in the intron of an unannotated gene, *C9ORF5*. The hsa-mir-32 sequence is initially flanked by intron sequence, shown as Ns. Processing by an enzyme called Drosha trims this flank, leaving just the hairpin. Further processing by Dicer generates the mature 22-nucleotide sequence.

The hairpin structure varies with the sequence of different miRNA genes. **Figure 10.6** shows the hairpin structures of members of the human mir-25 family. Within each family member structure, the mature sequence is underlined and the seven-nucleotide "seed" sequence held in common between all family members is highlighted. Notice that the first (5P) nucleotide of the mature miRNA is not included in the seed sequence.

The human mir-25 family also illustrates another important feature of miRNAs which is the mature sequence can come from either the top (5P) or bottom (3P) of the hairpin. For example, in mir-25, the mature sequence is on the bottom of the hairpin between nucleotides 52 and 73. In mir-32, the mature sequence is between nucleotides 6 and 27. To add to the complication of miRNA genesis and function, some hairpin structures produce mature miRNAs from both the 5P and 3P regions of the hairpin. In these cases, one is referred to as "mature" or the form most characterized, and the other is referred to as "minor" meaning that the expression level or impact is low or not known.

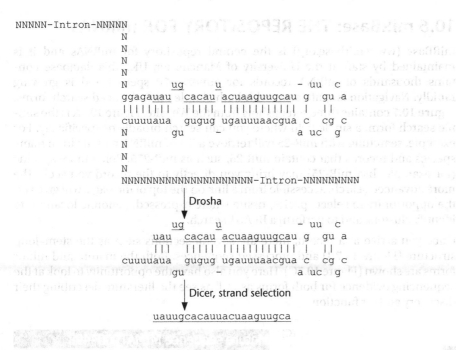

```
NNNNN-Intron-NNNNN
                  N
                  N
                  N
                  N
                  N       ug     u        -  uu  c
          ggagauau  cacau  acuaaguugcau  g    gu   a
          ||||||||  |||||  ||||||||||||  |    ||
          cuuuuaua  gugug  ugauuuaacgua  c    cg  c
          N       gu     a        -  a  uc  g
                  N
                  N
                  N
          NNNNNNNNNNNNNNNNNNNNNNNNN-Intron-NNNNNNNNNN
```

Figure 10.5 miRNA processing of the hsa-mir-32 gene. Shown is the miRNA gene within an intron, depicted as Ns. The mature miRNA sequence is underlined. Processing by a protein called Drosha releases the stem–loop and further processing by another protein, Dicer, generates the mature miRNA.

↓ Drosha

```
          ug     u        -  uu  c
     uau  cacau  acuaaguugcau  g    gu   a
          |||    ||||||  ||||||||||||  |    ||
     cuuuuaua  gugug  ugauuuaacgua  c    cg  c
          gu     a        -  a  uc  g
```

↓ Dicer, strand selection

uauugcacauuacuaaguugca

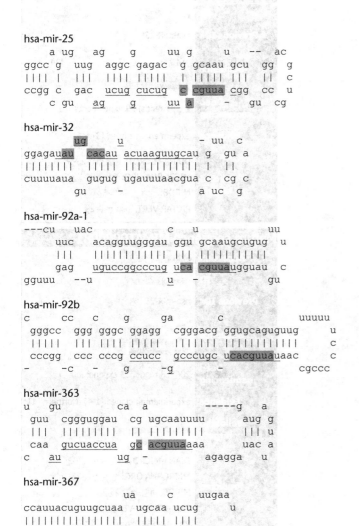

hsa-mir-25
```
       a  ug   ag    g    uu g     u   --  ac
ggcc g  uug  aggc gagac  g gcaau gcu  gg  g
|||| |  |||  |||| |||||  | ||||| |||  ||  c
ccgg c  gac  ucug cucug  c cguua cgg  cc  u
     c  gu   ag    g     uu a     -  gu  cg
```

hsa-mir-32
```
          ug     u        -  uu  c
ggagauau  cacau  acuaaguugcau  g    gu   a
||||||||  |||||  ||||||||||||  |    ||
cuuuuaua  gugug  ugauuuaacgua  c    cg  c
          gu     -          a  uc  g
```

hsa-mir-92a-1
```
---cu  uac         c  u       uu
    uuc  acagguugggau  ggu  gcaaugcugug  u
    |||  |||||||||||  |||  |||||||||||  u
    gag  uguccggcccug  uca  cguuaugguau  c
gguuu  --u         u          gu
```

hsa-mir-92b
```
c    cc   c   g    ga        c          uuuuu
 gggcc  ggg gggc ggagg  cgggacg ggugcaguguug   u
 |||||  ||| |||| |||||  ||||||| ||||||||||||   c
 cccgg  ccc cccg ccucc  gcccugc ucacguuauaac   c
 -     -c  -   g    -g      -             cgccc
```

hsa-mir-363
```
u  gu         ca a     -----g  a
 guu  cgggguggau  cg ugcaauuuu      aug  g
 |||  |||||||||  || |||||||||      |||  u
 caa  gucuaccua  gc acguuaaaa      uac  a
c  au         ug  -         agagga  u
```

hsa-mir-367
```
            ua     c    uugaa
ccauuacuguugcuaa  ugcaa  ucug    u
||||||||||||||||  |||||  ||||
gguagugguaacgauu  acguu  aggu    a
            uc     a    uaaau
```

Figure 10.6 Predicted structures of the hsa-mir-25 gene family. Although the predicted folding structure of the hairpins or stem–loops differs, the seed sequence AUUGCAC (shaded) is in common between family members. After processing, the single-stranded mature sequence (underlined) remains. In each case, the 5P end of the RNA begins at the left side of the top strand.

10.6 miRBase: THE REPOSITORY FOR miRNAs

miRBase (www.mirbase.org) is the central repository for miRNAs and it is maintained by staff at the University of Manchester, UK. This database contains thousands of miRNA records for almost 170 species and is growing rapidly. Navigation is quite easy, with both simple and advanced search forms. **Figure 10.7** contains three screenshots from the Website. Figure 10.7A is the simple search form, a single field where you can search broadly or specifically. For example, searching with miR-25 will retrieve a list of miRNA records from many species and records that contain miR-25, such as miR-255. Being more specific (for example, hsa-miR-25) may bring you directly to the record you seek. The more advanced search, accessible from a link on the top of the page, will give you the opportunity to select species, tissue where expressed, genomic location, to identify clusters, and to perform a BLAST search.

Once you arrive at a record, there are several sections such as the stem–loop structure (Figure 10.7B) and processed sequences. Both the mature and minor forms are shown (Figure 10.7C). Here you also have the opportunity to look at the sequencing evidence for both forms as well as see the literature describing their discovery and/or function.

(A)

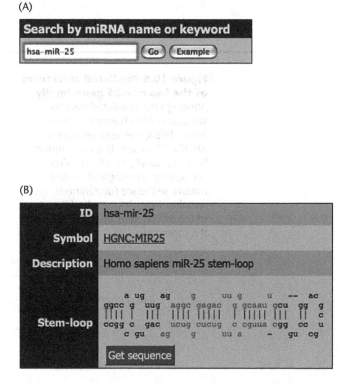

(B)

(C)

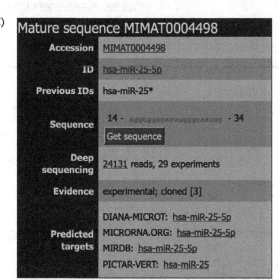

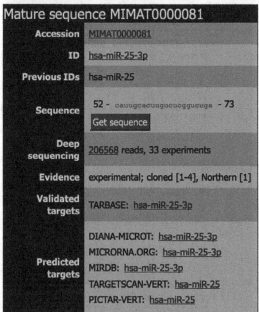

Figure 10.7 miRBase, the central repository of miRNA genes, structures, and mature sequences.
(A) Searching is straightforward through a Web form.
(B) Predicted stem–loop structures are available for every miRNA. (C) The sequences for mature miRNAs are provided. Some miRNA genes may have a second product referred to as the minor miRNA, often labeled with an asterisk, *. It appears that MIMAT0004498 was considered a minor miRNA (Previous ID) and is still found at considerably lower levels than MIMAT0000081 (Deep Sequencing). The nucleotide coordinates appear at the ends of the miRNA products (for example, 52 and 73), and those sequences have a different color in the stem–loop figure.

Before leaving this section on miRBase, another visit to nomenclature is necessary. As mentioned above, mature miRNAs can also originate from the opposite side of the hairpin and these minor species are often designated with a "*" symbol. In some miRNAs, mature sequences from both sides of the hairpin are known to be active miRNAs and are designated "-5p" or "-3p"; for example, mmu-miR-590-5p and mmu-miR-590-3p. Rather than refer to them as "mature" and "minor miR*," miRBase calls them both "Mature."

Finally, data can be easily gathered from miRBase search results. For example, the aligned sequences in Figure 10.4A originated from a miRBase BLASTN search, accessed from the tab across the top of the miRBase Website. Here, the mature sequence of hsa-miR-25 was entered and a search against the processed ("mature") sequences was launched. When the BLASTN results arrived, they were copied and pasted into a document and all non-subject lines were deleted.

10.7 NUMBERS AND LOCATIONS

As recently as 2008, approximately 500 human miRNA genes were known. But with deep sequencing of RNA libraries, over 1500 human miRNAs had been identified by early 2012, making it one of the largest gene families in the genome.

Some miRNA genes are present as singletons while others are clustered in pairs, triplets, or tens of miRNA genes. Their location is also variable. For example, **Figure 10.8A** shows that mir-155 is located within an exon of a noncoding RNA. Figure 10.8B shows a region of human DNA with six overlapping genes. One noncoding gene, *DLEU2*, has 11 exons and within one intron are two miRNA genes, mir-16-1 and mir-15a. *DLEU2* overlaps with another noncoding RNA and two protein-coding genes.

Figure 10.8C shows three miRNA genes (mir-25, mir-106b, and mir-93) within a single intron of a protein-coding gene, *MCM7*. *MCM7* also overlaps with another protein-coding gene, *AP4M1*. Having miRNA genes in the introns of protein-coding genes is quite common. In fact, the human plectin gene (studied in Chapter 9 to illustrate alternative splicing) has a miRNA gene in an intron in the 5P half of the gene. Another example of overlapping genes is shown in Figure 10.8D, with mir-191 and mir-425 found in the same intron of a protein-coding gene which overlaps with another gene. Figure 10.8E shows a large human cluster of 49 miRNA genes within 130,000 nucleotides on chromosome 19.

When looking up a miRNA gene in the NCBI Gene record, the view in the "Genomic regions, transcripts, and products" window will be zoomed in so the gene stretches from edge to edge in the window. To determine if the miRNA gene is within an intron, zoom out the view using the slider at the top of the window. It is important to remember that not all miRNA genes occur where protein-coding genes overlap or at otherwise complex genomic loci. The gene loci described here were chosen because they looked interesting.

10.8 LINKING miRNA ANALYSIS TO A BIOCHEMICAL PATHWAY: GASTRIC CANCER

The UTRs of many transcripts contain multiple and different miRNA target sites. Since numerous miRNAs are expressed in each cell, it is anticipated that a number of different miRNAs will participate in the regulation of the many biochemical pathways operating within a cell. Indeed, evidence is mounting that miRNAs are responsible for cooperatively down-regulating the entities of these pathways. The opposite is also true: by reducing the expression of miRNAs, mRNA repression would be eased and translation of those messages would increase. In this example, we will examine one section of a pathway and its regulation by a handful of miRNAs.

(A)

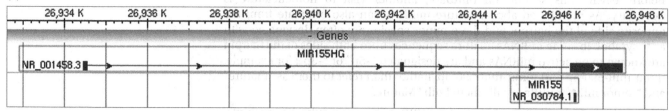

(B)

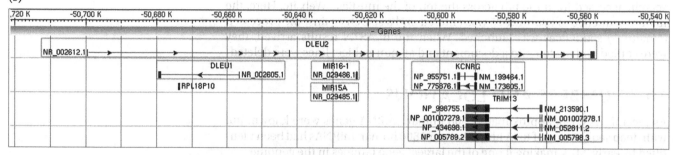

(C)

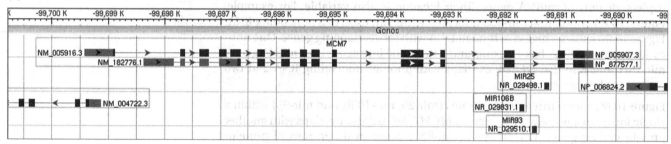

(D)

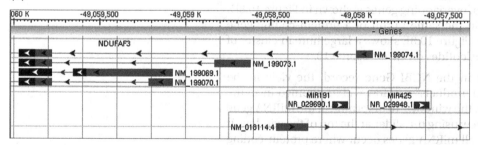

(E)

Figure 10.8 miRNA gene locations within the human genome. These examples are screenshots taken from the NCBI Gene records. (A) The miRNA sequence mir-155 is found in the exon of the noncoding gene *MIR155HG*. Like protein-coding genes, the introns are spliced out and further processing releases mir-155. (B) Two miRNA genes, mir-16-1 and mir-15a, in the intron of a noncoding gene, *DLEU2*. (C) mir-25, mir-106b, and mir-93 within the intron of a protein-coding gene, *MCM7*. (D) mir-425 in the intron of *DALRD3* (gene symbol not shown) and, nearby, mir-191 is found in the same intron as well as the intron of another protein-coding gene, *NDUFAF3*. (E) A very large cluster of 49 miRNA genes.

In 2009, Kim and colleagues published a paper concerning the impact of two clusters of miRNA genes on gastric cancer. Within an intron of human gene *MCM7* are three miRNA genes: mir-106b, mir-93, and mir-25. Expression of these three genes is up-regulated in gastric cancer tissues and the authors go on to show that these and transcripts from another miRNA cluster, mir-221 and mir-222, repress the translation of three genes involved with the cell division cycle. It is exciting to know that miRNA gene organization, in particular clustered miRNA genes, may prove to be a way to understand the large web of influence that multiple miRNAs have on multiple genes and pathways.

For the remainder of this chapter, we'll be using several different but complementary bioinformatics tools to identify or predict additional miRNAs and their targets within the cell division cycle. This will greatly expand our view of the regulation of this pathway, and identify potential avenues for laboratory research to verify these predictions.

10.9 KEGG: BIOLOGICAL NETWORKS AT YOUR FINGERTIPS

The Kyoto Encyclopedia of Genes and Genomes (KEGG) is a collection of databases which link and display a tremendous quantity of information on biochemical pathways, genes, diseases, drugs, and genomes. Text queries and analysis tools provide access to graphics and tables describing how individual proteins participate in large networks.

To find the network affected by the miRNA clusters described in the paper on gastric cancer, we'll perform a simple text query of the KEGG database. From the home page (www.genome.jp/kegg) navigate to the KEGG Table of Contents page (**Figure 10.9**). Here the gene symbol of one of the gene targets, *CDKN1B*, identified by Kim et al., can be used as a query.

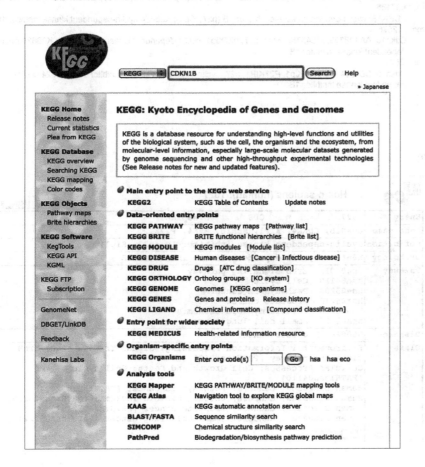

Figure 10.9 The KEGG database. The KEGG Table of Contents page offers many entry points into pathways, protein interactions, genes, chemicals and their inter-relationships. If you are not sure where to start, there is a search field at the top of the home page that allows queries of all databases at once. In this example, the gene symbol "CDKN1B" was entered in the Search field.

When the results are returned, *CDKN1B* appears in multiple places in the KEGG databases (**Figure 10.10A**). Under KEGG Genes, the hits are coded by species and *CDKN1B* is found in *Homo sapiens* (hsa), *Pan troglodytes* (ptr), *Macaca mulatta* (mcc), *Mus musculus* (mmu), and *Rattus norvegicus* (rno), among others.

Clicking on the human link, a KEGG Gene page is shown (Figure 10.10B). In addition to basic information like synonyms, we see that CDKN1B participates in six known pathways, four of which are linked to cancer: Pathways in cancer, Prostate cancer, Chronic myeloid leukemia, and Small cell lung cancer. Considering our original reason for this analysis was the link between miRNA and gastric cancer, this is interesting; there is no gastric cancer pathway in KEGG as yet. Kim et al. described the link between the target genes in their study and the cell cycle, so we'll investigate the "Cell cycle" pathway listed here.

Figure 10.10 KEGG database navigation. In Figure 10.9, "CDKN1B" was used as a query of the federation of KEGG databases. (A) There were hits in KEGG BRITE (classification within biology, for example, *CDKN1B* is under "cell growth and death" and a few others), KEGG ORTHOLOGY (orthologous sequences in sixteen species), and KEGG GENES. (B) Clicking on the "hsa:1027" (human) link shows details about this gene, some of which are shown here. Under the "Pathway" section, CDKN1B participates in at least six different pathways, one of which is the cell cycle.

(A)

Database: KEGG - Search term: CDKN1B

KEGG BRITE

ko00001
KO; KEGG Orthology (KO)

KEGG ORTHOLOGY

K06624
CDKN1B, P27, KIP1; cyclin-dependent kinase inhibitor 1B

KEGG GENES

hsa:1027
CDKN1B, CDKN4, KIP1, MEN1B, MEN4, P27KIP1; cyclin-dependent kinase inhibitor 1B (p27, Kip1); K06624 cyclin dependent kinase inhibitor 1B
ptr:466947
CDKN1B; cyclin-dependent kinase inhibitor 1B (p27, Kip1); K06624 cyclin-dependent kinase inhibitor 1B
mcc:697188
CDKN1B; cyclin-dependent kinase inhibitor 1B (p27, Kip1); K06624 cyclin-dependent kinase inhibitor 1B
mmu:12576
Cdkn1b, AA408329, AI843786, Kip1, p27, p27Kip1; cyclin-dependent kinase inhibitor 1B; K06624 cyclin-dependent kinase inhibitor 1B
mo:83571
Cdkn1b, CDKN4, Cdki1b, Kip1, P27KIP1, p27; cyclin-dependent kinase inhibitor 1B; K06624 cyclin-dependent kinase inhibitor 1B
• • • » display all

(B)

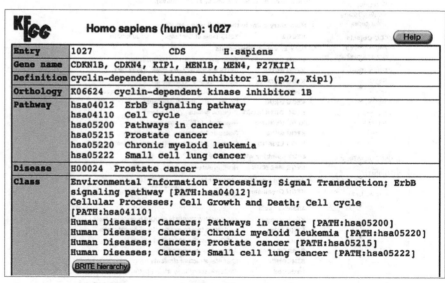

KE GG	Homo sapiens (human): 1027	Help
Entry	1027	CDS	H.sapiens
Gene name	CDKN1B, CDKN4, KIP1, MEN1B, MEN4, P27KIP1
Definition	cyclin-dependent kinase inhibitor 1B (p27, Kip1)
Orthology	K06624 cyclin-dependent kinase inhibitor 1B
Pathway	hsa04012 ErbB signaling pathway hsa04110 Cell cycle hsa05200 Pathways in cancer hsa05215 Prostate cancer hsa05220 Chronic myeloid leukemia hsa05222 Small cell lung cancer
Disease	H00024 Prostate cancer
Class	Environmental Information Processing; Signal Transduction; ErbB signaling pathway [PATH:hsa04012] Cellular Processes; Cell Growth and Death; Cell cycle [PATH:hsa04110] Human Diseases; Cancers; Pathways in cancer [PATH:hsa05200] Human Diseases; Cancers; Chronic myeloid leukemia [PATH:hsa05220] Human Diseases; Cancers; Prostate cancer [PATH:hsa05215] Human Diseases; Cancers; Small cell lung cancer [PATH:hsa05222]
	BRITE hierarchy

miRNAs in the cell cycle pathway

Figure 10.11 shows the KEGG cell cycle pathway. It is very challenging to display all the known interactions of a pathway, but the scientists at KEGG have organized them in a way that allows you to see the flow of signaling or change within a cell. The rectangular icons represent either individual proteins or protein complexes and are labeled with one or two common names. Floating your mouse over the icon will bring up a small box with the gene symbol(s). Clicking the icon for a single protein will take you to the corresponding gene record (for examples, Figure 10.10B). Should you click on a protein complex, multiple gene records are concatenated together so scroll down to see all the information. There are some nonrectangular icons and these usually represent places where the current pathway can intersect with another pathway. For example, in Figure 10.11, the MAPK signaling pathway, Apoptosis pathway, and Ubiquitin mediated proteolysis pathways all enter this pathway, but the other members of those pathways are not visible. Floating your mouse over these icons will reveal thumbnail views of those pathways and clicking on them will cause the navigation to those pathways.

The icons are joined by multiple lines, the style of which indicates the type of interaction between the entities joined by these lines. On each pathway Web page, there is a "Help" button on the upper right. Clicking there will give you a legend of all the types of interactions. For example, a solid arrow indicates an interaction, and with a "+p" added, that interaction means that one protein is phosphorylated by the other. Dashed lines mean indirect effects, and a line ending with a short perpendicular line indicates inhibition. For example, Figure 10.11 shows that Smad4 activates Cip1 while Skp2 inhibits Cip1.

Figure 10.11 The KEGG cell cycle pathway. This figure shows the known participants and relationships between factors in the cell cycle. Proteins are represented as rectangles and are labeled with common names, making these maps user-friendly. Floating your mouse over these icons may reveal that instead of a single protein, an icon may be a protein complex. For example, MCM and ORC are made up of six different proteins each, listed in the box in the lower left. Floating the mouse over icons also reveals the gene symbols of commonly named proteins. For example, the complex labeled "Kip1,2" (also known as p27 and p57) is CDKN1B and CDKN1C, respectively, and "Cip1," (also known as p21) is CDKN1A. Arrows show activation while lines ending with a perpendicular line indicate inhibition. Other symbols can be defined by clicking on the "Help" button at the top of the page (not pictured here). Icons for Kip1,2, Cip1, and MCM are highlighted in this figure.

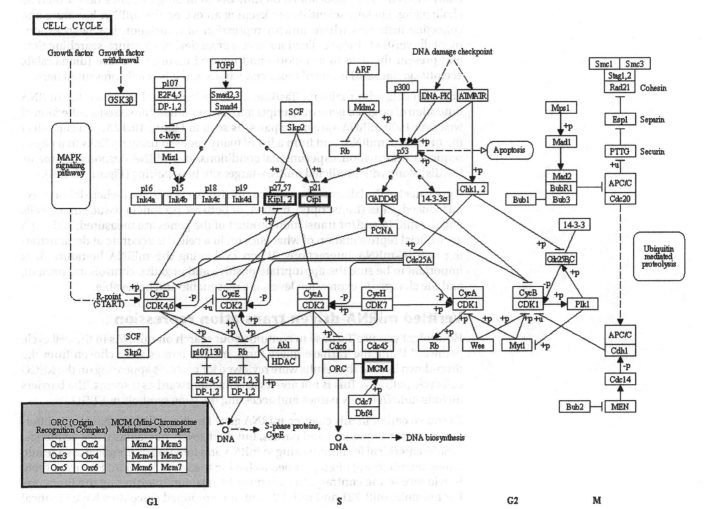

Three components of the pathway have been highlighted in Figure 10.11. Kip1, Kip2, and Cip1 represent the three genes identified by Kim et al. as the targets of miR-25, miR-93, miR-106b, and miR-221/222. Also highlighted on the KEGG pathway is MCM, a complex of various mini-chromosome maintenance proteins that are essential for the initiation of chromosomal DNA replication. As you recall, the gene for *MCM7* (found in this complex) harbors three miRNA genes analyzed in this study.

The presence of a protein in this pathway does not mean that this is its only function. When viewing the gene records, CDKN1C (Kip2) has only been identified as a participant in the cell cycle pathway but CDKN1B is found in six pathways (Figure 10.10B). Even within the same pathway, a protein can have multiple roles. CDKN2A, also known as ARF, inhibits Mdm2, but CDKN2A is also known as Ink4a, part of a larger complex that can inhibit CycD. Some proteins seem to have limited involvement in the pathway (for example, Rb), while others could be described as a "hub" of activity (for example, p53).

10.10 TarBase: EXPERIMENTALLY VERIFIED miRNA INHIBITION

We will now explore ways to try to expand the network of miRNAs that may be regulating the expression of proteins in the cell cycle pathway. Our first approach is to identify known examples where the five miRNAs in this study (miR-25, miR-93, miR-106b, miR-221, and miR-222) are repressing the translation of other gene transcripts from the cell cycle pathway. As stated earlier, there are literally thousands of publications on miRNAs so finding evidence here would be challenging. Luckily, scientists are keeping an eye on the miRNA literature and collecting instances where miRNA repression of translation has been experimentally verified. These collections save a great deal of literature-searching time and present the data in a uniform and focused manner. TarBase (diana.cslab. ece.ntua.gr/tarbase) is one of those collections and it is briefly presented here.

The scientists who maintain TarBase evaluate published evidence for miRNA inhibition of targeted gene transcripts and have manually amassed a collection of well over 1000 miRNA–target site pairs. As seen in **Figure 10.12A**, you can select the organism, miRNA, and from a list of many genes to bring up links to a paper, sequence information, experimental conditions, and miRNA Response Elements (MREs) along with predicted miRNA–target site base pairing (Figure 10.12B).

A great deal of the laboratory evidence consists of experiments where laboratory-introduced genes (transcripts), miRNA, or both are present in tissue culture cells and the mRNA and/or translation product of the genes are measured. Although an artificial representation of what goes on in a cell, it is accurate at demonstrating miRNA–mRNA interactions. When evaluating the miRNA literature, it is important to be sure the appropriate positive and negative controls are present, and the changes in expression levels are reasonable and believable.

Verified miRNA-driven translation repression

How might you use TarBase to continue your search on miRNAs in the cell cycle pathway? Using the TarBase query form, miRNA names were chosen from the drop-down list and the results were reviewed for proteins appearing on the KEGG cell cycle pathway. This is not always as straightforward as it seems. The barriers include miRNA family names and accessing the gene symbols in KEGG.

As shown earlier in this chapter, miRNA gene families will often have very different precursor sequences and names, but their seven-nucleotide seed sequence, absolutely critical for base-pairing to mRNA and function, is identical. As a result, family members are often grouped as having the same function. However, there is evidence to the contrary, so care must be taken in interpreting the literature. For example, miR-221 and miR-222 are often grouped since they have identical

(A)

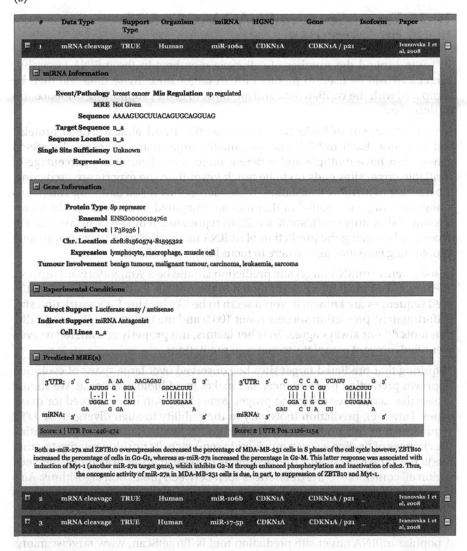

(B)

Figure 10.12 TarBase, a database of experimentally supported miRNA targets. There is a growing list of laboratory-verified miRNA target sites and TarBase has collected these into one Website. (A) TarBase can be queried by organism, miRNA, gene symbol, or any combination. Shown here is a query using *CDKN1A* (p21), also known as Cip1 (highlighted in Figure 10.11). (B) A list of hits appears, and clicking on them reveals additional information. This indicates that the human mRNA for *CDKN1A*/p21 has been shown to be repressed by miR-106a. At the bottom of this panel are two other experimentally supported translation repressions of *CDKN1A*/p21: miR-106b and miR-17-5P.

seed sequences and are clustered genomically. However, Kim et al. were careful to test both miRNAs in their assays and report similar activities. There are clear differences in the sequence as the following alignment shows (note that the identical seed sequences are underlined):

hsa-miR-221 AGCUACAUUGUCUGCUGGGUUUC

hsa-miR-222 AGCUACAUCUGGCUACUGGGU

When searching TarBase for miRNA names, be sure to use all the family names separately. For example, search with miR-221 and miR-222. Also, be sure to notice how many hits are returned; TarBase only shows 30 hits on a page and you may have to look at additional pages to see all the hits.

Figure 10.13 Experimentally verified miRNA targets of miR-25, miR-93, miR-106b, miR-221, and miR-222. These results are gathered from the Kim et al. paper and TarBase.

miRNA	Laboratory-verified target
miR-25	p57 (Kip2 or *CDKN1C*)
miR-93	p21 (Cip1 or *CDKN1A*)
miR-106b	p21 (Cip1 or *CDKN1A*)
miR-221	p27 (Kip1 or *CDKN1B*), p57 (Kip2 or *CDKN1C*)
miR-222	p27 (Kip1 or *CDKN1B*), p57 (Kip2 or *CDKN1C*)

Staying with the miRNAs used in assays from the Kim et al. paper and TarBase data, the miRNAs and targets are limited to those shown in **Figure 10.13**.

10.11 TargetScan: miRNA TARGET SITE PREDICTION

To try to expand the number of cell cycle members in the miRNA target list, TargetScan will be used to predict miRNA target sites. These data will then be compared with the verified data and attempts to gather supporting evidence will be described.

Over 1500 human miRNAs have been identified and almost 21,000 protein-coding genes. Each miRNA may potentially target many genes, and targeted genes often have multiple and different target sites. Even if the percentage of verifiable target sites ends up to be much lower than the experts are predicting, the number of possible translation–repression events is still daunting and paints a huge web of gene regulation that was not imagined until recently. The labor-intensive laboratory verification of miRNA repression of mRNAs will take years to accomplish, making the prediction of miRNA target sites an attractive approach to obtaining guidance as to where to focus laboratory efforts.

In some sense, miRNA target site prediction should be a straightforward process. The target sites must be a perfect match to the miRNA seed sequence. Since the seed sequences are known, it would seem to be like looking for a restriction site. Unfortunately, prediction success is not 100% and the results of different prediction tools do not always agree. So other factors, not properly considered or even identified, prevent complete accuracy in predictions.

Requiring that predicted target sites be conserved over large spans of evolution improves prediction performance, and makes good biological sense. We should expect that sequences critical to proper gene regulation be conserved for many genes. However, prediction tools vary in their ability to align divergent 3P UTR sequences and this undoubtedly contributes to prediction success. Calculating the chances of finding a seven-nucleotide sequence, its position relative to other predicted target sites, and the treatment of base pairing outside of the seed region all contribute to the differences seen between prediction algorithms. Also remember, if you require a great deal of evolutionary conservation, target sites that are, for example, human-specific will be overlooked. So we are only uncovering those sites in common between very distant species.

A popular miRNA target site prediction tool is TargetScan, www.targetscan.org. From the home page (see **Figure 10.14**), you can choose from 10 different species and enter gene symbols in a free text field. Or you may select miRNAs from drop-down menus.

It is important to understand how TargetScan and other applications treat related family members and their names. If miRNAs have the same seed sequence in the mature miRNA, they are grouped as a family. **Figure 10.15** shows a section of the page that is accessible from the TargetScan home page by clicking on the "mammalian miRNAs" link in the text below the search panel. Here you see that many "let" miRNAs as well as miR-98, from multiple species, all have the same GAGGUAG seed sequence and are grouped in a family called "let-7/98/4458/4500." The second family on this list is called "miR-1ab/206/613" and has miR-1, miR-206, and miR-613 miRNAs as family members. The seed sequence in this family is GGAAUGU. The mammalian miRNAs list is many pages

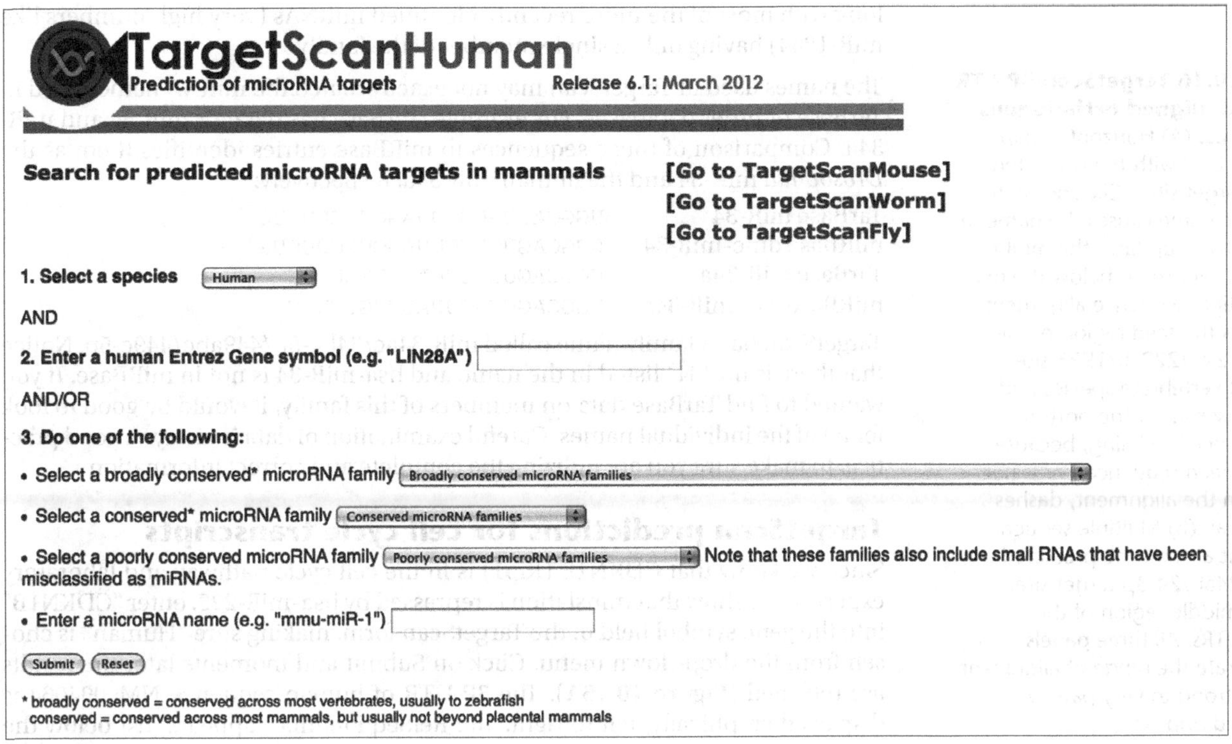

Figure 10.14 The TargetScan home page. Shown here is TargetScanHuman. Hyperlinks to mouse (*Mus musculus*), worm (*Caenorhabditis elegans*), and fly (*Drosophila melanogaster*) equivalents are on the right side of the page. Under the miRNA family drop-down menus are long lists of miRNAs.

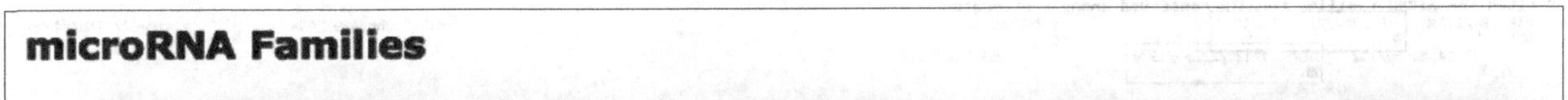
microRNA Families

microRNA family	Seed + m8	miRBase IDs in this Family
let-7/98/4458/4500	GAGGUAG	hsa-let-7a hsa-let-7b hsa-let-7c hsa-let-7d hsa-let-7e hsa-let-7f hsa-miR-98 mmu-let-7g mmu-let-7i mmu-let-7d hsa-let-7g hsa-let-7i mmu-let-7a mmu-let-7b mmu-let-7c mmu-let-7e mmu-let-7f mmu-miR-98 rno-let-7d rno-let-7a rno-let-7b rno-let-7c rno-let-7e rno-let-7f rno-let-7i rno-miR-98 gga-let-7i gga-let-7a gga-let-7b gga-let-7c gga-let-7g gga-let-7d gga-let-7f gga-let-7j gga-let-7k mml-miR-98 ptr-miR-98 bta-let-7f xtr-let-7b xtr-miR-98 xtr-let-7c xtr-let-7f xtr-let-7g xtr-let-7i xtr-let-7e xtr-let-7a bta-miR-98 bta-let-7d bta-let-7g bta-let-7a bta-let-7i mdo-let-7a mdo-let-7g mdo-let-7i mdo-let-7f mdo-let-7b mdo-let-7d bta-let-7b bta-let-7c bta-let-7e oan-let-7b oan-let-7e oan-let-7g oan-miR-98 oan-let-7f oan-let-7d mml-let-7a mml-let-7b mml-let-7c mml-let-7d mml-let-7e mml-let-7f mml-let-7g mml-let-7i cfa-let-7a cfa-let-7e cfa-let-7f cfa-let-7g cfa-let-7c cfa-miR-98 cfa-let-7j ptr-let-7a ptr-let-7b ptr-let-7c ptr-let-7d ptr-let-7e ptr-let-7f ptr-let-7g ptr-let-7i mmu-miR-1961 cfa-let-7b eca-let-7a eca-let-7e eca-let-7g eca-let-7d eca-let-7f eca-let-7c eca-miR-98 hsa-miR-4458 hsa-miR-4500
miR-1ab/206/613	GGAAUGU	mmu-miR-1a mmu-miR-206 hsa-miR-1 hsa-miR-206 rno-miR-206 gga-miR-1a gga-miR-206 gga-miR-1b rno-miR-1 hsa-miR-613 xtr-miR-1a xtr-miR-206 xtr-miR-1b mdo-miR-1 mdo-miR-206 oan-miR-1a oan-miR-206 oan-miR-1b mml-miR-1 mml-miR-206 cfa-miR-206 cfa-miR-1 ptr-miR-1 ptr-miR-206 ptr-miR-613 bta-miR-1 bta-miR-206 eca-miR-1 eca-miR-206

Figure 10.15 TargetScan miRNA family listings. This useful listing is found by clicking on the "mammalian miRNAs" link on the home page. Family names can be quite short (for example, "miR-9") or concatenations of the member names (for example, "let-7/98/4458/4500"). All members of a family share the same seed sequence. Clicking on the individual names will take you to the miRBase entry for that specific miRNA.

long with most of the more recently identified miRNAs (very high numbers like miR-1844) having only a single member of the family.

The names used in TargetScan may not exactly match the miRNA names used in TarBase or other databases. For example, TarBase has data for miR-34 and miR-34a. Comparison of those sequences to miRBase entries identifies them as the *Drosophila* miR-34 and the human miR-34a, respectively:

TarBase miR-34	UGGCAGUGUGGUUAGCUGGUUGUG
miRBase dme-miR-34	UGGCAGUGUGGUUAGCUGGUUGUG
TarBase miR-34a	UGGCAGUGUCUUAGCUGGUUGU
miRBase hsa-miR-34a	UGGCAGUGUCUUAGCUGGUUGU

TargetScan has a family name called miR-34ac/34bc-5p/449abc/449c-5p. Notice that there is no "34" listed in the name and hsa-miR-34 is not in miRBase. If you wanted to find TarBase data on members of this family, it would be good to look for all of the individual names. Careful examination of data is always a good practice, to make sure you are utilizing the complete and correct information.

TargetScan predictions for cell cycle transcripts

Since we know that CDKN1B (Kip1) is in the cell cycle pathway and laboratory experiments show that translation is repressed by hsa-miR-222, enter "CDKN1B" into the gene symbol field of the TargetScan form, making sure "Human" is chosen from the drop-down menu. Click on Submit and moments later the results are returned (**Figure 10.16A**). The 3P UTR of human sequence NM_004064 is displayed graphically, left to right. The RefSeq identifier appears just below the

Figure 10.16 TargetScan 3P UTR map and aligned orthologous sequences. (A) Horizontal map of the 3P UTR with the predicted miRNA target sites. Clicking on the target site name causes the name to be boxed and updates the multiple sequence alignment below the map. This multiple sequence alignment highlights the seed region of the miR-221/222/222ab/1928 site across 23 vertebrate species and has a consensus at the bottom. If a sequence is missing, because it hasn't been sequenced or is not present in the alignment, dashes (---) appear. (B) Multiple sequence alignment around the predicted miR-24/24ab/24-3p target site. (C) The middle region of the aligned UTRs. All three panels demonstrate the range of alignment quality: strong to very patchy (continued opposite).

(A)

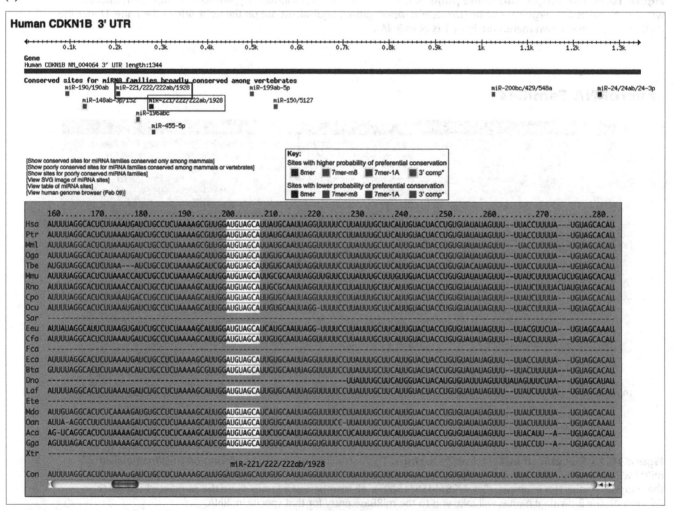

(B)

```
          .......1220......1230......1240......1250......1260...........1270........1280......1290.........1300......1310.........1320.
Hsa  --UUUUGUUAAAUAAUAUGGCUAUGCUUAAAAAGGUUGCAUACUGAGCCAAG------UAUAA-UUUUU--UGUAAUGUGUGAAAAAGAUG-CCAAU---UAUUGUUACACAUUAAGUAAUC----AAUAA
Ptr  --UUUUGUUAAAUAAUAUGGCUAUGCUUAAAAAGGUUGCAUACUGAGCCAAG------UAUAA-UUUUU--UGUAAUGUGUGAAAAAGAUG-CCAAU---UAUUGUUACACAUUAAGUAAUC----AAUAA
Mml  --UUUUGUUAAAUAAUAUGGCUAUGCUUAAAAAGAUUGCAUACUGAGCCAAG------UAUAA-UUUUU--UGUAAUGUGUGAAAAAGAUG-CCAAU---UAUUGUUAUACAUUAAGCAGUC----AAUAA
Oga  --UUCGUUAAAUAAUAUGGCUAUGCUU-AAAGGUUGCAUACUGAGCCAAG------UAUAA-UUUUU--UGUAAUGUGUGAAAAAGAUG-CCAAU---UAUUGUUACACAUUAAGCAAUC----AAUAA
Tbe  -------------------------------------------------------------------------------------------------------------------------
Mmu  --CUUUGUUAAAUAGUAUGGCUGUGCUUAAAAAGGUUGCAUACUGAGCCAAG------UAUAA--UUUU--UGUAAUGUGUGAAAAAGAUG-CCAAU---UAUUGUUACACAUCAAGCAAUC----AAUAA
Rno  --UUUUGUUAAAUAAUAUGGCUGUGCUUAAAAAGGUUGCAUACUGAGCCAAG------UAUAA--UUUU--UGUAAUGUGUGAAAAAGAUG-CCAAU---UAUUGUUACACAUCAAGCAAUC----AAUAA
Cpo  --UUUUGUUAAAUAAUAUGGCUAUGCUUAAAAAGGUUGCAUACUGAGCCAAG------UAU----GUUU--UGUAAUGUGUGAAAA-AUG-CCAAU---UAUUGUUACACUUCAAACAAUC----AAUAA
Ocu  --UUCGUUAAAUAAUAUGGCUGUGCUUAAAAAAUUGCAUACUGAGCCAAG------UAUAA-UUUUU--UGUAAUGUGUGAAAAAGAUU-CCAGU---UAUUGUUACACAUCAUGCAAUC----AAUAA
Sar  -------------------------------------------------------------------------------------------------------------------------
Eeu  --UUAGUUAAAUAAUAUGGCUAUGCUUAAAAAGUUUGCAUACUGAGCCAAG--------UAUAAUUUUUU--UGUAAUGUGUGAAAAAAAAUCCAAU---UAUUGUUACACA-CAAGAAUC----AAUAA
Cfa  --UUCGUUAAAUAAUAUGGCUAUGCUUAAAAAGGUUGCAUACUGAGCCAAG------UAUAA-UUUUU--UGUAAUGUGUGAAAAAGAUG-CCAAU---UAUUGUUACACAUCAAGCAAUC----AAUAA
Fca  -------------------------------------------------------------------------------------------------------------------------
Eca  --U--UGUAAGUAAUAUGGCUAUGCUUAAAAAGGUUGCAUACUGAGCCAAG------UAUAA-UUUUU--UGUAAUGUGUGAAAAAGAUG-CCAAU---UAUUGUUACACAUAAAGUAAUC----AAUAA
Bta  --UUCGUUAAAUAAUAUGGCUAUGCUUAAAAAGGUUGCAUACUGAGCCAAG------UAUAA-UUUUU--UGUAAUGUGUGAAAAAGAUG-CCAAU---UAUUGUUACAUGUCAAGUAAUC----AAUAA
Dno  UUUUUCGUUAAAUAAUAUGGCUAUGCUUAAAAGAUUGCAUACUGAGCCAAG------UAUAA-UUUUU--UGUAAUGUGUGAAAAAGAUG-ACAAU---UAUUGUUACACAUCAAGCAAUC----AAUAA
Laf  --UUUUGUUAAAUAAUAUGGCUAUGCUUAAAAAGAUUGCAUACUGAGCCAAG------UAUAA-UUUUU--UGUAAUGUGUGAAAAAGAUG-CCGAU---UAUUGUUACACAUCAAGCAAUC----AAUAA
Ete  UGUCCUGUU-AAUAAUAUGGCUAUGCUUAAAAAGAUUGCAUACUGAGCCAAG------UAUAA-UUUUU--UGUGAUGUGUGAAAAA-UG-C---U---UAUUGUUACACAUCAAGCAAUC----AAUAA
Mdo  --UUCGUUAAAUAAUAUGGCUGUGCUUAAAGAAUUGCCUACUGAGCCAAG------UAUAACUUUUU--UGUAAUGUGUA--------------AUUGUCCCUAUCAAUCAAUC----AAUCA
Oan  --UUUCGUUCA---AAUAUGGCUGUGCUUAAAAGUUUGCAUACUGAGCCAAG------UUUAA-UUUUUUUAUGUAAUGUGUGAAAAAGAUG-CCGAU---CAUUGUUUCACAUCCAGCAAUC----AAUAA
Aca  --GUUUUGUAAUUAAUAUGGCUGUGCUUAAAUGAUUGCAGACUGAGCCAAU------U-UAA-UUUU---GUGUG-UGUGUGAAGACAUUG-CCAAUUGAUAUUGUUUCCACAUCAAGCAAUCAAUAAAUAA
Gga  -------------------------------------------------------------------------------------------------------------------------
Xtr  --UUUUUUU---------GGCUA---UUAAUGGAUUGCAGACUGAGCCCAACGCAUUUGUAA-AUCUU---UUUACAGUUUAAAAAAAA----------------------------------
                                                  miR-24/24ab/24-3p
Con  ..UUcUGUUAAAUAAUAUGGCUAUGCUUAAAAAGgUUGCAUACUGAGCCAAG......UAUAA.UUUUU..UGUAAUGUGUGAAAAAGAUG.CCAAU...UAUUGUUACACAUCAAGCAAUC....AAUAA
```

(C)

```
          ..........600.................610.......620.....................................630.........640..........
Hsa  ------GCAUGUGGC-----------UU--UUUUUAAAAAAGCA-----A------C-------------AGAAAC--------------C-UAUC---CUCA---CUGCCCUCC-CC-----
Ptr  ------GCAUGUGGC-----------UU--UUUUUAAAAAAGCA-----A------C-------------AGAAAC--------------C-UAUC---CUCA---CUGCCCUCC-CC-----
Mml  ------GCAUUAUGGC-----------UU--UUUUUAAAAAAGCA-----A------C-------------AAAAAC--------------C-UAUC---CUCA---GUGCCCUCC-CC-----
Oga  ------GCACAUGGC-----------UU---UU----AACACAAA-----A------C-------------AAAAAU--------------C-UAUC---CUUA---GUGCUC-CU-CC-----
Tbe  ------GCAUUUUUU-----------UU--UU--AAAAAACA-----AAAAAAACCA-------------AAAAAC--------------U-UAUC---CUUGUCUGUGCUC-CU-CC-----
Mmu  ------AGAAAUGGC-----------UU-------A-----A-----U--------------------------------------------------------UUU-----------
Rno  ------AGAAAUGGC-----------UU------AAAAA-----A-----U----------------------------------------------------UUU----------
Cpo  ------ACAAAUGGC---------UU----UUAAAA------A-----U----------CAAAAA------------ACUAUC---C-UU---UUUUGCU-----------
Ocu  ------GCAAAUGGCGUGUUUUUUUUUUUUUU----UUUAAAGCA-----A------C-------------AAAA---------------UAUC---C-CA---GUGCUCCC-CC-----
Sar  -------------------------------------------------------------------------------------------------------------------
Eeu  ------GCAAAUGGC-----------UU--UUU-ACAAUAAAA-----A------C-----CAAAACACAAAAGCGAAAAAAAA-------C-ACUCUUUCUCA---GUGCCUUCCCA-----
Cfa  ------AAAAAGGGU-----------UU--UUUU-----AACC-----A------A------------AAAAAAUUUUUUUGGUUCCCC-CCCC---CCCC---CCCCCCCC--CC-----
Fca  -------------------------------------------------------------------------------------------------------------------
Eca  ------GCAAAUGGU-----------UU--UUUU----AAACA-----A------C-------------AGCAAA--------------C-UACC---CUCA---GUGCUCC--CA-----
Bta  ------GCAAAUGGC-----------UU--UUU--AACAACA-----A------C-------------AAAAAA--------------C-UAUC---CUCA---GUGCUCCCACU-----
Dno  ------AAAGGCU-----------UU-----------AAA-----A------A-------------AAAAAA--------------A-UACCCCUCUCA---GUGCCCUCC-CA-----
Laf  ------GUAAAUGGC-----------UU--UUUUAAAAGAACA-----A------U-------------AAAAAA--------------C-UGCCACUCUCA---GCGUCCCCU-CU-----
Ete  -------------------------------------------------------------------------------------------------------------------
Mdo  ------GCAAAGGCU-----------UU--UUUC-------------------------------------------------------------------------AAAGG
Oan  ------GCAAAAGGC-----------UU--UUUU--------------------------------------------------------------------------------
Aca  UAUAUGUAGAAAAGC-----------AG---AUUC---AUGGGACCUUAUG--------C-------------AAGAGG-----------G----C---GAUU--GUGUCC----UA-----
Gga  -------------------------------------------------------------------------------------------------------------------
Xtr  -------------------------------------------------------------------------------------------------------------------
Con  ......GCAAAUGGC..........UU....U.......AacA....A........c.........AGAAA...............C...C.ca...gUGccC...C......
```

scale, on the left. Note that if you choose another species from the drop-down menu (for example, chicken) the RefSeq sequence identifier will still be the human sequence. Remember, these sites are predicted based on being conserved between species so this is not a mistake. However, the UTR length will be adjusted on the results page; the human *CDKN1B* 3P UTR is 1344 nucleotides long while the chicken sequence is 431 nucleotides in length. See the TargetScan "Frequently Asked Questions" for more of an explanation.

From the result in Figure 10.16A, you can see predicted miRNA target sites as boxes below the thick horizontal line. Shown are two sites for miR-221/222/222ab/1928, and a number of other sites. miR-221, miR-222, miR-222a, miR-222b, and miR-1928 all have the same seed sequence (GCUACAU) so TargetScan combines their names to create a single site. The results page also shows the multiple sequence alignment of UTRs from 23 species plus a consensus on the bottom.

The rows, from top to bottom, reflect evolutionary distance; closest to human is on the top, farthest is on the bottom. A "Species key" link is just below the multiple sequence alignment horizontal scroll bar. Some species lines are blank (for example, Sar [shrew]) reflecting that the 3P UTR and/or gene for *CDKN1B* has not been identified or does not align in this region. By clicking on the miRNA family name (for example, miR-221/222/222ab/1928 in Figure 10.16A) that name becomes boxed and the multiple sequence alignment shifts to show the highlighted predicted target sequence to you.

The alignment of the predicted miR-221/222/222ab/1928 target site and the surrounding UTR sequence is impressive. Remember that the criterion for considering this site for prediction is the degree of conservation. This region is also only 200 nucleotides away from the end of translation so perhaps you might expect the homology to be high. However, it is still striking that within the UTR, under less evolutionary constraint than the coding region, these seven nucleotides have been conserved from the time the chicken and human ancestors diverged, over 300 million years ago. This site must be important. Compare this with the sequences you see in Figure 10.16B, the region of a predicted miR-24 site. This region is over 1200 nucleotides away from the end of translation and there are more differences and gaps in this region. However, this "island" of similarity around the predicted miR-24 target site is very good. A scroll across the UTR shows that other regions are very patchy (Figure 10.16 C) and the alignment here is barely detectable.

Scroll down the TargetScan results page and the alignments between the miRNAs and their predicted targets are displayed along with eight calculations performed by the TargetScan algorithm. As seen before, clicking on the miRNA name in the figure at the top of the results page changes what alignments and calculations are shown in this section. For example, clicking on miR-221/222/222ab/1928 at the top of the page refreshes the details below the multiple sequence alignment (**Figure 10.17**). Each miRNA of the family (for example, miR-221 and miR-222) is individually aligned to the predicted target site and both predicted sites (positions 201–208 and 274–281) are displayed.

The calculations reveal what is important to the TargetScan algorithm, including base-pairing contribution of the sequence 3P to the seed match, the local nucleotide composition (local AU contribution), and the position within the UTR (is it adjacent to other predicted sites and is it located near the beginning or end of the UTR?). Collectively, these values contribute to the "context score" which is expressed both as a value (for example, –0.33) and a percentile so the score can be compared with other predictions. The three additional base pairing 3P to the miR-221 seed match at position 201–208 had a small effect on the context score percentile; 94th versus 93rd percentile for the two alignments at position 201–208. The context score for the predicted miR-24 site is at the 99th percentile, putting this prediction in the top 1%. To read more about the calculations used by TargetScan, there are several links to published papers on the TargetScan Website.

Figure 10.17 Conserved TargetScan predicted sites for CDKN1B and algorithm scoring. Below the multiple sequence alignment (as seen in Figure 10.16) is this table, featuring pairwise alignments between miRNAs and the predicted sites, along with scoring parameters of the TargetScan algorithm.

Conserved

predicted consequential pairing of target region (top) and miRNA (bottom)	seed match	site-type contribution	3' pairing contribution	local AU contribution	position contribution	TA contribution	SPS contribution	context+ score	context+ score percentile	conserved branch length	PCT
Position 201-208 of CDKN1B 3' UTR 5' ...CUCUAAAAGCGUUGGAUGUAGCA... ||| ||||||| hsa-miR-221 3' CUUUGGGUCGUCUGUUACAUCGA	8mer	-0.247	-0.018	0.012	-0.063	-0.018	0.005	-0.33	94	2.443	0.60
Position 201-208 of CDKN1B 3' UTR 5' ...CUCUAAAAGCGUUGGAUGUAGCA... ||||||| hsa-miR-222 3' UGGGUCAUCGGUCUACAUCGA	8mer	-0.247	-0.008	0.012	-0.063	-0.018	0.005	-0.32	93	2.443	0.60
Position 274-281 of CDKN1B 3' UTR 5' ...UAGUUUUUACCUUUUAUGUAGCA... ||||||| hsa-miR-222 3' UGGGUCAUCGGUCUACAUCGA	8mer	-0.247	0.003	-0.047	-0.045	-0.018	0.005	-0.35	95	2.497	0.60
Position 274-281 of CDKN1B 3' UTR 5' ...UAGUUUUUACCUUUUAUGUAGCA... ||||||| hsa-miR-221 3' CUUUGGGUCGUCUGUUACAUCGA	8mer	-0.247	0.045	-0.047	-0.045	-0.018	0.005	-0.31	92	2.497	0.60

10.12 EXPANDING miRNA REGULATION OF THE CELL CYCLE USING TarBase AND TargetScan

This analysis path began in Section 10.8, with five miRNAs regulating three proteins (Cip1 or CDKN1A, Kip1 or CDKN1B, and Kip2 or CDKN1C) of the cell cycle, as verified by laboratory experiments (see Figure 10.13). Using additional data from TarBase and predictions from TargetScan, we will now try to expand the network of miRNA regulation of the cell cycle.

If you take the gene symbols of these three proteins and search TargetScanHuman for predicted miRNAs you can generate a list (**Figure 10.18**). Just concentrating on the first three genes of this figure, you notice that miR-24 is in common between Kip1 and Kip2. Could this be another miRNA regulating these transcripts? This is absent from TarBase. But could miR-24 be regulating the translation of any other cell cycle proteins? According to TarBase, miR-24 has been shown to regulate the translation of CDKN2A, also known as INK4A, ARF, or p16. This protein is on the cell cycle diagram from KEGG. So let's add this gene and target sites to the list in Figure 10.18. Other miRNAs or genes of interest can be gathered by systematically using the predicted miRNA target sites as queries in TarBase. The TarBase list of genes can, in turn, be used as queries in KEGG to identify members of the cell cycle pathway. Here are some findings using this approach.

1 The miRNA shown to target Cip1, miR-17-5P/93/106, is predicted to target E2F1 and *CCND1* (CycD). Both are also shown to be repressed by miR-17-5P according to TarBase. According to TarBase, miR-106a was shown to repress RB1 translation.

2. TarBase also shows that miR-17-5P represses the translation of *RBL2* or p107. TargetScan also predicts that *RBL2* is targeted by miR-25, one of the original miRNAs in this study.

3. According to TarBase, let-7, predicted to target Cip1, represses *MYC* translation.

4. According to TarBase, *E2F3*, *CDK6*, and *CCND1* are targeted by miR-34a. Both *E2F3* and *CDK6* are predicted to be targets of this miRNA by TargetScan. *CCND1* is also targeted by miR-let-7b according to TarBase.

TargetScan predictions allowed us to make directed searches of TarBase to identify proteins of interest in the KEGG cell cycle pathway. These proteins, all verified to be targets of miRNAs discussed here, are highlighted in **Figure 10.19**. The exception is *MCM7* but it is highlighted because its expression is tied to the

Gene	Verified and predicted miRNA target sites
p21 (Cip1 or *CDKN1A*)	miR-17-5P/93/106, let-7, miR-221/222/222ab/1928, miR-130, miR-132
p27 (Kip1 or *CDKN1B*)	miR-221, miR-222, miR-196ab, miR-200bc, miR-24
p57 (Kip2 or *CDKN1C*)	miR-221, miR-222, miR-25, miR-199, miR-129, miR-26a, miR-24, miR-34a
RB1	let-7, miR-26ab, miR-17-5P/93/106, miR-132, miR-7
CDKN2A (INK4A, ARF, or p16)	miR-24, (no conserved sites predicted)
E2F1	miR-205, miR-17-5P/93/106
CCND1 (CYCD)	miR-17-5P/93/106, let-7, miR-34a, miR-33, miR-15, miR-96, miR-503, miR-425, miR-365, miR-23ab, miR-193ab, miR-19, miR-155, miR-1
p107 (*RBL2*)	miR-17-5P/93/106, miR-23ab, miR-25
MYC	let-7, (no conserved sites predicted)
E2F3	miR-34a, miR-15, miR-138, miR-499, miR-30a, miR-25, miR-200bc, miR-203, miR-141, miR-10
CDK6	miR-34a, miR-33, miR-138, miR-29abc, miR-26ab, miR-218, miR-191, miR-145, miR-139-5P, miR-137

Figure 10.18 Genes with verified and predicted miRNA target sites. Experimentally verified miRNA-targeted transcripts are highlighted. All other target sites are predicted by TargetScan. Genes above the dashed line were the original genes identified by the gastric cancer study. TargetScan groups miR-106 and miR-93 because they are in the same family. Also grouped by TargetScan are miR-221 and miR-222.

Figure 10.19 KEGG cell cycle pathway mRNAs targeted by miRNAs. This is a close-up of the KEGG cell cycle pathway, seen in full in Figure 10.11, with highlighted proteins indicating where TarBase and TargetScan data were combined to link the proteins to a network of miRNA regulation.

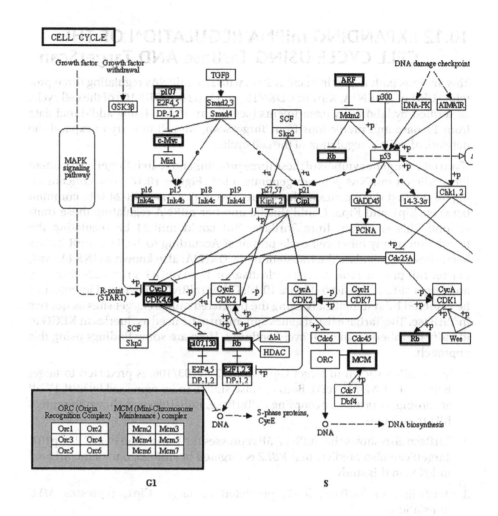

expression of the mir-25, mir-93, and mir-106b genes: an intron of *MCM7* has these miRNA genes. The number of pathway entities of interest, perhaps cooperatively under the control of a handful of miRNAs, has gone from three in Figure 10.11 to fourteen in Figure 10.19. Laboratory experiments could be designed around these miRNAs and target mRNAs to investigate possible links to gastric cancer. For example:

- miR-24 was shown to repress *CDKN2A*, and is predicted to target p27 and p57.

- miR-25 was shown to repress p57, and is predicted to target p107 and *E2F3*.

- miR-34a was shown to repress *CCND1*, *E2F3*, and *CDK6*, and is predicted to target p57.

- let-7 was shown to repress *MYC* and *CCND1*, and is predicted to target p21, *RB1*, and *CCND1*.

Returning to the TargetScan home page, the query form (shown in Figure 10.14) allows a search for all predicted target sites for individual miRNAs. Reset the page to be sure no other choices are made and clear the "Gene symbol" field of any entry. Choose miR-221/222/222ab/1928 from the broadly conserved drop-down menu and hit Submit. A table listing predicted target sites is shown, a portion of which is seen in **Figure 10.20**. One of our genes of interest, *CDKN1B*, is near the top of the list that is sorted by the column "Total context score." Because there are two predicted miR-221/222/222ab/1928 sites in *CDKN1B*, the two best context scores (–0.32 and –0.35 of miR-222) are added together. It is interesting that there are two predicted target sites from the same miRNA in this single UTR. If both are targeted by miRNAs, perhaps it is very important that the translation of this transcript is reduced by these miRNAs.

Human | miR-221/222/222ab/1928
446 conserved targets, with a total of **465** conserved sites and **154** poorly conserved sites.
Table sorted by total context score   [Sort table by aggregate P$_{CT}$]
Genes with only poorly conserved sites are not shown   [View top predicted targets, irrespective of site conservation]
The table shows at most one transcript per gene, selected for having the highest aggregate P$_{CT}$ (or the one with the longest 3' UTR, in case of a tie).   [Show all transcripts]

Target gene	Representative transcript	Gene name	Conserved sites				Poorly conserved sites				Representative miRNA	Total context+ score	Aggregate P$_{CT}$	Previous TargetScan publication(s)	Links to sites in UTRs
			total	8mer	7mer-m8	7mer-1A	total	8mer	7mer-m8	7mer-1A					
SNX4	NM_003794	sorting nexin 4	1	1	0	0	1	1	0	0	hsa-miR-222	-0.74	<0.1	2009	Sites in UTR
RGS6	NM_001204416	regulator of G-protein signaling 6	2	2	0	0	1	0	0	1	hsa-miR-222	-0.87	0.48	2009	Sites in UTR
CDKN1B	NM_004064	cyclin-dependent kinase inhibitor 1B (p27, Kip1)	2	2	0	0	0	0	0	0	hsa-miR-222	-0.67	0.74	2005, 2007, 2009	Sites in UTR
CXCL12	NM_000609	chemokine (C-X-C motif) ligand 12	1	1	0	0	1	0	1	0	hsa-miR-222	-0.67	0.27	2009	Sites in UTR
TCF12	NM_003205	transcription factor 12	1	1	0	0	1	0	1	0	hsa-miR-221	-0.63	0.58	2005, 2007, 2009	Sites in UTR
RFX7	NM_022841	regulatory factor X, 7	2	2	0	0	0	0	0	0	hsa-miR-222	-0.63	0.77	2007, 2009	Sites in UTR
OSTM1	NM_014028	osteopetrosis associated transmembrane protein 1	1	1	0	0	1	1	0	0	hsa-miR-222	-0.61	<0.1	2009	Sites in UTR
SEC62	NM_003262	SEC62 homolog (S. cerevisiae)	2	0	1	1	1	0	0	1	hsa-miR-221	-0.61	0.46	2005, 2007, 2009	Sites in UTR
ZNF181	NM_001029997	zinc finger protein 181	1	1	0	0	1	1	0	0	hsa-miR-222	-0.60	0.12	2009	Sites in UTR
MYBL1	NM_001080416	v-myb myeloblastosis viral oncogene homolog (avian)-like 1	1	1	0	0	2	0	1	1	hsa-miR-222	-0.57	0.36	2007, 2009	Sites in UTR
GABRA1	NM_000806	gamma-aminobutyric acid (GABA) A receptor, alpha 1	2	1	1	0	1	0	0	1	hsa-miR-222	-0.56	0.53	2005, 2007, 2009	Sites in UTR
BEAN1	NM_001136106	brain expressed, associated with NEDD4, 1	1	1	0	0	1	0	0	1	hsa-miR-221	-0.56	0.31		Sites in UTR
HECTD2	NM_182765	HECT domain containing 2	1	1	0	0	1	0	1	0	hsa-miR-222	-0.55	0.52	2005, 2007, 2009	Sites in UTR
PHACTR4	NM_001048183	phosphatase and actin regulator 4	1	1	0	0	1	0	0	1	hsa-miR-221	-0.55	0.12	2009	Sites in UTR
KIT	NM_000222	v-kit Hardy-Zuckerman 4 feline sarcoma viral oncogene homolog	1	1	0	0	1	1	0	0	hsa-miR-222	-0.55	0.44	2007, 2009	Sites in UTR
VGLL4	NM_001128219	vestigial like 4 (Drosophila)	2	1	1	0	0	0	0	0	hsa-miR-222	-0.54	0.23	2005, 2007, 2009	Sites in UTR
KCNQ3	NM_001204824	potassium voltage-gated channel, KQT-like subfamily, member 3	2	1	1	0	1	0	0	1	hsa-miR-222	-0.54	0.29		Sites in UTR
WSB2	NM_018639	WD repeat and SOCS box containing 2	1	1	0	0	1	0	0	1	hsa-miR-222	-0.53	0.46	2009	Sites in UTR
WDR35	NM_001006657	WD repeat domain 35	1	1	0	0	0	0	0	0	hsa-miR-222	-0.50	0.21	2009	Sites in UTR
PAIP1	NM_006451	poly(A) binding protein interacting protein 1	1	1	0	0	1	0	0	1	hsa-miR-221	-0.50	0.25	2005, 2007, 2009	Sites in UTR
CPEB3	NM_001178137	cytoplasmic polyadenylation element binding protein 3	2	1	1	0	0	0	0	0	hsa-miR-222	-0.49	0.40	2009	Sites in UTR
TP53BP2	NM_001031685	tumor protein p53 binding protein, 2	1	0	0	1	2	0	1	1	hsa-miR-222	-0.49	0.31	2009	Sites in UTR
NAP1L5	NM_153757	nucleosome assembly protein 1-like 5	1	1	0	0	1	0	1	0	hsa-miR-222	-0.49	0.23	2005, 2007, 2009	Sites in UTR
ZNF25	NM_145011	zinc finger protein 25	1	1	0	0	1	0	0	1	hsa-miR-221	-0.48	<0.1		Sites in UTR

The table in Figure 10.20 shows how many potential target sites exist for this single miRNA family. They are not all from genes involved in the cell cycle. Visually scanning the list, genes from nerves, muscles, and reproductive tissue are present. Exploring the gene symbol links (they are linked to NCBI Gene records) you find proteins involved in many functions including signal transduction, protein breakdown, and in cell structures.

Figure 10.20 TargetScan predicted target sites for human miR-221/222/222ab/1928. These miRNAs, acting on the same target site sequence, are predicted to repress hundreds of transcripts. This search was launched from the TargetScan home page, under "Broadly conserved miRNA families."

10.13 MAKING SENSE OF miRNAs AND THEIR MANY PREDICTED TARGETS

If you use TargetScan to find the predicted targets for our growing list of miRNAs (see **Figure 10.21**), there are thousands of target sites predicted. If only 10% of these were real sites, this would still leave hundreds of transcripts being regulated by miRNAs. And this is only for six miRNAs—remember there are hundreds! With all these potential translation-repression events taking place, how is *any* translation taking place? Can sense be made of it all?

First, you should remember that we are at the beginning of our understanding of miRNA action, coordination, and function so we are far from understanding how to predict their target sites accurately. In addition, the details of well-studied molecular functions such as transcriptional regulation are still being worked out so we have a long way to go before miRNA regulation of gene expression is well defined. However, a number of statements can be made to try to simplify this emerging complex world of miRNAs.

1. In order for translation repression to take place, both the mature miRNA and targeted mRNA must be present in the same cell. This is obvious, but remembering this statement will unclutter the possibilities of regulation when faced with many hundreds of predicted targets. For example, a liver-specific transcript is most probably not found in lung tissue. Then again, if that gene is transcribed at a low level in the wrong tissue, perhaps miRNAs function at keeping "inappropriate" translations to a minimum.

miRNA	Conserved targets
hsa-miR-25	893
hsa- miR-17-5P/93/106	1220
hsa- miR-221/222/222ab/1928	446
hsa-miR-24	631
hsa-miR-34a	655
let-7	1072

Figure 10.21 Total TargetScan predictions for individual miRNAs. Each miRNA is predicted to target hundreds of mRNAs.

2. There are transcription factors controlling the transcription of both the miRNAs and their potential mRNA targets. Besides tissue specificity, mentioned above, genes have to be transcribed at the correct time in development, in the cell cycle, and in response to stimuli such as hormones and metabolites. Therefore, transcription plays a major role in orchestrating the potential miRNA–mRNA interactions.

3. MicroRNAs can repress translation through the breakdown of mRNA and/or the reduction in translation. These should not be thought of as "all or nothing" events. A chemist's view of biology says that dramatic things can happen if you change the quantities of protein rather than eliminate them all together. Perhaps reducing or increasing a protein's level by 10 or 20% can make the difference between something important happening in a cell or not.

10.14 miRNAs ASSOCIATED WITH DISEASES

Since their discovery, miRNAs have been linked to many diseases and their affected pathways, and the literature on these links keeps growing. To briefly touch on this topic, a simple search at a Website dedicated to miRNAs and human disease will be performed. Specifically, we will search for links between miRNAs and gastric cancer to complement the work performed above.

The miR2Disease Website (www.mir2disease.org) gives you the opportunity to search by miRNA name, disease name, and target gene (**Figure 10.22**). There are over 160 diseases and over 300 miRNAs represented in the database. Over one-third of the diseases in this database are forms of cancer, most probably reflecting the number of cancers identified (there are over 200) and the abundance of cancer research. Many miRNA targets are identified and verified using tissue culture cells, and there is a large number of cancer cell lines that are amenable to these miRNA studies.

"Stomach cancer" was entered as a search term in the "Search by disease name" box (not shown) and the results were examined. Published studies are listed, describing associations between miRNA and gastric cancer. This list includes

Figure 10.22 The miR2Disease database search page.
This database is a large collection of miRNA–human disease associations based on published findings. By selecting "Search" from the left panel on the home page, you may search on the name of miRNAs, targeted mRNA, or disease.

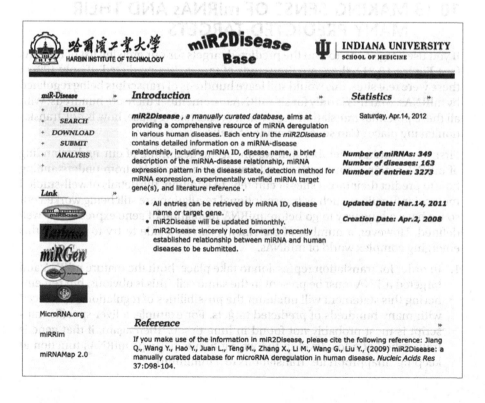

miR-24 and miR-34a, two miRNAs that were identified as interesting in the above TarBase/TargetScan predictions. Clicking on the miRNA name in any row brings up a new results list containing all diseases associated with that particular miRNA. For example, clicking on miR-103 brings up a listing of cardiac hypertrophy, neurodegeneration, liver disease, and seven forms of cancer.

The "Detail" hypertext in the table takes you to a summary of the study: methods, expression patterns, and validated targets along with a convenient link to the paper. For example, miR-1 was shown to be part of a five-miRNA "signature" of miRNAs found in the blood serum of patients with gastric cancer. This miRNA was elevated in the serum of patients with the disease. The authors argue that this signature could possibly be used to diagnose or follow the progression of the disease in patients. According to this miR2Disease record, the authors found possible gene targets for miR-1 in TarBase and links to these data also appear in this record.

Finally, in a 2007 paper by Saunders, Liang, and Li, many laboratory-verified and hundreds of predicted human miRNA target sites were found to be at the location of single nucleotide polymorphisms (SNPs), possibly altering or eliminating the recognition by their cognate miRNAs. Hundreds more SNPs were predicted to create new target sites. This raises the very interesting possibility that natural sequence variation in the form of SNPs could be having subtle to profound effects on gene expression, leading to functional variation or disease. As you will see in the Exercises at the end of this chapter, there is experimental evidence that the recent creation of a novel miRNA target site has occurred in nature, leading to a significant change in the phenotype of animals. Considering the many hundreds of miRNA genes and the many thousands of miRNA target sites, it is safe to assume that this type of variation has occurred in either modern human or our ancestors.

10.15 SUMMARY

The focus of this chapter is a single genetic entity: miRNA. These very short RNAs have a profound effect on gene regulation and miRNA function, structure, conservation, and diversity were examined. As miRNAs regulate the expression of biochemical pathways, we explored KEGG, a Website dedicated to the collection and display of biological networks. We then investigated a small network of miRNAs that regulate protein members of the cell cycle. Using a database of laboratory-verified mRNA targets, TarBase, and a Website which predicts miRNA target sites, TargetScan, we attempted to expand the network of miRNAs affecting the cell cycle. Finally, miR2Disease Base was introduced as an extensive collection of diseases linked to miRNAs.

EXERCISES

GDF8

GDF8 is a secreted protein and acts as a negative regulator of skeletal muscle growth. There are several spectacular examples of naturally occurring *GDF8* mutations where, in the absence of this negative regulation, there is hypertrophy of muscle tissue and body fat is decreased. The result is a very muscular animal, larger and stronger than related members of the species. You are encouraged to search the Internet for images of the animals described below.

The Belgian Blue cow was first documented to be "double-muscled" in the early 1800s. The *GDF8* gene of these animals has an 11-nucleotide deletion within the coding region of the third exon. The resulting frameshift in translation generates a C-terminally truncated protein. Belgian Blue cattle have twice the number of muscle fibers, very low body fat, and up to 20% more muscle tissue. Although favorable for the meat industry, the calves are often so large that births are difficult.

The Whippet dog breed is known for its speed and has been measured to run at over 35 miles per hour. A common variation within whippets is the "bully whippet" which is a larger and very muscular dog. Similar to the Belgian Blue cow, bully whippets have a two-nucleotide deletion within the third exon of *GDF8*, resulting in a frameshift. Heterozygotes for this mutated gene are stronger in appearance and, as a group, run faster than whippets not carrying the mutation.

There is at least one documented case of GDF8 deficiency in humans resulting in increased strength. In this case, there is a G-to-A mutation in an intron splice donor site and the resulting mis-splice generates a truncated protein. The young child born with this base change is observed to have significant muscle size and strength and no health issues have been reported thus far. A number of groups have proposed that pharmaceutically reducing GDF8 levels may provide treatments of muscle-wasting disorders such as muscular dystrophy.

Texel sheep are known for their heavily muscled size and lean body mass, making them a popular breed for the meat industry. In 2006, Clop et al. reported that these sheep possessed a single nucleotide change, G to A, within the 3P UTR of *GDF8* and this mutation was responsible for the increase in muscle mass. Further analysis showed that this small sequence change creates a miR-1 target site at this location resulting in translation inhibition of *GDF8*. Has nature experimented with miRNA-mediated changes in muscle mass before? Since miRNAs are known to work in networks, involving multiple targets and multiple miRNAs, this exercise will require you to find the pathway where GDF8 is thought to act and then identify possible targets for miR-1 and other miRNAs.

1. Using the NCBI Gene page, identify synonyms for human *GDF8*.

2. Using references found in PubMed, identify the receptor for GDF8. Be sure you have identified the correct receptor.

3. Using either the *GDF8* synonyms or the name of the receptor, identify the pathway in KEGG where GDF8 is thought to act.

4. This pathway is large so in the following steps, concentrate on the restricted part of the pathway that is most directly involved with GDF8 action. For the molecules in this part of the pathway, identify their gene symbols by floating your mouse over the KEGG icons and record your findings in a list. Be sure to list all the separate entities in protein complexes. This will be referred to as your "gene list."

5. Using TargetScan, identify all members of this gene list having predicted miR-1 sites.

6. Using TarBase, identify the transcripts, if any, repressed by miR-1. Since you are investigating only a handful of genes, a good approach would be to search for those genes in the gene drop-down menu and then look at their associated miRNAs.

7. Is miR-1 transcribed as a precursor that is associated with any other genes? If so, try to identify a possible link between these linked gene(s) and *GDF8*, its pathway, or its action on skeletal muscle.

8. Using PubMed and the miR2Disease database, try to find links to skeletal muscle tissue and the miRNAs you have worked with in this exercise.

FURTHER READING

Allen DL, Bandstra ER, Harrison BC et al. (2009) Effects of spaceflight on murine skeletal muscle gene expression. *J. Appl. Physiol.* 106, 582–595. An interesting article, first because it involved gene expression in spaceflight. Nicely written and contains many procedures and data associated with their findings. Describes the effects on GDF8 expression.

Bartel DP (2009) MicroRNAs: target recognition and regulatory functions. *Cell* 136, 215–233. A fantastic article written by an expert in the field.

Clop A, Marcq F, Takeda H et al. (2006) A mutation creating a potential illegitimate microRNA target site in the myostatin gene affects muscularity in sheep. *Nat. Genet.* 38, 813–818. This very interesting finding is the subject of the exercise at the end of the chapter.

Elkasrawy MN & Hamrick MW (2010) Myostatin (GDF-8) as a key factor linking muscle mass and bone structure. *J. Musculoskelet. Neuronal Interact.* 10, 56–63. Muscles and the underlying support, bone, should develop simultaneously and this article describes this association linked by GDF8.

Grimson A, Farh KK, Johnston WK et al. (2007) MicroRNA targeting specificity in mammals: determinants beyond seed pairing. *Mol. Cell* 27, 91–105. One of the TargetScan articles, describing details of the algorithm.

Gupta A, Gartner JJ, Sethupathy P et al. (2006) Anti-apoptotic function of a microRNA encoded by the HSV-1 latency-associated transcript. *Nature* 442, 82–85. This article describes how viruses utilize miRNAs to attack us.

Jiang Q, Wang Y, Hao Y et al. (2009) miR2Disease: a manually curated database for microRNA deregulation in human disease. *Nucleic Acids Res.* 37, D98–D104. This article provides information on a collation of literature associated with miRNA and disease.

Kanehisa M, Goto S, Furumichi M et al. (2010) KEGG for representation and analysis of molecular networks involving diseases and drugs. *Nucleic Acids Res.* 38, D355–D360. Biochemical pathways along with a tremendous amount of associated information.

Kim YK, Yu J, Han TS et al. (2009) Functional links between clustered microRNAs: suppression of cell-cycle inhibitors by microRNA clusters in gastric cancer. *Nucleic Acids Res.* 37, 1672–1681. We perform extensive analysis in the chapter based on the original findings described in this article.

Kozomara A & Griffiths-Jones S (2011) miRBase: integrating microRNA annotation and deep-sequencing data. *Nucleic Acids Res.* 39, D152–D157. miRBase is the reference miRNA database used worldwide.

Lewis BP, Shih IH, Jones-Rhoades MW et al. (2003) Prediction of mammalian microRNA targets. *Cell* 115, 787–798. This is an early publication about the development of TargetScan.

Mosher DS, Quignon P, Bustamante CD et al. (2007) A mutation in the myostatin gene increases muscle mass and enhances racing performance in heterozygote dogs. *PLoS Genet.* 3: e79 (DOI:10.1371/journal.pgen.0030079). Why do these dogs run so fast? Scientists identify the gene associated with this characteristic in dogs.

Papadopoulos GL, Reczko M, Simossis VA et al. (2009) The database of experimentally supported targets: a functional update of TarBase. *Nucleic Acids Res.* 37, D155–D158. TarBase is a well-known collection of experimental links between miRNAs and translation repression.

Rodriguez A, Griffiths-Jones S, Ashurst JL & Bradley A (2004) Identification of mammalian microRNA host genes and transcription units. *Genome Res.* 14, 1902–1910. A great article describing miRNA gene locations.

Saunders MA, Liang H & Li WH (2007) Human polymorphism at microRNAs and microRNA target sites. *Proc. Natl Acad. Sci. USA* 104, 3300–3305. As the exercise at the end of this chapter demonstrates, one base change can lead to the gain or loss of miRNA repression. There are many cases of this in human.

Xu J & Wong C (2008) A computational screen for mouse signaling pathways targeted by microRNA clusters. *RNA* 14, 1276–1283. This paper links clusters of miRNAs to genes that are coordinately expressed.

ACAAGGGACTAGAGAAACCAAAA
AGAAACCAAAACGAAAGGTGCAGAA
AACGAAAGGTGCAGAAGGGGAAACAGATGCAGA
GAAGGGGAAACAGATGCAGAAAGCAT
AGAAAGCAT
ACAAGGGACTAGAGAAACCAAAACGAAAGGTGCAGAAGGGGAAACAGATGCAGAAAGCAT
AGAAACCAAAACGAAAGGTGCAGAA
AACGAAAGGTGCAGAAGGG
GAAGGG

CHAPTER 11

Multiple Sequence Alignments

Key concepts

- Aligning sequences from the NCBI BLAST tool
- ClustalW alignments from the ExPASy Website
- ClustalW alignments from the EBI Website
- Comparison of ClustalW, MUSCLE, and COBALT alignments
- Aligning isoforms
- Aligning paralogs, orthologs, and domains
- Manual editing of multiple sequence alignments

11.1 INTRODUCTION

In earlier chapters you viewed pairwise sequence alignments generated by BLAST. The various versions of BLAST (for example, BLASTN or BLASTP) aligned the query to the subject and you were able to see the individual differences between the two sequences. But if you had multiple hits to many paralogs, for example, it was difficult to identify sequence held in common between all the sequences without scrolling the browser window up and down to look at the alignment details. In this chapter, through the use of multiple sequence alignments, you will be able to easily see the relationships between sequences: their similarities, highly conserved regions, hotspots for change, and the relationships between isoforms.

There is no "best" way to generate multiple sequence alignments. You will use a common set of Web tools in this chapter, but you should be prepared to vary the parameters and change tools as you systematically hunt for the right way to solve your particular problem.

11.2 MULTIPLE SEQUENCE ALIGNMENTS THROUGH NCBI BLAST

The NCBI BLAST page and other Websites offer the ability to generate a single alignment between the hits and the query. To explore this feature, we will use human beta globin protein as a BLASTP query against human RefSeq proteins. Because of the strong similarity between the globin proteins, this will not be a challenging alignment to generate or interpret but will serve as a good beginning for multiple sequence alignment viewing and analysis.

Before generating the multiple sequence alignment, you should first explore the pairwise alignments generated by the BLAST and choose the sequences you wish

Accession	Description	Max score	Total score	Query coverage	△ E value	Max ident	Links
NP_000509.1	hemoglobin subunit beta [Homo sapiens] >gi|55635219|ref|XP_5082¢	301	301	100%	7e-108	100%	U G M
NP_000510.1	hemoglobin subunit delta [Homo sapiens]	284	284	100%	3e-101	93%	U G M
NP_005321.1	hemoglobin subunit epsilon [Homo sapiens]	240	240	100%	5e-84	76%	U G M
NP_000175.1	hemoglobin subunit gamma-2 [Homo sapiens]	235	235	100%	5e-82	73%	U G M
NP_000550.2	hemoglobin subunit gamma-1 [Homo sapiens]	232	232	100%	8e-81	73%	U G M
NP_000508.1	hemoglobin subunit alpha [Homo sapiens] >ref|NP_000549.1| hemogl	114	114	97%	7e-35	43%	U G M
NP_005323.1	hemoglobin subunit zeta [Homo sapiens]	100	100	97%	2e-29	36%	U G M
NP_005322.1	hemoglobin subunit theta-1 [Homo sapiens]	97.1	97.1	97%	4e-28	39%	U G M
NP_001003938.1	hemoglobin subunit mu [Homo sapiens]	88.6	88.6	97%	6e-25	35%	U G M
NP_599030.1	cytoglobin [Homo sapiens]	72.8	72.8	86%	1e-18	29%	U G M
NP_056025.2	coiled-coil domain-containing protein 165 [Homo sapiens]	28.9	28.9	31%	0.005	36%	U G M
NP_001017526.1	rho GTPase-activating protein 8 isoform 1 [Homo sapiens]	27.7	27.7	57%	0.010	27%	G M
NP_004186.1	tumor necrosis factor receptor superfamily member 18 isoform 1 precu	26.2	26.2	34%	0.031	34%	U G M
NP_060101.3	probable ATP-dependent RNA helicase DDX60 [Homo sapiens]	26.2	26.2	10%	0.032	63%	U G M
NP_001099038.1	kinesin-like protein KIF13A isoform d [Homo sapiens]	26.2	26.2	28%	0.033	36%	U G M
NP_001099037.1	kinesin-like protein KIF13A isoform c [Homo sapiens]	26.2	26.2	28%	0.033	36%	U G M
NP_001099036.1	kinesin-like protein KIF13A isoform b [Homo sapiens]	26.2	26.2	28%	0.033	36%	U G M
NP_071396.4	kinesin-like protein KIF13A isoform a [Homo sapiens]	26.2	26.2	28%	0.033	36%	U G M

Figure 11.1 BLASTP search of human RefSeq proteins with human beta globin, NP_000509, as a query.

to align. It doesn't make sense to consider insignificant hits to your query, and any multiple sequence alignment including these weak hits will obscure any value gained from seeing all the significant sequences aligned. The BLASTP results (**Figure 11.1**) indicate that globin-like proteins range from E values 7e–108 to 1e–18. The rest of the hits are in the insignificant range for E value and the alignments are weak. For this example, we will focus on the hits from the top to alpha globin, at 7e–35.

First, we'll filter these results, only showing the described significant hits. To do so, click on the "Formatting options" link near the top of the BLAST results page. A number of options can be chosen (**Figure 11.2**). Near the bottom of this window are two fields where you can enter E value minimum and maximum values, for example "0" and "7e–34." Clicking on the "Reformat" button, seen in the upper right of Figure 11.2, applies this filter to your results and scrolling down reveals that the last hit on the results list is now hemoglobin subunit alpha.

To generate a multiple sequence alignment of just these hits we'll now use the "Alignment View" drop-down menu in this same "Formatting options" window. The default for BLAST is "Pairwise," the view we have commonly seen with BLAST results. Clicking on the down arrow for this menu reveals a number of choices, one of which is "Query-anchored with letters for identities." Choosing this and clicking on the "Reformat" button again generates an alignment (**Figure 11.3**).

Figure 11.2 NCBI BLASTP formatting options. The NCBI BLAST tools allow you to change the format of the output.

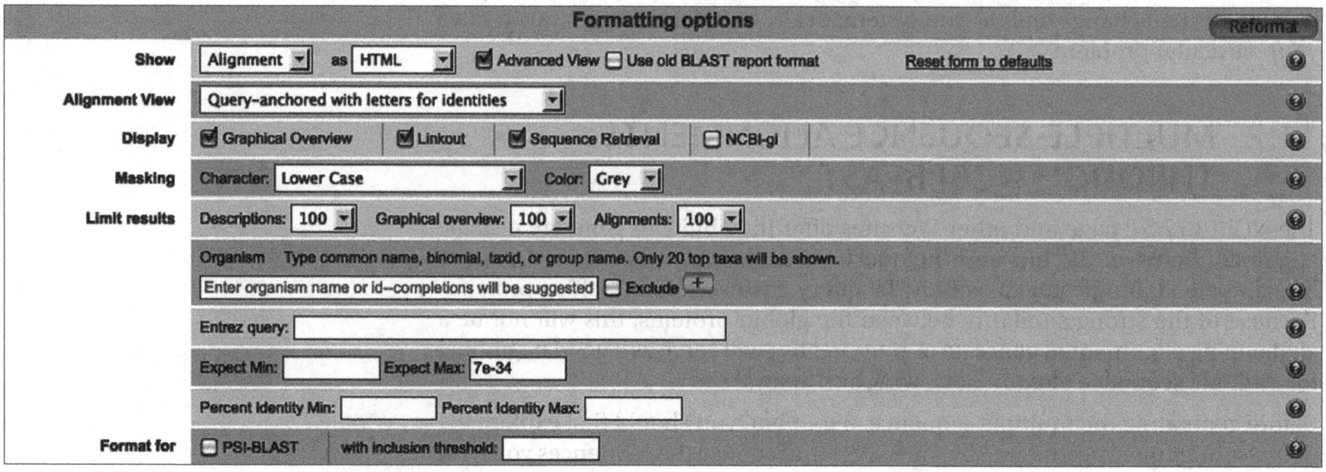

Now, with little effort, you can see all these sequences aligned to each other. Many positions are invariant, for example there is lysine (K) in position nine of the query

	Formatting options		Reformat
Show	Alignment ▾ as HTML ▾ ☑ Advanced View ☐ Use old BLAST report format	Reset form to defaults	❓
Alignment View	Query-anchored with letters for identities ▾		❓
Display	☑ Graphical Overview ☑ Linkout ☑ Sequence Retrieval ☐ NCBI-gi		❓
Masking	Character: Lower Case ▾ Color: Grey ▾		❓
Limit results	Descriptions: 100 ▾ Graphical overview: 100 ▾ Alignments: 100 ▾		❓
	Organism Type common name, binomial, taxid, or group name. Only 20 top taxa will be shown.		
	Enter organism name or id--completions will be suggested ☐ Exclude ⊞		❓
	Entrez query:		❓
	Expect Min: Expect Max: 7e-34		❓
	Percent Identity Min: Percent Identity Max:		❓
Format for	☐ PSI-BLAST with inclusion threshold:		❓

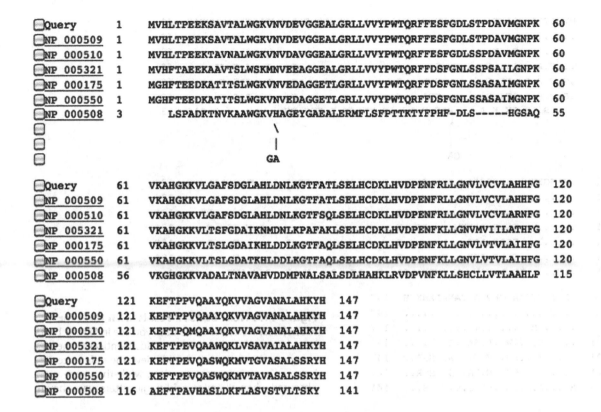

Query	1	MVHLTPEEKSAVTALWGKVNVDEVGGEALGRLLVVYPWTQRFFESFGDLSTPDAVMGNPK	60
NP_000509	1	MVHLTPEEKSAVTALWGKVNVDEVGGEALGRLLVVYPWTQRFFESFGDLSTPDAVMGNPK	60
NP_000510	1	MVHLTPEEKTAVNALWGKVNVDAVGGEALGRLLVVYPWTQRFFESFGDLSSPDAVMGNPK	60
NP_005321	1	MVHFTAEEKAAVTSLWSKMNVEEAGGEALGRLLVVYPWTQRFFDSFGNLSSPSAILGNPK	60
NP_000175	1	MGHFTEEDKATITSLWGKVNVEDAGGETLGRLLVVYPWTQRFFDSFGNLSSASAIMGNPK	60
NP_000550	1	MGHFTEEDKATITSLWGKVNVEDAGGETLGRLLVVYPWTQRFFDSFGNLSSASAIMGNPK	60
NP_000508	3	LSPADKTNVKAAWGKVHAGEYGAEALERMFLSFPTTKTYFPHF-DLS-----HGSAQ	55

GA

Query	61	VKAHGKKVLGAFSDGLAHLDNLKGTFATLSELHCDKLHVDPENFRLLGNVLVCVLAHHFG	120
NP_000509	61	VKAHGKKVLGAFSDGLAHLDNLKGTFATLSELHCDKLHVDPENFRLLGNVLVCVLAHHFG	120
NP_000510	61	VKAHGKKVLGAFSDGLAHLDNLKGTFSQLSELHCDKLHVDPENFRLLGNVLVCVLARNFG	120
NP_005321	61	VKAHGKKVLTSFGDAIKNMDNLKPAFAKLSELHCDKLHVDPENFKLLGNVMVIILATHFG	120
NP_000175	61	VKAHGKKVLTSLGDAIKHLDDLKGTFAQLSELHCDKLHVDPENFKLLGNVLVTVLAIHFG	120
NP_000550	61	VKAHGKKVLTSLGDATKHLDDLKGTFAQLSELHCDKLHVDPENFKLLGNVLVTVLAIHFG	120
NP_000508	56	VKGHGKKVADALTNAVAHVDDMPNALSALSDLHAHKLRVDPVNFKLLSHCLLVTLAAHLP	115

Query	121	KEFTPPVQAAYQKVVAGVANALAHKYH	147
NP_000509	121	KEFTPPVQAAYQKVVAGVANALAHKYH	147
NP_000510	121	KEFTPQMQAAYQKVVAGVANALAHKYH	147
NP_005321	121	KEFTPEVQAAWQKLVSAVAIALAHKYH	147
NP_000175	121	KEFTPEVQASWQKMVTGVASALSSRYH	147
NP_000550	121	KEFTPEVQASWQKMVTAVASALSSRYH	147
NP_000508	116	AEFTPAVHASLDKFLASVSTVLTSKY	141

sequence (the coordinates are displayed on the left and right of the blocks), while position ten can be an S, T, or A. The alpha globin sequence (NP_000508) in this alignment does not begin with a methionine like the other globins and appears one amino acid short of the others at the C-terminus, too. These amino acids are missing because there was insufficient similarity at the ends of the sequences for BLAST; they were absent from the pairwise BLASTP alignment too.

The alignment of alpha globin also offers a view of how gaps and insertions are handled in this BLAST alignment. Like pairwise alignments, gaps are represented as "-," as seen in the latter part of the first block. Notice the gaps introduced into NP_000508 (hemoglobin subunit alpha). Without those gaps at that location, sequence past this point would have been out of alignment, hiding the obvious similarity between the alpha globin protein and the other globins. This is routine for BLAST alignments and appears in this multiple sequence alignment that is based on those BLAST results. In this case, the placement of the "DLS" trio of amino acids in the gapped area is the best choice. The alpha globin protein is missing a glycine, found in all the other proteins, and the single gap aligns DLS to the DLS found in the first three proteins (the query, NP_000509, and NP_000510).

Insertions are handled differently. Because this alignment is "Query-anchored," the query sequence is uninterrupted. Therefore, the insertion of two amino acids, GA, between amino acids V and H in alpha globin is represented as diagonal and vertical lines ending with inserted amino acids.

There is another view which is definitely different from that seen in Figure 11.3, and that is "Query-anchored with dots for identities." This view, seen in **Figure 11.4**, uses a dot where there is agreement with the query sequence. Here, the differences between sequences are clearly visible because the amount of information you must survey with your eyes has been greatly reduced. Similarities between sequences are plainly seen, such as NP_000175 (gamma-2 globin) and NP_000550 (gamma-1 globin). The alignment shows that these two sequences are quite similar to each other, and have identical differences from the query and some of the other sequences. Looking at these two lines within the alignment, amino acids G.F.E.D.ATI.S.EDA and others show the near identity of these

Figure 11.3 NCBI BLASTP result with letters for identities.
In this view, called "Query-anchored with letters for identities," the hits are aligned to the query sequence in the first row. The alignments are derived from the BLAST query–hit relationships and the coordinates for these areas of alignment appear on both sides of the sequences. The accession numbers of these sequences appear on the far left. There is an insertion of "GA" in between a valine and a histidine in the first line of NP_000508.

```
Query       1   MVHLTPEEKSAVTALWGKVNVDEVGGEALGRLLVVYPWTQRFFESFGDLSTPDAVMGNPK   60
NP_000509   1   ............................................................   60
NP_000510   1   .........T..N........A..............................S.......   60
NP_005321   1   ...F.A...A...S...S.M..E.A...................D...N..S.S.IL....   60
NP_000175   1   .G.F.E.D.ATI.S.......EDA...T...............D...N..SAS.I.....   60
NP_000550   1   .G.F.E.D.ATI.S.......EDA...T...............D...N..SAS.I.....   60
NP_000508   3    .S.AD.TN.K.A....HAG.Y.A...E.MFLSF.T.KTY.PH.-...-----H.SAQ   55
                                    \
                                     |
                                     GA


Query      61   VKAHGKKVLGAFSDGLAHLDNLKGTFATLSELHCDKLHVDPENFRLLGNVLVCVLAHHFG   120
NP_000509  61   ............................................................   120
NP_000510  61   .........................SQ...........................RN..   120
NP_005321  61   .........TS.G.AIKNM....PA..K................K.....M.II..T...   120
NP_000175  61   .........TSLG.AIK...D......Q...............K.......T...I...   120
NP_000550  61   .........TSLG.ATK...D......Q...............K.......T...I...   120
NP_000508  56   ..G.....AD.LTNAV..V.DMPNALSA..D..AH..R...V..K..SHC.LVT..A.LP   115


Query     121   KEFTPPVQAAYQKVVAGVANALAHKYH   147
NP_000509 121   ...........................   147
NP_000510 121   .....QM....................   147
NP_005321 121   .....E....W...L.SA..I.......   147
NP_000175 121   .....E...SW..M.T...S..SSR..   147
NP_000550 121   .....E...SW..M.TA..S..SSR..   147
NP_000508 116   A....A.H.SLD.FL.S.STV.TS..   141
```

Figure 11.4 NCBI BLASTP result with dots for identities. Here, the same data you see in Figure 11.3 have been reduced and amino acids identical to the query are shown as dots. In this view, the differences between sequences are very clear, while they were obscured when all amino acid single letters were shown.

lines to each other. In fact, a single alanine is the only visible difference between gamma-1 globin and gamma-2 globin (can you find it?). This easy browsing of the sequence comparison would not have been possible with letters for identities.

One of the goals of the field of bioinformatics is **data reduction**. With the literal deluge of data, the only way for us to possibly interpret the information is that the data quantity is reduced to manageable levels. Viewing the multiple sequence alignment with dots for identities is a simple but powerful way that demonstrates how the data we look at can be filtered to the essence of what is needed.

11.3 ClustalW FROM THE ExPASy WEBSITE

When viewing UniProtKB records at the ExPASy Website, there is access to an alignment tool called ClustalW (pronounced "clust-al double-you"). This multiple sequence alignment tool is widely used throughout the world and many Websites run their own customized version. It has undergone over 20 years of development, with incremental improvements of the algorithm and the addition of new features. The version at the ExPASy Website can take advantage of the wealth of annotation that appears in many of the UniProtKB records and this will be demonstrated here.

First, let's retrieve the same globin sequences that we used in the above example. From the UniProtKB query window, enter "sapiens globin blood" and the results will have all the major human blood proteins near the top of the list, as seen in **Figure 11.5**. Check the boxes to the left of the UniProtKB accession numbers (also shown in this figure) and then on the bottom-right corner of the Web page, click on the "Align" button. This automatically retrieves the sequences and launches ClustalW. You may also access the align tool via the "Align" tab found on the top of the UniProtKB records (seen at the top of Figure 11.5) and manually enter FASTA formatted files.

When the page refreshes, a multiple sequence alignment is visible (**Figure 11.6**). This multiple sequence alignment is not "query-anchored" (there is no query!)

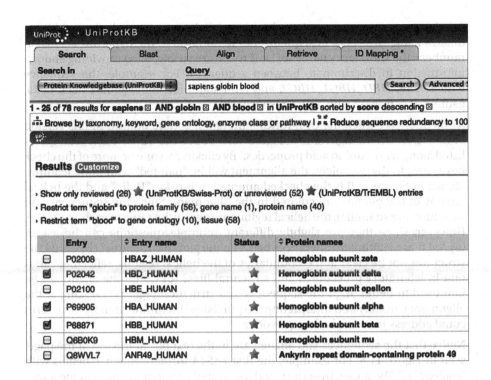

Figure 11.5 UniProtKB hit list from text query. After entering "sapiens globin blood" into the query window, a list of the major human blood proteins are near the top of the list. Checking the boxes adjacent to the UniProtKB accession numbers allows the alignment of these selected sequences.

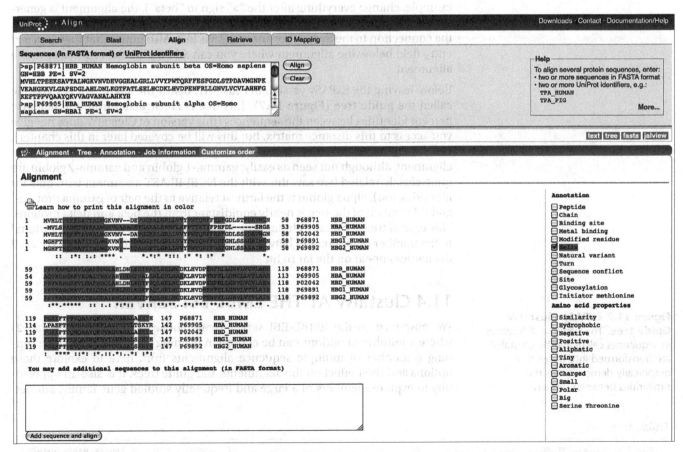

Figure 11.6 UniProtKB multiple sequence alignment. A multiple sequence alignment of UniProtKB files of human globin proteins. This version of ClustalW allows the "painting" of the alignment based on features found in the UniProtKB files (top half of features listed on the right) or amino acid properties (bottom half of choices, on right). Simply clicking the check boxes colors the alignment and featured here are the regions annotated as helical.

so, when necessary, gaps are introduced into all the sequences. Amino acid coordinates flank the blocks of sequence but in addition to the hypertext accession numbers, the short entry names of UniProtKB are visible, often providing enough information to discriminate between sequences. For example, the sequence names *HBB*, *HBD*, *HBG1*, *HBG2*, and *HBA* stand for hemoglobins beta, delta, gamma-1, gamma-2, and alpha, respectively.

As mentioned earlier, the UniProtKB annotation can be utilized in this version of ClustalW. On the right side of Figure 11.6, notice the "Annotation" categories listed along with "Amino acid properties." By clicking on one or more of the check boxes next to these choices, the alignment will be "painted" with these features. Shown in Figure 11.6 is the checked annotation box for "Helix" and the helical regions of the globins are colored in the alignment. Since the sequences and structures are so similar, the helical regions are nearly the same in each sequence. However, since they are slightly different, multiple questions can be asked: (a) Are these legitimate differences between the globin structures? (b) Could the structures not be clear at the boundaries of the helical regions, reflecting variation in laboratory procedures and structural files? (c) Could the alignment be improved to better align the sequences and structural regions? Looking at the alignment carefully and interpreting the published literature on these structures could address these and other questions.

Notice that the sequences you aligned are in the text box at the top of the page, seen in Figure 11.6. For example, the first FASTA format sequence begins with "sp|P68871|." The name, hypertext, and associated annotation found in the alignment are derived from this Swiss-Prot (sp) accession number. Should you modify the text after the ">" for the purposes of changing the names of the sequence (for example, change everything after the ">" sign to "beta"), the alignment is generated but you lose the ability to paint the annotation features on the alignment; the connection to the Swiss-Prot annotation is lost. Also notice that there is a text entry field below the alignment where you can add additional sequences to the alignment.

Before leaving the ExPASy version of ClustalW, below the alignment is a graphic called the **guide tree** (**Figure 11.7**). This tree is constructed from the pairwise percent identities between the sequences (this version of ClustalW does not give you access to this distance matrix, but this will be covered later in this chapter). However, the tree does show what can be gathered from the multiple sequence alignment, although not seen as easily: gamma-1 globin and gamma-2 globin are quite closely related (we saw this with the NCBI BLAST alignment with dots for identities, too), alpha globin is the farthest relative to the pair of gamma proteins, and this branch of the tree is nearly equidistant from the beta and delta proteins. This type of tree is called a **cladogram**. The branch lengths are not proportional to the number of differences between sequences, but have been elongated so all the names appear on the far right.

11.4 ClustalW AT THE EMBL-EBI SERVER

We now turn to the EMBL-EBI server (www.ebi.ac.uk/Tools/msa/clustalw2) where a number of options can be chosen in running ClustalW. We will be running a number of multiple sequence alignments from here to explore these options and their effect on the alignments and guide trees. It is also an opportunity to explore members of a large and frequently studied gene family, kinases.

Figure 11.7 ExPASy ClustalW Guide tree. The pairwise differences in sequences calculated by ClustalW are transformed into a tree that graphically demonstrates the similarities between sequences.

Guide tree

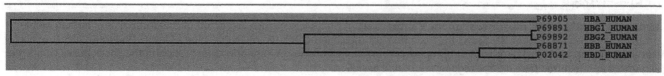

These will be the subjects of many of the alignments, below. Note that the EBI asks "If you plan to use these services during a course please contact us." To answer this request, use the email link under the words 'contact us.'

MARK1 kinase

First, we'll look at orthologous proteins of MARK1 (MAP/microtubule affinity-regulating kinase1). MARK1 proteins and related family members regulate microtubule interactions with other proteins and participate in the formation and disruption of microtubule structures and transport. MARK1 phosphorylates the brain protein tau, changing its affinity for microtubules, leading to the formation of tau aggregates, a widely seen characteristic of Alzheimer's disease. MARK1 is highly conserved in evolution, and we'll see this quite clearly by performing a multiple sequence alignment of proteins from a wide range of species.

Using the form from the EMBL-EBI ClustalW server, a multiple sequence alignment was performed using the default settings. The sequences that were aligned were from nine different species, from human to fish, and the N-terminal residues of the alignment appear in **Figure 11.8**. Notice that sequences are labeled on the left, and by editing the input FASTA file names, these were customized for this figure. ClustalW has a consensus line at the bottom indicating the conservation in each position. The asterisk (*) represents 100% identity, the colon (:) indicates that conservative substitutions are present, and a single dot (.) shows

```
Rattus       MSARTPLPTVNERDTEN-------------------------------------HTSVDGY  24
Mus          MSARTPLPTVNERDTEN-------------------------------------HTSVDGY  24
Canis        -----------------------MNLW--------------------------HTSVDGY  11
Oryctolagus  MEAEMCAFELSGSGVEYVRHRILGFAKQIWRRGRKAGELECLRGNGYSKGSWPHTSVDGY  60
Homo         MSARTPLPTVNERDTEN-------------------------------------HTSVDGY  24
Bos          MSARTPLPTVNERDTEN-------------------------------------HTSVDGY  24
Gallus       MSTRTPLPTVNERDTEN-------------------------------------HTSVDGY  24
Xenopus      MSARTPLPTVNERDTEN-------------------------------------HTSLDGY  24
Danio        MSTRTPLPTVNERDAEN-------------------------------------HTSVDGY  24
                                                                   ***:***

Rattus       TETHIPPTKSSSRQNIPRCRNSITSATDEQPHIGNYRLQKTIGKGNFAKVKLARHVLTGR  84
Mus          TETHIPPAKSSSRQNLPRCRNSITSATDEQPHIGNYRLQKTIGKGNFAKVKLARHVLTGR  84
Canis        TEPHIQPTKSSSRQNIPRCRNSITSATDEQPHIGNYRLQKTIGKGNFAKVKLARHVLTGR  71
Oryctolagus  TEPHIQPTKSSSRQNIPRCRNSITSATDEQPHIGNYRLQKTIGKGNFAKVKLARHVLTGR  120
Homo         TEPHIQPTKSSSRQNIPRCRNSITSATDEQPHIGNYRLQKTIGKGNFAKVKLARHVLTGR  84
Bos          TEPHIQPTKSSSRQNIPRCRNSITSATDEQPHIGNYRLQKTIGKGNFAKVKLARHVLTGR  84
Gallus       TEPHVQPIKSSSRQNIPRCRNSITSTNEEHPHIGNYRLLKTIGKGNFAKVKLARHVLTGR  84
Xenopus      PEPPVPPTKSSSRQSLPRCRNSIVSMTDEQPHIGNYRLLKTIGKGNFAKVKLARHVLTGR  84
Danio        TDTPAAPTKSSSRQSLPRSRNSVASITDEQPHVGNYRLLKTIGKGNFAKVKLARHVLTGR  84
              .: *.******.:**.***:.*..:*:**:*****.********************

Rattus       EVAVKIIDKTQLNPTSLQKLFREVRIMKILNHPNIVKLFEVIETEKTLYLVMEYASGGEV  144
Mus          EVAVKIIDKTQLNPTSLQKLFREVRIMKILNHPNIVKLFEVIETEKTLYLVMEYASGGEV  144
Canis        EVAVKIIDKTQLNPTSLQKLFREVRIMKILNHPNIVKLFEVIETEKTLYLVMEYASGGEV  131
Oryctolagus  EVAVKIIDKTQLNPTSLQKLFREVRIMKILNHPNIVKLFEVIETEKTLYLVMEYASGGEV  180
Homo         EVAVKIIDKTQLNPTSLQKLFREVRIMKILNHPNIVKLFEVIETEKTLYLVMEYASGGEV  144
Bos          EVAVKIIDKTQLNPTSLQKLFREVRIMKILNHPNIVKLFEVIETEKTLYLVMEYASGGEV  144
Gallus       EVAVKIIDKTQLNPTSLQKLFREVRIMKILNHPNIVKLFEVIETEKTLYLVMEYASGGEV  144
Xenopus      EVAVKIIDKTQLNPTSLQKLFREVRIMKILNHPNIVKLFEVIETEKTLYLIMEYASGGEV  144
Danio        EVAVKIIDKTQLNPTSLQKLFREVRIMKVLNHPNIVKLFEVIETEKTLYLIMEYASGGEV  144
              ************************************:*******************:********
```

Figure 11.8 ClustalW alignment of vertebrate MARK1 proteins. The row below the sequences indicates the conservation in the positions of the alignment. Only part of the alignment is shown and the sequences are as follows: *Rattus* NP_446399, *Mus* NP_663490, *Canis* XP_536123, *Oryctolagus* XP_002717438, *Homo* NP_061120, *Bos* NP_001179204, *Gallus* XP_419403, *Xenopus* NP_001085126, and *Danio* NP_001107948.

moderate conservation. If there is no conservation or a sequence is missing from a column, the bottom line will be blank in that position even if all the other amino acids in that position are identical.

As can be seen with the alignment, MARK1 orthologs are very highly conserved with some notable differences from dog (*Canis*) and rabbit (*Oryctolagus*). Both of these proteins were predictions and it is possible that further exploration of the dog and rabbit genomic sequences and transcripts would reveal problems with the predictions or sequences. Nevertheless, the remainder of the alignment shows remarkable conservation of sequence and length considering the evolutionary time represented by these species.

Additional results are available and the choices appear as tabs across the top of the results page (see **Figure 11.9**). As seen earlier at the ExPASy Website, ClustalW generates trees to show the relationships between sequences. **Figure 11.10** shows the default **phylogram** (Figure 11.10A) as well as the optional cladogram (Figure 11.10B). Notice that the branch points are in the same locations for both trees, but the branch lengths have all been stretched in the cladogram so the species names all appear on the far right of the graph. In the phylogram, the lengths of the branches are proportional to the evolutionary distances (sequence differences) calculated by ClustalW. As expected, the rat and mouse sequences are quite similar; they appear on the same larger branch, and have short branch lengths, indicating very few differences.

Moving to another part of the tree on the phylogram, look at the relationships between human and rabbit sequences, which share a branch. If the human and rabbit sequences are a certain distance apart, why not have two equal-length branches leading to their names? Why is the rabbit branch longer than the human branch? The answer is that the relationships conveyed here are not just between those nearest to each other. Although the human and rabbit sequences are some distance apart from each other, the rabbit sequence is a greater distance from the other sequences on the tree than the rabbit sequence is to the human sequence. So the rabbit branch is longer.

It is important to remember that this phylogram is an extremely limited view of evolution and the relationships between organisms. This is one gene from nine organisms. Some relationships (for example, the mouse and rat) would hold up to many different kinds of studies, including basing evolutionary relationships on body characteristics (morphology). We'll see later in this chapter that another ortholog study rearranges the phylogram branches we see here with MARK1. What this phylogram shows is the similarities between these MARK1 sequences in this alignment; how they are grouped suggests evolutionary relationships.

The "Result Summary" tab includes the pairwise comparisons generated by ClustalW during the calculations for the alignment (**Figure 11.11**). Here you see the methodical pairing of sequence 1 aligned with sequence 2, 1 aligned with 3, 1 aligned with 4, and so on until all the possible pairwise comparisons have been performed. Notice that the mouse and rat sequences are 97% identical (which may not be surprising) and that the human and rabbit sequences are 96% identical (which may be surprising). These percentage identities are among the highest in the table and these pairs of species ended up on the same branches. But the cow and human sequences are 97% identical; shouldn't the cow and human sequences share a branch? In this case, cow and dog (96%), cow and rabbit (95%), and human and rabbit (96%) identities must be part of the calculations. Fortunately for us, ClustalW considers all the pairwise comparisons and generates a tree that is consistent with all the data.

Figure 11.9 Output generated by the EBI ClustalW server. These tabs are across the top of the results page.

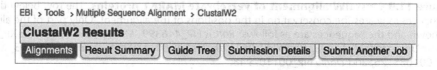

EBI › Tools › Multiple Sequence Alignment › ClustalW2

ClustalW2 Results

| Alignments | Result Summary | Guide Tree | Submission Details | Submit Another Job |

(A)

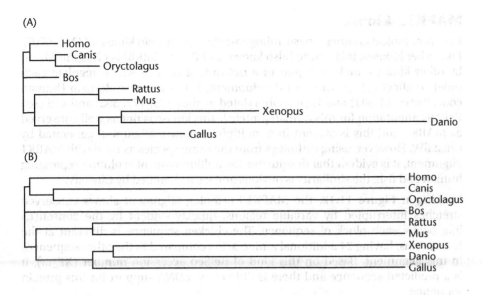

(B)

Figure 11.10 ClustalW guide trees of MARK1 ortholog alignment. Two kinds of trees are available in ClustalW. (A) Phylogram: the lengths of branches in the tree represent the number of differences between sequences. (B) Cladogram: the branch points are identical to those seen in the phylogram but branch lengths vary to align the sequence labels on the far right.

Of course, additional steps are needed to merge these pairwise comparisons into a coherent multiple sequence alignment. The calculated pairwise identities, discussed above, are used to construct the guide trees. The trees are then used to guide the alignment of all the sequences, taking into account local regions of similarity and divergence, and the consequences of introducing gaps. This parameter will be discussed below.

SeqA	Name	Length	SeqB	Name	Length	Score
1	Homo	795	2	Canis	782	97.0
1	Homo	795	3	Bos	795	97.0
1	Homo	795	4	Rattus	793	94.0
1	Homo	795	5	Mus	795	94.0
1	Homo	795	6	Oryctolagus	831	96.0
1	Homo	795	7	Xenopus	792	89.0
1	Homo	795	8	Gallus	794	92.0
1	Homo	795	9	Danio	772	82.0
2	Canis	782	3	Bos	795	96.0
2	Canis	782	4	Rattus	793	93.0
2	Canis	782	5	Mus	795	94.0
2	Canis	782	6	Oryctolagus	831	97.0
2	Canis	782	7	Xenopus	792	88.0
2	Canis	782	8	Gallus	794	91.0
2	Canis	782	9	Danio	772	80.0
3	Bos	795	4	Rattus	793	94.0
3	Bos	795	5	Mus	795	94.0
3	Bos	795	6	Oryctolagus	831	95.0
3	Bos	795	7	Xenopus	792	89.0
3	Bos	795	8	Gallus	794	92.0
3	Bos	795	9	Danio	772	82.0
4	Rattus	793	5	Mus	795	97.0
4	Rattus	793	6	Oryctolagus	831	92.0
4	Rattus	793	7	Xenopus	792	87.0
4	Rattus	793	8	Gallus	794	89.0
4	Rattus	793	9	Danio	772	81.0
5	Mus	795	6	Oryctolagus	831	92.0
5	Mus	795	7	Xenopus	792	86.0
5	Mus	795	8	Gallus	794	89.0
5	Mus	795	9	Danio	772	80.0
6	Oryctolagus	831	7	Xenopus	792	87.0
6	Oryctolagus	831	8	Gallus	794	90.0
6	Oryctolagus	831	9	Danio	772	80.0
7	Xenopus	792	8	Gallus	794	90.0
7	Xenopus	792	9	Danio	772	82.0
8	Gallus	794	9	Danio	772	83.0

Figure 11.11 ClustalW Result Summary of MARK1 multiple sequence alignment. To generate a multiple sequence alignment, ClustalW calculates the percent identity ("Score") between pairs of sequences and uses these calculations to guide the subsequent alignment steps. These calculations are available under the "Result Summary" tab. Although there are 450 million years of evolution separating the rabbit (*Oryctolagus*) and the fish (*Danio*), the MARK1 protein sequence is still 80% identical.

MAPK15 kinase

Let's now look at another kinase, mitogen-activated protein kinase 15 (MAPK15). Like other kinases, this kinase (also known as ERK7 or ERK8) is phosphorylated by other kinases, making it part of a network of kinases which regulate each other to affect cell growth and development. It has a kinase domain (human coordinates 13–304) and is phosphorylated at sites 175, 177, 392, and 415 (see ExPASy annotation for this human protein). This kinase is not as well conserved as MARK1 and this is evident in a multiple sequence alignment generated by ClustalW. However, using orthologs from the same species as used with MARK1 alignment, it is evident that despite the 450 million years of evolution separating human and fish, the similarity is obvious and easily handled by ClustalW.

Looking at **Figure 11.12**, the MAPK15 ortholog alignment shows conserved stretches interrupted by variable regions, quickly evident by the consensus line below each block of sequence. The chicken sequence is different at the N-terminus, having 24 additional amino acids compared to the other sequences in this alignment. Based on this kind of RefSeq accession number (XP_...), it is a predicted sequence and there is little or no cDNA support for this protein sequence.

Compare the result summary for the MAPK15 alignment (**Figure 11.13**) to that of the MARK1 alignment (Figure 11.11). The highest MAPK15 alignment score is 93% (mouse and rat), with the rest of the percent identities ranging from 80% down to 45%. Except for two data points, the MAPK15 identities do not overlap with the MARK1 lowest scores. This comparison nicely illustrates the differences in evolutionary rates for genes, even within the same larger family (kinases).

Figure 11.12 ClustalW alignment of vertebrate MAPK15 proteins. MAPK15 is considered a more rapidly evolving kinase than MARK1, although this portion of the full-length ClustalW multiple sequence alignment shows good conservation of length and sequence. The sequences are as follows: *Mus* NP_808590, *Rattus* NP_775453, *Bos* NP_001039575, *Canis* XP_539201, *Homo* NP_620590, *Oryctolagus* XP_002724352, *Danio* NP_001018581, *Xenopus* NP_001089435, and *Gallus* XP_423954.

```
Mus          ------------------------MCAAEVDRHVAQRYLIKRRLGKGAYGIVWKAMDRRT  36
Rattus       ------------------------MCAAEVDRHVSQRYLIKRRLGKGAYGIVWKAMDRRT  36
Bos          ------------------------MCTAEVDRHVAQRYLLKRRLGKGAYGIVWKAVDRRT  36
Canis        ------------------------MCAAEVDDHVAQRYLLKRRLGKGAYGIVWKAVDRRT  36
Homo         ------------------------MCT-VVDPRIVRRYLLRRQLGQGAYGIVWKAVDRRT  35
Oryctolagus  ------------------------MCAAEVDAHIARRFRLQRRLGKGAYGIVWKAVDRST  36
Danio        ------------------------MNITEVEEHISSKYEIKRRLGKGAYGIVWKAVDRKS  36
Xenopus      ------------------------MSGPEVEDHISQKYDIKKRLGKGAYGIVWKAIDRKS  36
Gallus       MRSLGLLAHALCVASVSTATEMSAMGEPEVDAAVAEKFEMKRRLGKGAYGIVWKAINRRT  60
                                     *     *:   :  :: ::::**:*********::* :

Mus          GEVVAIKKIFDAFRDQIDAQRTFREIMLLKEFGGHPNIIRLLDVIPAKNDRDIYLVFESM  96
Rattus       GEVVAIKKIFDAFRDQTDAQRTFREIMLLREFGGHPNIIRLLDVIPAKNDRDIYLVFESM  96
Bos          GEVVAIKKIFDAFKDKTDAQRTFREITLLQEFGDHPNIVRLLDVIPAENDRDIYLVFESM  96
Canis        GEVVAIKKIFDAFRDKTDAQRTFREITLLQELGDHPNIIRLLDVIRAENDRDIYLVFESM  96
Homo         GEVVAIKKIFDAFRDKTDAQRTFREITLLQEFGDHPNIISLLDVIRAENDRDIYLVFEFM  95
Oryctolagus  GEVVAIKKIFDAFRDQTDAQRTFREIVLLQEFGGHPNIIRLLDVIRAENDRDIYLVFESM  96
Danio        GETVAVKKIFDAFRNRTDAQRTFREIMFLQEFGDHPNIIKLLNVIRAQNDKDIYLIFEFM  96
Xenopus      GEIVAVKKIFDAFRNRTDAQRTFREIMFLQEFGEHPNIIKLLNVIRAQNDKDIYLVFEHM  96
Gallus       GEIVAVKKIFDAFRNRTDAQRTFREIMFLQEFGEHPNIIKLLDVIRAQNNKDIYLVFESM 120
             ** **:*********::: ********* :*:*:* ****: **:** *:*::****:** *

Mus          DTDLNAVIQKGRLLKDIHKRCIFYQLLRATKFIHSGRVIHRDQKPANVLLDSACRVKLCD 156
Rattus       DTDLNAVIQKGRLLEDIHKRCIFYQLLRATKFIHSGRVIHRDQKPANVLLDAACRVKLCD 156
Bos          DTDLNAVICKGTLLKDTHKRYIFYQLLRATKFIHSGRVIHRDQKPSNVLLDASCLVKLCD 156
Canis        DTDLNAVICKGRLLRDVHKRYIFYQLLRATKYIHSGRVIHRDQKPSNILLDSSCVVKLCD 156
Homo         DTDLNAVIRKGGLLQDVHVRSIFYQLLRATRFLHSGHVVHRDQKPSNVLLDANCTVKLCD 155
Oryctolagus  DTDLHAVIEKGTLLKDIHKRYIFYQLLRATQFLHSGHVVHRDQKPSNVLLDANCLVKLCD 156
Danio        DTDLHAVIKKGNLLKDIHKRYVMYQLLKATKYLHSGNVIHRDQKPSNILLDSDCFVKLCD 156
Xenopus      ETDLHAVIKKGNLLKDIHMRYILYQLLKATKFIHSGNVIHRDQKPSNILLDGDCLVKLCD 156
Gallus       ETDLHAVIKKGNLLKDIHKCYILYQLLKATKFIHSGNVIHRDQKPSNILLDADCFVKLCD 180
             :***:*** ** **.* *   ::****:**::.***.*:****:*:***. * *****
```

SeqA	Name	Length	SeqB	Name	Length	Score
1	Danio	524	2	Gallus	600	58.0
1	Danio	524	3	Xenopus	586	60.0
1	Danio	524	4	Oryctolagus	547	52.0
1	Danio	524	5	Mus	549	49.0
1	Danio	524	6	Rattus	547	52.0
1	Danio	524	7	Bos	536	51.0
1	Danio	524	8	Canis	559	51.0
1	Danio	524	9	Homo	544	49.0
2	Gallus	600	3	Xenopus	586	53.0
2	Gallus	600	4	Oryctolagus	547	48.0
2	Gallus	600	5	Mus	549	50.0
2	Gallus	600	6	Rattus	547	51.0
2	Gallus	600	7	Bos	536	50.0
2	Gallus	600	8	Canis	559	49.0
2	Gallus	600	9	Homo	544	48.0
3	Xenopus	586	4	Oryctolagus	547	45.0
3	Xenopus	586	5	Mus	549	51.0
3	Xenopus	586	6	Rattus	547	48.0
3	Xenopus	586	7	Bos	536	50.0
3	Xenopus	586	8	Canis	559	48.0
3	Xenopus	586	9	Homo	544	49.0
4	Oryctolagus	547	5	Mus	549	68.0
4	Oryctolagus	547	6	Rattus	547	69.0
4	Oryctolagus	547	7	Bos	536	72.0
4	Oryctolagus	547	8	Canis	559	68.0
4	Oryctolagus	547	9	Homo	544	68.0
5	Mus	549	6	Rattus	547	93.0
5	Mus	549	7	Bos	536	73.0
5	Mus	549	8	Canis	559	74.0
5	Mus	549	9	Homo	544	70.0
6	Rattus	547	7	Bos	536	74.0
6	Rattus	547	8	Canis	559	75.0
6	Rattus	547	9	Homo	544	69.0
7	Bos	536	8	Canis	559	80.0
7	Bos	536	9	Homo	544	71.0
8	Canis	559	9	Homo	544	72.0

Figure 11.13 ClustalW Result Summary of the vertebrate MAPK15 alignment. With the exception of the mouse (*Mus*)–rat (*Rattus*) 93% identity, the highest identity is 80%, which was the lowest identity seen earlier with the MARK1 alignment from the same species.

Perhaps surprising is the rearrangement of the branches in the tree for MAPK15 compared to MARK1. Comparing Figures 11.10A and **11.14**, we see that the human and rabbit branches are now separated onto two different branches for MAPK15. Mouse and rat are still close on the tree, with 93% identity as seen in Figure 11.13, but the fish, frog, and chicken branches are about the same length. This illustrates the hazard of basing evolutionary relationships on one or a few genes. In a spectacular example of solving this problem, Perelman et al. sequenced a total of eight Megabases (eight million bases) from 186 primates to collect enough sequence information to derive the molecular phylogeny of these very closely related species. Primate kinases, for example, are very similar in sequence and it would be difficult to define evolutionary relationships based on such close identities. However, this effort gained statistical power by having such a large dataset to analyze.

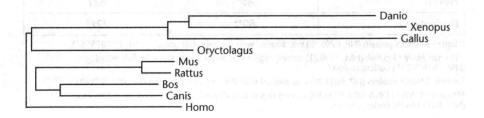

Figure 11.14 ClustalW phylogram of MAPK15. The evolutionary relationships between these vertebrates, as defined by the differences in the MAPK15 orthologs. Comparing the MARK1 (Figure 11.10A) and MAPK15 phylograms, the branch lengths are different (for example, the *Danio*, *Xenopus*, and *Gallus* branches) and several branches are different (for example, human and rabbit share a branch in the MARK1 tree but are on separate branches in the MAPK15 tree).

DNA versus protein identities

MAPK15 and MARK1 show different levels of protein sequence conservation, but what about DNA sequence? To briefly touch upon this topic, let's compare the human–zebrafish percent identities. To make a fair and direct comparison to protein sequence, only the coding regions of the cDNAs should be used since the 5P and 3P UTRs are under different evolutionary constraints. Using ClustalW to perform pairwise alignments, the percent identities for protein and DNA are shown in **Table 11.1**. Notice that in both cases, the *MARK1* sequences are more conserved than those of *MAPK15*. However, the differences are more striking when comparing protein identities (82% versus 49%) rather than nucleotide identities (72% versus 60%). In fact, the nucleotide sequence of *MAPK15* is more conserved than the protein sequence. Remember, there are only four choices in nucleotides while there are 20 in amino acids. This will affect the differences seen, especially in regions of low similarity.

In an interesting paper published in 1996, Makalowski, Zhang, and Boguski compared the 5P UTRs, 3P UTRs, coding regions, and protein sequences of 1196 human–mouse orthologs. They identified a number of genes where the coding sequence was more highly conserved than the protein sequence. But the averages were 85% for both the coding regions and the protein sequences. The percent identities of 3P UTRs (69%) were slightly higher than those of 5P UTRs (67%), perhaps reflecting the presence of regulatory sequences such as miRNA target sites. It is interesting that protein conservation ranged from 36% to 100%. Clearly, there are stories behind every gene, each acted upon by different evolutionary forces.

11.5 MODIFYING ClustalW PARAMETERS

There is a handful of ClustalW parameters that can be modified to try to improve multiple sequence alignments. Perhaps one of the easiest to understand and illustrate is the penalty applied to open a gap in alignments. Another is to vary the "method" applied to the alignment process. Both will be briefly described here. You are encouraged to explore these and other methods, easily accessible from the ClustalW interface. For more information on modifying ClustalW parameters, help is available on the EBI server and elsewhere on the Web.

Gap-opening penalty

Just as BLAST programs keep track of gap creations and extensions, ClustalW also has penalties for these two activities to constrain the alignment possibilities. Without these constraints, sequences that fail to align easily would have many gaps introduced, in many possible combinations, with sequences reaching far away in search of alignment, and accomplishing little except making you wait for, in many cases, biological nonsense. With penalties in place, ClustalW tries to find something that will make sense, and delivers a good estimate quickly. It is

Table 11.1 Comparison of DNA and protein pairwise similarities from human–zebrafish ClustalW alignment

	Percentage identity	
	MAPK15	*MARK1*
Protein	49*	82†
Nucleotide	60**	72††

*Human MAPK15 protein (NP_620590) was aligned with the zebrafish ortholog (NP_001018581).

**Human *MAPK15* cDNA (NM_139021) coding region was aligned with the zebrafish orthologous (NM_001020745) coding region.

†Human MARK1 protein (NP_061120) was aligned with the zebrafish ortholog (NP_001107948).

††Human *MARK1* cDNA (NM_018650) coding region was aligned with the zebrafish orthologous (NM_001114476) coding region.

up to you to interpret the results and try different parameters, or even a different alignment tool, to achieve the best outcome. Multiple sequence alignment tools should often not be used alone. BLAST results should be examined, other features like exon distributions should be considered, and sequences should be examined by eye.

A common challenge for multiple sequence alignment tools is the alignment of protein isoforms. When aligning the products of alternative splicing, otherwise perfect alignments can be easily disrupted by one or more protein sequences originating from these alternative exons. The mouse kinase Sgk1 has multiple isoforms that differ at their N-termini due to alternative splicing (**Figure 11.15A**). After these 5P exons, all the proteins share the same exons and sequence. How do the ClustalW defaults handle these sequences? In Figure 11.15B, the amino acids of the various N-termini are scattered about, as ClustalW attempts to align these regions to each other. None of these alignments make good sense.

If there were nothing in common between these N-termini, what would you prefer to see? One possibility is to keep each N-terminus adjacent to the regions that do align perfectly and show the various N-termini intact so you could compare their length and sequence. You would need to accept that the "alignment" of these N-termini is meaningless but more easily interpretable, as you will soon see. On the ClustalW home page, there are numerous options to choose, including gap penalties. By increasing the multiple sequence alignment gap-opening penalty to a very high number (for example, from 10 to 50) you prevent ClustalW from opening many gaps and the sequences are held together (Figure 11.15C). Very quickly you can see the relative lengths of the N-terminal sequences from the different isoforms.

The clustering method

Within ClustalW there are two clustering methods to choose from for generating the multiple sequence alignment. The Unweighted Pair Group Method with Arithmetic Mean (UPGMA) is a simple method that uses the calculated pairwise identities to find the closest sequences and then treats that pair as a single entity to make comparisons to other pairwise identities. This builds the alignment, and guide tree, based on these pairwise identities.

The default algorithm is called Neighbor-Joining (NJ) and is designed to build a tree with the lowest branch-length sum possible. Rather than solely depending on the pairwise identities, total branch lengths are calculated based on different pairs acting as the first group. Once one pairing generates the lowest branch-length sum, that group is treated as one branch and different pairings are again used to calculate the total branch length. Every time a group is chosen, the number of comparisons drops until the final guide tree is determined. These calculations are considered more rigorous than UPGMA; however, both algorithms have value.

These methods will often generate different alignments and trees. In this example, the kinase domains of orthologous epithelial discoidin domain-containing receptor 1 sequences from six species are aligned using either NJ or UPGMA clustering. Figures **11.16A** and 11.16B show portions of the full alignments using NJ and UPGMA methods, respectively. These alignments are quite similar, showing few differences because of sequence and length conservation. There is one region (underlined), near coordinates 100–125, where the lengths vary and the two methods differ in how the gaps are handled. With NJ, a gap is opened in all six sequences, with *Branchiostoma* and *Strongylocentrotus* sequences each having two gaps.

With UPGMA, only one gap is opened in three of the sequences. Are there too many gaps with NJ, or not enough gaps with UPGMA? Only careful analysis and, perhaps, more information will assist in this decision. The handling of the individual amino acids may be very important to you. For example, is this region

(A)

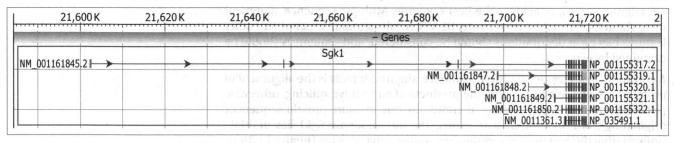

(B)

```
NP_001155322.1|  ------------------------------------------------------------MGEM    4
NP_001155320.1|  ------------------------------------------------------------        
NP_035491.1|     ---------------------------------------------------------MTV    3
NP_001155317.2|  MVNKDMNGFPVKKCSAFQFFKKRVRRWIKSPMVSVDKHQSPNLKYTGPAGVHLPPGESDF   60
NP_001155321.1|  ------------------------------------------------------------        


NP_001155322.1|  QG---------ALARARLESLLRPRHKKRAEAQKRSE---SVLLS-----------GLAF   41
NP_001155320.1|  ------------------------------------------------------------        
NP_035491.1|     KA---------EAARSTLT----------YSRMRGM---VAIL-------------IAF   27
NP_001155317.2|  EAMCQSCLGDHAFQRGMLPPEESCSWEIQPGCEVKEQCNHANILTKPDPRTFWTNDDAAF  120
NP_001155321.1|  ---------------------------------------MKEETLRSPWK----AF   13


NP_001155322.1|  MKQRRMGLNDFIQKIASNTYACKHAEVQSILKMSHPQEPELMNANPSPPPSPSQQINLGP  101
NP_001155320.1|  MKQRRMGLNDFIQKIASNTYACKHAEVQSILKMSHPQEPELMNANPSPPPSPSQQINLGP   60
NP_035491.1|     MKQRRMGLNDFIQKIASNTYACKHAEVQSILKMSHPQEPELMNANPSPPPSPSQQINLGP   87
NP_001155317.2|  MKQRRMGLNDFIQKIASNTYACKHAEVQSILKMSHPQEPELMNANPSPPPSPSQQINLGP  180
NP_001155321.1|  MKQRRMGLNDFIQKIASNTYACKHAEVQSILKMSHPQEPELMNANPSPPPSPSQQINLGP   73
                 ************************************************************
```

(C)

```
NP_001155322.1|  ------------------------------------------------------------        
NP_001155320.1|  ------------------------------------------------------------        
NP_035491.1|     ------------------------------------------------------------        
NP_001155317.2|  MVNKDMNGFPVKKCSAFQFFKKRVRRWIKSPMVSVDKHQSPNLKYTGPAGVHLPPGESDF   60
NP_001155321.1|  ------------------------------------------------------------        


NP_001155322.1|  ------------------MGEMQGALARARLESLLRPRHKKRAEAQKRSESVLLSGLAF   41
NP_001155320.1|  ------------------------------------------------------------        
NP_035491.1|     --------------------------MTVKAEAARSTLTYSRMRGMVAILIAF   27
NP_001155317.2|  EAMCQSCLGDHAFQRGMLPPEESCSWEIQPGCEVKEQCNHANILTKPDPRTFWTNDDAAF  120
NP_001155321.1|  -----------------------------------------------MKEETLRSPWKAF   13


NP_001155322.1|  MKQRRMGLNDFIQKIASNTYACKHAEVQSILKMSHPQEPELMNANPSPPPSPSQQINLGP  101
NP_001155320.1|  MKQRRMGLNDFIQKIASNTYACKHAEVQSILKMSHPQEPELMNANPSPPPSPSQQINLGP   60
NP_035491.1|     MKQRRMGLNDFIQKIASNTYACKHAEVQSILKMSHPQEPELMNANPSPPPSPSQQINLGP   87
NP_001155317.2|  MKQRRMGLNDFIQKIASNTYACKHAEVQSILKMSHPQEPELMNANPSPPPSPSQQINLGP  180
NP_001155321.1|  MKQRRMGLNDFIQKIASNTYACKHAEVQSILKMSHPQEPELMNANPSPPPSPSQQINLGP   73
                 ************************************************************
```

Figure 11.15 Handling of alternative N-termini by multiple sequence analysis tools.
(A) NCBI Gene view of the alternative splicing of mouse kinase *Sgk1*. (B) Each isoform has a different
N-terminus as shown by the scattered and weak alignment in the first 120 amino acids of the output
(as defined by NP_001155317). After this coordinate, the proteins are 100% identical and only 60
amino acids of this 100% alignment are shown in the figure. (C) Increasing the multiple sequence
alignment gap-opening penalty from 10 to 50 keeps the N-terminal sequences together.

(A)

```
Homo             DFLKEVKIMSRLKDPNIIRLLGVCVQDDPLCMITDYMENGDLNQFLS--AHQLEDKAAEG 116
Mus              DFLKEVKIMSRLKDPNIIRLLGVCVQDDPLCMITDYMENGDLNQFLS--ARQLENKATQG 116
Branchiostoma    DFFKEVKILARLRDPNIVRLLGVCTRDEPLCMIVEYMENGDLNQYLF--KHEFEGAVPS- 115
Ciona            DFLKEVRVMSQMQDPNIVHLIAVCTNDEPYAMITEYMENGDLNQYLRSCADKMNSDPSS- 118
Strongylocentrotus DFMKEMKIMSQLRDPNIVRLLAACTEDEPYCMIVEYMENGDLNQFLYE-REGFEVGLNN- 117
Pediculus        TFRQQVKVLSKIKDSNIVRVLGASLDHDPIFVVIEYMDYGDLYQFLQDHVADTTSPLPT- 114
                  *  ::::::::::*.**::::. ..:* .:: :**: *** *:*
```

```
Homo             APGDGQAAQGPTISYPMLLHVAAQIASGMRYLATLNFVHRDLATRNCLVGENFTIKIADF 176
Mus              LSGDTESDQGPTISYPMLLHVGAQIASGMRYLATLNFVHRDLATRNCLVGENFTIKIADF 176
Branchiostoma    ------ASNAPMLGLGALLYMAAQIASGMKYLSSLNFVHRDLATRNCLVGPRHSVRIADF 169
Ciona            ------KENSLFSSNEVLINMCVQIASGMRYLSSHKFVHRDLATRNCLVDSNHSIRIADF 172
Strongylocentrotus CNANLSNCNKSSISVGALVYMITQIASGMKYLSNMNFVHRDLATRNCLVGKSYTIRIADF 177
Pediculus        --------NAKTLSYGCLIYMATQIASGMKYLESLNFVHRDLAARNCLVGTGYKIKITDI 166
                   :   . *: : .*******:**   .:*******:*****.  ..::*:*:
```

(B)

```
Homo             DFLKEVKIMSRLKDPNIIRLLGVCVQDDPLCMITDYMENGDLNQFLSAHQLEDKAAEGAP 118
Mus              DFLKEVKIMSRLKDPNIIRLLGVCVQDDPLCMITDYMENGDLNQFLSARQLENKATQGLS 118
Branchiostoma    DFFKEVKILARLRDPNIVRLLGVCTRDEPLCMIVEYMENGDLNQYLFKHEFEGAVPS--- 115
Strongylocentrotus DFMKEMKIMSQLRDPNIVRLLAACTEDEPYCMIVEYMENGDLNQFLYEREGFEVGLNNCN 119
Ciona            DFLKEVRVMSQMQDPNIVHLIAVCTNDEPYAMITEYMENGDLNQYLRSCADKMNSDP--- 116
Pediculus        TFRQQVKVLSKIKDSNIVRVLGASLDHDPIFVVIEYMDYGDLYQFLQDHVADTTSPL--- 112
                  *  :::::::::*.**:::::.. .:* .:: :**: *** *:*
```

```
Homo             GDGQAAQGPTISYPMLLHVAAQIASGMRYLATLNFVHRDLATRNCLVGENFTIKIADFGM 178
Mus              GDTESDQGPTISYPMLLHVGAQIASGMRYLATLNFVHRDLATRNCLVGENFTIKIADFGM 178
Branchiostoma    ----ASNAPMLGLGALLYMAAQIASGMKYLSSLNFVHRDLATRNCLVGPRHSVRIADFGM 171
Strongylocentrotus ANLSNCNKSSISVGALVYMITQIASGMKYLSNMNFVHRDLATRNCLVGKSYTIRIADFGM 179
Ciona            --SSKENSLFSSNEVLINMCVQIASGMRYLSSHKFVHRDLATRNCLVDSNHSIRIADFGM 174
Pediculus        ----PTNAKTLSYGCLIYMATQIASGMKYLESLNFVHRDLAARNCLVGTGYKIKITDIGM 168
                    :   . *: : .*******:**   .:*******:*****.  ..::*:*:**
```

(C)

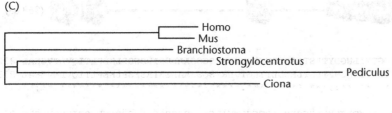

(D)

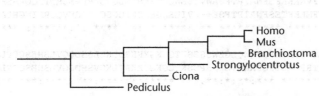

Figure 11.16 Small changes can greatly influence tree relationships. (A) Alignments generated by the Neighbor-Joining (NJ) clustering method. (B) Alignments generated by the UPGMA clustering method. (C) Guide tree for the NJ method. (D) Guide tree for the UPGMA method. The kinase domain sequences were obtained from the following: *Homo* NP_001945, *Mus* NP_001185762, *Branchiostoma* XP_002596288, *Strongylocentrotus* XP_001202828, *Ciona* XP_002123027, and *Pediculus* XP_002431712.

an unstructured loop, or is it an important structure such as an alpha helix? What are the consequences of and constraints on introducing gaps within these structures?

The resulting trees are different as well. Figure 11.16C shows the "unrooted" tree of the NJ clustering method while the "rooted" tree of UPGMA is shown in Figure 11.16D. In a rooted tree, a clear but unknown ancestor gives rise to the entire tree, but for the unrooted tree, the ancestral relationship between the species is not clear. The *Strongylocentrotus* and *Pediculus* sequences share a branch with the NJ method while they are on separate branches with UPGMA. Clearly, the small difference in the alignments had consequences on the tree structure.

11.6 COMPARING ClustalW, MUSCLE, AND COBALT

So far, we have introduced an NCBI multiple sequence alignment tool coupled with BLAST and the more sophisticated ClustalW. There is a handful of other multiple sequence alignment applications, each with different strengths and approaches to solving the problem. It is common to vary the parameters within a single tool and, when unable to achieve the desired results, to try another algorithm. In this section of the chapter, we will compare ClustalW to two applications: MUSCLE and COBALT.

MUSCLE (MUltiple Sequence Comparison by Log-Expectation), www.ebi.ac.uk/Tools/msa/muscle, was developed by R.C. Edgar at drive5.com and uses an iterative algorithm. Through building many multiple sequence alignments and guide trees, MUSCLE quickly achieves accurate alignment of benchmark protein sequences. COBALT (COnstraint Based ALignment Tool), www.ncbi.nlm.nih.gov/tools/cobalt, was developed at the NCBI and searches the conserved domain and PROSITE motif databases to obtain guidance for iterative construction of multiple sequence alignments.

To challenge all three algorithms, we'll be trying to align thrombospondin (TSP1) domains within ADAMTS1, a gene family we studied in Chapter 6. There are three thrombospondin domains in ADAMTS1 (**Figure 11.17A**). A human and mouse pairwise alignment in the vicinity of these second and third repeats shows good identity across species (Figure 11.17B), yet the second and third domains are approximately 21% identical to each other in both mouse (Figure 11.17C) and human (Figure 11.17D).

How do the algorithms handle an alignment of the second and third human thrombospondin repeats to the mouse sequences? Very well, if the human sequences are aligned one at a time. As seen in Figures **11.18A** and 11.18B, the second human TSP repeat (Hs-2) aligns with the second mouse thrombospondin repeat (mouse coordinates 855–911), and the third human thrombospondin

Figure 11.17 ADAMTS1 thrombospondin domain similarities. (A) Vertebrates have three thrombospondin (TSP1) domains in ADAMTS1, as shown in the ScanProsite graphic for mouse Adamts1. (B) Alignment between the human and mouse ADAMTS1 protein sequences in the adjacent second and third thrombospondin domains. (C) Despite their common name, ClustalW shows that the second and third mouse thrombospondin domains of Adamts1 are only 21% identical. (D) A similar identity is seen between the human ADAMTS1 thrombospondin domains. The mouse coordinates for domains two and three are amino acids 855–911 and 912–968, respectively. The human coordinates for domains two and three are amino acids 854–905 and 908–967, respectively.

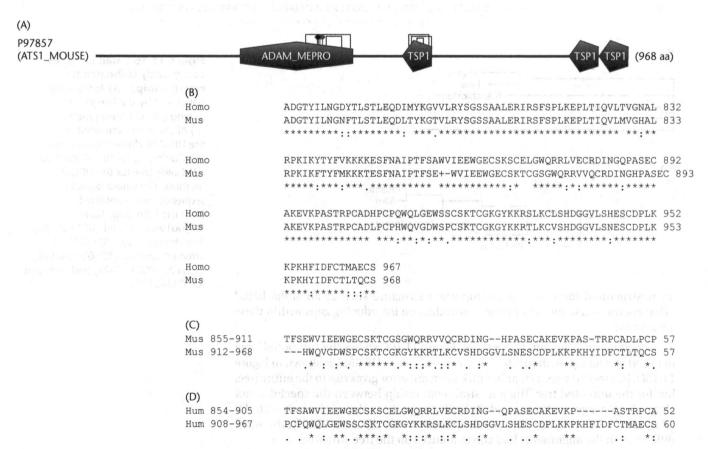

(A)

P97857
(ATS1_MOUSE) ────────────── ADAM_MEPRO ────── TSP1 ────────── TSP1 TSP1 (968 aa)

(B)
```
Homo        ADGTYILNGDYTLSTLEQDIMYKGVVLRYSGSSAALERIRSFSPLKEPLTIQVLTVGNAL 832
Mus         ADGTYILNGNFTLSTLEQDLTYKGTVLRYSGSSAALERIRSFSPLKEPLTIQVLMVGHAL 833
            ********* :********* : *** :*****************************:**:**

Homo        RPKIKYTYFVKKKKESFNAIPTFSAWVIEEWGECSKSCELGWQRRLVECRDINGQPASEC 892
Mus         RPKIKFTYFMKKKTESFNAIPTFSE+-WVIEEWGECSKTCGSGWQRRVVQCRDINGHPASEC 893
            *****:***:*** .********* ***********:*   *****:*:*******:*****

Homo        AKEVKPASTRPCADHPCPQWQLGEWSSCSKTCGKGYKKRSLKCLSHDGGVLSHESCDPLK 952
Mus         AKEVKPASTRPCADLPCPHWQVGDWSPCSKTCGKGYKKRTLKCVSHDGGVLSNESCDPLK 953
            ************** ***:**:*:** .***********:***:********* :*******

Homo        KPKHFIDFCTMAECS 967
Mus         KPKHYIDFCTLTQCS 968
            ****:*****:::.:**
```

(C)
```
Mus  855-911   TFSEWVIEEWGECSKTCGSGWQRRVVQCRDING--HPASECAKEVKPAS-TRPCADLPCP 57
Mus  912-968   ---HWQVGDWSPCSKTCGKGYKKRTLKCVSHDGGVLSNESCDPLKKPKHYIDFCTLTQCS 57
                   .*  : :*. ****** .*:::*.::* .  :*    ...* **     *: *.
```

(D)
```
Hum  854-905   TFSAWVIEEWGECSKSCELGWQRRLVECRDING--QPASECAKEVKP------ASTRPCA 52
Hum  908-967   PCPQWQLGEWSSCSKTCGKGYKKRSLKCLSHDGGVLSHESCDPLKKPKHFIDFCTMAECS 60
                 . .* : **..***:* *:::*. .**  : :.  .* .  **   *:.. *:
```

(A)
```
Hs-2        --------------TFSAWVIEEWGECSKSCELGWQRRLVECRDINGQPASECAKEVKPA 46
Mm-Adamts1  YFMKKKTESFNAIPTFSEWVIEEWGECSKTCGSGWQRRVVQCRDINGHPASECAKEVKPA 900
                          *** ***********:*   *****:*:******:***********

Hs-2        STRPCA------------------------------------------------------ 52
Mm-Adamts1  STRPCADLPCPHWQVGDWSPCSKTCGKGYKKRTLKCVSHDGGVLSNESCDPLKKPKHYID 960
            ******
```
(B)
```
Hs-3        ------------------------------------------------------------
Mm-Adamts1  YFMKKKTESFNAIPTFSEWVIEEWGECSKTCGSGWQRRVVQCRDINGHPASECAKEVKPA 900

Hs-3        --------PCPQWQLGEWSSCSKTCGKGYKKRSLKCLSHDGGVLSHESCDPLKKPKHFID 52
Mm-Adamts1  STRPCADLPCPHWQVGDWSPCSKTCGKGYKKRTLKCVSHDGGVLSNESCDPLKKPKHYID 960
                    ***:**.*:**.*********:***:***:********.*************

Hs1-3       FCTMAECS 60
Mm-Adamts1  FCTLTQCS 968
            ***::**
```
(C)
```
Hs-2        --------------TFSAWVIEEWGECSKSCELGWQRRLVECRDING--QPASECAKEVK 44
Mm-ADAMTS1  YFMKKKTESFNAIPTFSEWVIEEWGECSKTCGSGWQRRVVQCRDING--HPASECAKEVK 898
Hs-3        --------------PCPQWQLGEWSSCSKTCGKGYKKRSLKCLSHDGGVLSHESCDPLKK 46
                          . . *  :  **..***:*  *:::*  ::* . :*   . . .*      *

Hs-2        P------ASTRPCA---------------------------------------------- 52
Mm-ADAMTS1  P------ASTRPCADLPCPHWQVGDWSPCSKTCGKGYKKRTLKCVSHDGGVLSNESCDPL 952
Hs-3        PKHFIDFCTMAECS---------------------------------------------- 60
            *           . .:   . *:

Hs-2        ------------------ 
Mm-ADAMTS1  KKPKHYIDFCTLTQCS 968
Hs-3        ------------------
```
(D)
```
Hs-2        ------------------------------------------------------------
Hs-3        ------------------------------------------------------------
Mm-ADAMTS1  YFMKKKTESFNAIPTFSEWVIEEWGECSKTCGSGWQRRVVQCRDINGHPASECAKEVKPA 900

Hs-2        --------TFSAWVIEEWGECSKSCELGWQRRLVECRDINGQPASE--CAKEVKP-----
Hs-3        --------PCPQWQLGEWSSCSKTCGKGYKKRSLKCLSHDGGVLSHESCDPLKKPKHFID
Mm-ADAMTS1  STRPCADLPCPHWQVGDWSPCSKTCGKGYKKRTLKCVSHDGGVLSNESCDPLKKPKHYID 960
                    .     . : *  :.*****:* :*::.. *     :      *       **

Hs-2        -ASTRPCA
Hs-3        FCTMAECS
Mm-ADAMTS1  FCTLTQCS 968
             .:    *:
```
(E)
```
1  1    -------------------------------------------------TFSAWVIEEWGECSKSCELGWQRRLV 26
2      ------------------------------------------------------------------------
3  801 RYSGSSAALERIRSFSPLKEPLTIQVLMVGHALRPKIKFTYFMKKKTESFNAIPTFSEWVIEEWGECSKTCGSGWQRRVV 880

1  27  ECRDINGQPASECAKEVKPASTRPCA---------------------------------------------------- 52
2  1   -------------------------PCPQWQLGEWSSCSKTCGKGYKKRSLKCLSHDGGVLSHESCDPLKKPKHFID 52
3  881 QCRDINGHPASECAKEVKPASTRPCADLPCPHWQVGDWSPCSKTCGKGYKKRTLKCVSHDGGVLSNESCDPLKKPKHYID 960

1      --------
2  53  FCTMAECS 60
3  961 FCTLTQCS 968

1=Hs-2
2=Hs-3
3=Mm-ADAMTS1
```

Figure 11.18 Influence of sequences on the final alignment. (A) Using either ClustalW or MUSCLE, the human thrombospondin domain two aligns with domain two of the mouse full-length sequence. (B) Using either ClustalW or MUSCLE, the human thrombospondin domain three aligns with domain three of the mouse full-length sequence. (C) Simultaneously aligning all three sequences with ClustalW results in misplaced alignments, with human domains two and three aligning with the mouse domain two. (D) MUSCLE makes a different displacement, with both human domains placed on top of mouse domain three. (E) COBALT alignment of these sequences. Note that the COBALT labeling of the lines is "1," "2," and "3" and the legend for the row labels appears at the bottom. Using the default settings, the two human thrombospondin domains are properly aligned to the two domains of the mouse Adamts1 protein. Sequences were obtained from the mouse (P97857) and human (Q9UHI8) Swiss-Prot records.

repeat (Hs-3) aligns with the third mouse thrombospondin repeat (mouse coordinates 912–968), respectively. Both ClustalW and MUSCLE gave the same results. To follow the changes in the alignment, it might be easier if you focus on the first few amino acids of the human domains. For example, in Figure 11.18A, the N-terminal human amino acids TFS align with the mouse amino acids TFS. In Figure 11.18B, human amino acids PCP are aligning with mouse amino acids PCP. However, when trying to align both human TSP repeats to the mouse sequence simultaneously, both human sequences were aligned to the second mouse thrombospondin repeat with ClustalW (Figure 11.18C). Notice amino acids TFS being aligned with PCP in the third human domain. In the same test, MUSCLE aligns both human TSP repeats to the third mouse TSP repeat (Figure 11.18D). However, COBALT successfully aligns the second and third human TSP repeats to their respective mouse repeats (Figure 11.18E).

Does this mean that COBALT is the best tool to use for all alignments? No. The lesson here is that three different tools gave three different results and that having them all, and others, at your disposal when confronted with a challenge is a must. Getting lots of experience with one tool has its advantages but you should be prepared to venture elsewhere for answers when needed.

11.7 ISOFORM ALIGNMENT PROBLEM: INTERNAL SPLICING

Unlike the example seen earlier in this chapter of isoforms differing at the N-terminal, this section will address isoforms where the sequences derived from internal exons do not align properly. **Figure 11.19** shows the seven isoforms of mouse doublecortin-like kinase 1. Two shaded regions correspond to sequence encoded by separate exons that are found in different locations within the different isoforms. Notice how the first, darkly shaded exon (VNGTP..) can be found in all seven isoforms and in three different locations within the protein: very near the N-terminus, near the center of the protein, and near the C-terminus. The second, lightly shaded region (EESEE...) can be found near the N-terminus, near the center, or absent, being present in only five of the isoforms. Together, these two regions in these isoforms present a number of challenges to multiple sequence alignment tools.

Figure 11.20A is a ClustalW alignment (default settings) of these isoforms, focusing on the vicinity of the two shaded regions. Sequence from the darkly shaded region (VNGTP...) is aligned in five of the seven isoforms, with the two missing regions driven elsewhere in the alignment by ClustalW and absent from this figure. The lightly shaded region (EESEE...) is properly aligned in all five isoforms, remembering that it is only present in those five. Note the improper alignment to the sequence of NP_001104523.1. Changing the clustering method of ClustalW from the default NJ to UPGMA improves the alignment (Figure 11.20B), now showing all seven of the darkly shaded sequences aligned. The second region (EESEE...) is also aligned in all five isoforms, although the sequence VGDSV from NP_001182469 and NP_001104523 now appears in the middle of the lightly shaded block. This most probably does not make good biological sense.

Figure 11.21A shows the same regions of the proteins aligned with MUSCLE. In this case, sequences from all seven of the darkly shaded regions (VNGTP...) and all five of the lightly shaded regions are aligned properly. The only exception is the appearance of another short sequence, "DLY," from NP_001182469 and NP_001104523, improperly aligned to the lightly shaded region sequence.

Finally, Figure 11.21B shows the alignment generated by COBALT. In this case the sequences from the darkly and lightly shaded regions are all aligned and there are no short sequences brought into the picture as seen earlier with ClustalW and MUSCLE. Perhaps parameter changes for these two tools can be found that generate the same outcome as that seen with COBALT. Although it is impressive that the default conditions of COBALT did the best in this challenging alignment, the lack of line labeling makes it more difficult to interpret the alignment and the absence of the consensus symbols (*) requires more work to identify regions of identity.

```
>NP_001182469.1| serine/threonine-protein kinase DCLK1 isoform 7 [Mus musculus]
MLELIEVNGTPGSQLSTPRSGKSPSPSPTSPGSLRKQRDLYRPLSSDDLDSVGDSV
>NP_001182468.1| serine/threonine-protein kinase DCLK1 isoform 6 [Mus musculus]
MLELIEVNGTPGSQLSTPRSGKSPSPSPTSPGSLRKQRISQHGGSSTSLSSTKVCSSMDENDGPEGDEL
GRRHSLQRGWRREESEEGFQIPATITERYKVGRTIGDGNFAVVKECIERSTAREYALKIIKKSKCRGKEH
MIQNEVSILRRVKHPNIVLLIEEMDVPTELYLVMELVKGGDLFDAITSTSKYTERDASGMLYNLASAIKY
LHSLNIVHRDIKPENLLVYEHQDGSKSLKLGDFGLATIVDGPLYTVCGTPTYVAPEIIAETGYGLKVDIW
AAGVITYILLCGFPPFRGSGDDQEVLFDQILMGQVDFPSPYWDNVSDSAKELINMMLLVNVDQRFSAVQV
LEHPWVNDDGLPENEHQLSVAGKIKKHFNTGPKPSSTAAGVSVIATTALDKERQVFRRRRNQDVRSRYKA
QPAPPELNSESEDYSPSSSETVRSPNSPF
>NP_001182467.1| serine/threonine-protein kinase DCLK1 isoform 5 [Mus musculus]
MSFGRDMELEHFDERDKAQRYSRGSRVNGLPSPTHSAHCSFYRTRTLQTLSSEKKAKKVRFYRNGDRYFK
GIVYAISPDRFRSFEALLADLTRTLSDNVNLPQGVRTIYTIDGLKKISSLDQLVEGESYVCGSIEPFKKL
EYTKNVNPNWSVNVKTTSASRAVSSLATAKGGPSEVRENKDFIRPKLVTIIRSGVKPRKAVRILLNKKTA
HSFEQVLTDITDAIKLDSGVVKRLYTLDGKQVMCLQDFFGDDDIFIACGPEKFRYQDDFLLDESECRVVK
STSYTKIASASRRGTTKSPGPSRRSKSPASTSSVNGTPGSQLSTPRSGKSPSPSPTSPGSLRKQRISQHG
GSSTSLSSTKVCSSMDENDGPGEEESEEGFQIPATITERYKVGRTIGDGNFAVVKECIERSTAREYALKI
IKKSKCRGKEHMIQNEVSILRRVKHPNIVLLIEEMDVPTELYLVMELVKGGDLFDAITSTSKYTERDASG
MLYNLASAIKYLHSLNIVHRDIKPENLLVYEHQDGSKSLKLGDFGLATIVDGPLYTVCGTPTYVAPEIIA
ETGYGLKVDIWAAGVITYILLCGFPPFRGSGDDQEVLFDQILMGQVDFPSPYWDNVSDSAKELINMMLLV
NVDQRFSAVQVLEHPWVNDDGLPENEHQLSVAGKIKKHFNTGPKPSSTAAGVSVIATTALDKERQVFRRR
RNQDVRSRYKAQPAPPELNSESEDYSPSSSETVRSPNSPF
>NP_001104523.1| serine/threonine-protein kinase DCLK1 isoform 4 [Mus musculus]
MSFGRDMELEHFDERDKAQRYSRGSRVNGLPSPTHSAHCSFYRTRTLQTLSSEKKAKKVRFYRNGDRYFK
GIVYAISPDRFRSFEALLADLTRTLSDNVNLPQGVRTIYTIDGLKKISSLDQLVEGESYVCGSIEPFKKL
EYTKNVNPNWSVNVKTTSASRAVSSLATAKGGPSEVRENKDFIRPKLVTIIRSGVKPRKAVRILLNKKTA
HSFEQVLTDITDAIKLDSGVVKRLYTLDGKQVMCLQDFFGDDDIFIACGPEKFRYQDDFLLDESECRVVK
STSYTKIASASRRGTTKSPGPSRRSKSPASTSSVNGTPGSQLSTPRSGKSPSPSPTSPGSLRKQRDLYRP
LSSDDLDSVGDSV
>NP_001104522.1| serine/threonine-protein kinase DCLK1 isoform 3 [Mus musculus]
MLELIEVNGTPGSQLSTPRSGKSPSPSPTSPGSLRKQRISQHGGSSTSLSSTKVCSSMDENDGPGEEESE
EGFQIPATITERYKVGRTIGDGNFAVVKECIERSTAREYALKIIKKSKCRGKEHMIQNEVSILRRVKHPN
IVLLIEEMDVPTELYLVMELVKGGDLFDAITSTSKYTERDASGMLYNLASAIKYLHSLNIVHRDIKPENL
LVYEHQDGSKSLKLGDFGLATIVDGPLYTVCGTPTYVAPEIIAETGYGLKVDIWAAGVITYILLCGFPPF
RGSGDDQEVLFDQILMGQVDFPSPYWDNVSDSAKELINMMLLVNVDQRFSAVQVLEHPWVNDDGLPENEH
QLSVAGKIKKHFNTGPKPSSTAAGVSVIALDHGFTIKRSGSLDYYQQPGMYWIRPPLLIRRGRFSDEDAT
RM
>NP_001104521.1| serine/threonine-protein kinase DCLK1 isoform 2 [Mus musculus]
MLELIEVNGTPGSQLSTPRSGKSPSPSPTSPGSLRKQRISQHGGSSTSLSSTKVCSSMDENDGPGEEESE
EGFQIPATITERYKVGRTIGDGNFAVVKECIERSTAREYALKIIKKSKCRGKEHMIQNEVSILRRVKHPN
IVLLIEEMDVPTELYLVMELVKGGDLFDAITSTSKYTERDASGMLYNLASAIKYLHSLNIVHRDIKPENL
LVYEHQDGSKSLKLGDFGLATIVDGPLYTVCGTPTYVAPEIIAETGYGLKVDIWAAGVITYILLCGFPPF
RGSGDDQEVLFDQILMGQVDFPSPYWDNVSDSAKELINMMLLVNVDQRFSAVQVLEHPWVNDDGLPENEH
QLSVAGKIKKHFNTGPKPSSTAAGVSVIATTALDKERQVFRRRRNQDVRSRYKAQPAPPELNSESEDYSP
SSSETVRSPNSPF
>NP_064362.1| serine/threonine-protein kinase DCLK1 isoform 1 [Mus musculus]
MSFGRDMELEHFDERDKAQRYSRGSRVNGLPSPTHSAHCSFYRTRTLQTLSSEKKAKKVRFYRNGDRYFK
GIVYAISPDRFRSFEALLADLTRTLSDNVNLPQGVRTIYTIDGLKKISSLDQLVEGESYVCGSIEPFKKL
EYTKNVNPNWSVNVKTTSASRAVSSLATAKGGPSEVRENKDFIRPKLVTIIRSGVKPRKAVRILLNKKTA
HSFEQVLTDITDAIKLDSGVVKRLYTLDGKQVMCLQDFFGDDDIFIACGPEKFRYQDDFLLDESECRVVK
STSYTKIASASRRGTTKSPGPSRRSKSPASTSSVNGTPGSQLSTPRSGKSPSPSPTSPGSLRKQRISQHG
GSSTSLSSTKVCSSMDENDGPEGDELGRRHSLQRGWRREESEEGFQIPATITERYKVGRTIGDGNFAVV
KECIERSTAREYALKIIKKSKCRGKEHMIQNEVSILRRVKHPNIVLLIEEMDVPTELYLVMELVKGGDL
DAITSTSKYTERDASGMLYNLASAIKYLHSLNIVHRDIKPENLLVYEHQDGSKSLKLGDFGLATIVDGPL
YTVCGTPTYVAPEIIAETGYGLKVDIWAAGVITYILLCGFPPFRGSGDDQEVLFDQILMGQVDFPSPYWD
NVSDSAKELINMMLLVNVDQRFSAVQVLEHPWVNDDGLPENEHQLSVAGKIKKHFNTGPKPSSTAAGVSV
IATTALDKERQVFRRRRNQDVRSRYKAQPAPPELNSESEDYSPSSSETVRSPNSPF
```

Figure 11.19 Multiple isoforms for mouse doublecortin-like kinase 1. This mouse gene is alternatively spliced to generate seven protein isoforms pictured here. The two shaded regions are featured in Figures 11.20 and 11.21. These sequences were obtained from RefSeq.

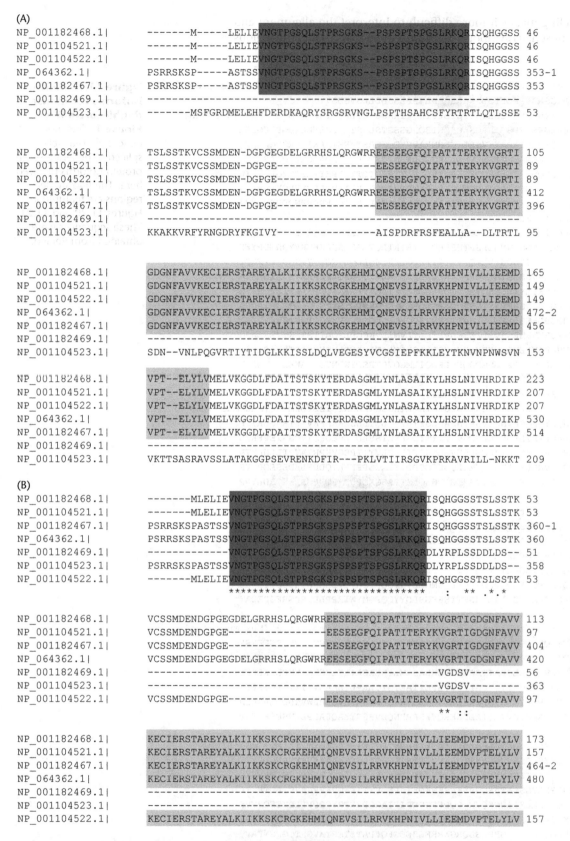

Figure 11.20 ClustalW alignments of the mouse doublecortin-like kinase 1. (A) A section of a ClustalW alignment in the vicinity of the amino acids shaded in Figure 11.19. The default settings of ClustalW fail to align the first region of darkly shaded amino acids from all seven isoforms. The lightly shaded region is aligned with a sequence in NP_001104523.1, with little obvious similarity. (B) Another ClustalW alignment of the same sequences, but in this case the clustering parameter was changed from the default NJ to UPGMA. Note that the darkly shaded region is now aligned properly. The alignment of the lightly shaded sequence has also improved from panel (A).

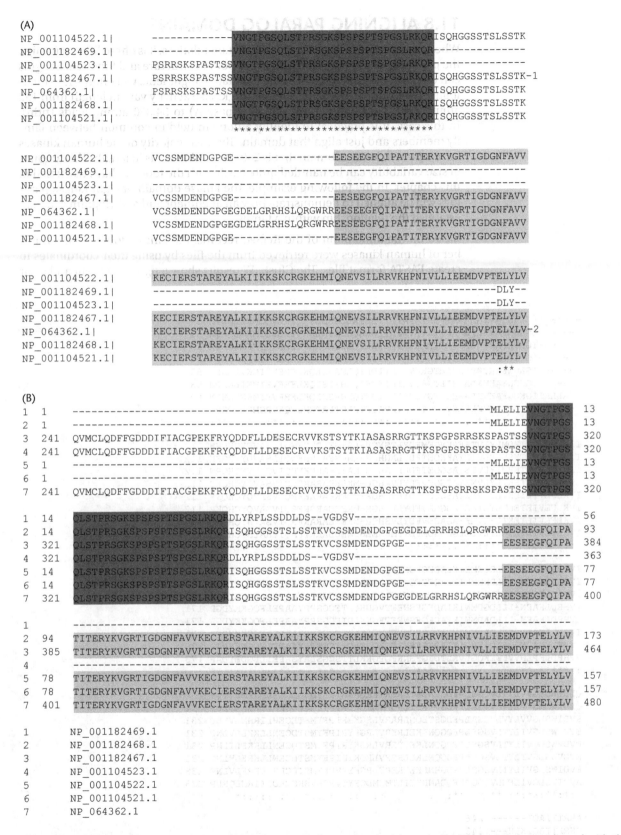

Figure 11.21 MUSCLE and COBALT alignments. (A) The mouse doublecortin-like kinase 1 isoforms aligned with the default settings of MUSCLE. Note that the darkly shaded region in all sequences is properly aligned. The lightly shaded sequence is a tight alignment, with only three amino acids, "DLY," appearing out of place. (B) The alignment by COBALT with the legend for the numbered lines immediately below the alignment. The default settings for COBALT properly align both shaded regions.

11.8 ALIGNING PARALOG DOMAINS

When working with a large family with distant relatives, it is often difficult to align the paralogs because they have diverged both in structure and function. Lengths can often vary considerably and the generated alignments will often strain credibility. For example, human proteins with kinase activity vary in length from 152 amino acids (nucleotide diphosphate kinase A) to 34,350 amino acids (titin). In this case, it is practical to identify a domain held in common between family members and just align that domain. The vast majority of the human kinases have a protein kinase domain that is 240–300 amino acids in length. This amount of size variability can be handled quite easily with the tools we have discussed in this chapter. In the following example, the kinase domain of eight kinase paralogs will be aligned to demonstrate how an alignment and a phylogram can help establish the relationship within a family.

Based on the annotation of the RefSeq records, the kinase domains of a number of human kinases were retrieved from the files by using their coordinates to create FASTA format files. The ClustalW results show a tight alignment and clear

(A)

```
SIK1    -YDIERTLGKGNFAVVKLARHRVTKTQVAIKIIDKTRLDSSNLEKIYREVQLMKLLNHPH 59
SIK2    -------LGKGNFAVVKLGRHRITKTEVAIKIIDKSQLDAVNLEKIYREVQIMKMLDHPH 53
SIK3    -------IGKGNFAVVKRATHLVTKAKVAIKIIDKTQLDEENLKKIFREVQIMKMLCHPH 53
MARK1   -------IGKGNFAKVKLARHVLTGREVAVKIIDKTQLNPTSLQKLFREVRIMKILNHPN 53
MARK2   -------IGKGNFAKVKLARHILTGKEVAVKIIDKTQLNESSLQKLFREVRIMKVLNHPN 53
MARK3   -------IGKGNFAKVKLARHILTGREVAVKIIDKTQLNPTSLQKLFREVRIMKILNHPN 53
MARK4   NYRLLRTIGKGNFAKVKLARHILTGREVAIKIIDKTQLNPSSLQKLFREVRIMKGLNHPN 60
SNRK    -------LGRGHFAVVKLARHVFTGEKVAVKVIDKTKLDTLATGHLFQEVRCMKLVQHPN 53
           :*:*:**:**  .  *  .*  :**:*:***::*:    :::::**: ** : **:

SIK1    IIKLYQVMETKDMLYIVTEFAKNGEMFDYLTSNGH-LSENEARKKFWQILSAVEYCHDHH 118
SIK2    IIKLYQVMETKSMLYLVTEYAKNGEIFDYLANHGR-LNESEARRKFWQILSAVDYCHGRK 112
SIK3    IIRLYQVMETERMIYLVTEYASGGEIFDHLVAHGR-MAEKEARRKFKQIVTAVYFCHCRN 112
MARK1   IVKLFEVIETEKTLYLVMEYASGGEVFDYLVAHGR-MKEKEARAKFRQIVSAVQYCHQKY 112
MARK2   IVKLFEVIETEKTLYLVMEYASGGEVFDYLVAHGR-MKEKEARAKFRQIVSAVQYCHQKF 112
MARK3   IVKLFEVIETEKTLYLIMEYASGGEVFDYLVAHGR-MKEKEARSKFRQIVSAVQYCHQKR 112
MARK4   IVKLFEVIETEKTLYLVMEYASAGEVFDYLVSHGR-MKEKEARAKFRQIVSAVHYCHQKN 119
SNRK    IVRLYEVIDTQTKLYLILELGDGGDMFDYIMKHEEGLNEDLAKKYFAQIVHAISYCHKLH 113
           *::*::*:::*:   :*:: *  .. *::**::   :  .: *. *:  * **: *: :**

SIK1    IVHRDLKTEN-LLLDGNMDIKLADFGFGNFYKSGEPLSTWCGSPPYAAPEVFEGKEYEGP 177
SIK2    IVHRDLKAEN-LLLDNNMNIKIADFGFGNFFKSGELLATWCGSPPYAAPEVFEGQQYEGP 171
SIK3    IVHRDLKAEN-LLLDANLNIKIADFGFTPGQLLKTWCGSPPYAAPELFEGKEYDGP 171
MARK1   IVHRDLKAEN-LLLDGDMNIKIADFGFSNEFTVGNKLDTFCGSPPYAAPELFQGKKYDGP 171
MARK2   IVHRDLKAEN-LLLDADMNIKIADFGFSNEFTFGNKLDTFCGSPPYAAPELFQGKKYDGP 171
MARK3   IVHRDLKAEN-LLLDADMNIKIADFGFSNEFTVGGKLDTFCGSPPYAAPELFQGKKYDGP 171
MARK4   IVHRDLKAEN-LLLDAEANIKIADFGFSNEFTLGSKLDTFCGSPPYAAPELFQGKKYDGP 178
SNRK    VVHRDLKPENVVFFEKQGLVKLTDFGFSNKFQPGKKLTTSCGSLAYSAPEILLGDEYDAP 173
          :******:**  ::::   :    :*::****.* :    *   * * *** .*:***:: *.:*::.*

SIK1    QLDIWSLGVVLYVLVCGSLPFDGPNLPTLRQRVLEGRFRIPFFMSQDCESLIRRMLVVDP 237
SIK2    QLDIWSMGVVLYVLVCGALPFDGPTLPILRQRVLEGRFRIPYFMSEDCEHLIRRMLVLDP 231
SIK3    KVDIWSLGVVLYVLVCGALPFDGSTLQNLRARVLSGKFRIPFFMSTECEHLIRHMLVLDP 231
MARK1   EVDVWSLGVILYTLVSGSLPFDGQNLKELRERVLRGKYRIPFYMSTDCENLLKKLLVLNP 231
MARK2   EVDVWSLGVILYTLVSGSLPFDGQNLKELRERVLRGKYRIPFYMSTDCENLLKKFLILNP 231
MARK3   EVDVWSLGVILYTLVSGSLPFDGQNLKELRERVLRGKYRIPFYMSTDCENLLKRFLVLNP 231
MARK4   EVDIWSLGVILYTLVSGSLPFDGHNLKELRERVLRGKYRVPFYMSTDCESILRRFLVLNP 238
SNRK    AVDIWSLGVILFMLVCGQPPFQEANDSETLTMIMDCKYTVPSHVSKECKDLITRMLQRDP 233
           :*:**:**:*: **.*  **:  .        ::  :: :*  .:*  :*:  ::  ::*    :*

SIK1    ARRITIAQI------ 246
SIK2    SKRLTIAQIKEH--- 243
SIK3    NKRLSMEQICKH--- 243
MARK1   IKRGSLEQIMKDRWM 246
MARK2   SKRGTLEQIMKDRWM 246
MARK3   IKRGTLEQIMKDRW- 245
MARK4   AKRCTLEQIMK---- 249
SNRK    KRRASLEEIENHPW- 247
           :*  :: :*
```

Figure 11.22 Multiple sequence alignment of paralogous kinase domains. (A) The alignment was performed with ClustalW using the default settings. The Prosite ATP-binding signature is shaded.

(B)

SeqA	Name	Length	SeqB	Name	Length	Score
1	MARK1	246	2	MARK3	245	95.0
1	MARK1	246	3	MARK2	246	95.0
1	MARK1	246	4	MARK4	249	88.0
1	MARK1	246	5	SIK3	243	68.0
1	MARK1	246	6	SIK1	246	60.0
1	MARK1	246	7	SIK2	243	62.0
1	MARK1	246	8	SNRK	247	48.0
2	MARK3	245	3	MARK2	246	94.0
2	MARK3	245	4	MARK4	249	90.0
2	MARK3	245	5	SIK3	243	68.0
2	MARK3	245	6	SIK1	246	60.0
2	MARK3	245	7	SIK2	243	63.0
2	MARK3	245	8	SNRK	247	48.0
3	MARK2	246	4	MARK4	249	89.0
3	MARK2	246	5	SIK3	243	68.0
3	MARK2	246	6	SIK1	246	60.0
3	MARK2	246	7	SIK2	243	62.0
3	MARK2	246	8	SNRK	247	47.0
4	MARK4	249	5	SIK3	243	68.0
4	MARK4	249	6	SIK1	246	63.0
4	MARK4	249	7	SIK2	243	63.0
4	MARK4	249	8	SNRK	247	47.0
5	SIK3	243	6	SIK1	246	67.0
5	SIK3	243	7	SIK2	243	73.0
5	SIK3	243	8	SNRK	247	46.0
6	SIK1	246	7	SIK2	243	80.0
6	SIK1	246	8	SNRK	247	45.0
7	SIK2	243	8	SNRK	247	43.0

(C)

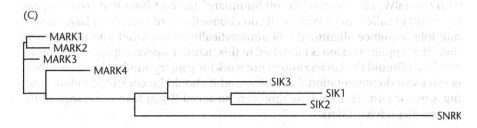

Figure 11.22 Multiple sequence alignment of paralogous kinase domains. (B) The "Result Summary" output from ClustalW. Listed here are the pairwise comparisons of the sequences. The "Score" is the percent identity between pairs of sequences, ranging from 95% (MARK1 and MARK2, or MARK3) to 43% (SIK2 and SNRK). (C) The guide tree, based on these values. The lengths of the branches in this tree are proportional to the number of amino acid differences between sequences. The human kinase domains were obtained from the following full-length sequences: MARK1 NP_061120, MARK2 NP_001034558, MARK3 NP_001122392, MARK4 NP_001186796, SIK1 NP_775490, SIK2 NP_056006, SIK3 NP_079440, and SNRK NP_060189.

identities in many positions (**Figure 11.22A**). These results could have been expected by the similarity in names (for example, *MARK1, MARK2, MARK3*) but remember that until a protein is well studied, name similarity may reflect the historical discovery rather than the function and membership to a family.

The alignment also shows the "Protein kinases ATP-binding region signature" as defined by Prosite, and it is shaded in Figure 11.22A. This pattern (PS00107) is

```
[LIV]-G-{P}-G-{P}-[FYWMGSTNH]-[SGA]-{PW}-[LIVCAT]-{PD}-x-
[GSTACLIVMFY]-x(5,18)-[LIVMFYWCSTAR]-[AIVP]-[LIVMFAGCKR]-K
```

(Note this is one signature wrapped to two lines of text.) Square brackets indicate "or" (for example, L or I or V), the curly brackets indicate "not" (for example, not P), x indicates "any amino acid" may be in that position, and parentheses indicate a range (for example, 5–18 of the preceding amino acid). The last position, lysine, binds the ATP. Based on the alignment, it appears that SIK1 and MARK4 kinase domains may be longer than the others at the N-terminus.

Paralogs related by name can vary in their conservation. Notice that MARK1, MARK2, and MARK3 are 94–95% identical based on the Result Summary in Figure 11.22B, while the identity of MARK4 to these other family members is "only" 88–90%. Still, these members are quite close to each other. The SIK family identities are more distant: the pairs range from 67 to 80% identity. They are also 60–68% identical to the MARK family. The odd kinase in this alignment, SNRK, has a maximum identity of 48% to the others.

These relationships are reflected in the guide tree (Figure 11.22C). MARK1, MARK2, and MARK3 are clustered on short branches, while MARK4 is separated from the others. The SIK family members are more spread out on another branch, and SNRK is far away from all the other kinases.

For an interesting view of the entire human kinase family, a paper published in 2002 by Manning et al. described the entire human kinase family or "kinome" and it has a wonderful figure of the entire tree (so large, in fact, it is referred to as a poster). The members shown in Figure 11.22 are on a single branch of this huge tree. Note that the figure in the poster has outdated names for many of the kinases.

11.9 MANUALLY EDITING A MULTIPLE SEQUENCE ALIGNMENT

It should be evident by now that there will be times where a perfect multiple sequence alignment will not be achievable. Even if you are experienced with one tool, able to employ different parameters with ease, or comfortable with trying multiple tools, you will encounter situations where the minor or major problems with alignments will not be solved but you could see with your eyes how to fix them.

Jalview

Luckily, there is a nice editing tool available called Jalview (www.jalview.org). You can get to Jalview from the ClustalW home page (on the sidebar of www.ebi. ac.uk/Tools/msa/clustalw2) or, after submitting a multiple sequence alignment with ClustalW, click on the "Result Summary" tab and from that page you can see a button called "Start Jalview." If you choose this route to launch Jalview, your multiple sequence alignment will automatically be imported into the application. The "applet" version is launched in this manner, operating within a browser window. (Should the Jalview applet not work for you, try another browser.) There is extensive documentation for Jalview and it should be consulted before starting a major edit. Tremendous time can be saved if you take advantage of the approaches Jalview offers.

Jalview can accept six different formats, including ClustalW. ClustalW is one of the output formats for MUSCLE, so alignments from this second tool can be edited in Jalview as well. After running COBALT, there is a "Download" link at the top of the results page. Click here and you are given the opportunity to save the file in ClustalW format. So the output from all three of these tools can be edited in Jalview.

Figure 11.23 shows a simple session with Jalview. In this example, a number of gaps are removed from an alignment in order to force two amino acids to be adjacent to each other. The Jalview session was launched within ClustalW. The editing window is quite colorful, showing different colors for the amino acids. Figure 11.23A (and color plates) shows the alignment to be edited, the colors evident in this figure as different shades (see the color plates for the color version). Here, the first methionine in three rows is separated from a block of sequence by five gaps. Highlighting the gaps with the mouse draws a box (Figure11.23B). Under the "Edit" menu in Jalview, use the "Cut" command to eliminate these highlighted gaps. The result looks good except that every sequence row which had the gaps removed is now moved to the left by five positions, throwing the alignment off all the way downstream (Figure 11.23C). To correct this, the three gaps to the left of the methionines are highlighted, drawing a box (Figure 11.23D). This time, tap the keyboard right arrow key five times to push the sequences in those rows back over to the right, restoring the former alignment of these sequences (Figure 11.23E). On occasion, the amino acids lose their coloring. Using the "Edit" menu again, click "Undo" and then "Redo" to refresh the view and the colors return (Figure 11.23F).

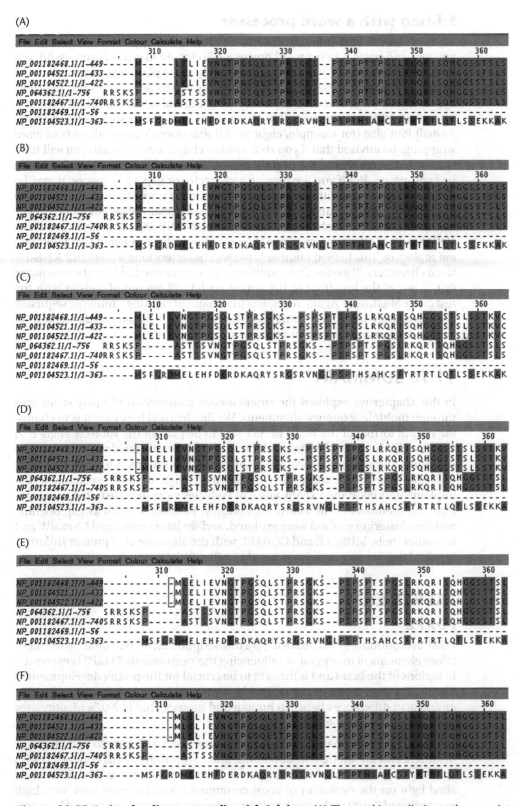

Figure 11.23 A simple alignment edit with Jalview. (A) The goal is to eliminate the gaps introduced by ClustalW between the methionines and leucine–glutamic acid pairs in the vicinity of alignment coordinate 310. (B) The gaps to be eliminated are highlighted by clicking and dragging the mouse. To eliminate the highlighted gaps, use the "Cut" command from the Jalview "Edit" menu, upper left in the Jalview window. Note that the sequences are now misaligned (C). (D) To realign the residues, highlight the gaps to the left of the methionines. (E) Using the keyboard right arrow key, "push" the sequences to the right by five taps. Should you overshoot, use the left arrow key. Although the sequences are now realigned, the colored columns (grayscale in this figure) may not show the strong alignment displayed in (A) and (B). (F) To make the coloring reappear, use the Jalview Edit command "Undo" followed by a "Redo" to refresh the column coloring. See color plates for a color version of this figure.

Editing with a word processor

If your edits are very minor, you can also choose to copy the alignment out of any tool and paste it into a word processor. To insure that this is successful, use "paste special" in your word processor to only paste plain text. If you paste the default format, it may be in html that certainly has its advantages (for example, hypertext will often transfer), but html formatting will also transfer and the text is sometimes difficult to manipulate. Use a monospaced font such as Courier and a small font size (for example, eight or six) that doesn't cause all sorts of word wrapping. Be advised that if you shift around characters too much you will have to change the content of the next block, and the next block, and so on! This can be error-prone. But if you are moving a character or gap a few spaces, it may be easy. Moving your multiple sequence alignment to a word processor gives you the ability to edit the names of the lines. This is especially important for COBALT alignments which are labeled "1, 2, 3... ". Instead of NP_001182469, you can call it "DCLK1-iso7." Of course, you have to be mindful of the number of characters you are replacing. This RefSeq number is twelve characters long while "DCLK1-iso7" is ten characters. Therefore, two additional spaces are needed after the new name or the rest of the line (that is, the amino acids) will get out of register with the rest of the alignment. Again, recognize that manually editing a multiple sequence alignment may introduce errors. See the Appendix 1 for more information on editing sequences using a word processor.

11.10 SUMMARY

In this chapter we explored the simultaneous comparison of many sequences through multiple sequence alignments. We first learned how easy it is to change the output format of the NCBI BLAST tool to display all the subjects aligned to a query. By changing the identical amino acids to dots, the data were reduced to a form allowing easy recognition of variable positions. ClustalW was then introduced as a tool specifically designed to align multiple sequences. The implementations at the ExPASy and EBI Websites were demonstrated using globins and members of the kinase superfamily. Parameters such as gap opening and the clustering method were explored, and we later challenged ClustalW and two other tools, MUSCLE and COBALT, with the alignments of protein isoforms. Finally, Jalview was briefly introduced to edit multiple sequence alignments.

EXERCISES

FOXP2

FOXP2 (pronounced "fox pea two") is a transcription factor that binds to the regulatory elements of many genes, influencing their expression. FOXP2 is expressed in regions of the brain and is thought to be crucial for the proper development of that organ. The FOXP2 protein is highly conserved in evolution, with only three amino acid differences between human and mouse. That is 99.6% identity over 90 million years of evolution!

Mutations in the human *FOXP2* gene are associated with speech and language disorders and are thought to impair the fine motor movements connected with speech. The differences between species have been closely studied in order to shed light on the evolution of vocal communication. Canaries have very high expression of Foxp2 in the brain striatum, a region associated with vocal learning. Furthermore, expression in canaries varies seasonally as does their song. Easily manipulated experimentally, the mouse *Foxp2* gene has been "humanized" to encode a sequence identical to the human protein. Interestingly, newborn mice carrying these mutations have different vocalizations.

Since the discovery of Neanderthal bones, scientists have been intrigued by the Neanderthal hominids who coexisted with modern humans up to approximately

25,000 years ago. Considering their physical attributes, large brain, and extensive tools, it has been a mystery as to why they perished while we thrived. One speculation is that, although quite similar to modern man, they were missing one or more key skills that prevented them from competing in the rapidly changing world. One skill that is proposed as missing is the ability to communicate at a very high level. Therefore, when the Neanderthal *FOXP2* gene was sequenced from carefully isolated DNA obtained from fossilized bone, a compelling analysis was to compare the sequence of this gene to that of modern human. The result: Neanderthal protein is identical to that of the modern human.

In this exercise, you will examine the sequence of this interesting protein in animals using multiple sequence alignments. Be sure to examine the alignments carefully to determine if the alignment parameters or tools have to be changed to optimize the alignment.

1. Retrieve the six human FOXP2 protein isoform sequences from RefSeq and align them.

2. Retrieve the five rat Foxp2 protein isoform sequences from RefSeq and align them.

3. Align the human and rat sequences in one multiple sequence alignment. How do the rat and human isoforms compare?

4. Retrieve the RefSeq protein records for FOXP2 from human (isoform 4), mouse, rat (isoform 1), baboon, gorilla, and macaque and align them.

5. Ignoring differences between isoforms, what are the amino acid differences between human and mouse FOXP2?

6. Based on the alignments generated in this exercise (4), there are two amino acid positions that are uniquely human. What are they?

FURTHER READING

Abe MK, Saelzler MP, Espinosa R et al. (2002) ERK8, a new member of the mitogen-activated protein kinase family. *J. Biol. Chem.* 277, 16733-16743. This paper provides an introduction to the *MAPK15* family of kinases, used as examples in this chapter.

Adams JC & Tucker RP (2000) The thrombospondin type 1 repeat (TSR) superfamily: diverse proteins with related roles in neuronal development. *Dev. Dyn.* 218, 280-299. A nice review article, with beautiful and colorful figures.

Burgess HA & Reiner O (2002) Alternative splice variants of doublecortin-like kinase are differentially expressed and have different kinase activities. *J. Biol. Chem.* 277, 17696-17705. A detailed and thorough scientific study on this protein, featured in this chapter.

Chenna R, Sugawara H, Koike T et al. (2003) Multiple sequence alignment with the Clustal series of programs. *Nucleic Acids Res.* 31, 3497-3500. The reference for ClustalW.

Drewes G, Ebneth A, Preuss U et al. (1997) MARK, a novel family of protein kinases that phosphorylate microtubule-associated proteins and trigger microtubule disruption. *Cell* 89, 297-308. A nice article on the early characterization of the MARK family of proteins, featured in this chapter.

Edgar RC (2004) MUSCLE: multiple sequence alignment with high accuracy and high throughput. *Nucleic Acids Res.* 32, 1792-1797. The reference for MUSCLE.

Enard W, Przeworski M, Fisher SE et al. (2002) Molecular evolution of FOXP2, a gene involved in speech and language. *Nature* 418, 869-872. A paper describing the first links between mutations in this protein and speech disorders.

Krause J, Lalueza-Fox C, Orlando L et al. (2007) The derived FOXP2 variant of modern humans was shared with Neandertals. *Curr. Biol.* 17, 1908-1912. The discovery stated in the title suggests that we shared a common ancestor who had these human-specific amino acids important for speech.

Makałowski W, Zhang J & Boguski MS (1996) Comparative analysis of 1196 orthologous mouse and human full-length mRNA and protein sequences. *Genome Res.* 6, 846-857. A nice and early large-scale comparison of orthologs.

Manning G, Whyte DB, Martinez R et al. (2002) The protein kinase complement of the human genome. *Science* 298, 1912–1934. A massive work based on the multiple sequence alignment of hundreds of proteins. See also the human kinome poster: kinase.com/human/kinome.

Papadopoulos JS & Agarwala R (2007) COBALT: constraint-based alignment tool for multiple protein sequences. *Bioinformatics* 23, 1073–1079. The reference for COBALT.

Perelman P, Johnson WE, Roos C et al. (2011) A molecular phylogeny of living primates. *PLoS Genet.* 7, e1001342. Our family tree based on large-scale sequencing.

Saitou N & Nei M (1987) The neighbor-joining method: a new method for reconstructing phylogenetic trees. *Mol. Biol. Evol.* 4, 406–425. An early description of Neighbor Joining, a widely used method of multiple sequence alignments.

Sneath PHA & Sokal RR (1973) Numerical Taxonomy, pp 230–234. WH Freeman, San Francisco, CA. A very early description of the UPGMA algorithm for generating multiple sequence alignments.

Stockand JD (2005) Preserving salt: in vivo studies with Sgk1-deficient mice define a modern role for this ancient protein. *Am. J. Physiol. Regul. Integr. Comp. Physiol.* 288, R1–R3. A relatively short review of the physiological and molecular biological issues of this protein.

Vogel W (1999) Discoidin domain receptors: structural relations and functional implications. *FASEB J.* 13 (Suppl), S77–S82. A nice review of these membrane proteins which connect the outside and inside worlds of the cell.

Waterhouse AM, Procter JB, Martin DM et al. (2009) Jalview Version 2—a multiple sequence alignment editor and analysis workbench. *Bioinformatics* 25, 1189–1191. A short description of this powerful editor.

Zambelli F, Pavesi G, Gissi C et al. (2010) Assessment of orthologous splicing isoforms in human and mouse orthologous genes. *BMC Genomics* 11, 534. An interesting example of large-scale database mining.

ACAAGGGACTAGAGAAACCAAAA
AGAAACCAAAACGAAAGGTGCAGAA
AACGAAAGGTGCAGAAGGGGAAACAGATGCAGA
GAAGGGGAAACAGATGCAGAAAGCAT
AGAAAGCAT
ACAAGGGACTAGAGAAACCAAAACGAAAGGTGCAGAAGGGGAAACAGATGCAGAAAGCAT

CHAPTER 12

Browsing the Genome

Key concepts

- Chromosome structure and statistics
- Synteny and comparing the genomes between species
- Navigating the genome with the UCSC Genome Browser and exploring gene details

12.1 INTRODUCTION

In many of the previous chapters we have studied the details of DNA and protein sequences. From these fine details we now step back to get a larger view, using genome browsers. These wonderful tools allow the visualization of large portions of the genome, yet reduce the data into forms and numbers that are manageable and interpretable.

12.2 CHROMOSOMES

The human genome consists of 22 autosomes and two sex chromosomes, X and Y. Together they contain over three billion nucleotides and are stored in a microscopic cell structure, the nucleus. While not engaged in cell division activities, these chromosomes are in an elongated state, often evoking an image of tangled and disorganized strands crammed into a tiny compartment. Remarkably, the chromosomes are quite organized and sequestered to distinct and reproducible sections of the nucleus. In this state, nonadjacent DNA strands are brought into close proximity and some forms of gene expression are regulated through these interactions.

In the cell division stage called metaphase, the chromosomes become tightly coiled and structured. With proper staining, scientists can then observe them under a light microscope, revealing variable sizes and their distinct and differentiating banding patterns (**Figure 12.1**). Figure 12.1 was obtained from the Ensembl Genome Browser Website, www.ensembl.org (Ensembl is pronounced "on som bull," as in a group of musicians). After navigating to this page, click on the human link and then click on "Karyotype" on the left sidebar. The chromosomes are numbered according to their sizes as seen under the microscope. Various stains show bands that allow chromosomes nearly identical in size to be specified (for example, see chromosomes 10 and 11 in Figure 12.1). Each metaphase chromosome has a visual constriction point called the centromere. Here, specialized proteins and structures attach and pull apart paired chromosomes to either daughter cell during cell division. These centromeres act as visual landmarks, dividing many of the chromosomes into two distinct regions. For example, in Figure 12.1 chromosome 2 has a short arm (p, for petite), the centromere (seen as a constricted point in the cartoon), and then the large arm (q).

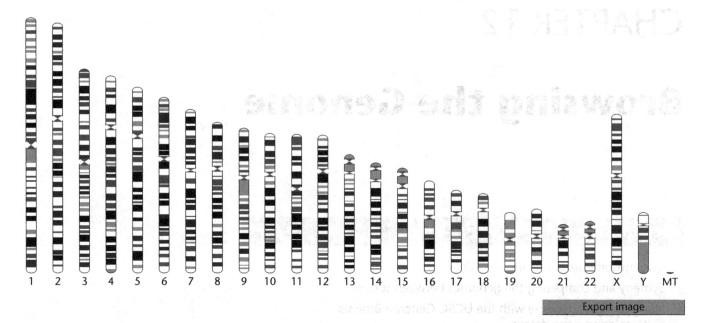

Export image

Figure 12.1 Karyotype view of the human genome from the Ensembl Genome Browser. There are 22 autosomes and two sex chromosomes. The mitochondrial genome is the small figure on the far right. These figures represent metaphase chromosomes, stained to reveal dark and light bands and viewed by light microscopy. Chromosome banding is used for chromosome identification as well as the discovery of chromosome rearrangements, frequently associated with genetic disorders.

For years, those skilled in **karyotyping** studied these microscopic images and both discovered and diagnosed many genetic disorders caused by chromosomal abnormalities such as rearrangements and chromosome number variations. Today, the molecular consequences of many of these abnormalities are now defined or studied, and these genetic elements are the focus in this book.

Human chromosome statistics

Since the completion of the Human Genome Project, there has been a continuous refinement of the sequence as well as improvement in gene prediction, gene discovery, definition of genetic elements, and addition of new data. As a result, you should always consider this and any genome a "version," subject to change.

At the Ensembl Genome Browser Website, statistical summaries are available for the latest "build" or version of the genome. **Figure 12.2** presents some of this version and gene count information. These data are found on the same page as the karyotype seen in Figure 12.1. Along with important information such as version and release date, the size and gene numbers are presented. As novel genes are identified (for example, they go from "predicted gene" to "liver kinase 123") you can expect the "Novel" category to shrink and the "Known" gene number to grow. Other discoveries can lead to changes in the listed categories. For example, there could be two adjacent gene predictions but with a new transcript that links these exons together as one gene, the predicted gene count would decrease.

Figure 12.2 Version information on the Ensembl Human Genome Browser.

Version
Assembly: GRCh37.p6, Feb 2009
Database version: 66.37
Base Pairs: 3,286,906,305
Golden Path Length: 3,101,804,739
Genebuild by: Ensembl
Genebuild released: Apr 2011
Genebuild last updated/patched: Feb 2012

Gene counts
Known protein-coding genes: 20,563
Novel protein-coding genes: 536
Pseudogenes: 15,520
RNA genes: 11,960
Gene exons: 673,807
Gene transcripts: 190,053

Clicking on the chromosomes seen in Figure 12.1 brings you to summary data for those individual chromosomes and statistics collected from these pages are collated in **Table 12.1**. The length (in base pairs) of the chromosomes largely confirms the rank ordering and naming of the chromosomes that were determined decades ago by scientists using light microscopy, which is a testament to their

Table 12.1 Human chromosome statistics

Chromosome	1	2	3	4	5	6	7	8
Length (bp)	249,250,621	243,199,373	198,022,430	191,154,276	180,915,260	171,115,067	159,138,663	146,364,022
Known protein-coding genes	2040	1220	1055	719	868	1021	895	675
Novel protein-coding genes	63	66	34	56	36	53	59	76
Pseudogene genes	1123	942	717	694	675	733	795	568
miRNA genes	134	115	99	92	83	81	90	80
rRNA genes	66	40	29	24	25	26	24	28
snRNA genes	221	161	138	120	106	111	90	86
snoRNA genes	145	117	87	56	61	73	76	52
Other RNA genes	106	93	77	71	68	67	70	42
SNPs	2,396,450	2,462,825	1,991,453	1,989,953	1,790,533	1,866,905	1,677,602	1,526,823

Chromosome	9	10	11	12	13	14	15	16
Length (bp)	141,213,431	135,534,747	135,006,516	133,851,895	115,169,878	107,349,540	102,531,392	90,354,753
Known protein-coding genes	812	753	1276	1032	329	630	621	868
Novel protein-coding genes	22	30	76	34	17	13	19	33
Pseudogene genes	712	495	771	357	321	204	120	104
miRNA genes	69	64	63	72	42	92	78	52
rRNA genes	19	32	24	27	16	10	13	32
snRNA genes	66	87	74	106	45	65	63	53
snoRNA genes	51	56	76	62	34	97	136	58
Other RNA genes	55	56	53	69	36	46	39	34
SNPs	1,336,410	1,430,521	1,408,765	1,358,610	1,010,455	932,828	848,069	959,552

Chromosome	17	18	19	20	21	22	X	Y
Length (bp)	81,195,210	78,077,248	59,128,983	63,025,520	48,129,895	51,304,566	155,270,560	59,373,566
Known protein-coding genes	1173	275	1428	542	233	449	827	41
Novel protein-coding genes	54	10	25	11	9	25	38	14
Pseudogene genes	247	54	174	211	148	304	779	322
miRNA genes	61	32	110	57	16	31	128	15
rRNA genes	15	13	13	15	5	5	22	7
snRNA genes	80	51	29	46	21	23	85	17
snoRNA genes	71	36	31	37	19	23	64	3
Other RNA genes	46	25	15	34	8	23	52	2
SNPs	822,500	793,335	647,509	708,569	459,053	436,162	831,295	145,120

Data from the chromosome summary pages of the Ensembl Genome Browser Website for the GRCh37 genebuild (March 2009). For each chromosome, the number of the following genetic elements is listed: known and novel protein-coding genes, pseudogenes, microRNA (miRNA) genes, ribosomal RNA (rRNA) genes, small nuclear RNA (snRNA) genes, small nucleolar RNA (snoRNA) genes, genes for other RNAs, and single nucleotide polymorphisms (SNPs). bp, base pairs.

fine skills. Except for the four smallest autosomes, chromosome 1 is larger than chromosome 2, which is larger than chromosome 3, and so on. Find photographs of metaphase chromosomes on the Internet and you will see how challenging the original naming must have been.

With the total number of protein-coding genes for the human genome being over 21,000, this translates to two-thirds of the chromosomes only having gene counts in the hundreds and novel genes in the dozens. Pseudogenes are a separate line in these statistics and there are over 15,000 of these genes without apparent protein-coding capacity. However, recent work suggests that pseudogene transcripts and short translation products may play important roles in gene regulation.

Not to be ignored are the numerous small RNA genes with no protein product, such as the miRNAs presented in Chapter 10. The study of these RNAs is a very active area and will most likely increase the number and importance of these genes dramatically. Finally, single nucleotide polymorphisms (SNPs) number in the millions and have been critical in identifying connections with diseases and defining populations around the world.

For another source of gene listings for each chromosome, www.genenames.org is the Website for the HUGO Gene Nomenclature Committee. Here you will see similar chromosome **ideograms** and clicking on them brings up similar statistics. In addition, this Website gives you the opportunity to download gene lists in the form of large tables that you can customize for content.

Chromosome details and comparisons

From the karyotype view of the human genome seen in Figure 12.1, the Ensembl Genome Browser Website allows you to click on the individual chromosomes and choose "chromosome summary view," bringing up more detail about landmarks and genes, in addition to the statistics seen in Table 12.1. As seen in **Figure 12.3**, these chromosome ideograms clearly show the p and q arms, along with the coordinates (for example, p34.3) based on the band positions. Also visible are the gross phenotypic differences (for example, chromosome 1 versus chromosome 22). It is important to remember the scale in this figure; chromosome 1 is 249 million nucleotides long while chromosome 22 is 51 million nucleotides long, a fifth of the size.

Gene distribution is represented as horizontal bars, the length of each being proportional to the number of genes at that location. Looking across all four panels of Figure 12.3, representing four different human chromosomes, the gene

Figure 12.3 Human chromosome appearance, gene location, and gene density. Shown here are the individual chromosome maps and gene displays of chromosomes 1, 22, X, and Y obtained from the Ensembl Genome Browser. Genes located on the short (p) and long (q) arms of chromosomes are represented as horizontal bars to the right of the banded cartoon of the chromosome, with dark gray representing known genes.

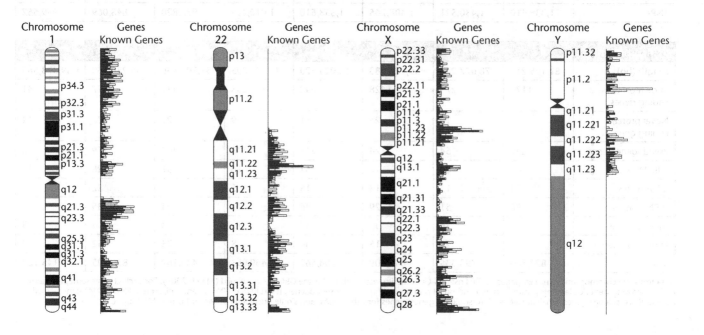

distribution varies considerably. There are regions where the distribution of genes is relatively uniform (for example, chromosome 1 p34.3), gene-dense loci (for example, chromosome 22 q11.22), and regions devoid of genes (for example, chromosome 22 p13). It is notable that the X chromosome has numerous and evenly distributed genes while the Y chromosome has few genes and a large region that has none.

12.3 SYNTENY

As species evolved, genes were lost and gained and the genomes underwent tremendous upheavals in gross structure. Chromosomes grew and shrunk through gene and whole-segment duplication and deletion. Chromosomes also fragmented, swapping pieces in both even and uneven exchanges. Through it all (or by selection), some adjacent genes managed to travel together and, despite millions of years from a common ancestor, we see genome fragments today that are quite similar to each other even though they are from completely different species. This is the essence of **synteny**.

A wonderful tool for visualizing the myriad of exchanges is found on the Ensembl Genome Browser Website. While viewing the detail of a single chromosome, such as that seen for chromosome 1 in Figure 12.3, click on the "Synteny" link on the left sidebar and a new page comes up (**Figure 12.4**). Drop-down menus at the lower right allow you to choose the human chromosome and comparison

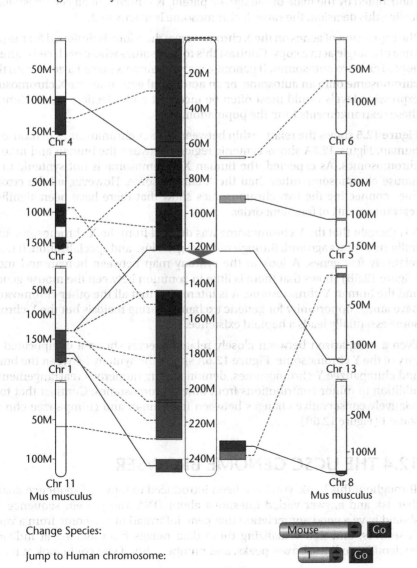

Figure 12.4 Synteny between human chromosome 1 and mouse chromosomes. If you look at the conserved regions of human and mouse chromosomes (Chr), orthologous blocks from eight different mouse chromosomes (flanking images) map to human (center) chromosome 1. Brown lines (shown here as dashed) indicate regions that have flipped in orientation. The Ensembl Genome Browser colorfully displays these and 15 other syntenic relationships for all human chromosomes.

Calico cats

The domestic cat has a gene for coat color on the X chromosome and this gene has two alleles. One version of the gene will produce black pigmentation while the other will result in an orange coat color. In early female cat development, the cells responsible for hair color start spreading throughout the embryo and then undergo random X inactivation. The individual cells continue to divide becoming growing islands of cells, all with the same inactivated X chromosome. The result is the characteristic coat color seen in calico cats: blotches of black and orange.

species. Displayed is a colorful figure showing that parts of human chromosome 1 (center of figure) correspond to eight different mouse chromosomes: 1, 3, 4, 5, 6, 8, 11, and 13 (the flanks of the figure). Lines between species connect the centers of each block. The lines are color-coded (not visible in Figure 12.4) indicating whether the mouse synteny is in the same orientation or flipped. Note that there is a section near the centromere (labeled 140M for the 140 million nucleotides coordinate) that has no detected mouse counterpart.

Synteny of the sex chromosomes

The sex chromosomes are under different and interesting constraints in evolution. The X chromosome exists as two copies in females and only one copy in males. Since subtle differences in gene expression can influence gene pathways and, ultimately, the survival of the animal, how does this difference in the number of X chromosomes affect males and females? Are X-linked genes in males expressed in double amounts to match the levels seen in females? If so, how could nature preferentially double the activity of the hundreds of gene promoters? Or does the male just cope with half the X chromosome gene expression levels seen in females? Nature's solution is to turn genes off. Early in female development, one of the X chromosomes is inactivated to turn off all the gene contributions of that one chromosome, a situation that persists in the ensuing generations of somatic cells. The result is that both males and females have just one "active" X chromosome. Interestingly, the choice of which X chromosome is inactivated, contributed by the male or the female parent, is random. In all cell divisions following this decision, the same X chromosome is inactivated.

The expression of genes on the X chromosome, therefore, is designed for the presence of a single active copy. Contrast this to autosomes where each cell expresses genes from two autosomes. If genomic rearrangements placed a gene from the X chromosome onto an autosome, or an autosomal gene onto the X chromosome, expression levels would most often be incorrect and selection would eliminate these rearrangements from the population.

Figure 12.5 shows the relationship between the sex chromosomes of mouse and human. Figure 12.5A shows syntenic regions between the human and mouse X chromosomes. As expected, the human X chromosome is not syntenic to any mouse chromosome other than the X chromosome. However, all the crossing lines connecting the two chromosomes show that there have been significant rearrangements of fragment order.

It is thought that the Y chromosome was derived from the X chromosome many millions of years ago and, through massive gene loss and specialization, it retains relatively few genes. A look at the synteny map between human and mouse (Figure 12.5B) shows that there is little in common between the mouse genome and the human Y chromosome. It is interesting that all the other chromosomes have ample opportunity for genetic exchange during meiosis but the Y chromosome essentially leads a haploid existence.

Even a comparison between closely related species shows a complicated history of the Y chromosome. **Figure 12.6A** shows the synteny between the human and chimpanzee Y chromosomes, demonstrating numerous rearrangements in addition to minor contributions from other chromosomes. Contrast that to the relatively conservative changes between the human and chimpanzee chromosome 1 (Figure 12.6B).

12.4 THE UCSC GENOME BROWSER

Throughout this book, you have been introduced to topics of sequence analysis that ask and answer varied questions about DNA and protein sequence. You should have a good appreciation that gene information can come from a variety of sources, and that visualizing these data ranges from looking at individual nucleotides to lines, boxes, peaks, and numbers. But if you could look at most of

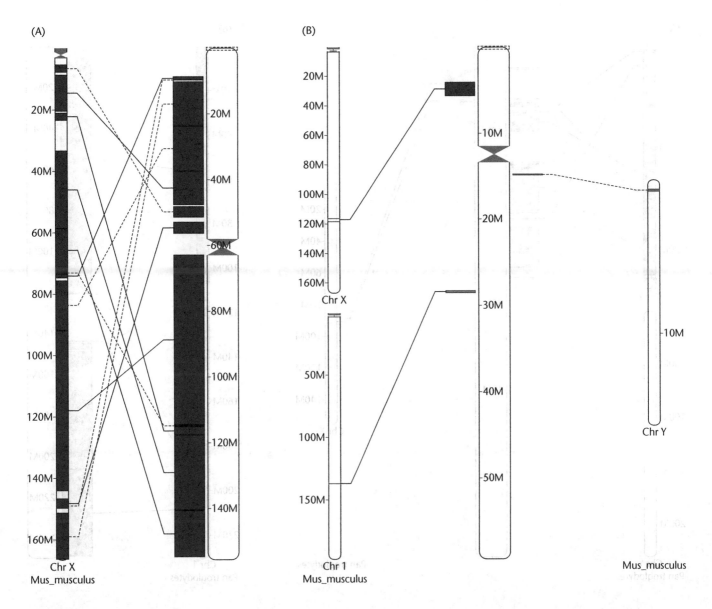

these data simultaneously, in an organized and customizable fashion, you might literally see the big picture more easily, come to conclusions more rapidly, or formulate new questions more readily. From these ideas and promise came the concept of a genome browser. Above, only a fraction of the Ensembl Genome Browser was explored and you are encouraged to explore its capabilities on your own. Now we will examine some of the capabilities of the other great browser, the UCSC Genome Browser from the University of California at Santa Cruz.

To navigate to this browser, go to the UCSC Genome Informatics Web page (genome.ucsc.edu) and along the top and left side of this page are links to many of their projects and tools that they offer. The "Genome Browser" link is at the top-left corner. Clicking there brings you to their Gateway page (**Figure 12.7**), displaying a simple interface to gain access to genome browsers for over 50 different species. Through several drop-down menus, you can access these species and specific genes directly. As expected, the information for human and common model organisms is very extensive compared to other species on the list.

OPN5: a sample gene to browse

To explore the functions of the UCSC Genome Browser, let's browse the human *OPN5* gene locus. Opsin-5 is a member of the G-protein-coupled receptor family and most probably acts as a light-sensitive protein in vertebrates. If you enter

Figure 12.5 Synteny between human and mouse sex chromosomes. (A) The Ensembl syntenic map between the human (right) and mouse (left) X chromosomes. Although there have been a number of rearrangements, orthologous blocks have remained within the same chromosome. (B) The unusual syntenic map of the human (center) and mouse (right) Y chromosomes. There are relatively few genes on the human and mouse Y chromosomes, and only a tiny fraction of sequence is in common between species.

(A)

(B)

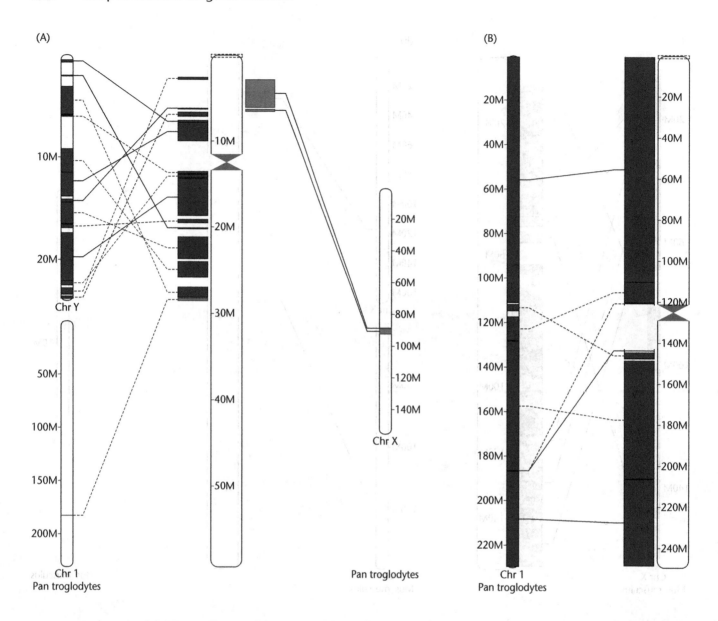

Figure 12.6 Synteny between human and chimpanzee chromosomes. Even between closely related species, the syntenic maps of (A) human (center) and chimpanzee (flank) Y chromosomes show a complex evolution compared to that of (B) an autosome [chromosome 1of human (right) and chimpanzee].

opsin or G-protein-coupled receptor in the "position or search term" box (Figure 12.7), you are presented a long list of hits that may be interesting to explore but is not the most direct route to *OPN5*. The best approach for quick navigation to your gene is to use the official gene symbol in the "gene" text box (Figure 12.7). Enter "OPN5" into this field and you are taken directly to the browser view quite similar to that seen in **Figure 12.8**.

It is hard to ignore all the information in the main browser window (the big white window), so before going into the other details of the window, here is a very brief overview. The *OPN5* gene stretches from left to right, 5P to 3P, respectively. The gene's exons and introns are near the top of the browser window with boxes for exons and thin lines for introns. UTRs are shorter in height than the taller coding regions. Below the gene are rows of different kinds of information represented by peaks, boxes, and shaded coloration. These rows are labeled both on the left and in the middle of the window.

Near the top of the browser window is a row of navigation buttons. Movement of the entire window along the chromosome is controlled by buttons with "<" and ">" symbols, with more symbols controlling greater movement. You may also click and hold the mouse in the window and drag left or right. Zooming in or out is controlled by two sets of buttons, and the amount of zoom is indicated by the numbers 1.5×, 3×, and 10×. Clicking on the "base" button will zoom in until individual nucleotides are seen.

Next is a row containing the "position/search" and "gene" text entry fields seen earlier. The "position/search" window in Figure 12.8 is populated with the genomic coordinates of human chromosome 6 that correspond to the *OPN5* gene. Since we are at the genomic scale, the coordinates are in millions

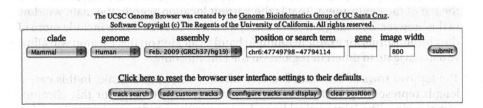

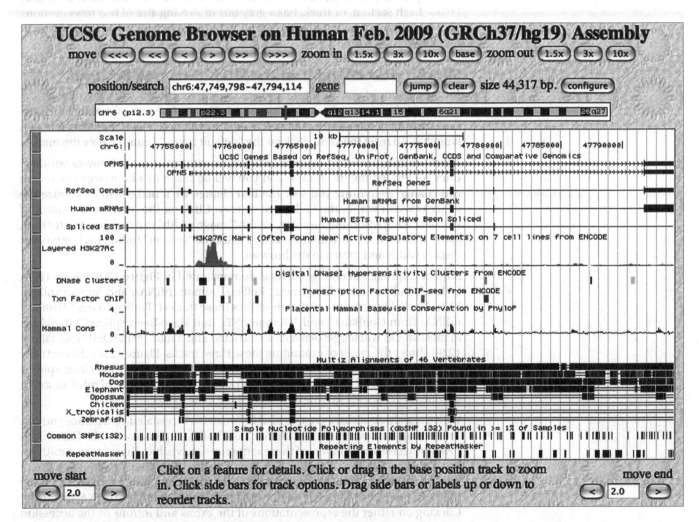

Figure 12.8 The primary interface of the UCSC Genome Browser. Although it is easy to suffer data overload, the browser offers convenient navigation and varying levels of data display. Navigation buttons at the top of this window direct scrolling to the left or right as well as zooming in or out to view less and more of a genomic region, respectively. A karyotype view shows where you are at all times, indicated by a rectangle on top of the banded chromosome figure. The main browser (white) window is organized by rows or tracks with the scale appearing at the very top. In this default view, different representations of the gene features appear as boxes (exons) joined by lines (introns). Small arrows on the top lines of gene *OPN5* point 3P-ward. Transcriptional elements or evidence is shown along with a graphical representation of sequence conservation between animals. At the bottom are common single nucleotide polymorphisms along with mapped repeat elements. All tracks can be expanded to show more detail or hidden completely. A host of other features are not shown here but are available through menu choices below this interface.

of nucleotides. Changing the chromosome and coordinates in this window will change the view in the white window. You can navigate to places that have no genes, for example, by using this coordinates field. The "jump" button functions as a "submit" button should you enter anything in the two text entry fields; "clear" will clear these two fields.

The next row is a view of the chromosome, in this case chromosome 6. Note the small dark gray box (red on the screen) crossing the p arm of the chromosome to the left of the centromere. That is the current location seen in the main window of the browser. Zooming out (seen later) would widen this box to include more of the chromosome, for example multiple bands. Clicking on this cartoon will allow you to navigate to different regions on the chromosome.

The top two rows of the main browser window contain the scale, in this case a length representing 10,000 nucleotides, and the coordinates for this chromosomal region. The faint vertical lines throughout the window assist you in seeing how objects align. Also note that the left boundaries of the white window are gray boxes. Each section, or track, has a gray box of varying size of two rows or more. These gray boxes are actually configuration buttons that allow you to configure or hide these rows entirely.

Simple view changes in the UCSC Genome Browser

Note that should you be working on a shared computer, you may not see the tracks of information shown in Figure 12.8. The preferences of the previous session with the genome browser are saved and will take effect the next time you launch it. To reset the view, click on the "default tracks" button below the image.

There are many ways to customize the view within the main browser window. Click on the "configure" button and you will be taken to a long page where you can change everything you see. The first change we'll make is to customize the window width within the Web page display. Click on the "configure" button. If you have a big computer monitor, set the "image width" to a number greater than 800 to take up the entire monitor width. Click the "submit" button to change your display width and return to the browser.

A quick approach to change views is demonstrated in **Figure 12.9**. Figure 12.9A is the default view for three tracks: RefSeq, human mRNAs, and human spliced ESTs. By clicking on the text that describes these tracks ("RefSeq Genes," "Human mRNAs from GenBank," and "Human ESTs That Have Been Spliced") the views in these tracks will toggle between a restricted and a more expanded view. Figure 12.9B shows the result of expanding these three tracks. Figure 12.9A showed only one RefSeq sequence, one human mRNA, and a consensus of exons from spliced ESTs, but Figure 12.9B shows two RefSeq sequences, a number of mRNAs, and a number of ESTs with varying coverage of the gene exons.

With these three clicks, a great deal of information is revealed. At a glance you can see the supporting data for the RefSeq consensus sequences. On the left of each row, the accession numbers of these sequences are visible. There are two mRNAs, BC042544 and BX647224, which help confirm the shorter RefSeq sequence. There are two mRNAs, HQ258199 and AY288419, which are missing the 3P UTRs. Are these unique splice forms or do they just represent incomplete cDNA synthesis? Clicking on either the representations of the exons and introns or the accession numbers will take you to a page with more detail on those sequences and links elsewhere. For the ESTs, they cover the 5P exons very well but the 3P exons have little EST coverage. Sequence DB043324 looks interesting in that it appears to continue off the 3P end of the view, meaning that there is at least one additional exon, at least for this single EST, 3P to the known 3P UTR for this gene.

In the next few rows (**Figure 12.10A**), two related representations of data are present with the default view of *OPN5*. One of the many pre-calculation activities of the genome browser database is the alignment of the human genomic sequence to the sequences of 45 other vertebrates. Looking at the lower rows of

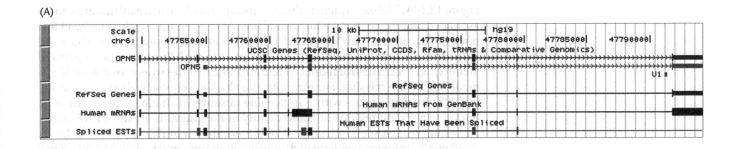

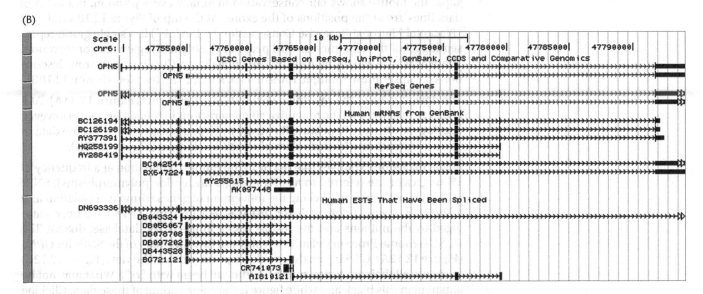

Figure 12.9 Expanding the UCSC Genome Browser data displays with a "click." Clicking the description lines, for example "Human mRNAs from GenBank," allows you to display more or less data. (A) The RefSeq Genes, Human mRNAs, and Human ESTs sections of *OPN5* appear as they did in Figure 12.8. (B) These tracks have all been expanded.

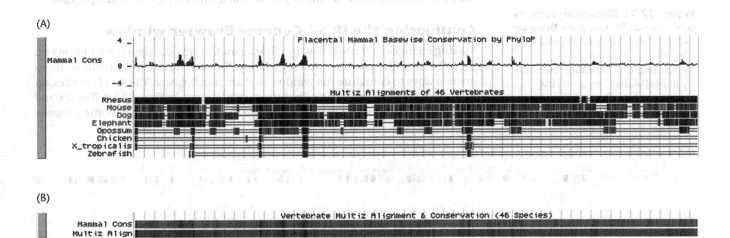

Figure 12.10 Displaying alignments with the UCSC Genome Browser. (A) Showing the nucleotides of a multiple sequence alignment at this scale (for example, 44,000 nucleotides across) is impossible so a graphical display is used instead. Peaks in the consensus track and a dark shade in the species lines indicate the highest conservation and often occur near exons, but not exclusively. Smaller peaks and lighter shades reflect varying degrees of conservation while no conservation shows a flat consensus line and no shading in the species tracks. (B) Clicking on the text "Vertebrate Multiz Alignment & Conservation (46 Species)" collapses the multiple lines of data seen below this label in (A) to two shaded tracks.

Figure 12.10A, Multiz sequence alignments for a handful of organisms are shown as shaded lines or boxes on a horizontal line. At this genomic scale (that is, 44,000 nucleotides per window) it would be impossible or impractical to show the individual nucleotides so the data are reduced to dark lines for good regions of identity, lighter shades for reduced identities, and no lines where there is no alignment or data are missing. For example, the chicken sequence aligns very poorly to the human sequence with a noticeable absence of shaded lines in the row, but there are tight regions of good identity in the columns for the *OPN5* exons. The exons are not seen in Figure 12.10 but are shown in Figure 12.8. The rhesus monkey alignment shows very high conservation from end to end, with just a few gaps. The mouse shows fair conservation in almost every position, but areas of dark lines are at the positions of the exons. At the top of Figure 12.10A (labeled "Placental Mammal Basewise Conservation by PhyloP") is a single graph representing all of the alignments, with peaks at the highest regions of conservation. These peaks are usually at the exons, but not always. Clicking on the row descriptions collapses these multiple rows of data, simplifying the view (Figure 12.10B).

The default view of the repetitive elements is a single row (**Figure 12.11A**). At a glance, you can see all the repetitive element regions in this single gene. However, clicking on the text "Repeating Elements by RepeatMasker" expands the data to show a separate row for each class of repetitive element (Figure 12.11B).

Single nucleotide changes that occur in the human population at a frequency of 1% or greater are referred to as SNPs (single nucleotide polymorphisms). SNPs offer a powerful way of tracking the movement of genes within a population and associations with certain traits or diseases. For humans, SNPs have been cataloged by the millions and the results reside in the NCBI database, dbSNP. The UCSC Genome Browser default view shows a row summary of the SNPs for *OPN5* (**Figure 12.12A**). Clicking on the row description expands the view (Figure 12.12B) to expose all their accession numbers (they all begin with "rs"). What may not be apparent in this black and white figure is the color-coding of these data. Clicking on the gray bar on the left brings up a Web page, a portion of which is shown in Figure 12.12C, describing and allowing the changing of the color-coding.

Finally, you can change the order of the tracks. Move your mouse to the left of the data but to the right of the gray configuration buttons and the color of the entire track changes. The mouse arrow may also change in appearance. At this point you can click your mouse and hold it down, then drag that track up or down in the main browser window and let go. The track will then move to that position.

Configuring the UCSC Genome Browser window

Making changes to the window by clicking on the gray configuration buttons on the left side of the genome browser window for each track can be tedious. There are two ways to configure the window that are much easier. The first is to click on the "configure" button, visible in the top-right corner in Figure 12.8. The second is to use the multitude of menus below the main window of the genome browser.

Figure 12.11 Displaying repeats with the UCSC Genome Browser.
(A) The mapped repeating elements identified by RepeatMasker.
(B) Clicking on "Repeating Elements by RepeatMasker" expands the section to show the types of repeats along with their location.

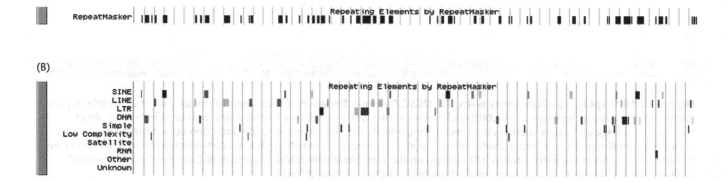

(A)

(B)

(A)

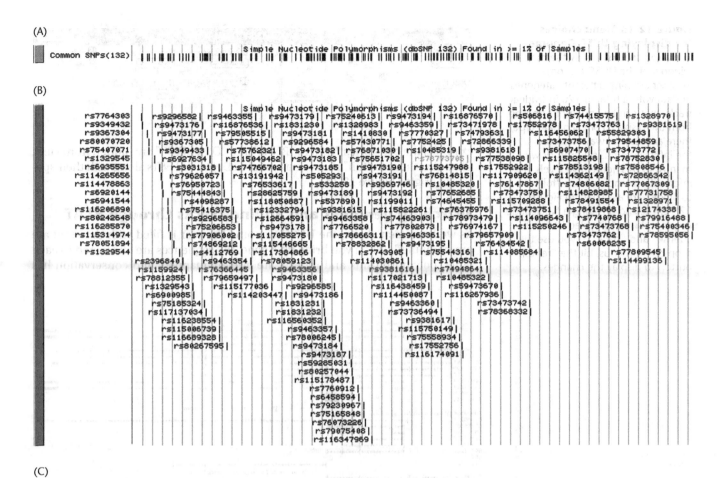

(B)

(C)

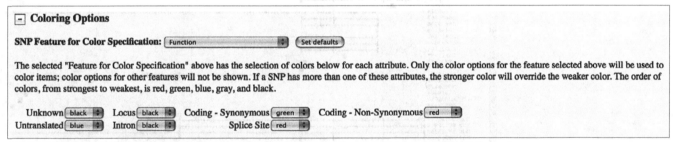

Figure 12.12 Displaying SNPs with the UCSC Genome Browser. (A) A single track of SNP locations. (B) Clicking on the text describing this track reveals the names and locations of each SNP. These names are hypertext and will take you to additional information on the individual SNP. (C) The SNPs are color-coded to reflect their location and impact on the gene. SNPs occurring in noncoding regions are blue, SNPs in introns are black, those found in coding regions but which have no impact on the amino acid sequence (synonymous changes) are green, while those that change the amino acid sequence (non-synonymous) or splice sites are red. This color scheme can be varied with the drop-down menus.

Scroll down and for each data category (for example, mRNA and EST Tracks) all the possible tracks are shown and, for each, a drop-down menu for the level of data shown. This is much like what is seen when you click on the "configure" button mentioned above. Click on the hypertext name of these track names and you will be taken to an explanation of their content. **Figure 12.13** shows an example drop-down menu and the UCSC naming convention for these varying levels of data portrayal. The names "hide," "dense," "squish," "pack," and "full" are in order from lowest to greatest data exposure. There are times when there will be no perceived difference between levels.

As seen in Figure 12.13, there are several types of EST data that were not shown by the default settings and were absent from earlier figures in this chapter. For

Figure 12.13 Menu choices for displaying UCSC Genome Browser tracks. Below the main window of the UCSC Genome Browser are nine different categories of data with a total of 120 individual tracks that can be added or removed from the display. For each data type, there are varying ways to reveal the information using drop-down menus. In this figure, five levels of information ranging from "full" to "hide" are available. In Figure 12.12, clicking on "Single Nucleotide Polymorphisms" toggled between "dense" and "pack." If you use the drop-down menus, hit the "refresh" button to display the changes.

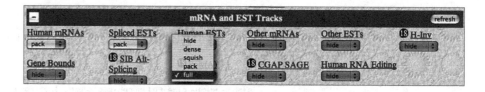

example, the categories "Other mRNAs" and "Other ESTs" are hidden from view. Changing these to "full" shows many more cDNAs and ESTs along with their species of origin (**Figure 12.14**).

Searching genomes and adding tracks through BLAT

As we have seen above, the UCSC Genome Browser is filled with data that are pre-computed. When you bring up a gene, the sequences are all aligned to each other, polymorphisms and repeats are already mapped, and conservation has

Figure 12.14 "Full" display of mRNA and EST tracks. Many genes have a tremendous amount of supporting data from various sources (for example, mRNAs, ESTs) and different species. By choosing the "full" display of data from the drop-down menus, transcripts from these sources can all be displayed.

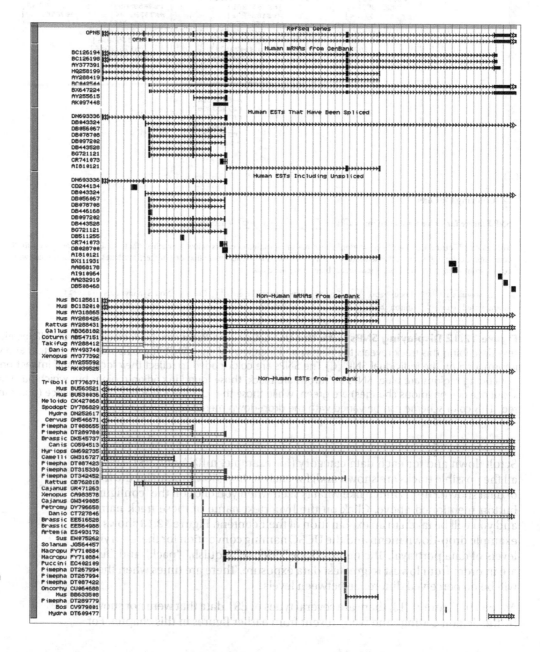

already been calculated. But there may be times when you will wish to add your own data to the system. You may have a sequence you have just generated or obtained that is not represented in the Genome Browser. Under these circumstances, a very convenient way to add your sequence and customize the view is through a tool called BLAT. The Website states "BLAT on DNA is designed to quickly find sequences of 95% and greater similarity of length 25 bases or more." This description is not adequate; BLAT is *very* fast and will give you your hits within moments of submission. It accomplishes this great feat by generating an index of the genome and keeping it, but not the sequence, in memory.

The following example will illustrate the capabilities of BLAT, which is available on the top toolbar of the main browser window. If you know something about your sequence already (for example, it is a transcript from a receptor called *OPN5*) you can navigate to that page using the gene symbol, as discussed earlier in this section. Or you can let BLAT take you there. In **Figure 12.15A**, a sequence has been pasted into the BLAT window. PGR12 is a synonym (an alternative name) for *OPN5* and BLAT will bring us to the same locus we have been studying.

Upon clicking on "submit," the page refreshes and four hits are presented (Figure 12.15B). Notice that the "START" and "END" coordinates for the first hit indicate that the entire query of 300 nucleotides (QSIZE) aligns at 100%. The other hits are shorter, hit other chromosomes (CHRO), and span shorter distances on the genome (SPAN). The first hit spans 3595 nucleotides, much larger than the query size, indicating that it is spliced and has long introns.

From this list you have the opportunity to examine the alignment by clicking on "details." This new page contains your sequence highlighted on the genomic sequence as well as the alignment (Figure 12.15C). Based on this alignment, the query is split between two exons, coordinates 1–42 (00000001–00000042) and 43–300. Hit the back button and you can now click on "browser." There is now a new track above all the other sequences (Figure 12.15D) and it is named after the query, AY255615.1. This is not quite like the RefSeq sequences (the middle exon is missing) and so it deserves a closer look.

(A)

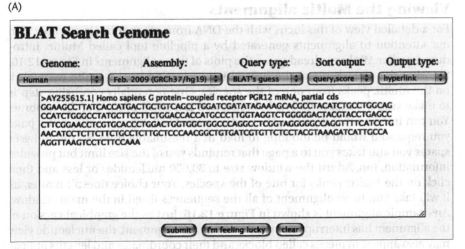

(B)

Figure 12.15 BLAT searches and incorporating your sequence into the UCSC Genome Browser. You can add customized tracks to the UCSC Genome Browser via the BLAT tool. (A) Entering your FASTA sequence, cDNA AY255615, and hitting "submit" launches an extremely fast search, generating a list of hits (B) (continued overleaf).

BLAT Search Results

ACTIONS	QUERY	SCORE	START	END	QSIZE	IDENTITY	CHRO	STRAND	START	END	SPAN
browser details	AY255615.1|	299	1	300	300	100.0%	6	+	47759642	47763236	3595
browser details	AY255615.1|	23	215	239	300	96.0%	3	+	130355558	130355582	25
browser details	AY255615.1|	22	74	96	300	100.0%	8	+	13368272	13368296	25
browser details	AY255615.1|	20	63	82	300	100.0%	7	+	73012093	73012112	20

Figure 12.15 BLAT searches and incorporating your sequence into the UCSC Genome Browser. (C) The "details" of the first hit displays two alignments, most probably two exons identical to genomic sequence. (D) Clicking on the "browser" link shown in (B) brings you to the genome browser main window with your sequence, AY255615, on top and aligned with all other data the browser makes available to you. Not all data are shown in this figure.

(C)

```
00000001  ggaagccttatcaccatgactgctgtcagcctggatcgatat  00000042
>>>>>>>>  ||||||||||||||||||||||||||||||||||||||||||  >>>>>>>>
47759642  ggaagccttatcaccatgactgctgtcagcctggatcgatat  47759683


00000043  agaaagcacgcctacatctgcctggcagccatctgggcctatgcttccttt  00000092
>>>>>>>>  ||||||||||||||||||||||||||||||||||||||||||||||||||  >>>>>>>>
47762979  agaaagcacgcctacatctgcctggcagccatctgggcctatgcttcctt  47763028

00000093  ctggaccaccatgcccttggtaggtctgggggactacgtacctgagccct  00000142
>>>>>>>>  ||||||||||||||||||||||||||||||||||||||||||||||||||  >>>>>>>>
47763029  ctggaccaccatgcccttggtaggtctgggggactacgtacctgagccct  47763078

00000143  tcggaacctcgtgcaccctggactggtggctggcccaggcctcggtaggg  00000192
>>>>>>>>  ||||||||||||||||||||||||||||||||||||||||||||||||||  >>>>>>>>
47763079  tcggaacctcgtgcaccctggactggtggctggcccaggcctcggtaggg  47763128

00000193  ggccaggttttcatcctgaacatcctcttcttctgcctcttgctcccaac  00000242
>>>>>>>>  ||||||||||||||||||||||||||||||||||||||||||||||||||  >>>>>>>>
47763129  ggccaggttttcatcctgaacatcctcttcttctgcctcttgctcccaac  47763178

00000243  ggctgtgatcgtgttctcctacgtaaagatcattgccaaggttaagtcct  00000292
>>>>>>>>  ||||||||||||||||||||||||||||||||||||||||||||||||||  >>>>>>>>
47763179  ggctgtgatcgtgttctcctacgtaaagatcattgccaaggttaagtcct  47763228
```

(D)

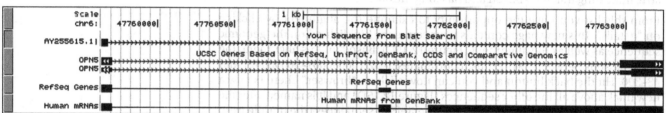

Viewing the Multiz alignments

For a detailed view of this locus with the DNA from other animals, we now turn our attention to alignments generated by a pipeline tool called Multiz, introduced earlier. We have already seen the plots of these alignments in Figure 12.10. Showing the alignments at the individual nucleotide level is quite easy; click on the Multiz peaks from one of the species. One potentially confusing step is to make sure the main window being viewed is a small portion of the genome. You can imagine that if you had a window size of 500,000 nucleotides, the page you requested would be too long to load in a reasonable time. So the browser spares you and takes you to a page that reminds you of the size limit but provides information, too. Adjust the window size to 30,000 nucleotides or less and then click on the Multiz peaks for one of the species. Your choice doesn't matter, as it will take you to an alignment of all the sequences listed in the main window. An example alignment is shown in **Figure 12.16.** Just as the graphical version of the alignment has interruptions where there is no alignment, the nucleotide view may also appear in pieces called blocks and their coordinates and lengths appear above each block. The default configuration for this track is to have blue exons in uppercase and all other sequence in black.

To configure the Multiz alignments, click on the left configuration button and a window much like **Figure 12.17** appears. Here, you can choose the species sequence to align with the genome you are browsing (the "reference" sequence) as well as make a few adjustments to the nucleotide alignment settings.

There is an alternative way to look at the aligned nucleotides that is quite powerful because it is within the main browser window. This affords the simultaneous viewing of the other data while you are examining the sequence. In **Figure 12.18A,**

```
Alignment block 2 of 54 in window, 47759664 - 47759719, 56 bps
B D        Human      CTGTCAGCCTGGATCGATATTTGAAAATCTGCTATTTATCTTATGgtaagttggc---a
B D        Rhesus     CTGTCAGCCTGGATCGATATTTGAAAATCTGCTATTTATCTTATGgtaagttggc---a
B D        Mouse      CTGTCAGCCTGGACCGCTATCTGAAGATCTGTTATCTGTCTTATGgtaagttgga---a
B D        Dog        CTGTCAGCCTGGATCGATATCTGAAAATCTGCTATTTATCTTATGgtaagtacga---a
B D        Elephant   CTGTCAGCCTGGATCGATATCTGAAAATCTGCTATTTATCTTATGgtaagta-ga---a
B D        Opossum    CTGTCAGCCTGGACAGATACTTGAAAATCTGCCACTTGTCTTACGgtaagttgga---a
B D        Chicken    CCGTCAGCCTGGATAGATACTTAAAAATCTGTCACTTGGCTTATGgtaaggaaga---g
B D  X. tropicalis    TAGTGAGCCTGGACAGGTACCTGAAGATCTGCCATCTGCGATACGgtaag----------
B D        Zebrafish  TTGTCAGCTTTGACCGCTACTTGAAAATTTGTCATTTAAGATATGgtaa-----------

Inserts between block 2 and 3 in window
B D        Chicken 12bp
B D  X. tropicalis 621bp
B D        Zebrafish 4144bp

Alignment block 3 of 54 in window, 47759720 - 47759724, 5 bps
B D        Human      gg----t----------------tt
B D        Rhesus     gg----t----------------tt
B D        Mouse      gg----t----------------cc
B D        Dog        gg----t----------------tt
B D        Elephant   gg----t----------------tt
B D        Opossum    gg----ctcctgaagaagtgtcatt
B D        Chicken    -----------------gggatgtt
B D  X. tropicalis    =========================
B D        Zebrafish  =========================

Alignment block 4 of 54 in window, 47759725 - 47759847, 123 bps
B D        Human      ctcatt---------------------ccctgacagtt-aaagctagg---tggattgag-------tgt
B D        Rhesus     ctcatt---------------------ccctgacagtt-aaagctagg---tggattgag-------tgt
B D        Mouse      cttgtt---------------------ccctgatagg--aaagttaga---tgatcggag-------tgt
B D        Dog        ctcatt---------------------ccctgatagtc-aaaacta-g---tggat----------tgt
B D        Elephant   ctcggt---------------------ccctgatagga-aaagctagg---tggattgag-------tgt
B D        Opossum    ctcctt---------------------tcttgatagC-agggctgaagattggtttgggaggataaagt
B D        Chicken    ctcattctgcctcctctccccc-----tcctgcccaag-aaggc---a---ttggt-----------cat
B D  X. tropicalis    ======================================================================
B D        Zebrafish  ======================================================================

             Human      tagaactgt----acatattt----tat--gca--ggt--aataaacaaagat-tatgaataacc--tca
             Rhesus     tagaactgt----acatgttt----tat--gca--ggt--aataaacaaagat-tatgaataacc--tca
             Mouse      cagagctgt----gcatattt----tct--aca--g------taaataaaggt-tattaatgact--tca
             Dog        cag-actgt----aaatactg----tactgata--ggt--gatgaacaaagat-tactaataacc--tag
             Elephant   cagagctgc----acatgctt----tac---cct-tgt--aataaacaaaaat-gactgctaacc--tca
             Opossum    caga----------------------tac--ata--gac--cttgaa-------------------------a
             Chicken    tagggttgt----gaacaatcagggcaa--gga--aacaaaacaattagaaa-------atattt--cct
     X. tropicalis      ====================================================================
         Zebrafish      ====================================================================

             Human      gagat-----------caagaa-tatttctatttatagcc-tatc
             Rhesus     gagat-----------caagaa-tatttctatttatagcc-tatc
             Mouse      gtgac-----------caagaa-cccttctattcatagcc-ta---
             Dog        gacac-----------caagag-tatctccatttctagcc-ta---
             Elephant   gagac-----------caagaa-tatttctatttatagcc-t----
             Opossum    tggca-----------tctgtg-taatcccacatagattc------
             Chicken    gaaat-----------aaa------------------------
     X. tropicalis      ==========================================
         Zebrafish      ==========================================
```

an exon of *OPN5* is near the center with introns flanking the sequence. Figure 12.18B shows the same region after zooming in. We can now see the translation of the small exon in the Multiz alignment tracks while the flanking introns are individual nucleotides. Visible now is the conservation of the amino acid sequence and substitutions appear as different colors. In Figure 12.18C, the configuration of the Multiz track (using the gray configuration button) was set to "no codon translation," revealing the nucleotides of the exons. To keep the translations in view, an additional track was added, the UCSC Genes, and this shows the amino acid sequence of the exon. Also visible is the PhyloP plot of conservation for the individual nucleotides. Note the peaks, valleys, and even the negative values for the third position of several of the codons, indicating there has been good conservation of the first two nucleotides but the wobble base has had substitutions.

Figure 12.16 UCSC Genome Browser Multiz alignment of orthologous sequences. If you display a small-enough region of a chromosome (for example, 30,000 nucleotides or less), clicking on the bands within the Multiz Alignment graphics panel (shown in Figure 12.10A) will take you to the multiple sequence alignment blocks to show the conservation across species. Coding regions are in uppercase while noncoding sequence is lowercase.

Figure 12.17 Multiz alignment tree. To configure the display of orthologous sequences, click on the gray configuration bar to the left of the Multiz Alignment section in the main browser window. Species can be chosen, guided by the evolutionary tree displayed on the right of this configuration window.

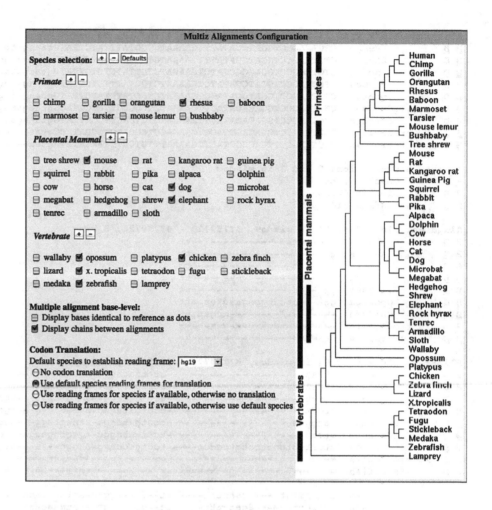

Zooming out: seeing the big picture

Now we'll take a look at what is in the "neighborhood" of the human *OPN5* gene. For **Figure 12.19**, a minimalist view in the browser was chosen to speed up the refresh rate and keep the information overload to a minimum. In Figure 12.19A, only the scale, RefSeq genes, and spliced EST tracks are visible for the default magnification of *OPN5*. This window is roughly the size of the gene, 44,317 nucleotides wide. In Figure 12.19B, we've zoomed out and now the window is over 132,000 nucleotides wide. Besides being smaller, we now see that there is a spliced EST with exons beyond the RefSeq 3P exons, as mentioned earlier in this chapter. The ESTs are being seen in a "dense" view (one line) and expanding to "full" would reveal their accession numbers for further analysis. There is another spliced EST that has come into view 5P of *OPN5*. This EST has no RefSeq record so it would be interesting to study this sequence. The fact that splicing is occurring suggests something other than a "random" transcription product, but this would have to have more supporting evidence such as conservation in other species.

In Figure 12.19C, we have zoomed out to a window size of almost 400,000 nucleotides and a handful of other RefSeq genes are now visible. To the 3P side of *OPN5* is an uncharacterized gene called *C6orf138*; this is open reading frame 138 on chromosome 6 and may someday receive a more informative name. To the 5P side of *OPN5* are a few genes that are related based on their names; *GPR111* and *GPR115* are G-protein-coupled receptors, as is *OPN5*, but it is a large superfamily and little similarity may be seen.

Zooming out to a window size of over 10 million nucleotides (Figure 12.19D) shows a large number of RefSeq transcripts. The increase in window size is also reflected in the chromosome cartoon near the top of the window. In

(A)

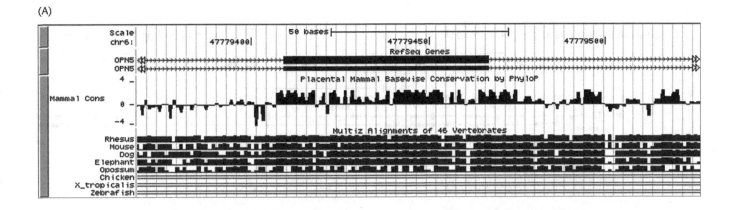

(B)

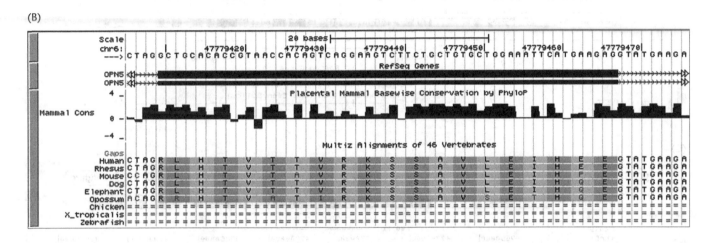

(C)

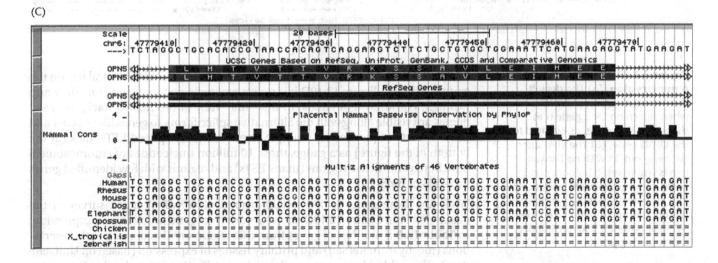

Figure 12.18 "Zoom in" function for sequence display. As you zoom in your view using the UCSC Genome Browser, details down to the sequence can be revealed. (A) The view of an *OPN5* exon. (B) Zooming in (note the scale is 20 bases) shows the nucleotides across the top and the graphical display of similarity to other species has been transformed into the translation flanked by nucleotides. Also notice that the Mammalian Consensus (Mammal Cons) row now shows the conservation in each nucleotide position as calculated by PhyloP. This reveals that the third position of some codons is not well conserved, with the graph even going below the midline in the track. (C) The Multiz Alignment configuration was set to "no codon translation" and the UCSC Genes track has been added to show the translation simultaneously.

(A)

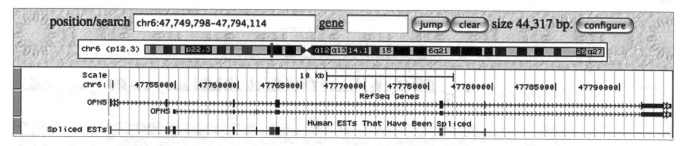

(B)

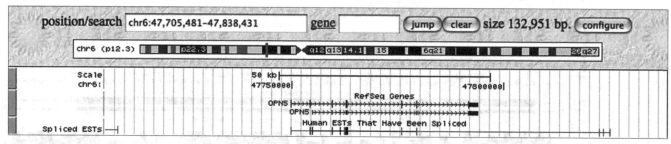

(C)

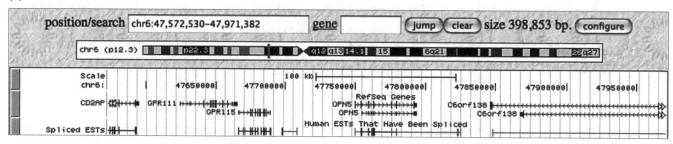

Figure 12.19 A larger view of a gene region. (A) The default view of the UCSC Genome Browser when searching for a gene is to show only that gene, as pictured for the *OPN5* gene. (B) The view after zooming out 3× with no flanking genes visible. (C) Zoom out 3× again and now adjacent genes come into view. Note that the window size has gone from 44,000 nucleotides (A) to almost 400,000 nucleotides (continued opposite).

Figures 12.19A–C the region being viewed appeared as a thick vertical bar on the cartoon while in Figure 12.19D it is now a rectangle encompassing one dark and two light chromosomal bands. In this "full" view of the RefSeq track, the gene names are visible along with the alternative splice forms so genes like *VEGFA* and *CRISP2* show sequences on multiple lines. Using the spliced ESTs track (at the bottom of the figure) as a gauge, these 10 million nucleotides of chromosome 6 have both gene-dense regions (many ESTs) and regions with few identified genes (no ESTs or RefSeqs).

What are these genes in the neighborhood of *OPN5*? A quick survey of the genes visible in Figure 12.19D shows a variety of gene function and expression. In **Table 12.2** a number of genes from this region are listed along with descriptions (mostly from RefSeq) and primary tissues of expression (based on UniGene data). This table shows genes related by family (for example, the G-protein-coupled receptors), function (for example, transcription factors), and patterns of expression (for example, intestine) or location (for example, mitochondria). Some family members are clustered (for example, *DEFB** and *GSTA**).

Very large genes: dystrophin and titin

Dystrophin and titin are among the largest genes in the human genome. Both have numerous exons, encoding huge proteins, and both are expressed in muscle cells. Despite these similarities, the genome browser reveals striking differences between these two genes.

(D)

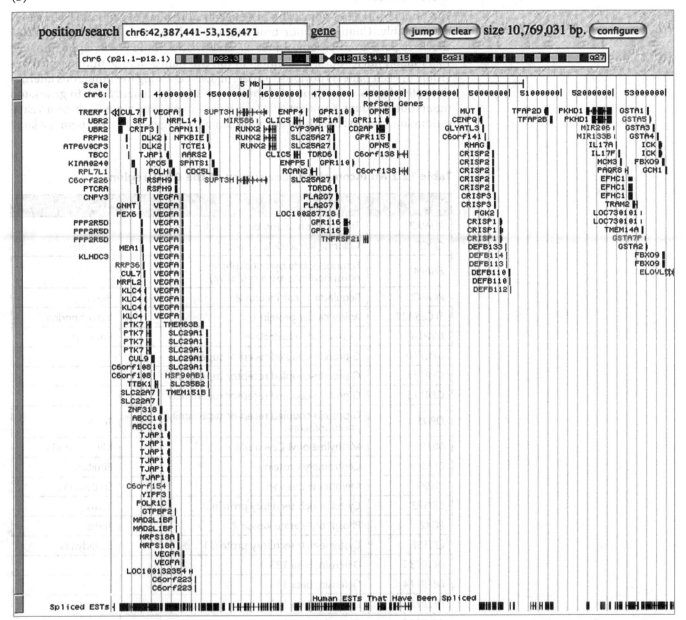

Figure 12.19 A larger view of a gene region. (D) Zoom out a few more times and now the window is almost 11 million nucleotides. The *OPN5* gene is near the center (top row) but many genes, along with their numerous splice forms, are now showing the gene density of this region. *VEGFA* is over 16,000 nucleotides long but in this view is reduced to a single vertical line. The exons of *C6orf138* are still clearly visible because this gene is over 190,000 nucleotides long.

Mutations in dystrophin can cause muscular dystrophy. The gene is over 2.2 million nucleotides long and has 85 exons (**Figure 12.20A**). To compare sizes, the entire window in Figure 12.19D, filled with genes, would only be able to hold five dystrophin genes. Numerous splice forms have been identified and there are thousands of SNPs visible in Figure 12.20A. The SNP distribution does not seem to be uniform based on the shape of the SNP track graphic. There are also thousands of repetitive elements within the introns of the gene from the six major classes of repeats.

By comparison, titin is much smaller, "only" 281,000 nucleotides long (Figure 12.20B). But it has over 300 exons and these encode the largest protein of 33,000 amino acids. The tightly packed exons of titin leave little room for repetitive elements. But there are two overlapping genes at this locus, a predicted gene *LOC100506866* and *MIR548N*, the status of which is not definite.

Gene density

The number of genes appearing in equal-sized regions of genomic DNA varies considerably. Continuing our browsing of the human genome, **Figure 12.21** shows regions from chromosomes 17 and 5. The window size, approximately 13.5 million nucleotides, is identical in the two panels yet the number of genes visible in each panel is obviously different. Even ignoring the numerous alternative splice forms, this region of chromosome 17 is particularly rich in genes, too numerous to count. Contrast this to a region in chromosome 5 where a visual estimation suggests that there are approximately 50 genes. In fact, there is a large

Table 12.2 An incomplete gene survey of the *OPN5* region

Gene	Description	Expression
RUNX2	Transcription factor, bone development	Bone
CLIC5	Chloride intracellular channel	Placental
ENPP4	Ectonucleotide pyrophosphatase/ phosphodiesterase 4 (putative)	
RCAN2	Regulator of calcineurin 2	Brain
SLC25A27	Anion carrier protein	Mitochondria
MEP1A	PABA peptide hydrolase	Intestine
GPR110	G-protein-coupled receptor superfamily	
GPR111	G-protein-coupled receptor superfamily	
GPR115	G-protein-coupled receptor superfamily	
OPN5	G-protein-coupled receptor superfamily, photoisomerase	Testis
MUT	Methylmalonyl CoA mutase	Mitochondria
CENPQ	Centromere protein	Pituitary
RHAG	Membrane channel	Erythrocyte
CRISP2	Cysteine-rich secretory protein 2	Testis
PGK2	Phosphoglycerate kinase 2	Testis
CRISP1	Cysteine-rich secretory protein 1	Epididymis
DEFB133	Defensin, beta 133	
DEFB114	Defensin, beta 114	
DEFB113	Defensin, beta 113	
DEFB110	Defensin, beta 110	
DEFB112	Defensin, beta 112	
TFAP2D	Transcription factor	
TFAP2B	Transcription factor	Eye
GSTA1	Glutathione S-transferase alpha 1	
GSTA5	Glutathione S-transferase alpha 5	
GSTA3	Glutathione S-transferase alpha 3	
GSTA4	Glutathione S-transferase alpha 4	
ICK	Intestinal cell (MAK-like) kinase	Intestine
FBXO9	Ubiquitination subunit	
GCM1	DNA binding protein	

Some of the genes within the *OPN5* gene region, shown in Figure 12.19D, are listed here. The first column contains the gene symbols, the second contains information easily gathered from the Gene or RefSeq records, and the third column contains information about gene expression, again gathered from sequence annotation.

(A)

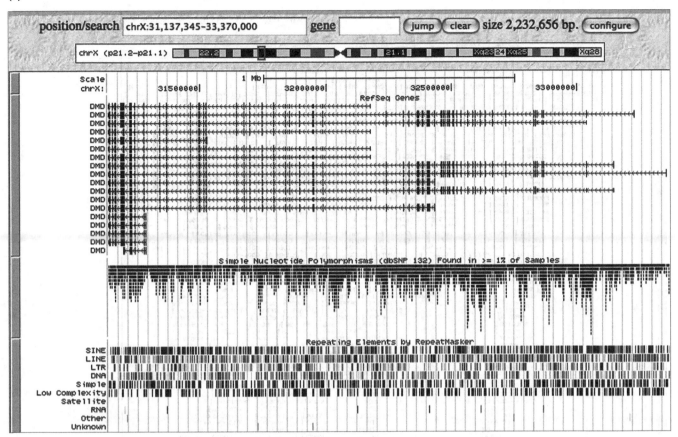

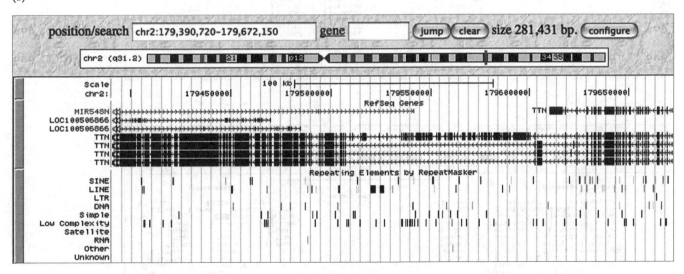

(B)

Figure 12.20 Very large genes. (A) The human dystrophin gene has been described as the longest in nature, at over 2.2 million nucleotides. In addition to the known splice forms and numerous exons, also visible are the many SNPs and repeat elements. According to the NCBI Gene annotation, the dystrophin gene has eight, independent, tissue-specific promoters and 85 exons. (B) The titin gene encodes the largest protein at over 33,000 amino acids and has over 300 exons, although it is eightfold smaller (approximately 281,000 nucleotides) than the dystrophin gene. Note that titin has many large exons and fewer repeats. Also pictured are two overlapping genes, one of which is a gene tentatively identified as *MIR548*. Both dystrophin and titin are expressed in muscle and mutations in dystrophin are responsible for Duchenne muscular dystrophy.

(A)

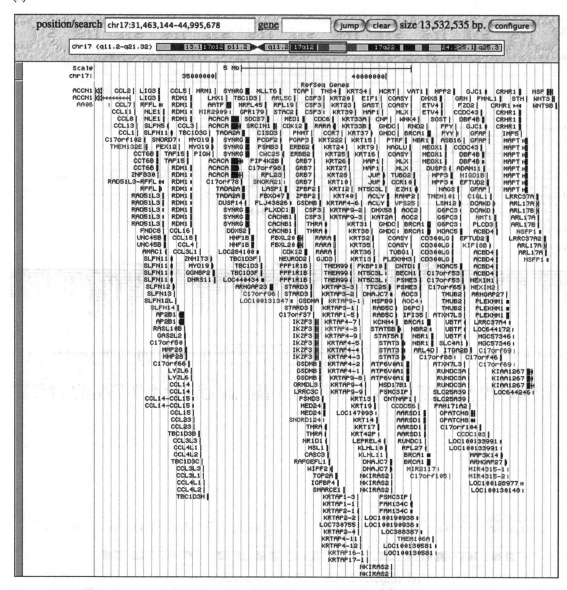

(B)

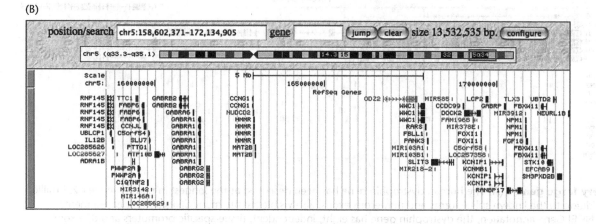

Figure 12.21 Gene-dense and gene-desert regions. Two different regions of the human genome, both at the same scale (a 13,532,535-nucleotide window). (A) A gene-rich region of chromosome 17; the RefSeq genes extend considerably down the page. (B) A gene desert on chromosome 5; a region with relatively few genes and a large expanse with no identifiable protein-coding genes. Note the 5 Mb scale.

gap near the center of the region from chromosome 5, with few genes present for almost 5 million nucleotides. This and other regions of the genome with few genes appearing are generally referred to as **gene deserts**.

This variation in gene density is even seen at the chromosome level. Going back to Table 12.1, chromosome 17 has 1173 known protein-coding genes, 54 novel protein-coding genes, and 247 pseudogenes. This adds up to 1474 genes from these three classes. Chromosome 18, just three million nucleotides smaller than chromosome 17, has only 339 genes of the same classes. Chromosome 5, a portion of which is seen in Figure 12.21B, is more than twice the size of chromosome 17, yet has almost the same number of known, novel, and pseudogenes (1579).

Interspecies comparison of genomes

We'll now make a comparison of gene structures between species. Dr. Sydney Brenner and colleagues established *Takifugu rubripes* as a model organism for genomic studies. It was discovered that *Takifugu*, also known as *Fugu* (rhymes with voodoo) or the Japanese pufferfish, had a genome one-eighth the size of the human and mouse genomes. Brenner argued that if you want to study a vertebrate genome, why not save time, effort, and expense by choosing a small one like *Fugu*? It may not be a mammal but will nevertheless have many if not all of the major features of development and gene regulation shared between vertebrates. As a result, the *Fugu* genome was among the first large genomes to be sequenced.

In 2001, W.P. Yu published a paper comparing the human and *Fugu PTEN* loci and this comparison is featured in **Figure 12.22**. PTEN is a phosphatase that normally

Figure 12.22 Human and *Fugu* genome comparison. This figure compares the *PTEN* genes of human and *Fugu*. (A) The human gene; the repeats track indicates a large number of repetitive elements. (B) The orthologous gene in *Fugu*. This gene is 16-fold smaller than the human ortholog. Noticeably absent is the large number of repetitive elements. Note: the *Fugu* view was reversed (configuration buttons are on the right side instead of left) so the gene orientation, 5P to 3P, left to right, is the same in (A) and (B). This was accomplished by clicking the "reverse" button, seen on the bottom right.

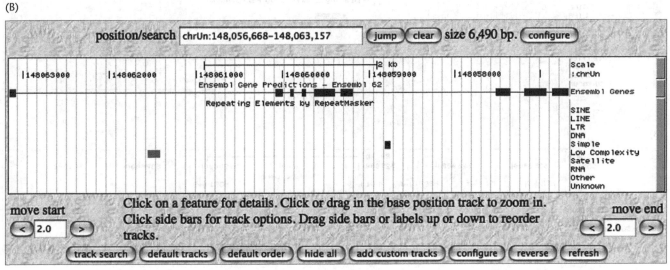

acts as a tumor suppressor and is mutated in a number of cancers. Figure 12.22A shows the human locus, a 105,000-nucleotide region of chromosome 10. Note the large number of repetitive elements and the large introns often seen between the nine exons. Contrast this to the *Fugu* locus for *PTEN* (Figure 12.22B). This window is only 6490 nucleotides and is 16-fold smaller than the human gene. The exons appear larger in Figure 12.22B because the window size is small. Note the near absence of repetitive elements, a feature contributing to the small size of the *Fugu* genome. Also note the chromosome number in the *Fugu* window: "chrUn." The *Fugu* genome has not been fully assembled to resolve into individual chromosomes, so this and the other *Fugu* genes are mapped to an uncharacterized chromosome.

The beta globin locus

Throughout this book, beta globin has appeared as an example gene for analysis. One of the first genes cloned, beta globin is compact, uncomplicated with only a few exons, and part of a gene family whose members have high identity. Here we will take one last look at beta globin and the clustered family members with the UCSC Genome Browser. By entering the gene symbol, *HBB*, you quickly navigate to the 1600-nucleotide gene (**Figure 12.23A**). At this scale, the slightly narrower 5P and 3P UTRs are visible on the first and last exons, respectively. There are two repetitive elements in the first intron (Figure 12.23A).

Zooming out to a 60,000-nucleotide window, the entire locus is seen, showing all five beta globin genes: *HBE1*, *HBG2*, *HBG1* (*HBBP1* is a pseudogene), *HBD*, and *HBB* (Figure 12.23B). Remarkably, these genes appear in order of their expression throughout development. In fact, if you rearrange the gene order, their expression order changes as well. As the embryo grows and matures, these different globins take over the responsibility of assisting alpha globin to bind oxygen in our blood cells. This expression is orchestrated by the locus control region which lies upstream of the *HBE1* gene (the far right of Figure 12.23B). This region is sensitive to nucleases, reflecting an open conformation that is most probably caused by the binding of transcription factors. Other nuclease-sensitive sites, often at the 5P

Figure 12.23 The human beta globin locus viewed by the UCSC Genome Browser. (A) Human beta globin is encoded by a gene approximately 1600 nucleotides in length. Compare this to the 2.2 million-nucleotide dystrophin gene in Figure 12.20A. (B) Zooming out the view, family members of this beta globin locus are visible. The locus control region, a set of DNase I-sensitive sites upstream of *HBE1*, is critical to the timing and control of expression for all of these genes.

(A)

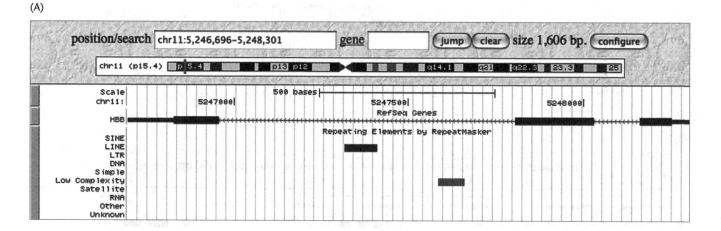

(B)

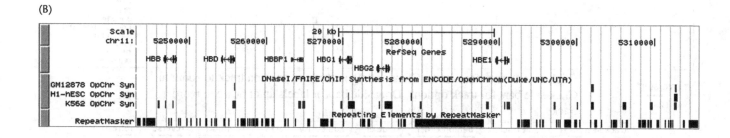

ends of the genes, appear and disappear with that gene's expression and silencing, respectively. These DNase I-sensitive sites appear as tracks in Figure 12.23B.

12.5 SUMMARY

In this chapter, we stepped back from the minute details of sequence analysis and looked at a big view of genome content. Focusing on the human genome, we learned about the structures and diversity of chromosomes. Throughout evolution, significant genome rearrangements have taken place, but these still leave obvious evidence of past and present relationships, which are visible with the synteny viewer at the Ensembl Genome Browser. The UCSC Genome Browser was extensively explored, allowing us to easily see the collation and display of data about individual genes, loci, and larger chromosomal regions.

EXERCISES

Olfactory genes

Our sense of smell starts with the interaction between the multitude of odorants in the world and our millions of olfactory receptors. Located on the outside of our olfactory sensory neurons within our nose, these G-protein-coupled receptors bind odorants and trigger a cascade of molecular and cognitive events that lead to an almost instantaneous recognition of the foreign, familiar, fragrant, foul, friend, foe, or food. Minute quantities of chemicals can be detected, yet our sensitivity diminishes rapidly under constant exposure. Certain smells can mask others, or blend to produce no smell at all.

There are 855 olfactory receptor genes in the human genome, which makes them members of one of the largest protein-encoding gene superfamilies. There are 18 families and 300 subfamilies, making their study challenging for bioinformatics problems ranging from sequence similarity to multiple sequence alignments (and associated phylograms), and for nomenclature. But the story behind the olfactory receptor gene number is more complicated because approximately 465 of these genes are pseudogenes, no longer capable of encoding a full-length protein. As we evolved, the importance of our sense of smell to our survival diminished and the number of our functional olfactory receptor genes has dropped. Contrast our olfactory receptor gene numbers to those of the mouse, which absolutely depends on a keen sense of smell and has over 1200 functional genes. Although the dog may have "only" 800 olfactory receptor genes, their nasal epithelium is extremely large and one-third of their brain is dedicated to the sense of smell, making them powerhouses for this sense.

Studies of 25 individuals from three different populations demonstrate that the olfactory receptor repertoire varies between people. Since there are so many genes under relatively weak selection, SNPs are seen that inactivate the receptors, having little apparent effect on the survival of the individuals. Also seen are changes in copy number, resulting in sometimes higher but mostly lower numbers of functional olfactory receptor genes, suggesting that we are witnessing the shrinkage of our olfactory receptor gene family.

Human olfactory receptor proteins are encoded by genes with single exons. Some are organized in clusters while others are scattered to all the chromosomes except autosome 20 and the Y chromosome. In this set of exercises, you will examine orthologous clusters of olfactory receptor genes and determine the reasons why some genes are considered pseudogenes. In addition, you will investigate a nearby gene cluster in mouse and explore this family's gene number and structure in human.

1. Open the UCSC Genome Browser and navigate to human chromosome 11. Navigate to the q arm and the region immediately adjacent to the centromere.

Here you will find a cluster of olfactory receptor genes, the symbols of which begin with "OR" in the "RefSeq Genes" rows of the browser. Zoom in and/or adjust the coordinates in the "position/search" window until you see just the OR cluster and a few of the flanking non-OR genes. Keep this window open.

2. Open the Ensembl Genome Browser and navigate to the human genome, karyotype view, chromosome 11. Go to the synteny view of this chromosome and bring up the syntenic relationship to the mouse genome. Considering the position of the human OR cluster found with the UCSC Genome Browser, above, what mouse chromosome segment is syntenic to the region of human chromosome 11 with the OR cluster? Keep this window open.

3. Open up another UCSC Genome Browser window and navigate to the mouse chromosome that is syntenic with the human chromosome 11 region of interest. Navigate to the syntenic region containing the mouse OR clusters. Their gene symbols begin with "Olfr."

4. Zoom in and/or adjust the coordinates in the mouse genome "position/search" window until you see just the OR cluster and a few of the flanking non-OR genes. Keep this window open.

5. Comparing the human and the mouse clusters, what are the approximate sizes (in nucleotides) of the clusters?

6. Comparing the human and mouse clusters, what is the *approximate* number of genes in each cluster?

7. Both human and mouse have at least one non-OR gene in the middle of their respective OR clusters (ignore the flanks). Are these human and mouse non-OR genes related in function?

8. There is at least one pseudogene in the human cluster, *OR7E5P*. Using any bioinformatics tool you wish, can you provide evidence that it is not a functional gene?

9. The human pseudogene, above, had a "P" at the end of the symbol, unlike any other symbol in this human cluster. The mouse naming convention is different, but by scanning the names of the mouse olfactory receptor genes in this cluster, can you identify any pseudogenes?

10. Adjacent to the mouse olfactory receptor gene locus is a small cluster of predicted genes which we will now study. Navigate to the following coordinates with the UCSC Genome Browser: chr2:83,786,121-83,854,484.

11. Using the drop-down menus below the browser window, under "Genes and Gene Predictions Tracks," hide all other tracks but make visible (in 'Full' mode) the "Other RefSeq" and the "Ensembl Genes" tracks.

12. Under "mRNA and EST Tracks," hide all except "Mouse mRNAs" and show those in "Full."

13. Notice the unusual repetition of the genes as well as the repetition of the gene names. Using any bioinformatics tool or database, try to make sense of this gene cluster. Remember your most basic tool: pencil and paper.

FURTHER READING

Blanchette M, Kent WJ, Riemer C et al. (2004) Aligning multiple genomic sequences with the threaded blockset aligner. *Genome Res.* 14, 708–715. This paper discusses the challenging problem and solution of aligning very divergent sequences at the genomic scale.

Eickbush TH & Eickbush DG (2007) Finely orchestrated movements: evolution of the ribosomal RNA genes. *Genetics* 175, 477–485. Ribosomal RNA genes are excellent models for the evolution of tandemly arrayed genes.

Flicek P, Amode MR, Barrell D et al. (2011) Ensembl 2011. *Nucleic Acids Res.* 39, D800–D806. A brief article on the Ensembl Genome Browser.

Fujita PA, Rhead B, Zweig AS et al. (2011) The UCSC Genome Browser database: update 2011. *Nucleic Acids Res.* 39, D876–D882. A recent update to the genome browser featured prominently in this chapter.

Harju S, Navas PA, Stamatoyannopoulos G & Peterson KR (2005) Genome architecture of the human beta-globin locus affects developmental regulation of gene expression. *Mol. Cell. Biol.* 25, 8765–8778. A very interesting paper describing experiments where the gene order of the globin locus was rearranged, with concomitant changes to expression observed as a result.

Hasin Y, Olender T, Khen M et al. (2008) High-resolution copy-number variation map reflects human olfactory receptor diversity and evolution. *PLoS Genet.* 4, e1000249 (DOI: 10.1371/journal.pgen.1000249). This very interesting gene family is the subject of the exercises for this chapter.

Hughes JF, Skaletsky H, Pyntikova T et al. (2010) Chimpanzee and human Y chromosomes are remarkably divergent in structure and gene content. *Nature* 463, 536–539. A surprising degree of evolution is seen in the Y chromosome.

Kawaji H & Hayashizaki Y (2008) Exploration of small RNAs. *PLoS Genet.* 4, e22. (DOI: 10.1371/journal.pgen.0040022). A nice review article covering this very broad topic.

Kemkemer C, Kohn M, Cooper DN et al. (2009) Gene synteny comparisons between different vertebrates provide new insights into breakage and fusion events during mammalian karyotype evolution. *BMC Evol. Biol.* 9, 84. A detailed study of the chromosomal breakpoints and the re-creation of our ancestral eutherian genome. Colorfully illustrated.

Kent WJ (2002) BLAT–the BLAST-like alignment tool. *Genome Res.* 12, 656–664. BLAT, faster than lightning, appears in the UCSC Genome Browser.

Levings PP & Bungert J (2002) The human beta-globin locus control region. A center of attraction. *Eur. J. Biochem.* 269, 1589–1599. This review focuses on the regulatory elements that orchestrate the expression of the five beta globin genes.

Olender T, Lancet D & Nebert DW (2008) Update on the olfactory receptor (OR) gene superfamily. *Hum. Genomics* 3, 87–97. This review paper compares the olfactory receptor gene families between species. An impressive phylogram appears in this paper.

Ovcharenko I, Loots GG, Nobrega MA et al. (2005) Evolution and functional classification of vertebrate gene deserts. *Genome Res.* 15, 137–145. Gene deserts contain few genes but they are conserved in evolution. This paper discusses the topic.

Sherry ST, Ward MH, Kholodov M et al. (2001) dbSNP: the NCBI database of genetic variation. *Nucleic Acids Res.* 29, 308–311. A brief discussion on this huge and detailed database at the NCBI.

Tam OH, Aravin AA, Stein P et al. (2008) Pseudogene-derived small interfering RNAs regulate gene expression in mouse oocytes. *Nature* 453, 534–538. Pseudogenes are not necessarily dead! Some are transcribed and actively participate in gene regulation.

Yu WP, Pallen CJ, Tay A et al. (2001) Conserved synteny between the Fugu and human *PTEN* locus and the evolutionary conservation of vertebrate PTEN function. *Oncogene* 20, 5554–5561. A nice article discussing these orthologs.

APPENDIX 1

Formatting Your Report

A1.1 INTRODUCTION

The presentation of information in science is often challenging, and bioinformatics is no exception. Once you have analyzed your sequences, you will need to record the results in a format that lends itself to easy viewing, interpretation, and understanding. Much of your analysis will be done on the Internet, but saving the results and compiling them into other documents involves some skill and planning.

A1.2 FONT CHOICE AND PASTING ISSUES

A clear example of data presentation is a sequence. **Figure A1.1** is a view of a protein sequence from an NCBI file; note the orderly rows, the blocks of 10 amino acids, and the alignment of the numbers.

If you highlight this sequence on the Web page, copy it to the clipboard, and then paste it into your word processor document, you might be surprised to see something like that seen in **Figure A1.2**.

What has happened? First, there is a font difference. Most sequence records on the Web are displayed using a monospaced font in which each character uses the same amount of space on the line. A narrow "I" will take up the same amount of space as a wide "W." The default font for most word processors is likely to be one where the space assigned to letters is variable in width. So, when you copy and paste from the Web to your word processor document, the font changes and the file loses its orderly format.

Depending on the documents you generate, you may want to change the default font in your word processor to a monospaced font such as Courier. Alternatively, you can just highlight the sequence and change the font to Courier (**Figure A1.3**) or another monospaced font.

```
  1 mvagtrclla lllpqvllgg aaglvpelgr rkfaaassgr pssqpsdevl sefelrllsm
 61 fglkqrptps rdavvppyml dlyrrhsgqp gspapdhrle raasrantvr sfhheeslee
121 lpetsgkttr rfffnlssip teefitsael qvfreqmqda lgnnssfhhr iniyeiikpa
181 tanskfpvtr lldtrlvnqn asrwesfdvt pavmrwtaqg hanhgfvvev ahleekqgvs
241 krhvrisrsl hqdehswsqi rpllvtfghd gkghplhkre krqakhkqrk rlkssckrhp
301 lyvdfsdvgw ndwivappgy hafychgecp fpladhlnst nhaivqtlvn svnskipkac
361 cvptelsais mlyldenekv vlknyqdmvv egcgcr
```

1 mvagtrclla lllpqvllgg aaglvpelgr rkfaaassgr pssqpsdevl sefelrllsm   61 fglkqrptps rdavvppyml dlyrrhsgqp gspapdhrle raasrantvr sfhheeslee   121 lpetsgkttr rfffnlssip teefitsael qvfreqmqda lgnnssfhhr iniyeiikpa   181 tanskfpvtr lldtrlvnqn asrwesfdvt pavmrwtaqg hanhgfvvev ahleekqgvs   241 krhvrisrsl hqdehswsqi rpllvtfghd gkghplhkre krqakhkqrk rlkssckrhp   301 lyvdfsdvgw ndwivappgy hafychgecp fpladhlnst nhaivqtlvn svnskipkac   361 cvptelsais mlyldenekv vlknyqdmvv egcgcr

Figure A1.1 A protein sequence.
This is the formatting you will typically see in the GenBank format.

Figure A1.2 A protein sequence pasted directly from a Web page. The orderly rows and neat appearance of Figure A1.1 have now disappeared.

Figure A1.3 The text of Figure A1.2 changed from Times New Roman to Courier font.

```
1 mvagtrclla lllpqvllgg aaglvpelgr rkfaaassgr pssqpsdevl
sefelrllsm          61 fglkqrptps rdavvppyml dlyrrhsgqp
gspapdhrle raasrantvr sfhheeslee         121 lpetsgkttr
rfffnlssip teefitsael qvfreqmqda lgnnssfhhr iniyeiikpa
181 tanskfpvtr lldtrlvnqn asrwesfdvt pavmrwtaqg hanhgfvvev
ahleekqgvs         241 krhvrisrsl hqdehswsqi rpllvtfghd
gkghplhkre krqakhkqrk rlkssckrhp         301 lyvdfsdvgw
ndwivappgy hafychgecp fpladhlnst nhaivqtlvn svnskipkac
361 cvptelsais mlyldenekv vlknyqdmvv egcgcr
```

Figure A1.4 The text of Figure A1.3 changed to font size 10.

```
1 mvagtrclla lllpqvllgg aaglvpelgr rkfaaassgr pssqpsdevl sefelrllsm
61 fglkqrptps rdavvppyml dlyrrhsgqp gspapdhrle raasrantvr sfhheeslee
121 lpetsgkttr rfffnlssip teefitsael qvfreqmqda lgnnssfhhr iniyeiikpa
181 tanskfpvtr lldtrlvnqn asrwesfdvt pavmrwtaqg hanhgfvvev ahleekqgvs
241 krhvrisrsl hqdehswsqi rpllvtfghd gkghplhkre krqakhkqrk rlkssckrhp
301 lyvdfsdvgw ndwivappgy hafychgecp fpladhlnst nhaivqtlvn svnskipkac
361 cvptelsais mlyldenekv vlknyqdmvv egcgcr
```

Although there is some improvement, the format is still not right because the font size is too large to accommodate the five blocks of 10 amino acids, with numbering on the left margin, and spaces in between the blocks. When the sequence ran out of space on the line, it wrapped around to the next line with unpleasant results. Reducing the font size will restore the correct lines (**Figure A1.4**).

Figure A1.4 is much better, but still not quite right. Note that the numbers on the first two lines are not aligned, which throws off the blocks of sequence. The problem may be with the hidden commands that were pasted along with the sequence characters. Although the text on a Web page looks plain, it is specially formatted in a language called HyperText Markup Language (HTML). When viewing a Web page, you don't see any of the HTML formatting commands that define how the text appears in the Web document. But when you copy the contents of a Web page, you are copying these formatting commands as well. If you use the default paste command in word processors, it actually pastes the HTML, or some version of it. You might be able to fix your pasted sequence by adding spaces, but this does not often work. The better approach would be to use a paste option (often called "paste special") where what you paste is text without any formatting. It may still be in the incorrect font and size, but that can all be cleaned up with a few clicks. You may also have to clean up the first line by adding a few spaces in front of the number, but this time it will be easy and the format will closely resemble what you see on the Web (**Figure A1.5**).

Finally, you may also notice that it takes longer to paste in HTML. There is often a delay while your word processor is trying to interpret the hidden commands that are being pasted in. If you paste in unformatted text, it instantly appears on the page.

Figure A1.5 Text copied from the Web and pasted in as unformatted text. A small amount of cleaning up will make the sequence look like that seen on the Web, and in Figure A1.1.

```
  1 mvagtrclla lllpqvllgg aaglvpelgr rkfaaassgr pssqpsdevl sefelrllsm
 61 fglkqrptps rdavvppyml dlyrrhsgqp gspapdhrle raasrantvr sfhheeslee
121 lpetsgkttr rfffnlssip teefitsael qvfreqmqda lgnnssfhhr iniyeiikpa
181 tanskfpvtr lldtrlvnqn asrwesfdvt pavmrwtaqg hanhgfvvev ahleekqgvs
241 krhvrisrsl hqdehswsqi rpllvtfghd gkghplhkre krqakhkqrk rlkssckrhp
301 lyvdfsdvgw ndwivappgy hafychgecp fpladhlnst nhaivqtlvn svnskipkac
361 cvptelsais mlyldenekv vlknyqdmvv egcgcr
```

A1.3 FIND AND REPLACE

The multitudes of sequence files in online databases are given accession numbers. In order for you to maintain proper records and to allow others to reproduce your analysis, you must keep track of these unique identifiers for later reference. But for communication purposes you will often want to change these anonymous numbers into meaningful text which provides key information for the analysis to be interpreted. Who will care that NP_031579.2 and NP_058874.1 are nearly identical? But if these accession numbers were changed into "mouse" and "rat," then the reader of your reports can instantly relate to and better understand the data that are presented.

You may have already used your word processor's "find and replace" function to change all the occurrences of one word or phrase with a new word or phrase. In the example below, a multiple sequence alignment was generated for the bone morphogenetic protein 2 sequences from mouse, rat, pig, and human using the EBI tool ClustalW. In the output (**Figure A1.6**), note that all of the text is in Courier.

Rather than the RefSeq accession numbers, it would be helpful to label the rows by the common name of the animal. In this short example, you may choose to do this manually. However, a much larger alignment could be tedious and error-prone. The "find and replace" command can quickly and accurately replace the accession numbers with words you can recognize: MOUSE, RAT, PIG, and HUMAN.

There are some important considerations before proceeding. Most importantly, you want to maintain the alignment of the sequences. The words MOUSE, RAT, PIG, and HUMAN differ in character length. If you replace an eleven-character accession number, NP_001191.1, with a five-character name, HUMAN, you will shift the sequences in those human lines by six spaces (eleven minus five). This would be fine if the mouse, rat, and pig accession numbers were each eleven characters long and their names were all five characters long. They are not, so you will have to include spaces in the "replace" step in order to maintain the alignment of the sequences.

It may be helpful to generate a guide to provide the exact number of spaces needed. First, generate a list of the accession numbers followed by a list of common names after each. Be sure that your list is in a monospaced font to properly align the characters and, importantly, that the accession numbers and their corresponding common names are in the same order. With these groupings of the accession numbers and names, at a glance you can see the differences in length (**Figure A1.7**).

```
CLUSTAL 2.0.12 multiple sequence alignment

NP_001191.1       MVAGTRCLLALLLPQVLLGGAAGLVPELGRRKFAAASSGRPSSQPSDEVLSEFELRLLSM 60
XP_001928064.1    MVAGTRCLLALLLPQVLLGGAADLIPELGRRKFAAS-TGLSSSQPSDDVLSEFELRLLSM 59
NP_031579.2       MVAGTRCLLVLLLPQVLLGGAAGLIPELGRKKFAAASSRPLSR-PSEDVLSEFELRLLSM 59
NP_058874.1       MVAGTRCLLVLLLPQVLLGGAAGLIPELGRKKFAGAS-RPLSR-PSEDVLSEFELRLLSM 58
                  ********* .*********** * .*:*****:***.:      *  **:.:*********** 

NP_001191.1       FGLKQRPTPSRDAVVPPYMLDLYRRHSGQPGPSPAPDHRLERAASRANTVRSFHHEESLEE 120
XP_001928064.1    FGLKQRPTPSRDAVVPPYMLDLYRRHSGQPGAPAPDHRLERAASLANTVRSFHHEESLEE 119
NP_031579.2       FGLKQRPTPSKDVVVPPYMLDLYRRHSGQPGAPAPDHRLERAASRANTVRSFHHEEAVEE 119
NP_058874.1       FGLKQRPTPSKDVVVPPYMLDLYRRHSGQPGALAPDHRLERAASRANTVLSFHHEEAIEE 118
                  **********:* .****************** * ********** **** ******::** 
```

Figure A1.6 A multiple sequence alignment with the lines labeled with accession numbers.

```
NP_001191.1
XP_001928064.1
NP_031579.2
NP_058874.1
HUMAN
PIG
MOUSE
RAT
```

Figure A1.7 Accession numbers and names, in a monospaced font.

```
NP_001191.1XXX
XP_001928064.1
NP_031579.2XXX
NP_058874.1XXX
HUMAN
PIGXX
MOUSE
RATXX
```

Figure A1.8 Accession numbers and names equalized with spaces (shown here as Xs).

The longest accession number (XP_001928064.1) is 14 characters long while the others are 11. For "find and replace," all the accession numbers can be made the same length by the addition of spaces. In **Figure A1.8** these are shown as Xs so you can see these normally invisible characters. The common names can be equalized as well by adding two spaces to the PIG and RAT names.

So with Figure A1.8 as a guide, perform the "find and replace" command and provide spaces as needed. Replace NP_001191.1 (followed by three spaces) with HUMAN, replace XP_001928064.1 with PIG (followed by two spaces), and so on. Replace one occurrence and verify that you are getting the desired result. If it goes as expected, you may then choose to replace the text all at once instead of one occurrence at a time. Since spaces may be invisible in the "find and replace" window, you must be sure that the window fields do not have any extra spaces that you are not expecting. Note that in this example, the whole alignment shifts left. If this is not desired, additional spaces can be appended to the common names.

The last cleanup to be performed is the consensus line at the bottom of each block, under the columns of amino acids. It contains the characters "*", ":", and ".", indicating the agreement in the alignment (see Figure A1.6). This last cleanup can be done manually in this short example, but you can also use "find and replace." Many word processors have a feature where you can view the hidden formatting. This would reveal spaces, carriage returns, page breaks, etc. By doing so the 20 spaces that ClustalW uses to push the consensus line to the right are revealed. Manually, it takes the removal of nine spaces from the consensus line to bring that line under the sequences. Using the "find and replace" command, replace the 20 spaces with 11 spaces and the consensus lines should come into alignment with the amino acids (**Figure A1.9**). Although there are other blocks of spaces in the multiple sequence alignment (for example, to the right of the common names) the consensus line is the only place where there are 20 spaces.

Be careful and make sure you understand how to use the "find and replace" command. Some word processors will replace all the occurrences in the entire document unless you either click the starting point or highlight the area to be worked on. If you are not careful, you could end up performing replacements throughout your entire document instead of at the desired locations. In some common word processors, a dialog box asks if you want to perform the replace function on the remainder of the document. Be sure to answer "no" or it will go to the beginning of the document and start replacing. If you are preparing a report with many different sections of analysis, the safest approach is to open a separate document that only contains the text to be modified. You can later copy and paste the modified text back into the final report.

Changing file format

"Find and replace" can also be used to change the format of a document. **Figure A1.10A** is a tab-delimited display of three protein sequences. The first column contains UniProtKB accession numbers, and the second column has sequence from each accession number. Although this list of three rows can easily be

Figure A1.9 The multiple sequence alignment, seen in Figure A1.6, with accession numbers replaced with readable words.

```
HUMAN    MVAGTRCLLALLLPQVLLGGAAGLVPELGRRKFAAASSGRPSSQPSDEVLSEFELRLLSM 60
PIG      MVAGTRCLLALLLPQVLLGGAADLIPELGRRKFAAS-TGLSSSQPSDDVLSEFELRLLSM 59
MOUSE    MVAGTRCLLVLLLPQVLLGGAAGLIPELGRKKFAAASSRPLSR-PSEDVLSEFELRLLSM 59
RAT      MVAGTRCLLVLLLPQVLLGGAAGLIPELGRKKFAGAS-RPLSR-PSEDVLSEFELRLLSM 58
         ******** .************* .*:*****:***.:    *  **::***********

HUMAN    FGLKQRPTPSRDAVVPPYMLDLYRRHSGQPGSPAPDHRLERAASRANTVRSFHHEESLEE 120
PIG      FGLKQRPTPSRDAVVPPYMLDLYRRHSGQPGAPAPDHRLERAASLANTVRSFHHEESLEE 119
MOUSE    FGLKQRPTPSKDVVVPPYMLDLYRRHSGQPGAPAPDHRLERAASRANTVRSFHHEEAVEE 119
RAT      FGLKQRPTPSKDVVVPPYMLDLYRRHSGQPGALAPDHRLERAASRANTVLSFHHEEAIEE 118
         **********:*.*****************:  ********** **** ******::**
```

(A)

```
P01275   HDEFERHAEGTFTSDVSSYLEGQAAKEFIAWLVKGRG
P55095   HDEFERHAEGTFTSDVSSYLEGQAAKEFIAWLVKGRG
O42143   HAEGTFTSDVTQQLDEKAAKEFIDWLINGGPSKEIIS
```

(B)

```
P01275   →   HDEFERHAEGTFTSDVSSYLEGQAAKEFIAWLVKGRG¶
P55095   →   HDEFERHAEGTFTSDVSSYLEGQAAKEFIAWLVKGRG¶
O42143   →   HAEGTFTSDVTQQLDEKAAKEFIDWLINGGPSKEIIS¶
¶
```

(C)

```
>P01275   HDEFERHAEGTFTSDVSSYLEGQAAKEFIAWLVKGRG
>P55095   HDEFERHAEGTFTSDVSSYLEGQAAKEFIAWLVKGRG
>O42143   HAEGTFTSDVTQQLDEKAAKEFIDWLINGGPSKEIIS
>
>
```

(D)

```
>P01275
HDEFERHAEGTFTSDVSSYLEGQAAKEFIAWLVKGRG
>P55095
HDEFERHAEGTFTSDVSSYLEGQAAKEFIAWLVKGRG
>O42143
HAEGTFTSDVTQQLDEKAAKEFIDWLINGGPSKEIIS
```

Figure A1.10 Altering formatting using "find and replace." (A) Rows with accession numbers followed by sequence. (B) The same sequences but with formatting symbols (tab and paragraph marks) visible. (C) Each paragraph mark has been replaced by a paragraph mark followed by the symbol ">." (D) Each tab has now been replaced with a paragraph mark and the sequences are now in multi-FASTA format.

converted to FASTA format manually, what if the list were 50 rows long? "Find and replace" would definitely be the better choice of approaches as it would be faster and error-free.

As noted earlier, many word processors have the ability to show or hide the formatting symbols in a document. Figure A1.10B shows the same set of sequences with the formatting symbols visible: there is a paragraph mark after each sequence and a tab mark between the accession number and the sequence.

First, convert the paragraph marks into paragraph marks followed by a ">" symbol using "find and replace" (Figure A1.10C). You end up with extra ">" symbols at the end of the document, or missing one at the top, but these are easily deleted or added. Now you only have to replace the "tabs" with paragraph marks (Figure A1.10D).

A1.4 HYPERTEXT

When a BLAST search is performed and the results are returned, every hit in the database is hypertext to allow easy navigation to the individual sequence records. In addition, hypertext is used to navigate within the many pages of results, instantly transporting you from the tabular results at the top of the page to the specific pairwise alignments many pages below. When copying your results from the Web, why lose this power of hypertext? Luckily, this is one of the easiest features to maintain. Your results will vary depending on the Web pages of origin, but usually a default copy and paste will preserve many of the hypertext links.

In **Figure A1.11**, the default paste was hypertext. The links on the right (for example, 774) are no longer functional as these were only used to navigate to another results page in your Web browser. However, the hypertext accession numbers will still take you to the individual sequence records. As long as the site where you performed the search, in this example the Website of the DNA Data Bank of Japan, maintains its URL (uniform resource locator) and system for linking hypertext to records, this hypertext taken off the Web and placed into your word processor document should function.

Figure A1.11 The results table of a BLAST search, pasted from a Web page. The hypertext from the Web page is underlined here.

```
sp|P12643|BMP2_HUMAN RecName: Full=Bone morphogenetic protein 2;...    774   0.0
sp|O46564|BMP2_RABIT RecName: Full=Bone morphogenetic protein 2;...    743   0.0
sp|O19006|BMP2_DAMDA RecName: Full=Bone morphogenetic protein 2;...    712   0.0
sp|P49001|BMP2_RAT RecName: Full=Bone morphogenetic protein 2;  ...    710   0.0
sp|P21274|BMP2_MOUSE RecName: Full=Bone morphogenetic protein 2;...    709   0.0
sp|Q90751|BMP2_CHICK RecName: Full=Bone morphogenetic protein 2;...    596   e-170
sp|P25703|BMP2A_XENLA RecName: Full=Bone morphogenetic protein 2...    581   e-165
sp|P30884|BMP2B_XENLA RecName: Full=Bone morphogenetic protein 2...    575   e-163
sp|Q804S2|BMP2_TETNG RecName: Full=Bone morphogenetic protein 2;...    478   e-134
sp|Q90752|BMP4_CHICK RecName: Full=Bone morphogenetic protein 4;...    470   e-132
sp|P30885|BMP4_XENLA RecName: Full=Bone morphogenetic protein 4;...    468   e-131
sp|Q8MJV5|BMP4_SUNMU RecName: Full=Bone morphogenetic protein 4;...    462   e-129
sp|P12644|BMP4_HUMAN RecName: Full=Bone morphogenetic protein 4;...    462   e-129
sp|Q06826|BMP4_RAT RecName: Full=Bone morphogenetic protein 4;  ...    457   e-128
sp|Q2KJH1|BMP4_BOVIN RecName: Full=Bone morphogenetic protein 4;...    457   e-128
sp|O46576|BMP4_RABIT RecName: Full=Bone morphogenetic protein 4;...    455   e-127
sp|P21275|BMP4_MOUSE RecName: Full=Bone morphogenetic protein 4;...    455   e-127
```

Creating hypertext

Occasionally, it is useful to add your own hypertext to a report. This is easily done and full instructions can be found in the help document of your word processor. The critical piece of text to have is the Web page address (the URL). For example, if you were to navigate to the human bone morphogenetic protein 2 record at the DNA Data Bank of Japan, the URL of the Web page would look like this:

```
http://getentry.ddbj.nig.ac.jp/search/get_entry?mode=view&type=flatfil
e&database=sprot&accnumber=P12643
```

This URL could be copied to your clipboard and then pasted into the appropriate window as instructed by your word processor help document. You could then turn any text (for example, human BMP2) into a convenient link.

Selecting a column of text

In the list of BLAST hits in Figure A1.11, the first column of hypertext (for example, P12643) is functional while the second (for example, 774) is not. But the second column is still formatted (colored and underlined) as if it contained hyperlinks. Cleaning up this part of your document would reduce confusion for the reader. This can be done with your word processor by "clearing" any formatting or hypertext (this may appear under your "Edit" menu).

Some word processors allow you to select any column of text. For example, if you wanted to select the entire column of nonfunctional hypertext in Figure A1.11, you could try holding down the "Alt" key on your keyboard, clicking your mouse before the nonfunctional hypertext, and, while holding the mouse button down, highlighting the area. You can then clear the formatting all at once. **Figure A1.12** shows a column of text highlighted.

A1.5 SUMMARY

It is often easy to generate data faster than it can be assimilated. So don't forget to import your data into documents for the purposes of saving, later interpretation, presentation, and publication. You will need to record your analysis results, and this section presented basic strategies for working with Web page content.

```
sp|P12643|BMP2_HUMAN RecName: Full=Bone morphogenetic protein 2;...    774   0.0
sp|O46564|BMP2_RABIT RecName: Full=Bone morphogenetic protein 2;...    743   0.0
sp|O19006|BMP2_DAMDA RecName: Full=Bone morphogenetic protein 2;...    712   0.0
sp|P49001|BMP2_RAT RecName: Full=Bone morphogenetic protein 2;   ...   710   0.0
sp|P21274|BMP2_MOUSE RecName: Full=Bone morphogenetic protein 2;...    709   0.0
sp|Q90751|BMP2_CHICK RecName: Full=Bone morphogenetic protein 2;...    596   e-170
sp|P25703|BMP2A_XENLA RecName: Full=Bone morphogenetic protein 2...    581   e-165
sp|P30884|BMP2B_XENLA RecName: Full=Bone morphogenetic protein 2...    575   e-163
sp|Q804S2|BMP2_TETNG RecName: Full=Bone morphogenetic protein 2;...    478   e-134
sp|Q90752|BMP4_CHICK RecName: Full=Bone morphogenetic protein 4;...    470   e-132
sp|P30885|BMP4_XENLA RecName: Full=Bone morphogenetic protein 4;...    468   e-131
sp|Q8MJV5|BMP4_SUNMU RecName: Full=Bone morphogenetic protein 4;...    462   e-129
sp|P12644|BMP4_HUMAN RecName: Full=Bone morphogenetic protein 4;...    462   e-129
sp|Q06826|BMP4_RAT RecName: Full=Bone morphogenetic protein 4;   ...   457   e-128
sp|Q2KJH1|BMP4_BOVIN RecName: Full=Bone morphogenetic protein 4;...    457   e-128
sp|O46576|BMP4_RABIT RecName: Full=Bone morphogenetic protein 4;...    455   e-127
sp|P21275|BMP4_MOUSE RecName: Full=Bone morphogenetic protein 4;...    455   e-127
```

Figure A1.12 A column of highlighted text. Some word processors allow any column of text to be selected by using the "Alt" key on the keyboard.

ACAAGGGACTAGAGAAACCAAAA

AGAAACCAAAACGAAAGGTGCAGAA

AACGAAAGGTGCAGAAGGGGAAACAGATGCAGA

GAAGGGGAAACAGATGCAGAAAGCAT

AGAAAGCAT

ACAAGGGACTAGAGAAACCAAAACGAAAGGTGCAGAAGGGGAAACAGATGCAGAAAGCAT

ACCAAAA

AGAAACCAAAACGAAAGGTGCAGAA

AACGAAAGGTGCAGAAGGG

GAAGGG

APPENDIX 2

Running NCBI BLAST in "batch" Mode

When you have more than one sequence to analyze, and you will be performing the same analysis on each sequence, it is efficient to analyze them simultaneously. This is commonly referred to as doing things in "batch" mode. The NCBI recommends no more than 50 sequences be used in a single BLAST form at once. A column of numbers pasted into the BLAST form works fine:

```
NG_007393
NG_007394
etc.
```

What if you don't have accession numbers? Multiple sequences work as well but the sequences have to be in a certain form in order to run successfully: it is called a multi-FASTA format. This can be constructed manually or you can let the NCBI Website do it for you.

1) Using the Entrez form, with "EST" chosen from the drop-down menu (**Figure A2.1**), type or paste in a list of accession numbers and submit them by hitting "Search." A space- or comma-delimited list in one line will be accepted by the Web form when searching sequence databases.

2) The default format of the results is "Summary" (**Figure A2.2**). This is a good opportunity to look at the sequence information and verify that the list is correct. Once these records are converted into FASTA format, some annotation may no longer be visible or as readable.

Figure A2.1 NCBI Entrez query window, with "EST" chosen from the drop-down menu. Note that two EST accession numbers have been entered, with a comma in between them.

```
1. Ep_venom_10B06_M13F-20 Echis pyramidum leakeyi Venom Gland Gateway cDNA library Echis
pyramidum leakeyi cDNA clone Ep_venom_10B06 5-, mRNA sequence
390 bp linear mRNA
Accession:GR951114.1 GI:281560290
EST GenBank FASTA

2. Ep_venom_15B08_M13F-20 Echis pyramidum leakeyi Venom Gland Gateway cDNA library Echis
pyramidum leakeyi cDNA clone Ep_venom_15B08 5-, mRNA sequence
448 bp linear mRNA
Accession:GR951108.1 GI:281560284
EST GenBank FASTA
```

Figure A2.2 Summary format of EST records.

3) Click on the "Display Settings" menu and select "FASTA (text)." The format changes to individual FASTA formatted files which are numbered and have blank lines separating them (**Figure A2.3**).

```
>gi|281560290|gb|GR951114.1|GR951114 Ep_venom_10B06_M13F-20 Echis pyramidum leakeyi Venom
Gland Gateway cDNA library Echis pyramidum leakeyi cDNA clone Ep_venom_10B06 5', mRNA
sequence
CAAAAAGTGTGGGGTAACCCAGGCTAAATGGGAATCAGATGAGCCCATCAAAAAGGCCTCTAATTTAGTT
GCTACTTCTGAACAACAACGTTTTGACCCAAGATACATTCAGCTTGTCATAGTTGCAGATCACGCAATGA
TTACGAAAACAAAGGTGATTTAACTGCTGTAAGAACATGGGTACATCAAATTTTCAACGATATGACTGT
GATGTACAGAAATCTGAATATTCATATAACACTGGTTGCCATAGTAACTTGGAGCAAAAGAGATATGATT
ACTGTGACATCATCAGCACGTGGTACTTTGAAGTTATTTGGAACATGGAGAGAGACAAAGTTGCTGGAAA
AAATAAAGCATGATAATGCTCAGTTACTCACAGGCATTAA


>gi|281560284|gb|GR951108.1|GR951108 Ep_venom_15B08_M13F-20 Echis pyramidum leakeyi Venom
Gland Gateway cDNA library Echis pyramidum leakeyi cDNA clone Ep_venom_15B08 5', mRNA
sequence
TTATTCAACTGACGAACAAAACCCCTGCTAGCTACGTGTCTTACGGATGCTTCTGCGGGGGTGGGGACAA
AGGCAAGCCAAAGGACGCCACCGACCGCTGCTGCTTCGTGCATAGCTGCTGTTACGACACGCTGCCCGAC
TGCAGCCCCAAAACGGACCAATACAAATACAAATGGGAAAACGGGGAAATCATCTGTGAAAACAGCACCT
CGTGCAAGAAACGAATTTGTGAGTGTGACAAGGCTGTGGCAATCTGCTTGCGAGACAATCTGAACACATA
CAACAAGAAATATAGGATTTACCCGAACTTCTTATGCAGAGGAGATCCAGACAAATGCTAAGTCTCTGCA
GGCCGGGGGAAAAACCCTCCAATTACACAATTGTGGTTGTGTTACTCTATTATTCTGAATGCAATACTGA
GTAATAAACAGGTGCCAGCTTTGCACTC
```

Figure A2.3 FASTA format of EST records.

This is the NCBI's version of a multi-FASTA file and can be copied and pasted directly into the BLAST Web forms. Other Websites will require you to clean up the annotation further; for example, by removing the text preceding the GenBank accession number and making sure the final text is free of blank lines (**Figure A2.4**).

```
>GR951114.1|GR951114 Ep_venom_10B06_M13F-20 Echis pyramidum leakeyi Venom Gland Gateway
cDNA library Echis pyramidum leakeyi cDNA clone Ep_venom_10B06 5', mRNA sequence
CAAAAAGTGTGGGGTAACCCAGGCTAAATGGGAATCAGATGAGCCCATCAAAAAGGCCTCTAATTTAGTT
GCTACTTCTGAACAACAACGTTTTGACCCAAGATACATTCAGCTTGTCATAGTTGCAGATCACGCAATGA
TTACGAAAACAAAGGTGATTTAACTGCTGTAAGAACATGGGTACATCAAATTTTCAACGATATGACTGT
GATGTACAGAAATCTGAATATTCATATAACACTGGTTGCCATAGTAACTTGGAGCAAAAGAGATATGATT
ACTGTGACATCATCAGCACGTGGTACTTTGAAGTTATTTGGAACATGGAGAGAGACAAAGTTGCTGGAAA
AAATAAAGCATGATAATGCTCAGTTACTCACAGGCATTAA
>GR951108.1|GR951108 Ep_venom_15B08_M13F-20 Echis pyramidum leakeyi Venom Gland Gateway
cDNA library Echis pyramidum leakeyi cDNA clone Ep_venom_15B08 5', mRNA sequence
TTATTCAACTGACGAACAAAACCCCTGCTAGCTACGTGTCTTACGGATGCTTCTGCGGGGGTGGGGACAA
AGGCAAGCCAAAGGACGCCACCGACCGCTGCTGCTTCGTGCATAGCTGCTGTTACGACACGCTGCCCGAC
TGCAGCCCCAAAACGGACCAATACAAATACAAATGGGAAAACGGGGAAATCATCTGTGAAAACAGCACCT
CGTGCAAGAAACGAATTTGTGAGTGTGACAAGGCTGTGGCAATCTGCTTGCGAGACAATCTGAACACATA
CAACAAGAAATATAGGATTTACCCGAACTTCTTATGCAGAGGAGATCCAGACAAATGCTAAGTCTCTGCA
GGCCGGGGGAAAAACCCTCCAATTACACAATTGTGGTTGTGTTACTCTATTATTCTGAATGCAATACTGA
GTAATAAACAGGTGCCAGCTTTGCACTC
```

Figure A2.4 "Cleaned-up" EST records in multi-FASTA format.

4) You then copy the sequences (for example, everything in Figure A2.4) and paste them into an NCBI BLAST form, as usual. When the results come back (**Figure A2.5**), they are slightly different than that seen with a single sequence. The upper-left corner indicates the number of sequences used in the search. There is a new drop-down menu called "Results for:" and the sequence descriptions appear as separate lines in the menu. Choosing from this menu will cause the screen to refresh each time, and the different results will be displayed.

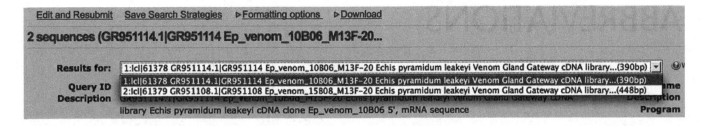

Figure A2.5 "Results for:" drop-down menu of the BLAST results.

ABBREVIATIONS

AC	accession numbers	LINE	long interspersed nuclear element
ACTH	adrenocorticotropic hormone	MAPK15	mitogen-activated protein kinase 15
BLAST	basic local alignment search tool	MAPP	multivariate analysis of protein polymorphism
BLASTN	nucleotide BLAST	MARK1	MAP/microtubule affinity-regulating kinase 1
BLASTP	protein BLAST	MeSH	Medical Subject Headings
bp	base pair	miRNA	microRNA
CBS	Center for Biological Sequence Analysis	MRE	miRNA Response Element
CCDS	Consensus CDS Protein Set	mRNA	messenger RNA
CDART	Conserved Domain Architecture Retrieval Tool	MSH	melanocyte-stimulating hormone
cDNA	complementary DNA	MUP	major urinary protein
CDS	coding sequence	MUSCLE	Multiple Sequence Comparison by Log-Expectation
CNV	copy number variation	NCBI	National Center for Biotechnology Information
COBALT	Constraint Based Alignment Tool	NIH	National Institutes of Health
DDBJ	DNA Data Bank of Japan	NJ	Neighbor-Joining
EBI	European Bioinformatics Institute	NPP	N-terminal peptide of POMC
EC	Enzyme Commission	OMIM	Online Mendelian Inheritance in Man
EGF	epidermal growth factor	ORF	open reading frame
EPO	European Patent Office	PCR	polymerase chain reaction
ER	endoplasmic reticulum	PDB	Protein Data Bank
ER	estrogen receptor	PIR	Protein Information Resource
ESTs	expressed sequence tags	PMID	PubMed Identifier
FD	familial dysautonomia	POMC	pro-opiomelanocortin
GABA	gamma-aminobutyric acid	RCSB	Research Collaboratory for Structural Bioinformatics
GeneRIF	Gene Reference into Function	RefSeq	Reference Sequence
HGNC	HUGO Gene Nomenclature Committee	rRNA	ribosomal RNA
HIV	human immunodeficiency virus	RuBisCO	ribulose bisphosphate carboxylase
HSPs	high-scoring subject pairs	SINE	short interspersed nuclear element
HSV-1	herpes simplex virus-1	SNP	single nucleotide polymorphism
HTML	HyperText Markup Language	SRS	Sequence Retrieval System
HUGO	Human Genome Organisation	STS	Sequence Tagged Site
IDA	Inferred from Direct Assay	TPM	transcripts per million
iHOP	information Hyperlinked Over Proteins	TSP	thrombospondin
ILGF	insulin-like growth factors	uORF	upstream open reading frame
IMAGE	Integrated Molecular Analysis of Genomes and their Expression	UPGMA	Unweighted Pair Group Method with Arithmetic Mean
IMP	Inferred from Mutant Phenotype	URL	uniform resource locator
indels	insertions or deletions	UTR	untranslated region
IUBMB	International Union of Biochemistry and Molecular Biology	VLDL	very low density lipoprotein
KEGG	Kyoto Encyclopedia of Genes and Genomes	WIPO	World Intellectual Property Organization
LAT	latency-associated transcript		

GLOSSARY

The chapter numbers refer to the first substantial discussion of the topic.

accession numbers
Unique identifiers of records in a database. Often a mix of letters and numbers. (Chapter 2)

alignments
The positioning of two or more DNA or protein sequences to demonstrate their similarities. (Chapter 3)

alpha helices
A regular protein structure, stabilized by hydrogen bonds between amino acids, and forming a spiral, rod-like structure. (Chapter 8)

alternatively spliced
The use of different combinations of exons when processing mRNA to its mature form. (Chapter 9)

beta sheets
A regular protein structure, stabilized by hydrogen bonds between adjacent strands, and forming a pleated surface. (Chapter 8)

bioinformatics
The generation, visualization, analysis, storage, and retrieval of large quantities of biological information. (Chapter 1)

BLAST
Basic Local Alignment Search Tool, used for DNA and protein sequence similarity searches. (Chapter 3)

BLASTN
A BLAST program that uses a DNA query to search a DNA database. (Chapter 3)

BLASTP
A BLAST program that uses a protein query to search a protein database. (Chapter 4)

BLASTX
A BLAST program that uses a DNA query to search a protein database. (Chapter 5)

C-terminus (carboxyl terminus)
The last amino acids of a protein or peptide. It can refer to a single amino acid or the region. (Chapter 4)

cDNA library
A collection of cDNAs that originate from a single pool of mRNAs. (Chapter 5)

CDS
An abbreviation for "coding sequence" in GenBank records. (Chapter 3)

Central Dogma
Coined by Dr Francis Crick, this term refers to the direct passage of information from DNA to RNA to protein. (Chapter 10)

cladogram
A tree-like graphic representing the relationships between related DNA or protein sequences. (Chapter 11)

coding region
The region of a gene that is translated into protein sequence. (Chapter 5)

codons
Triplets of nucleotides which are translated to specific amino acids by the ribosome. (Chapter 4)

copy number variations (CNVs)
Differences in gene number between individuals resulting from changes in genomic structure. (Chapter 1)

data reduction
A generic term referring to the transforming of complex or great quantities of data to forms more easily visualized and understood. (Chapter 11)

deep sequencing
A phrase describing the very extensive sequencing of a DNA to reach the most rare of sequences. (Chapter 1)

degeneracy
A term referring to the presence of different codons that can be translated to the same amino acid. (Chapter 4)

domain
A section of protein possessing a certain sequence, structure, and/or function. (Chapter 8)

E value
A statistical measure of a query–hit relationship in BLAST. E values, also labeled as "Expect," represent a numerical probability of encountering the observed identity. (Chapter 3)

ESTs (Expressed Sequence Tags)
Short cDNA sequences derived from high-throughput cDNA library sequencing efforts. (Chapter 1)

FASTA
A DNA or protein sequence file format consisting of a ">" sign immediately followed (no space) by a name or number, annotation (optional), then a carriage return. The lines that follow consist of nothing but sequence. Note that there is also a sequence database searching tool called FASTA which is not covered in this book. (Chapter 3)

gene annotation
The description of biological sequences, usually in the form of their origins, analysis results, possible functions, features, and associated literature. Annotation in bioinformatics refers to the information about a sequence, either gathered by people or generated automatically. Databases such as GenBank have a specific structure to their annotation: fields of information that are the same in each record. Examples include definition, accession number, and species of origin. (Chapter 1)

gene deserts
Regions of chromosomes with an apparent low density of genes. (Chapter 12)

gene ontology
The standardization of gene function and description through a controlled vocabulary and nomenclature. (Chapter 2)

genome browser
An application designed to reduce large stretches of genomic DNA and associated data to easily interpretable graphics. (Chapter 9)

genome sequence assembly
The organization and alignment of overlapping pieces of DNA sequence to reconstruct the intact original sequence. (Chapter 1)

gigabase
One billion nucleotides.

glycosylation
The covalent placement of one or more sugar groups onto a biomolecule. (Chapter 8)

guide tree
A graphical representation of the relationship between sequences. (Chapter 11)

hominin
Classification clade that includes *Homo sapiens* and extinct *Homo* species but not apes. (Chapter 1)

homologs
Sequences that are clearly related by descent from a common ancestor and may or may not perform the same function. (Chapter 3)

housekeeping genes
Genes thought to provide functional products used by all cell types in an organism. (Chapter 5)

Human Genome Project
A large, multi-year, and multinational effort to sequence the human genome. (Chapter 1)

identity
This word has two meanings: (a) the proper name for a sequence based on function or strong similarity to another sequence that is already described. For example, if you determine that an unknown sequence is identical to a known sequence, you have determined its identity; (b) the degree of similarity between two sequences, usually expressed as a percentage. (Chapter 3)

ideogram
Graphical representation of chromosomes showing gross details such as banding and centromeres. (Chapter 12)

indels
An abbreviation for "insertions or deletions," referring to differences seen between very similar sequences. (Chapter 4)

isoforms
Different translational products of a gene due to alternative splicing. (Chapter 4)

karyotype
Pictures or illustrations representing the chromosomes of an organism. (Chapter 12)

kilobase
One thousand nucleotides.

Kozak sequence
Named after Dr Marilyn Kozak, the ten nucleotides around the start codon of a gene; it has a consensus sequence of GCCRCCAUGG. (Chapter 9)

LINEs
Long interspersed nuclear elements; these eukaryotic repetitive elements are found in very large numbers in genomic DNA. (Chapter 6)

low complexity
A region of a protein where the sequence composition is limited to only a small number of amino acids. (Chapter 5)

Max Score
One of the statistical measures of similarity found by BLAST. (Chapter 3)

megabase
One million nucleotides.

metagenomics
The collection and study of genetic information from environmental or biomedical samples. (Chapter 1)

miRNA
MicroRNA, an approximately 22-nucleotide sequence that binds and represses the translation of mRNAs. (Chapter 10)

motif
A region of a protein sequence associated with a specific function, sometimes used interchangeably with "signature." (Chapter 8)

multimer
A term describing a protein complex. Single and two-protein complexes may be referred to as monomeric and dimeric, respectively, but larger numbers of proteins in a complex are often called multimers. (Chapter 10)

N-terminus (amino terminus)
The first amino acids of a protein or peptide. It can refer to a single amino acid or the region. (Chapter 4)

non-redundant (nr) database
A database of unique sequences. (Chapter 5)

open reading frame (ORF)
A DNA sequence of variable length bounded by termination codons, the end of DNA sequences, or a mixture of both. It is a region which could potentially encode a protein sequence, is free of other termination codons, and may have other characteristics consistent with being coding sequence. (Chapters 5 and 7)

orthologs
Sequences in different species which perform the same function in each organism and are clearly related by descent from a common ancestor. (Chapter 3)

paleogenetics
The study of DNA sequences or traits from very old or fossilized materials. (Chapter 1)

paralogs
Sequences within a species that are clearly related by descent from a common ancestor. (Chapter 3)

PCR (polymerase chain reaction)
A procedure for amplifying DNA sequences through successive rounds of priming, DNA synthesis, denaturing, and re-priming. (Chapter 7)

phylogram
A tree graphic showing the relationships between sequences with branch lengths proportional to the sequence differences. (Chapter 11)

pipeline
In the context of bioinformatics, the joining of multiple analytical tools to automatically perform a number of tasks. (Chapter 1)

poly(A) tail
The structure of multiple adenosines placed at the end of most eukaryotic mRNAs. (Chapter 1)

polylinker
A relatively short DNA sequence engineered to contain many restriction enzyme digestion sites to aid in cloning steps. (Chapter 7)

post-translational processing
Any modification of cellular proteins which occurs after translation. (Chapter 8)

primers
Short DNA sequences, often introduced to base-pair with RNA or DNA, which act as start sites for the commencement of DNA synthesis. (Chapter 7)

pseudogene
DNA sequence which is clearly related to a functional gene but contains features which are thought to make it nonfunctional, such as extra termination codons, truncations, and the absence of introns. (Chapter 1)

query
A text term or sequence used to ask a question or discover records in a database. (Chapter 3)

reading frame
One of six ways to translate a DNA sequence starting from base one, two, or three (the three forward reading frames) and the corresponding bases on the opposite DNA strand (the three reverse reading frames). (Chapters 5 and 7)

RefSeq
An abbreviation for "Reference Sequence." This sequence should represent a standard version of a sequence and is created and maintained at the NCBI. (Chapter 3)

repeats
In the context of bioinformatics, this refers to repetitive DNA or protein elements. (Chapter 6)

re-sequencing
A term referring to the determination and assembly of DNA sequence with the aid of previously generated versions or standards. (Chapter 1)

restriction enzymes
Enzymes produced by bacteria which recognize very specific DNA sequences and cleave at or near these sequences. (Chapter 7)

scoring matrix
A set of pairwise scores used by programs assessing the alignment between two sequences. (Chapter 4)

script
Short and/or simple groups of computer commands to perform a task. (Chapter 8)

signal peptide
A short sequence at the N-terminus of a protein that commits the protein for insertion into a membrane or for export from the cell. (Chapter 8)

sequence similarities
see similarity

similar or **similarity**
Used to describe how sequences are related to each other taking into account biochemically similar or evolutionarily conserved substitutions in addition to identical amino acid residues. (Chapter 4)

simple sequences
Usually refers to DNA sequences which have low complexity and are highly repetitive, for example CTCTCTCTCTCTC. (Chapter 6)

SINEs
Short interspersed nuclear elements; these eukaryotic repetitive elements are found in very large numbers in genomic DNA. (Chapter 6)

single-pass sequencing
Sequencing of DNA where each nucleotide's identity is determined once with no verification from other sequencing reads. (Chapter 5)

SNP (single nucleotide polymorphism)
These are single nucleotide differences from a reference sequence, or standard, which occur in 1% or more of the population. (Chapter 1)

Subject (Sbjct)
Also known as a "hit," it is the label given by BLAST to any sequence found by the program. (Chapter 3)

superfamily
A large set of sequences which are related by descent from a distant ancestor yet still retain a recognizable common function. (Chapter 4)

synteny
The evolutionary conservation of physical linkage between genes. (Chapter 12)

tandem arrays
Repetitive DNA sequences which are immediately adjacent to each other and often arranged in a "head-to-tail" fashion. (Chapter 6)

TBLASTN
A BLAST program that uses a protein query to search a DNA database. (Chapter 5)

transcription factor
A protein or protein complex which binds DNA elements to regulate the transcription of genes. (Chapter 9)

transcription factor binding sites
The specific short sequences of DNA which are bound by transcription factors while regulating transcription. (Chapter 9)

transcription variant
Different transcriptional products of a gene due to alternative splicing. (Chapter 3)

transmembrane
A term usually describing a domain of a protein which crosses a cellular membrane. (Chapter 8)

UTR (untranslated region)
Regions of mRNA transcripts which are not translated because they are upstream (5P UTR) or downstream (3P UTR) of the coding region. (Chapter 1)

words
In the context of database searching, sub-sequences of a larger DNA or protein sequence used to compare a query to a database of sequences, also broken into words. (Chapter 3)

X-ray crystallography
The use of an orderly arrangement of identical molecules allowing the determination of their structure through irradiation by X-rays. (Chapter 8)

WEB RESOURCES

Bionumbers
A large collection of biological facts and measurements. (Chapter 7)
bionumbers.hms.harvard.edu/default.aspx

CCDS
An abbreviation for Consensus CDS Protein Set database, it is a collection of cDNA and protein sequences, broken down by exon. (Chapter 6)
www.ncbi.nlm.nih.gov/projects/CCDS/CcdsBrowse.cgi

Center for Biological Sequence Analysis (CBS)
An extensive collection of web analytical tools. (Chapter 8)
www.cbs.dtu.dk/services

ClustalW
A widely used multiple sequence alignment tool. (Chapter 11)
www.ebi.ac.uk/Tools/msa/clustalw2

COBALT
An abbreviation for COnstraint Based ALignment Tool, it is a multiple sequence alignment tool. (Chapter 11)
www.ncbi.nlm.nih.gov/tools/cobalt

DotPlot
Sequence analysis tool that aligns sequences and portrays identities as dots on a graph. (Chapter 7)
www.vivo.colostate.edu/molkit/dnadot/index.html

EBI Jalview
European Bioinformatics Institute implementation of Jalview, a multiple sequence alignment viewer and editor. (Chapter 11)
www.ebi.ac.uk/Tools/msa/clustalw2

Ensembl Genome Browser
A large collection of genomics information and data visualization tools. (Chapter 12)
www.ensembl.org

Entrez
A text query-based interface to many of the databases maintained at the NCBI Website. (Chapter 2)
www.ncbi.nlm.nih.gov/gquery

European Patent Office (EPO)
A Website specializing in European patents. (Chapter 2)
www.epo.org

ExPASy
Proteomics Server of the Swiss Institute of Bioinformatics, hosting databases such as UniProtKB and PROSITE. (Chapter 4)
www.expasy.ch

fast DB
Alternative mRNA splicing and transcripts database. (Chapter 9)
www.fast-db.com

GenBank
A very large public DNA database maintained at the NCBI. It contains the sequences and annotation of genomic DNA, cDNA, and synthetic nucleotides. (Chapter 1)
www.ncbi.nlm.nih.gov/genbank

Gene
A gene-centered database at the NCBI Website, it contains extensive genomics data for many organisms. (Chapter 2)
www.ncbi.nlm.nih.gov/gene

Gene Ontology
Website specializing in the standardization of gene name and function. (Chapter 2)
www.geneontology.org

HUGO Gene Nomenclature Committee
Official human gene names and symbols. (Chapter 12)
www.genenames.org

iHOP
An abbreviation for Information Hyperlinked Over Proteins, it is a Website linking PubMed abstracts to protein symbols and synonyms. (Chapter 2)
www.ihop-net.org

Jalview
Official site for Jalview, a multiple sequence alignment viewer and editor. (Chapter 11)
www.jalview.org

Jmol
An open-source viewer of protein structures in 3D. (Chapter 8)
www.jmol.org

KEGG
An abbreviation for Kyoto Encyclopedia of Genes and Genomes, it is a collection of 16 databases which link and display a tremendous quantity of information on biochemical pathways, genes, diseases, drugs, and genomes. (Chapter 10)
www.genome.jp/kegg

miR2Disease
A large collection of miRNA-human disease associations based on published findings. (Chapter 10)
www.mir2disease.org

miRBase
The central repository for miRNAs, their structures, and associated annotation. (Chapter 10)
www.mirbase.org

MUSCLE
An abbreviation for MUltiple Sequence Comparison by Log-Expectation, it is a multiple sequence alignment tool. (Chapter 11)
www.ebi.ac.uk/Tools/msa/muscle

NCBI
An abbreviation for the National Center for Biotechnology Information, it is an extensive Website for bioinformatics

containing many biomedical databases and tools. (Chapter 2)
www.ncbi.nlm.nih.gov

NCBI ORF Finder
A tool for identifying open reading frames. (Chapter 7)
www.ncbi.nlm.nih.gov/gorf/gorf.html

NEBCutter tool
A restriction mapping tool at the New England Biolabs Website.
(Chapter 7)
tools.neb.com/NEBcutter2/index.php

NIH RePORTER
The National Institutes of Health Website for government funded
research projects. (Chapter 2)
projectreporter.nih.gov

OMIM
An abbreviation for Online Mendelian Inheritance in Man. This
database is a joint project between Johns Hopkins University and
the NCBI, containing records of human genes, phenotypes, and
genetic disorders. (Chapter 2)
www.ncbi.nlm.nih.gov/OMIM

Primer3
A widely used primer prediction program. (Chapter 7)
frodo.wi.mit.edu/primer3

ProPhylER
A data reduction and visualization tool that shows the freedom and constraints of evolution upon protein sequences.
(Chapter 8)
www.prophyler.org

PROSITE
A database and associated search tools that contains over one
thousand patterns and protein family signatures. (Chapter 8)
www.expasy.ch/prosite

Protein Data Bank (PDB)
The single worldwide repository of information about the 3D
structures of large biological molecules, including proteins and
nucleic acids, maintained by the Research Collaboratory for
Structural Bioinformatics (RCSB). (Chapter 8)
www.rcsb.org

Protein Information Resource (PIR)
A large Website focused on proteins. The URL points to a
Composition/Molecular Weight calculation form. (Chapter 7)
pir.georgetown.edu/pirwww/search/comp_mw.shtml

PubMed
A biomedical literature database maintained by the NCBI.
(Chapter 2)
www.ncbi.nlm.nih.gov

Sequence Manipulation Suite
A large collection of utilities for analyzing and manipulating
sequences. (Chapter 7)
www.bioinformatics.org/sms2/index.html

Sequence Retrieval System (SRS)
A relational database from the European Bioinformatics Institute
(EBI) Website, allowing the construction of very complex and
focused queries of sequence databases. (Chapter 7)
srs.ebi.ac.uk

SignalP
A program for predicting secretion signal sequences, from
the Center for Biological Sequence Analysis (CBS) Website.
(Chapter 8)
www.cbs.dtu.dk/services/SignalP

Splign
Pairwise alignment tool at the NCBI Website, specializing on
cDNA-genomic DNA alignments. (Chapter 9)
www.ncbi.nlm.nih.gov/sutils/splign/splign.cgi

TarBase
A curated database of miRNA-target site pairs, collected from
published evidence for miRNA inhibition of targeted gene transcripts. (Chapter 10)
diana.cslab.ece.ntua.gr/tarbase

TargetScan
A widely used miRNA target site prediction tool. (Chapter 10)
www.targetscan.org

TimeTree
A Website that provides evolutionary time divergence between
groups of organisms along with supporting published evidence.
(Chapter 5)
www.timetree.org

TMHMM
A tool for the prediction of transmembrane helices in proteins.
(Chapter 8)
www.cbs.dtu.dk/services/TMHMM

Transfac database
A curated collection of transcription factors, binding sites, and
associated data. (Chapter 9)
www.gene-regulation.com

UCSC Genome Browser
A large collection of genomics information and data visualization
tools. (Chapter 12)
genome.ucsc.edu

UniGene
An NCBI database containing cDNA sequences organized by
genetic loci. (Chapter 2)
www.ncbi.nlm.nih.gov/unigene

UniProtKB
A widely respected protein database with extensive and curated
annotation. (Chapter 4)
www.uniprot.org

U.S. Patent and Trademark Office
A Website specializing in U.S. patents and trademarks.
(Chapter 2)
www.uspto.gov

World Intellectual Property Organization (WIPO)
United Nations Website of intellectual property. (Chapter 2)
www.wipo.int

INDEX

Note: genes and their products are presented as single entries – thus "CDKN1A" includes references to both the *CDKN1A* gene and CDKN1A protein. Abbreviations following page numbers are: F, figure; T, table; B, box; and i, information box.

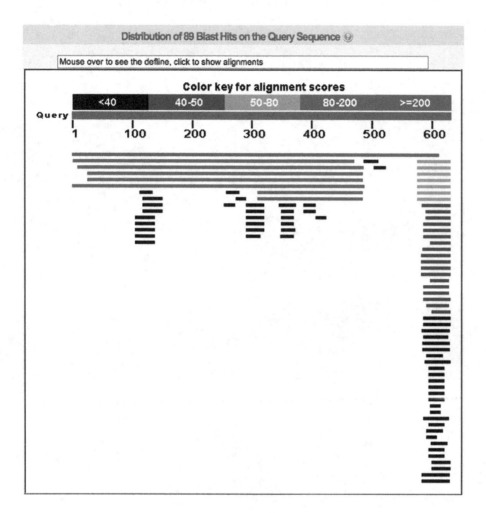

Figure 3.6 The graphic pane of the NCBI BLASTN results. The query coordinates and length correspond to the numbered scale across the top. Sequences found by BLAST, "hits," are represented as horizontal bars below this scale. These will vary in length, position, and color-coded scoring.

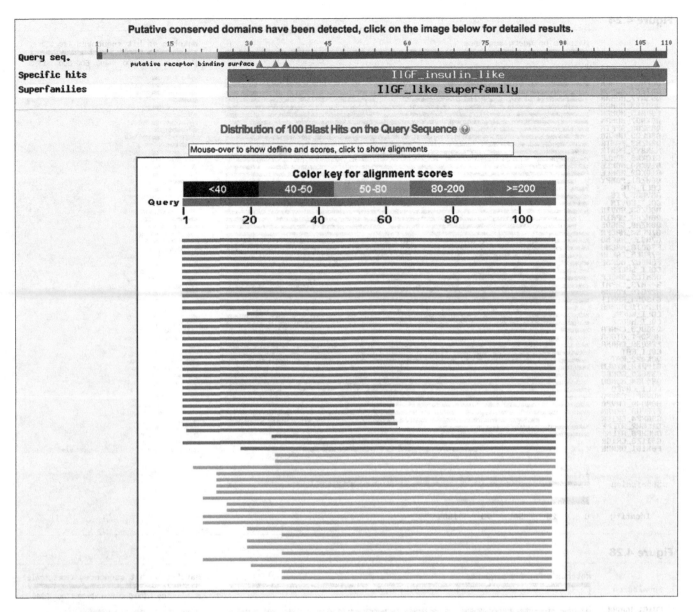

Putative conserved domains have been detected, click on the image below for detailed results.

Query seq.
Specific hits   IlGF_insulin_like
Superfamilies   IlGF_like superfamily

putative receptor binding surface

Distribution of 100 Blast Hits on the Query Sequence

Mouse-over to show defline and scores, click to show alignments

Color key for alignment scores

| <40 | 40-50 | 50-80 | 80-200 | >=200 |

Query

Figure 4.14 The graphic from the NCBI BLASTP results. The query is the human preproinsulin protein sequence NP_000198 and the database is RefSeq protein.

Figure 4.24 (opposite, top) The results graphic for ExPASy BLASTP. The POMC fragment, UniProtKB accession number Q53WY7, was used as a query against the mammalian division of the UniProtKB database. On the far left are sequence names for each row. The center panel depicts the query and the best areas of identity with the hits. The right panel shows the hits as gray bars with the sequence identities with the query colored.

Figure 4.28 (opposite, bottom) The results graphic from ExPASy BLASTP query Q9UMA6. The human division of UniProtKB was searched. Note the multiple regions of similarity between the query and the hits in the right panel.

Figure 4.24

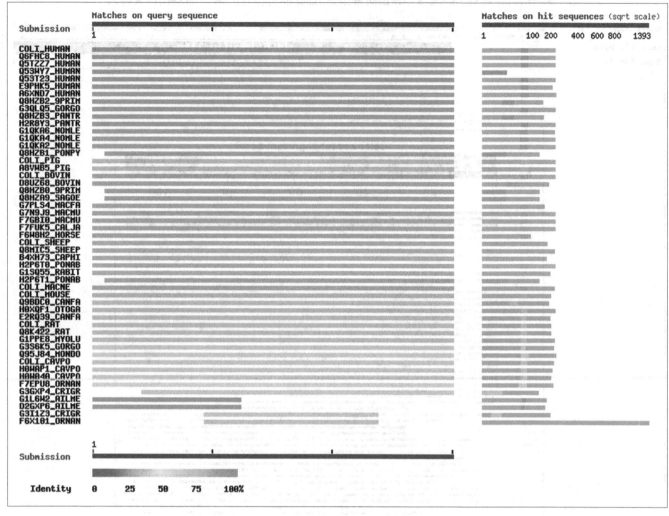

Figure 4.28

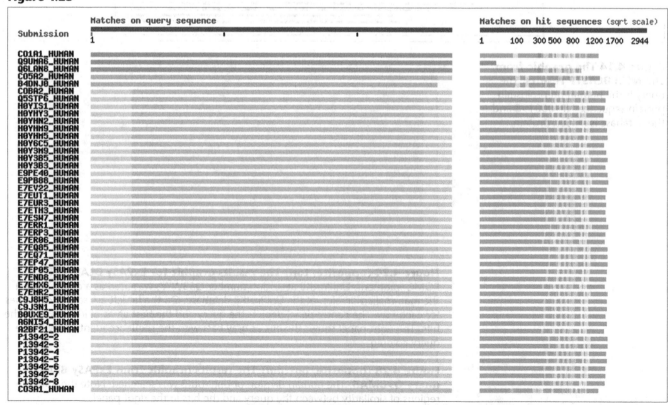

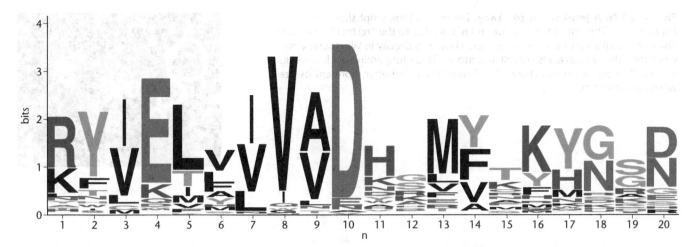

Figure 8.8 Logo for a section of the ADAM type metalloprotease profile, PS50215. The sequence is read from left to right. The variability in a position is represented by stacked one-letter amino acid codes, the height of each being proportional to the frequency found in that position.

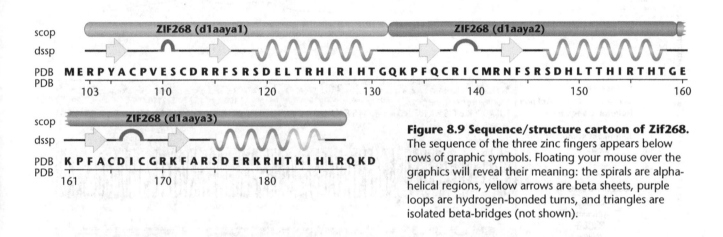

```
scop    ━━━━━━━━  ZIF268 (d1aaya1) ━━━━━━━━━━━━━   ━━━━━━━━━━━  ZIF268 (d1aaya2) ━━━━━━━━━━━━
dssp    ──▶──────◠──▶────〰〰〰〰〰〰〰─   ──▶──────◠──▶──〰〰〰〰〰〰〰─
PDB     M E R P Y A C P V E S C D R R F S R S D E L T R H I R I H T G Q K P F Q C R I C M R N F S R S D H L T T H I R T H T G E
PDB       103         110              120             130           140            150           160
```

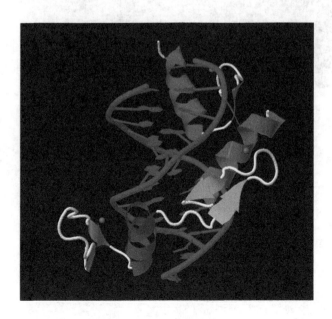

```
scop    ━━━━━━━━  ZIF268 (d1aaya3) ━━━━━━━━━
dssp    ──▶──────◠──▶──〰〰〰〰〰〰〰─
PDB     K P F A C D I C G R K F A R S D E R K R H T K I H L R Q K D
PDB     161         170            180
```

Figure 8.9 Sequence/structure cartoon of Zif268. The sequence of the three zinc fingers appears below rows of graphic symbols. Floating your mouse over the graphics will reveal their meaning: the spirals are alpha-helical regions, yellow arrows are beta sheets, purple loops are hydrogen-bonded turns, and triangles are isolated beta-bridges (not shown).

Figure 8.11 Zif268 PDB structure, 1aay. Each zinc finger consists of a helical region, interacting directly with the central DNA helix. External to these helices are two beta sheets that are parallel, joined by a loop (white strand). The zinc atoms (gray spheres) lie between the helix and loop. Note that the colors described and shown here may differ from what is seen on your screen.

Figure 8.12B A Jmol script for 1aay. The result of the script shown in Figure 8.12A. The rest of the structure is invisible due to the "restrict" command. The two histidines and two cysteines were chosen to display in Wireframe view, while the other amino acids are still in Cartoon. These four amino acids were also labeled. The zinc atom was changed to Spacefill view to better represent its size within the structure.

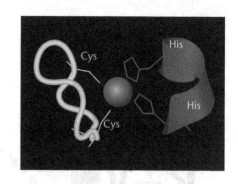

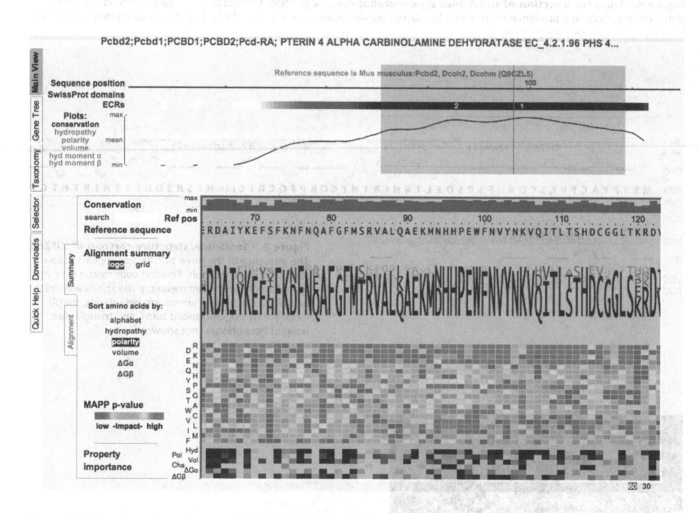

Figure 8.15 Interface view of mouse DCoH2 in ProPhylER.

Figure 8.16 "Cartoon" views of the DCoH2 dimer in ProPhylER.
(A) In this view, there is a clear divide between the two proteins in this dimer. Floating your mouse over a position will provide the identification of that amino acid as shown for tryptophan 66 on Chain B.
(B) Another view of the structure, showing how the beta sheets form an almost flat region while the helices are on the outside of the protein.
(C) The color-coding of the positions is based on the conservation of sequence revealed by a multiple sequence alignment.

(A)

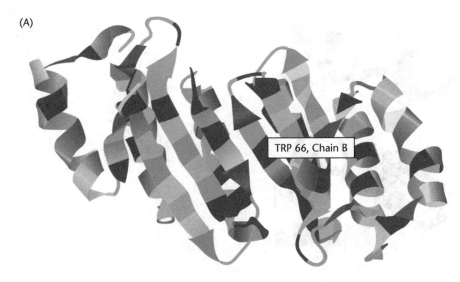

TRP 66, Chain B

(B)

(C)

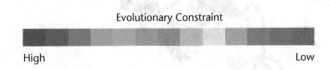

Evolutionary Constraint

High   Low

(A)

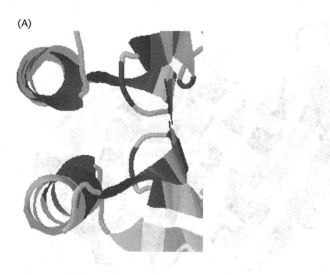

(B)

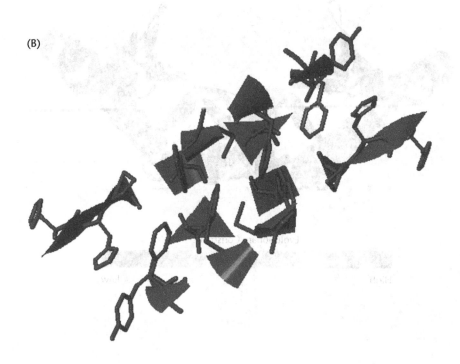

Figure 8.17 Close-up of the DCoH2 dimer structure. (A) Side view of conserved regions of both proteins. From this angle, you are looking down the helices that are in close proximity. The beta sheets are on edge, but their ends curve toward the helices. Although numbering in the opposite direction, this region appears like a mirror image, with one chain on top and the other on the bottom. (B) Isolated amino acids, showing side chains; Chain A is on top, Chain B at the bottom. The side chains at the far right and far left belong to histidine 62 of the two chains. All other groups are pointing toward the other chain.

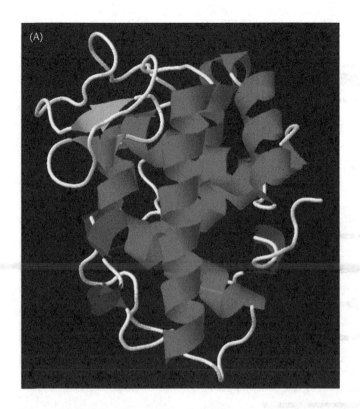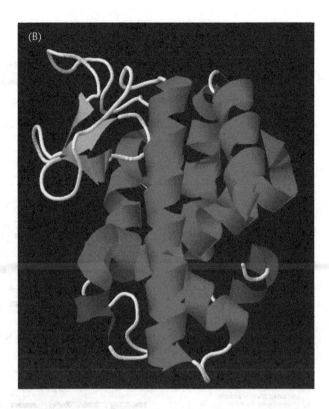

Figure 8.19 The swan and cod fish lysozyme structures. Both structures were generated by the RCSB Jmol viewer and are from (A) the swan (PDB identifier 1gbs) and (B) the cod fish (PDB identifier 3gxk).

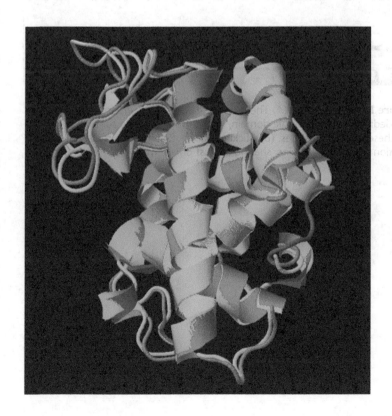

Figure 8.20 Structural alignment of the swan and cod fish lysozymes using the jFATCAT-rigid method. The swan and cod lysozyme structures are orange/dark gray and cyan/light gray, respectively. Gray regions signify where the two structures are too divergent to align.

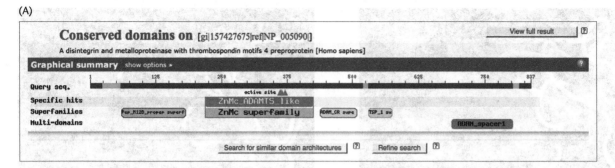

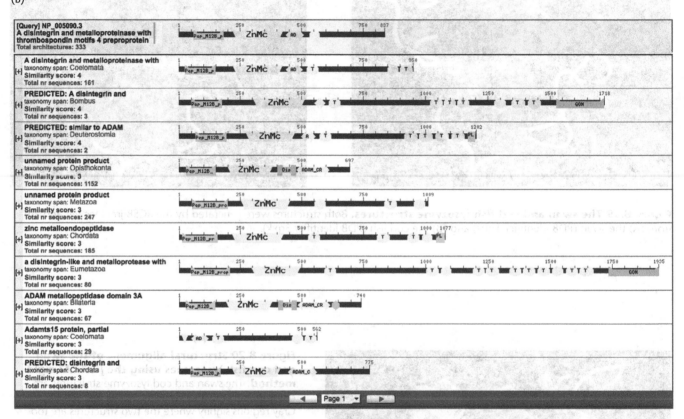

Figure 8.21 The NCBI Conserved Domain Architecture Retrieval Tool. (A) The conserved domains found in NP_005090, ADAMTS4. Clicking on these icons will take you to a page dedicated to that single domain, listing other proteins containing this domain and presenting a multiple sequence alignment of these many sequences. (B) Clicking on the "Search for similar domain architectures" button shown in (A) takes you to a presentation of related proteins, sharing at least one domain with your query.

Figure 8.22 SignalP prediction of secretion signal peptide cleavage sites. The SignalP-NN prediction, graphically showing a sudden drop in the probability for being part of the signal peptide (S score), and increases in the probability of being the new N-terminus after cleavage (C score) and the correlation between the two (Y score).

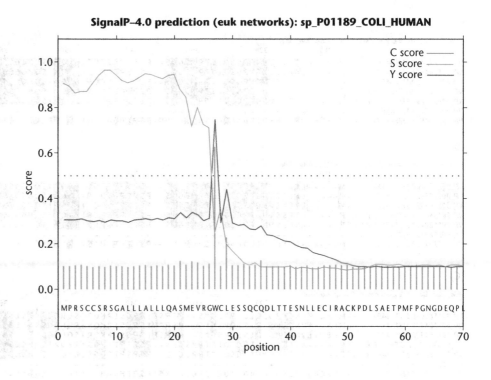

Figure 8.27A Transmembrane domain prediction using TMHMM. Within the span of 348 amino acids, human rhodopsin is predicted to pass through the cell outer membrane seven times. The result is four sections on the exterior of the cell and four in the interior.

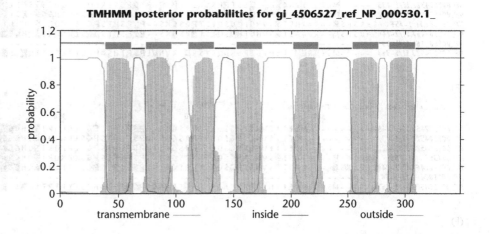

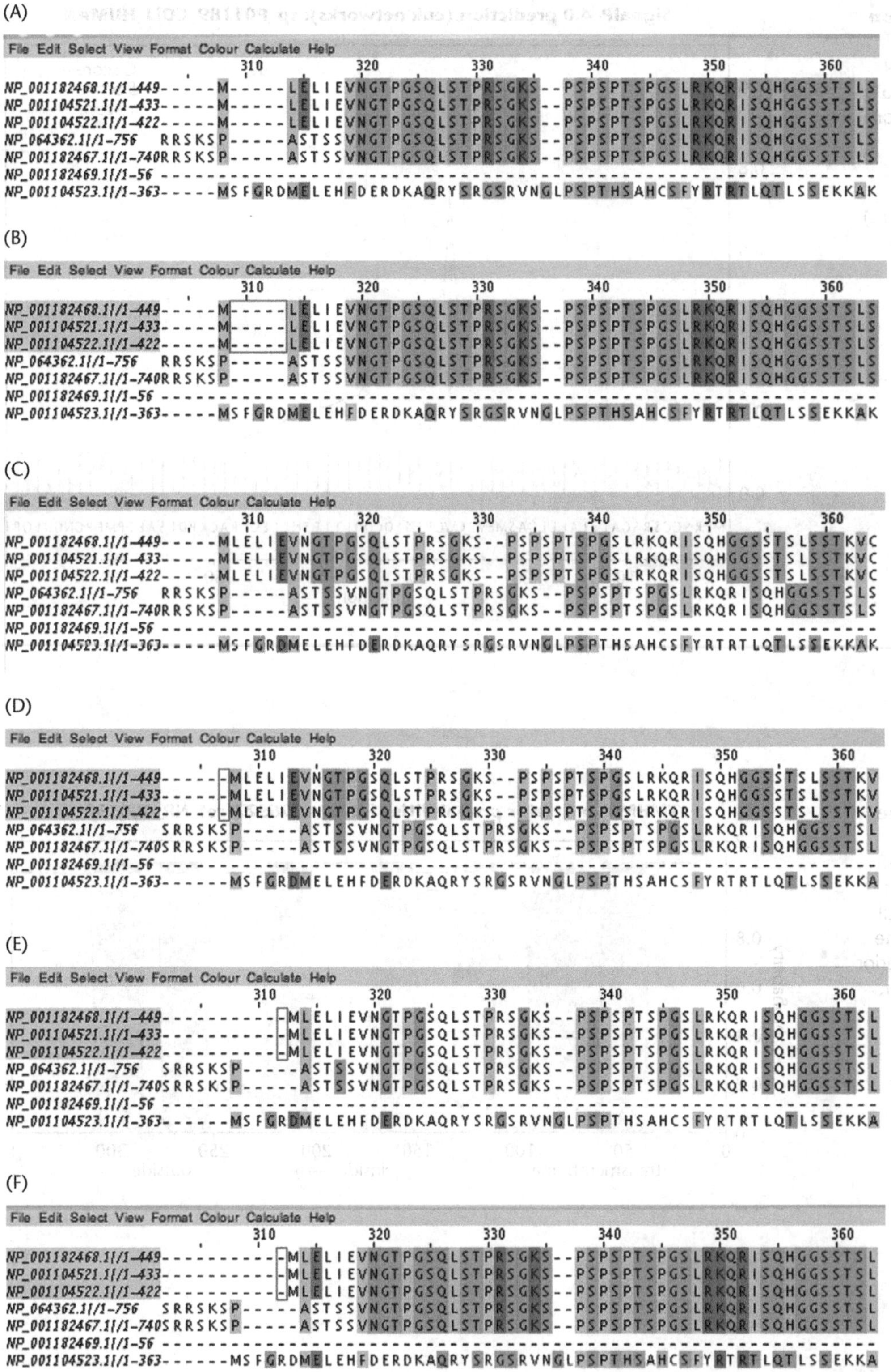

Figure 11.23 A simple alignment edit with Jalview. (A) The goal is to eliminate the gaps introduced by ClustalW between the methionines and leucine–glutamic acid pairs in the vicinity of alignment coordinate 310. (B) The gaps to be eliminated are highlighted by clicking and dragging the mouse. To eliminate the highlighted gaps, use the "Cut" command from the Jalview "Edit" menu, upper left in the Jalview window. Note that the sequences are now misaligned (C). (D) To realign the residues, highlight the gaps to the left of the methionines. (E) Using the keyboard right arrow key, "push" the sequences to the right by five taps. Should you overshoot, use the left arrow key. Although the sequences are now realigned, the colored columns may not show the strong alignment displayed in (A) and (B). (F) To make the coloring reappear, use the Jalview Edit command "Undo" followed by a "Redo" to refresh the column coloring.